T0140323

Springer INdAM Series

Volume 40

This series will publish textbooks, multi-authors books, thesis and monographs in English language resulting from workshops, conferences, courses, schools, seminars, doctoral thesis, and research activities carried out at INDAM - Istituto Nazionale di Alta Matematica, http://www.altamatematica.it/en. The books in the series will discuss recent results and analyze new trends in mathematics and its applications.
THE SERIES IS INDEXED IN SCOPUS

More information about this series at http://www.springer.com/series/10283

Valentina Barucci • Scott Chapman •
Marco D'Anna • Ralf Fröberg
Editors

Numerical Semigroups

IMNS 2018

Editors
Valentina Barucci
Matematica
Sapienza University of Rome
Roma, Italy

Scott Chapman
Department of Mathematics & Statistics
Sam Houston State University
Huntsville
TX, USA

Marco D'Anna
Dip di Matematica e Informatica
Università di Catania
Catania
Catania, Italy

Ralf Fröberg
Mathematics
Stockhom University
Stockholm, Sweden

ISSN 2281-518X ISSN 2281-5198 (electronic)
Springer INdAM Series
ISBN 978-3-030-40824-4 ISBN 978-3-030-40822-0 (eBook)
https://doi.org/10.1007/978-3-030-40822-0

This Springer imprint is published by the registered company Springer Nature Switzerland AG.
The registered company address is: Gewerbestrasse 11, 6330 Cham, Switzerland

Preface

During the week of September 3–7, 2018, 55 mathematical researchers from 15 different countries gathered at the "Il Palazzone" in Cortona, Italy, for the "International Meeting on Numerical Semigroups." This meeting has evolved into a biennial event, with the initial gatherings meeting under the title "The Iberian Meeting on Numerical Semigroups," in Porto, Portugal (2008), Granada, Spain (2010), and Vila Real, Portugal (2012). In 2014, the first meeting to use the "International" title was held in Cortona, followed by the second such meeting in 2016 in Levico Terme, Italy. The pages of this volume constitute the proceedings of the 2018 meeting in Cortona.

Talks were given at the conference by 41 participants. These talks centered on not only traditional types of numerical semigroups (such as Arf or symmetric) and their usual properties (such as the Frobenius number, genus, gap sets, and non-unique factorization), but also related types of semigroups (such as affine, Puiseux, Weierstrass, and primary) and their applications in other branches of algebra (including semigroup rings, coding theory, star operations, and Hilbert functions). The 21 papers in this Proceedings reflect the variety of the talks presented.

The meeting was organized by Marco D'Anna, University of Catania, Pedro A. García-Sánchez, University of Granada, and Vincenzo Micale, University of Catania. The Scientific Committee consisted of Valentina Barucci, Sapienza University of Rome; Scott Chapman, Sam Houston State University; Ralf Fröberg, Stockholm University; Pieter Moree, Max Planck Institute for Mathematics; and José Carlos Rosales, University of Granada. Marco D'Anna, assisted by Valentina Barucci, Scott Chapman, and Ralf Fröberg, edited these Proceedings.

The principal sponsor for the meeting was the "Istituto Nazionale di Alta Matematica 'Francesco Severi'" and additional support was received from various grants based in the Mathematics Departments at the Universities of Granada, Catania, and Cadiz. The organizers and Scientific Committee thank all involved for their generous support. We look forward to another International Meeting on Numerical Semigroups in 2020, which is currently in the planning stage.

In this book, we chiefly present research papers. Additionally, we present a few survey articles which collect results and examples which are difficult to find elsewhere. The book is intended for researchers and students who want to learn about recent developments in the theory of numerical semigroups. Our aim is to present the current status of research on numerical semigroups and to gather together papers on the topic from different areas, such as Semigroup Theory, Factorization Theory, Algebraic Geometry, Combinatorics, Commutative Algebra, and Coding Theory, which reflects how numerical semigroups arise in different research contexts.

Roma, Italy	Valentina Barucci
Huntsville, TX, USA	Scott Chapman
Catania, Italy	Marco D'Anna
Stockholm, Sweden	Ralf Fröberg

Contents

Counting Numerical Semigroups by Genus and Even Gaps via Kunz-Coordinate Vectors

Matheus Bernardini

Abstract We construct a one-to-one correspondence between a subset of numerical semigroups with genus g and γ even gaps and the integer points of a rational polytope. In particular, we give an overview to apply this correspondence to try to decide if the sequence (n_g) is increasing, where n_g denotes the number of numerical semigroups with genus g.

Keywords Numerical semigroup · Multiplicity · Even gaps · Genus · Kunz-coordinate vector

1 Introduction

A *numerical semigroup* S is a subset of $\mathbb{N}_0$ such that $0 \in S$, it is closed under addition and the set $G(S) := \mathbb{N}_0 \setminus S$, the set of *gaps* of S, is finite. The number of elements $g = g(S)$ of $G(S)$ is called the *genus* of S and the first non-zero element in S is called the *multiplicity* of S. If S is a numerical semigroup with positive genus g then one can ensure that all gaps of S belongs to $[1, 2g - 1]$; in particular, $\{2g + i : i \in \mathbb{N}_0\} \subseteq S$ and the number of numerical semigroups with genus g, denoted by n_g, is finite. Some excellent references for the background on numerical semigroups are the books [5] and [7].

Throughout this paper, we keep the notation proposed by Bernardini and Torres [1]: the set of numerical semigroups with genus g is denoted by $\mathcal{S}_g$ and has n_g elements and the set of numerical semigroups with genus g and γ even gaps is denoted by $\mathcal{S}_\gamma(g)$ and has $N_\gamma(g)$ elements.

M. Bernardini (✉)
Universidade de Brasília, Área Especial de Indústria Projeção A - UNB, Brasília, Brazil
e-mail: matheusbernardini@unb.br

V. Barucci et al. (eds.), *Numerical Semigroups*, Springer INdAM Series 40,
https://doi.org/10.1007/978-3-030-40822-0_1

In this paper we use the quite useful parametrization

$$\mathbf{x}_g : \mathcal{S}_\gamma(g) \to \mathcal{S}_\gamma, S \mapsto S/2, \tag{1}$$

where $S/2 := \{s \in \mathbb{N}_0 : 2s \in S\}$.

Naturally, the set $\mathcal{S}_\gamma(g)$ and the map $\mathbf{x}_g$ can be generalized. Let $d > 1$ be an integer. The set of numerical semigroups with genus g and γ gaps which are congruent to 0 modulo d is denoted by $\mathcal{S}_{(d,\gamma)}(g)$. There is a natural parametrization given by

$$\mathbf{x}_{g_d} : \mathcal{S}_{(d,\gamma)}(g) \to \mathcal{S}_\gamma, S \mapsto S/d,$$

where $S/d := \{s \in \mathbb{N}_0 : ds \in S\}$. This concept appears in [9], for instance.

In this paper, we obtain a one-to-one correspondence between the set $\mathbf{x}_g^{-1}(T)$ and the set of integer points of a rational polytope.

As an application of this correspondence, we give a new approach to compute the numbers $N_\gamma(g)$. Our main goal is finding a new direction to discuss the following question.

$$\text{Is it true that } n_g \leq n_{g+1}, \text{ for all } g? \tag{2}$$

The first few elements of the sequence (n_g) are $1, 1, 2, 4, 7, 12, 23, 39, 67$. Kaplan [6] wrote a nice survey on this problem and one can find information of these numbers in Sloane's On-line Encyclopedia of Integer Sequences [10].

Bras-Amorós [3] conjectured remarkable properties on the behaviour of the sequence (n_g):

1. $\lim_{g\to\infty} \frac{n_{g+1}+n_g}{n_{g+2}} = 1$;
2. $\lim_{g\to\infty} \frac{n_{g+1}}{n_g} = \varphi := \frac{1+\sqrt{5}}{2}$;
3. $n_{g+2} \geq n_{g+1} + n_g$ for any g.

Zhai [12] proved that $\lim_{g\to\infty} n_g \varphi^{-g}$ is a constant. As a consequence, it confirms that items (1) and (2) hold true. However, item (3) is still an open problem; even a weaker version, proposed at (2), is an open question. Zhai's result also ensures that $n_g < n_{g+1}$ for large enough g. Fromentin and Hivert [4] verified that $n_g < n_{g+1}$ also holds true for $g \leq 67$.

Torres [11] proved that $\mathcal{S}_\gamma(g) \neq \emptyset$ if, and only if, $2g \geq 3\gamma$. Hence,

$$n_g = \sum_{\gamma=0}^{\lfloor 2g/3 \rfloor} N_\gamma(g)\,. \tag{3}$$

In order to work on Question (2), Bernardini and Torres [1] tried to understand the effect of the even gaps on a numerical semigroup. By using the so-called t-translation, they proved that $N_\gamma(g) = N_\gamma(3\gamma)$ for $g \geq 3\gamma$ and also $N_\gamma(g) < N_\gamma(3\gamma)$ for $g < 3\gamma$. Although numerical evidence points out that $N_\gamma(g) \leq N_\gamma(g+$

1) holds true for all g and γ, their methods could not compare numbers $N_\gamma(g_1)$ and $N_\gamma(g_2)$, with $3\gamma/2 \leq g_1 < g_2 < 3\gamma$. Notice that if $N_\gamma(g_1) \leq N_\gamma(g_2)$, for $3\gamma/2 \leq g_1 < g_2 < 3\gamma$ then $n_g < n_{g+1}$, for all g.

2 Apéry Set and Kunz-Coordinate Vector

Let S be a numerical semigroup and $n \in S$. The Apéry set of S (with respect to n) is the set $Ap(S, n) = \{s \in S : s - n \notin S\}$. If $n = 1$, then $S = \mathbb{N}_0$ and $Ap(\mathbb{N}_0, 1) = \{0\}$. If $n > 1$, then there are $w_1, \ldots, w_{n-1} \in \mathbb{N}$ such that $Ap(S, n) = \{0, w_1, \ldots, w_{n-1}\}$, where $w_i = \min\{s \in S : s \equiv i \pmod{n}\}$.

Proposition 1 *Let S be a numerical semigroup with multiplicity m and $Ap(S, m) = \{0, w_1, \ldots, w_{m-1}\}$. Then*

$$S = \langle m, w_1, w_2, \ldots, w_{m-1}\rangle.$$

Proof It is clear that $am \in \langle m, w_1, w_2, \ldots, w_{m-1}\rangle, \forall a \in \mathbb{N}$. For $s \in S$, $m \nmid s$, there is $i \in \{1, \ldots, m-1\}$ such that $s = mk + i$. By minimality of w_i, there is $\tilde{k} \in \mathbb{N}_0$ such that $s = w_i + \tilde{k}m \in \langle m, w_1, w_2, \ldots, w_{m-1}\rangle$. On the other hand, $m, w_1, \ldots, w_{m-1} \in S$.

Let S be a numerical semigroup, $n \in S$ and consider $Ap(S, n) = \{0, w_1, \ldots, w_{n-1}\}$. There are $e_1, \ldots, e_{n-1} \in \mathbb{N}$ such that $w_i = ne_i + i$, for each $i \in \{1, \ldots, n-1\}$. The vector $(e_1, \ldots, e_{n-1}) \in \mathbb{N}_0^{n-1}$ is called the *Kunz-coordinate vector* of S (with respect to n). In particular, if m is the multiplicity of S, then the Kunz-coordinate vector of S (with respect to m) is in $\mathbb{N}^{m-1}$. This concept appears in [2], for instance.

A natural task is finding conditions for a vector $(x_1, \ldots, x_{m-1}) \in \mathbb{N}^{m-1}$ to be a Kunz-coordinate vector (with respect to the multiplicity m of S) of some numerical semigroup S with multiplicity m. The following examples illustrate the general method, which is presented in Proposition 2.

Example 1 Numerical semigroups with multiplicity 2 are $\langle 2, 2e_1 + 1\rangle$, where $e_1 \in \mathbb{N}$. Observe that e_1 is the genus of such numerical semigroup.

There is a one-to-one correspondence between the set of numerical semigroups with multiplicity 2 and the set of positive integers given by $\langle 2, 2e_1 + 1\rangle \mapsto e_1$.

Example 2 Let $S = \langle 3, 3e_1+1, 3e_2+2\rangle$ be a numerical semigroup with multiplicity 3 and genus g, where $e_1, e_2 \in \mathbb{N}$. By minimality of $w_1 = 3e_1+1$ and $w_2 = 3e_2+2$, (e_1, e_2) satisfies

$$\begin{cases} (3e_1 + 1) + (3e_1 + 1) \geq 3e_2 + 2 \\ (3e_2 + 2) + (3e_2 + 2) \geq 3e_1 + 1. \end{cases}$$

The set of gaps of S has $e_1 + e_2$ elements, since $G(S) = \{3n_1 + 1 : 0 \leq n_1 < e_1\} \cup \{3n_2+2 : 0 \leq n_2 < e_2\}$. Thus, $e_1+e_2 = g$. On the other hand, if $(e_1, e_2) \in \mathbb{N}^2$

is such that $2e_1 \geq e_2$, $2e_2+1 \geq e_1$ and $e_1+e_2=g$, then $\langle 3, 3e_1+1, 3e_2+2\rangle$ is a numerical semigroup with multiplicity m and genus g.

Hence, there is a one-to-one correspondence between the set of numerical semigroups with multiplicity 3 and genus g the vectors of $\mathbb{N}^2$ which are solutions of

$$\begin{cases} 2X_1 \geq X_2 \\ 2X_2+1 \geq X_1 \\ X_1+X_2 = g. \end{cases}$$

In order to give a characterization of numerical semigroups with fixed multiplicity and fixed genus, the main idea is generalizing Example 2. The following is a result due to Rosales et al. [8].

Proposition 2 *There is a one-to-one correspondence between the set of numerical semigroups with multiplicity m and genus g and the positive integer solutions of the system of inequalities*

$$\begin{cases} X_i + X_j \geq X_{i+j}, & for\ 1 \leq i \leq j \leq m-1;\ i+j < m; \\ X_i + X_j + 1 \geq X_{i+j-m}, & for\ 1 \leq i \leq j \leq m-1;\ i+j > m \\ \sum_{k=1}^{m-1} X_k = g. \end{cases}$$

Let $S = \langle m, w_1, \ldots, w_{m-1}\rangle$ be a numerical semigroup with multiplicity m and genus g, where $w_i = me_i + i$. The main idea of the proof is using the minimality of $w_1, \ldots, w_{m-1}$ and observing that $w_i + w_j \equiv i + j \pmod{m}$ and $G(S) = \bigcup_{i=1}^{m-1}\{mn_i + i : 0 \leq n_i < e_i\}$. For a full proof, see [8].

3 The Main Result and an Application to a Counting Problem

In [1], the calculation of $N_\gamma(g)$ was given by

$$N_\gamma(g) = \sum_{T \in \mathcal{S}_\gamma} \#\mathbf{x}_g^{-1}(T). \tag{4}$$

In this section, we present a new way for computing those numbers. In order to do this, we fix the multiplicity of $T \in \mathcal{S}_\gamma$.

First of all, we obtain a relation between the genus and the multiplicity of a numerical semigroup.

Proposition 3 *Let S be a numerical semigroup with genus g and multiplicity m. Then $m \leq g+1$.*

Proof If a numerical semigroup S has multiplicity m and genus g with $m \geq g+2$, then the number of gaps of S would be, at least, $g+1$ and it is a contradiction. Hence $m \leq g+1$.

Remark 1 The bound obtained in Proposition 3 is sharp, since $\{0, g+1, \ldots\}$ has genus g has multiplicity $g+1$.

If $\gamma = 0$, then $\mathcal{S}_0 = \{\mathbb{N}_0\}$ and $\mathbf{x}_g^{-1}(\mathbb{N}_0) = \{\langle 2, 2g+1\rangle\}$. Hence, $N_0(g) = 1$, for all g. If $\gamma > 0$, we divide the set $\mathcal{S}_\gamma$ into the subsets $\mathcal{S}_\gamma^m := \{S : g(S) = \gamma \text{ and } m(S) = m\}$, where $m \in [2, \gamma+1] \cap \mathbb{Z}$. We can write

$$\mathcal{S}_\gamma = \bigcup_{m=2}^{\gamma+1} \mathcal{S}_\gamma^m. \tag{5}$$

Putting (4) and (5) together, we obtain

$$N_\gamma(g) = \sum_{m=2}^{\gamma+1} \sum_{T \in \mathcal{S}_\gamma^m} \#\mathbf{x}_g^{-1}(T).$$

Thus, it is important to give a characterization for $T \in \mathcal{S}_\gamma^m$. We can describe T by its Apéry set (with respect to its multiplicity m) and write

$$T = \langle m, me_1+1, me_2+2, \ldots, me_{m-1}+(m-1)\rangle,$$

where $me_i + i = \min\{s \in S : s \equiv i \pmod m\}$.

The next result characterizes all numerical semigroups of $\mathbf{x}_g^{-1}(T)$, for $T \in \mathcal{S}_\gamma^m$. It is a consequence of Proposition 2.

Theorem 1 *Let $T = \langle m, me_1+1, \ldots, me_{m-1}+(m-1)\rangle \in \mathcal{S}_\gamma^m$. A numerical semigroup S belongs to $\mathbf{x}_g^{-1}(T)$ if, and only if,*

$$S = \langle 2m, 2me_1+2, \ldots, 2me_{m-1}+(2m-2), 2mk_1+1, 2mk_3+3, \ldots, 2mk_{2m-1}+(2m-1)\rangle,$$

where $(k_1, k_3, \ldots, k_{2m-1}) \in \mathbb{N}_0^m$ satisfies the system

$$\begin{cases} (*) \begin{cases} X_{2i-1} + e_j \geq X_{2(i+j)-1}, & \text{for } 1 \leq i \leq m; 1 \leq j \leq m-1; i+j \leq m; \\ X_{2i-1} + e_j + 1 \geq X_{2(i+j-m)-1}, & \text{for } 1 \leq i \leq m; 1 \leq j \leq m-1; i+j > m; \end{cases} \\ (**) \begin{cases} X_{2i-1} + X_{2j-1} \geq e_{i+j-1}, & \text{for } 1 \leq i \leq j \leq m; i+j \leq m; \\ X_{2i-1} + X_{2j-1} + 1 \geq e_{i+j-1-m}, & \text{for } 1 \leq i \leq j \leq m; i+j \geq m+2; \end{cases} \\ \sum_{i=1}^m X_{2i-1} = g - \gamma, \end{cases}$$

Proof The even numbers $2m, 2me_1+2, \ldots, 2me_{m-1}+2me_{m-1}+2(m-1)$ belongs to $Ap(2m, S)$. Let $2mk_1+1, 2mk_3+3, \ldots, 2mk_{2m-1}+(2m-1)$ be the odd numbers of $Ap(2m, S)$. Thus, $(e_1, k_1, e_2, k_3, \ldots, e_{m-1}, k_{2m-1}) \in \mathbb{N}_0^{m-1}$ is the Kunz-coordinate vector of S (with respect to $2m$).

Now, we apply Proposition 2. Inequalites given in $(*)$ come from sums of an odd element of $Ap(2m, S)$ with an even element of $Ap(2m, S)$, while inequalities given in $(**)$ come from sums of two odd elements of $Ap(2m, S)$. Since $(e_1, \ldots, e_{m-1})$ is the Kunz-coordinate vector of T (with respect to m), then the sum of two even elements of $Ap(2m, S)$ belongs to S. Finally, last equality comes from the fact that S has $g - \gamma$ odd gaps.

Remark 2 Some of the numbers k_i can be zero. Hence, it is possible that the multiplicity of S is not $2m$.

Example 3 Let $T = \langle 2, 2\gamma + 1\rangle \in \mathcal{S}_\gamma^2$, with $\gamma \in \mathbb{N}$. Theorem 1 ensures that if $S \in \mathbf{x}_g^{-1}(T)$, then

$$S = \langle 4, 4\gamma + 2, 4k_1 + 1, 4k_3 + 3\rangle,$$

where $(k_1, k_3) \in \mathbb{N}_0^2$ satisties

$$(\#)\begin{cases} (*)\begin{cases} -\gamma - 1 \leq X_3 - X_1 \leq \gamma \end{cases} \\ (**)\begin{cases} X_1 + X_1 \geq \gamma \\ X_3 + X_3 + 1 \geq \gamma \end{cases} \\ X_1 + X_3 = g - \gamma. \end{cases}$$

The set of integer points of the region in Figure 1 is in one-to-one correspondence with the set $\{S \in \mathcal{S}_\gamma(g) : S/2 \text{ has multiplicity } 2\}$.

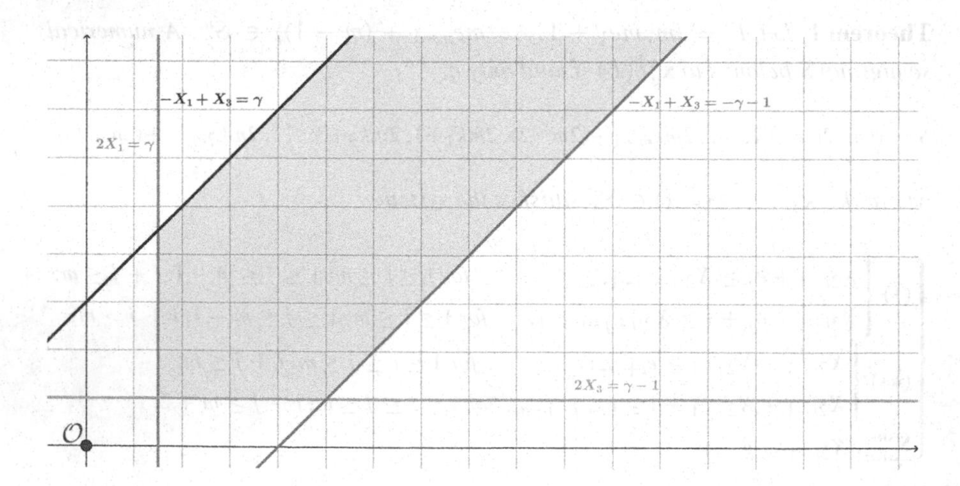

Fig. 1 Region in $\mathbb{R}^2$ given by inequalities $(*)$ and $(**)$

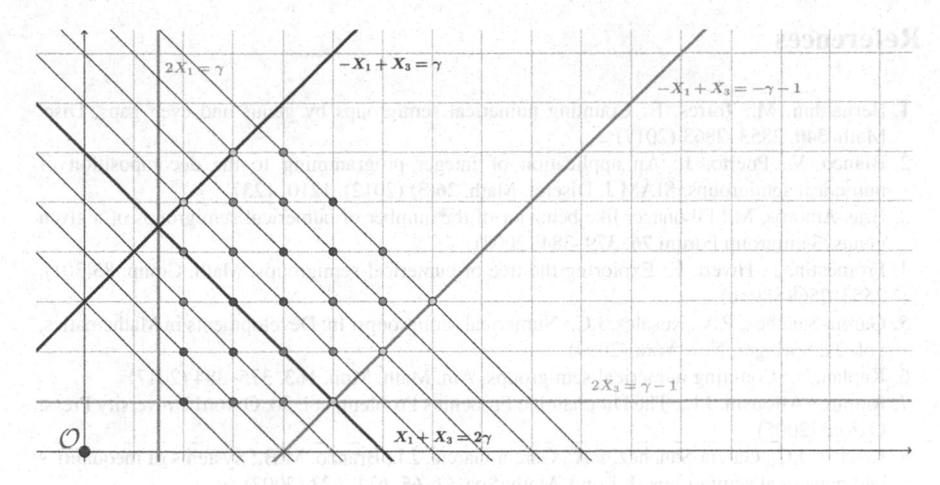

Fig. 2 For fixed g, the integer points in the line segment represent numerical semigroups of $\{S \in \mathcal{S}_\gamma(g)\colon S/2$ has multiplicity $2\}$

If g is fixed, then the set of points that satisfies the system (#) is a polytope (a line segment). We are interested in the set of integer points of this polytope. Figure 2 shows examples for some values of g. Each integer point represents a numerical semigroup of the set $\{S \in \mathcal{S}_\gamma(g) : S/2$ has multiplicity $2\}$.

Let $N_\gamma^m(g) = \sum_{T \in \mathcal{S}_\gamma^m} \#\mathbf{x}_g^{-1}(T)$. After some computations, we obtain

$$N_\gamma^2(g) = \begin{cases} 0, \text{ if } g < 2\gamma \\ k+1, \text{ if } g = 2\gamma + k \text{ and } k \in \{0, 1, \ldots, \gamma - 1\} \\ \gamma + 1, \text{ if } g \geq 3\gamma. \end{cases}$$

In particular, $N_\gamma^2(g) \leq N_\gamma^2(g+1)$. We leave the following open question.

$$\text{Let } \gamma \in \mathbb{N} \text{ and } m \in [2, \gamma+1] \cap \mathbb{Z}. \text{ Is it true that } N_\gamma^m(g) \leq N_\gamma^m(g+1), \text{ for all } g? \quad (6)$$

A positive answer to Question (6) implies a positive answer to Question (2).

Acknowledgements The author was partially supported FAPDF-Brazil (grant 23072.91.49580.29052018). Part of this paper was presented in the "INdAM: International meeting on numerical semigroups" (2018) at Cortona, Italy. I am grateful to the referee for their comments, suggestions and corrections that allowed to improve this version of the paper.

References

1. Bernardini, M., Torres, F.: Counting numerical semigroups by genus and even gaps. Disc. Math. **340**, 2853–2863 (2017)
2. Blanco, V., Puerto, J.: An application of integer programming to the decomposition of numerical semigroups. SIAM J. Discret. Math. **26**(3) (2012), 1210–1237
3. Bras-Amorós, M.: Fibonacci-like behavior of the number of numerical semigroups of a given genus. Semigroup Forum **76**, 379–384 (2008)
4. Fromentin, J., Hivert, F.: Exploring the tree of numerical semigroups. Math. Comp. **85**(301), 2553–2568 (2016)
5. García-Sánchez, P.A., Rosales, J.C.: Numerical semigroups. In: Developments in Mathematics, vol. 20. Springer, New York (2009)
6. Kaplan, N.: Counting numerical semigroups. Am. Math. Mon. **163**, 375–384 (2017)
7. Ramírez-Alfonsín, J.L.: The Diophantine Frobenius Problem, vol. 30. Oxford University Press, Oxford (2005)
8. Rosales, J.C., García-Sánchez, P.A., García-García, J.I., Branco, M.B.: Systems of inequalities and numerical semigroups. J. Lond. Math. Soc. (2) **65**, 611–623 (2002)
9. Rosales, J.C., García-Sánchez, P.A., García-Sánchez, J.I., Urbano-Blanco, J.M.: Proportionally modular Diophantine inequalities. J. Number Theory **103**, 281–294 (2003)
10. Sloane, N.J.A.: The on-line encyclopedia of integer sequences, A007323 (2009). http://www.research.att.com/~njas/sequences/
11. Torres, F.:On γ-hyperelliptic numerical semigroups. Semigroup Forum **55**, 364–379 (1997)
12. Zhai, A.: Fibonacci-like growth of numerical semigroups of a given genus. Semigroup Forum **86**, 634–662 (2013)

Patterns on the Numerical Duplication by Their Admissibility Degree

Alessio Borzì

Abstract We develop the theory of patterns on numerical semigroups in terms of the admissibility degree. We prove that the Arf pattern induces every strongly admissible pattern, and determine all patterns equivalent to the Arf pattern. We study patterns on the numerical duplication $S \bowtie^d E$ when $d \gg 0$. We also provide a definition of patterns on rings.

Keywords Numerical semigroup · Arf semigroup · Pattern on a numerical semigroup · Numerical duplication

1 Introduction

A numerical semigroup S is an additive submonoid of $\mathbb{N}$ with finite complement in $\mathbb{N}$. The set of values of a Noetherian, one-dimensional, analytically irreducible, local, domain is a numerical semigroup, therefore the study of numerical semigroups is related to the study of this class of rings. In [14], Lipman introduces and motivates the study of Arf rings, which constitute an important class of rings for the classification problem of singular curve branches. A good reference for the study of Arf rings in the analytically irreducible case is [1]. The value semigroup of an Arf ring is an Arf numerical semigroup. We say that a numerical semigroup S is Arf if for every $x, y, z \in S$ with $x \geq y \geq z$ we have $x + y - z \in S$. There are several works in the literature about Arf numerical semigroups, see for instance [11, 17]. Note that Arf semigroups are related to the polynomial $x + y - z$. In [6], Bras-Amóros and García-Sánchez generalize the definition of Arf semigroup to any linear homogeneous polynomial, introducing the theory of patterns on numerical semigroups [7, 19–21].

A. Borzì (✉)
University of Catania, Catania, Italy

Scuola Superiore di Catania, Catania, Italy
e-mail: alessio.borzi@studium.unict.it

V. Barucci et al. (eds.), *Numerical Semigroups*, Springer INdAM Series 40,
https://doi.org/10.1007/978-3-030-40822-0_2

In this manner, Arf numerical semigroups are the semigroups that admit the *Arf pattern* $x + y - z$. In addition, Arf numerical semigroups can be characterized in terms of their additive behaviour (see for instance [4, 5]). Therefore, one can translate similar characterizations for certain classes of patterns.

Given a numerical semigroup S we can consider the quotient of S by a positive integer $d \in \mathbb{N}$

$$\frac{S}{d} = \{x \in \mathbb{N} : dx \in S\}.$$

In [9], D'Anna and Strazzanti define a semigroup construction, called the numerical duplication, that is, in a certain sense, the reverse operation of the quotient by 2. If $A \subseteq \mathbb{N}$, the set of doubles is denoted by $2 \cdot A = \{2a : a \in A\}$ (note that $2 \cdot A \neq 2A = A + A$). Given a numerical semigroup S, a *semigroup ideal* of S is a subset $E \subseteq S$ such that $E + S \subseteq E$. If $d \in S$ is an odd integer, the numerical duplication of S with respect to the semigroup ideal E and d is

$$S \bowtie^d E = 2 \cdot S \cup (2 \cdot E + d).$$

The numerical duplication can be seen as the value semigroup of a quadratic quotient of the Rees algebra, see for instance [2, 3]. This construction generalizes Nagata's idealization and the amalgamated duplication (see [8]), and it is one of the main tools used in [15] to give a negative answer to a problem of Rossi [18].

In [3] it was characterized when the numerical duplication $S \bowtie^d E$ is Arf. The characterization is given in terms of the multiplicity sequence of the Arf semigroup S. A natural question is how this characterization can be generalized to any pattern. This paper deals with this question.

In particular, in Sects. 3 and 4 we develop the theory of patterns on numerical semigroups in terms of the admissibility degree, generalizing some results of [6] proved for Boolean patterns. Further, we prove that the Arf pattern induces every strongly admissible pattern and we determine the family of patterns equivalent to the Arf pattern. In Sect. 5 we characterize when the numerical duplication $S \bowtie^d E$ admits a monic pattern for $d \gg 0$ and give some examples of the general case. In Sect. 6 we give some observations and trace possible future work about pattern on rings.

Several computations are performed by using the GAP system [22] and, in particular, the NumericalSgps package [10].

2 Preliminaries

Let S be a numerical semigroup, the *multiplicity* of S is the integer $\mathrm{m}(S) = \min(S \setminus \{0\})$, the *conductor* of S is $\mathrm{c}(S) = \min\{x \in \mathbb{N} : x + \mathbb{N} \subseteq S\}$. If $E \subseteq S$ is a semigroup ideal of S, set $\mathrm{c}(E) = \min\{x \in \mathbb{N} : x + \mathbb{N} \subseteq E\}$. Note that, if $d \in S$ is an odd integer, from [9, Proposition 2.1] the conductor of the numerical duplication is $c(S \bowtie^d E) = 2\,c(E) + d - 1$.

A *pattern* $p(x_1, \ldots, x_n)$ of length n is a linear homogeneous polynomial in n variables with non-zero integer coefficients. The pattern of length zero is the zero polynomial $p = 0$. A numerical semigroup S *admits* a pattern p if for every $s_1, \ldots, s_n \in S$ with $s_1 \geq \cdots \geq s_n$ we have $p(s_1, \ldots, s_n) \in S$. The family of all numerical semigroups admitting p is denoted by $\mathscr{S}(p)$. Given two patterns p_1, p_2, we say that p_1 *induces* p_2 if $\mathscr{S}(p_1) \subseteq \mathscr{S}(p_2)$; we say that p_1 and p_2 are *equivalent* if they induce each other, or equivalently $\mathscr{S}(p_1) = \mathscr{S}(p_2)$. Let p be a pattern of length n, set

$$p(x_1, \ldots, x_n) = \sum_{i=1}^{n} a_i x_i,$$

and $b_i = \sum_{j \leq i} a_j$, we will keep this notation throughout. Note that we can write

$$\begin{aligned} p(x_1, \ldots, x_n) &= a_1 x_1 + \cdots + a_n x_n = \\ &= b_1(x_1 - x_2) + \cdots + b_{n-1}(x_{n-1} - x_n) + b_n x_n, \end{aligned}$$

we will use frequently this decomposition in the sequel. The pattern p is *admissible* if $\mathscr{S}(p) \neq \emptyset$, that is, p is admitted by some numerical semigroup. Set

$$p' = \begin{cases} p - x_1 & \text{if } a_1 > 1 \\ p(0, x_1, \ldots, x_{n-1}) & \text{if } a_1 = 1, \end{cases}$$

and define recursively $p^{(0)} = p$ and $p^{(i)} = (p^{(i-1)})'$ for $i \in \mathbb{N} \setminus \{0\}$. The *admissibility degree* of p, denoted by $\mathrm{ad}(p)$, is the least integer k such that $p^{(k)}$ is not admissible, if such integer exists, otherwise is ∞. If p' is admissible, p is *strongly admissible*. With this definitions, p is admissible if $\mathrm{ad}(p) \geq 1$, strongly admissibile if $\mathrm{ad}(p) \geq 2$.

Proposition 1 *[6, Theorem 12] For a pattern p the following conditions are equivalent*

1. *p is admissible,*
2. *$\mathbb{N}$ admits p,*
3. *$b_i \geq 0$ for all $i \in \{1, \ldots, n\}$.*

Corollary 1 *If p has admissibility degree* 1*, then there exists $i \in \{1, \ldots, n\}$ such that $b_i = 0$.*

Proof By hypothesis p' is not admissibile, then from Proposition 1 there exists i such that $(a_1 - 1) + \sum_{j=2}^{i} a_j = -1 \Rightarrow b_i = \sum_{j=1}^{i} a_j = 0$.

The *trivializing pattern* is $x_1 - x_2$, note that $\mathscr{S}(x_1 - x_2) = \{\mathbb{N}\}$, so from Proposition 1 it induces every admissibile pattern, in other words it induces every pattern p with $\mathrm{ad}(p) \geq 1$. The *Arf pattern* is $x_1 + x_2 - x_3$, it is equivalent to $2x_1 - x_2$ (see [6, Example 5]). The family $\mathscr{S}(x_1 + x_2 - x_3)$ is the family of Arf numerical semigroups. More in general, the *subtraction pattern* of degree k is the pattern $x_1 + x_2 + \cdots + x_k - x_{k+1}$. So the trivializing pattern and the Arf pattern are the subtraction patterns of degree 1 and 2. Note that the admissibility degree of a subtraction pattern is equal to its degree.

Following [16, Chapter 6], a *Frobenius variety* is a nonempty family $\mathscr{F}$ of numerical semigroups such that

1. $S, T \in \mathscr{F} \Rightarrow S \cap T \in \mathscr{F}$,
2. $S \in \mathscr{F} \setminus \{\mathbb{N}\} \Rightarrow S \cup \{F(S)\} \in \mathscr{F}$.

Proposition 2 *[16, Proposition 7.17] If p is a strongly admissible pattern, then $\mathscr{S}(p)$ is a Frobenius variety.*

Given a Frobenius variety $\mathscr{F}$, it is possible to define the closure of a numerical semigroup S as the smallest (with respect to set inclusion) numerical semigroup in $\mathscr{F}$ that contains S. From this idea, we can define the notion of system of generators with respect to the variety. In addition, we can construct a tree of all numerical semigroups in $\mathscr{F}$ rooted in $\mathbb{N}$ and such that T is a son of S if and only if $T = S \cup \{F(S)\}$.

From Proposition 2, these definitions generalize many notions given in [6], for instance p-closure or p-system of generators.

3 Patterns and Their Admissibility Degree

In [19] and [21] it was noted that a pattern p is strongly admissibile (i.e. $\mathrm{ad}(p) \geq 2$) if and only if $b_i \geq 1$ for all $i \in \{1, \ldots, n\}$. Of course if $b_i \geq k$ for all $i \in \{1, \ldots, n\}$ then $\mathrm{ad}(p) \geq k + 1$.

Proposition 3 *If a pattern p has admissibility degree at least $k + 1$, then $b_i \geq \min\{i, k\}$ for all $i \in \{1, \ldots, n\}$.*

Proof Let a'_i be the coefficients of p' and $b'_i = \sum_{j \leq i} a'_j$. We proceed by induction on k. The base case follows from Proposition 1. For the inductive step, firstly we assume that p is monic. For all $i \in \{1, \ldots, n-1\}$ we have $b_{i+1} = b'_i + 1$, then

$$\mathrm{ad}(p) \geq k + 1 \Rightarrow \mathrm{ad}(p') \geq k \Rightarrow b'_i \geq \min\{i, k-1\} \Rightarrow b_{i+1} \geq \min\{i+1, k\},$$

in addition $b_1 = 1 \geq \min\{1, k\}$. On the other hand, if p is not monic, for all $i \in \{1, \ldots, n\}$ we have $b_i = b'_i + 1$, then

$$\mathrm{ad}(p) \geq k + 1 \Rightarrow \mathrm{ad}(p') \geq k \Rightarrow$$
$$\Rightarrow b'_i \geq \min\{i, k-1\} \Rightarrow b_i \geq \min\{i+1, k\} \geq \min\{i, k\}.$$

Example 1 Proposition 3 cannot be inverted. For instance consider the pattern $p = x_1 + 3x_2 - x_3$, then $b_i \geq \min\{i, k\}$ for all $k \in \mathbb{N}$, but p has admissibility degree 4.

The next result generalizes [6, Lemma 42] and the proof is similar.

Lemma 1 *An admissible pattern p with finite admissibility degree can be written uniquely as*

$$p(x_1, \ldots, x_n) = H_p(x_1, \ldots, x_h) + C_p(x_h, \ldots, x_t) + T_p(x_{t+1}, \ldots, x_n),$$

where either $H_p = 0$ or all the coefficients of H_p are positive and their sum is equal to $\mathrm{ad}(p) - 1$*, C_p is admissible and the sum of all its coefficients is zero,* $\mathrm{ad}(T_p) > 1$.

Proof Set $\mathrm{ad}(p) = k + 1$, then p can be written uniquely as the sum

$$p(x_1, \ldots, x_n) = H_p(x_1, \ldots, x_h) + p^{(k)}(x_h, \ldots, x_n)$$

where H_p is a pattern with positive coefficients and their sum is equal to $k = \mathrm{ad}(p) - 1$, and $p^{(k)}$ is admissible with $\mathrm{ad}(p^{(k)}) = 1$. If a'_i are the coefficients of $p^{(k)}$, by Corollary 1 there exists an integer i such that $\sum_{j=h}^{i} a'_j = 0$, set t to be the largest of such integers. Set

$$C_p(x_h, \ldots, x_t) = \sum_{i=h}^{t} a'_i x_i, \quad T_p(x_{t+1}, \ldots, x_n) = \sum_{i=t+1}^{n} a'_i x_i.$$

By the choice of t it follows $\sum_{i=t+1}^{m} a'_i = \sum_{i=h}^{m} a'_i > 0$ for all $m \in \{t + 1, \ldots, n\}$, hence $\mathrm{ad}(T_p) > 1$.

If the pattern p has admissibility degree ∞, we set $H_p = p$ and $C_p = T_p = 0$. Therefore, we can write every pattern as

$$p(x_1, \ldots, x_n) = H_p(x_1, \ldots, x_h) + C_p(x_h, \ldots, x_t) + T_p(x_{t+1}, \ldots, x_n) \tag{1}$$

we will keep this notation throughout.

Definition 1 Let p be a pattern. With the notation of Lemma 1 we call H_p the *head*, C_p the *center* and T_p the *tail* of p. The decomposition (1) is the *standard decomposition* of p.

Example 2 Let $p = x_1 + 3x_2 + x_3 - 2x_4 + x_5 + x_6$, the admissibility degree of p is 4, the standard decomposition of p is

$$H_p(x_1, x_2) = x_1 + 2x_2,$$
$$C_p(x_2, x_3, x_4) = x_2 + x_3 - 2x_4,$$
$$T_p(x_5, x_6) = x_5 + x_6.$$

Corollary 2 *Any non-zero strongly admissible pattern p can be decomposed into the sum*

$$p = p_1 + q_1 + p_2 + q_2 + \cdots + p_m + q_m,$$

where the coefficients of the pattern p_i are positive, the pattern q_i is admissible and the sum of its coefficients is zero, for all $i \in \{1, \ldots, m\}$.

Proof It follows by recursively applying Lemma 1 on the tail of p.

Remark 1 Note that the head of every pattern of admissibility degree 1 is zero. Further, if p is an admissible pattern in which the sum of all coefficients is zero (i.e. $b_n = 0$), the tail of p is zero. In addition, by Proposition 3, the admissibility degree of p is 1, so the head of p is also zero, consequently p is equal to its center. Therefore, an admissible pattern is equal to its center if and only if the sum of all its coefficients is equal to zero.

The next result follows a similar idea of [21, Proposition 2.4].

Proposition 4 *Let p be an admissible pattern such that the sum of its coefficients is zero. A numerical semigroup S admits p if and only if the monoid generated by the integers $b_1, \ldots, b_n$ is a subset of S.*

Proof *Necessity* Let $i \in \{1, \ldots, n\}$ and $\lambda \in \mathbb{N}$ such that $\lambda, \lambda + 1 \in S$. Then

$$p(\underbrace{\lambda + 1, \ldots, \lambda + 1}_{i}, \lambda, \ldots, \lambda) = \sum_{j=1}^{n} a_j \lambda + \sum_{j=1}^{i} a_j = b_n \lambda + b_i = b_i \in S.$$

Sufficiency It is enough to write

$$\begin{aligned} p(x_1, \ldots, x_n) &= a_1 x_1 + \cdots + a_n x_n = \\ &= b_1(x_1 - x_2) + \cdots + b_{n-1}(x_{n-1} - x_n) + b_n x_n. \end{aligned}$$

Proposition 5 *If p has admissibility degree 1, then a numerical semigroup S admits p if and only if it admits C_p and T_p.*

Proof Sufficiency follows from $p = C_p + T_p$. For the necessity it is enough to write

$$\begin{aligned} p(x_1, \ldots, x_t, 0, \ldots, 0) &= C_p(x_1, \ldots, x_t), \\ p(x_{t+1}, \ldots, x_{t+1}, x_{t+2}, \ldots, x_n) &= T_p(x_{t+1}, \ldots, x_n), \end{aligned}$$

where t is the same index used in the proof of Lemma 1.

Corollary 3 *If p has admissibility degree 1, then a numerical semigroup S admits p if and only if S admits T_p and contains the monoid generated by $b_1, \ldots, b_t$.*

By iterating on the tail, the previous Corollary 3 with [6, Lemma 14] gives us an algorithm to determine if a numerical semigroup admits an admissible pattern. Further, the previous result allows us to extend Proposition 2 to (not necessarily strongly) admissible patterns.

Proposition 6 *If p is monic and has admissibility degree 2 with*

$$p(x_1, \ldots, x_n) = x_1 + C_p(x_2, \ldots, x_t) + T_p(x_{t+1}, \ldots, x_n),$$

then S admits p if and only if it admits $p_i(x_1, x_2, x_3) = x_1 + (b_i - 1)(x_2 - x_3)$ for all $i \in \{2, \ldots, n\}$, and $x_1 + T_p$.

Proof First, write

$$p(x_1, \ldots, x_n) = x_1 + \sum_{i=2}^{t}(b_i - 1)(x_i - x_{i+1}) + T_p(x_{t+1}, \ldots, x_n).$$

Necessity Let $i \in \{2, \ldots, n\}$, we have

$$p(x_1, \underbrace{x_2, \ldots, x_2}_{i-1}, \underbrace{x_3, \ldots, x_3}_{t-i}, 0, \ldots, 0) = x_1 + (b_i - 1)(x_2 - x_3),$$

$$p(\underbrace{x_1, \ldots, x_1}_{t}, x_{t+1}, \ldots, x_n) = x_1 + T_p(x_{t+1}, \ldots, x_n).$$

Sufficiency Let $\lambda_1, \ldots, \lambda_n \in S$ with $\lambda_1 \geq \ldots \lambda_n$. We can write

$$p(\lambda_1, \ldots, \lambda_t, 0, \ldots, 0) = \lambda_1 + \sum_{i=2}^{t}(b_i - 1)(\lambda_i - \lambda_{i+1}).$$

By hypothesis $\lambda_1 + (b_2 - 1)(\lambda_2 - \lambda_3) \in S$ and it is greater than λ_1. Thus also $\left(\lambda_1 + (b_2 - 1)(\lambda_2 - \lambda_3)\right) + (b_3 - 1)(\lambda_3 - \lambda_4) \in S$. By iterating this process we obtain $p(\lambda_1, \ldots, \lambda_t, 0, \ldots, 0) \in S$ and it is greater than λ_1. Finally, since S admits $x_1 + T_p$, we have

$$p(\lambda_1, \ldots, \lambda_n) = p(\lambda_1, \ldots, \lambda_t, 0, \ldots, 0) + T_p(\lambda_{t+1}, \ldots, \lambda_n) \in S.$$

4 Patterns Equivalent to the Arf Pattern

The next result is a straightforward generalization of [6, Proposition 34].

Lemma 2 *A numerical semigroup S admits every pattern of admissibility degree greater or equal than $\lceil \frac{c(S)}{m(S)} \rceil + 1$.*

Proof Write

$$p(x_1, \ldots, x_n) = H_p(x_1, \ldots, x_h) + C_p(x_{h+1}, \ldots, x_t) + T_p(x_{t+1}, \ldots, x_n),$$

and recall that the coefficients of H_p are positive and their sum is equal to $\mathrm{ad}(p) - 1 \geq \lceil \frac{\mathrm{c}(S)}{\mathrm{m}(S)} \rceil$. Let $s_1, \ldots, s_n \in S$ with $s_1 \geq \cdots \geq s_n$. If $s_{h+1} < \mathrm{m}(S)$, then $s_{h+1} = s_{h+2} = \cdots = s_n = 0$ and

$$p(s_1, \ldots, s_n) = \sum_{i=1}^{h} a_i s_i \in S.$$

On the other hand, if $s_{h+1} \geq \mathrm{m}(S)$, then $s_1 \geq \ldots s_h \geq \mathrm{m}(S)$, therefore

$$\begin{aligned} p(s_1, \ldots, s_n) \geq H_p(s_1, \ldots, s_h) \geq (\mathrm{ad}(p) - 1)\,\mathrm{m}(S) \geq \\ \geq \left\lceil \frac{\mathrm{c}(S)}{\mathrm{m}(S)} \right\rceil \mathrm{m}(S) \geq \mathrm{c}(S). \end{aligned}$$

Proposition 7 *If p has admissibility degree k, then there exists a numerical semigroup S that admits every pattern of admissibility degree $k+1$ but it does not admit p.*

Proof If $k = 0$ take $S = \mathbb{N}$. Assume $k \geq 1$. The sum of the coefficients of C_p is zero, therefore we can write

$$p(x_1, \ldots, x_n) = H_p(x_1, \ldots, x_h) + \sum_{i=h+1}^{t} c_i(x_i - x_{i+1}) + T_p(x_{t+1}, \ldots, x_n)$$

for some $c_i \in \mathbb{N}$. Note that there exists $r \in \{h+1, \ldots, t\}$ such that $c_r > 0$. Now let $q \in \mathbb{N}$ such that $q > c_r + k - 1$. Set $S = \langle q, q+1 \rangle \cup (kq + \mathbb{N})$, then

$$p(\underbrace{q+1, \ldots, q+1}_{r}, \underbrace{q, \ldots, q}_{t-r}, 0, \ldots, 0) = (k-1)(q+1) + c_r = \lambda,$$

with $(k-1)q + k - 1 < \lambda < kq$, therefore $\lambda \notin S$, so S does not admit p. Nonetheless, since $\mathrm{c}(S) = kq = k\,\mathrm{m}(S)$, from the preceding lemma S admits every pattern of admissibility degree $k+1$.

Corollary 4 *Let p and q be two patterns.*

1. *If p induces q, then* $\mathrm{ad}(p) \leq \mathrm{ad}(q)$.
2. *If p and q are equivalent, then* $\mathrm{ad}(p) = \mathrm{ad}(q)$.

Lemma 3 *The Arf pattern induces the pattern $x_1 + n(x_2 - x_3)$ for every $n \in \mathbb{N}$.*

Proof We prove this by induction on n. The case $n = 0$ is the pattern x_1, the case $n = 1$ is the Arf pattern itself. For the inductive step, suppose that the Arf pattern induces $x_1 + n(x_2 - x_3)$, then it is enough to write

$$x_1 + (n+1)(x_2 - x_3) = \big(x_1 + n(x_2 - x_3)\big) + x_2 - x_3.$$

Recall that a pattern p is strongly admissible if and only if it has admissibility degree at least 2.

Proposition 8 *The Arf pattern induces every strongly admissible pattern.*

Proof Let p be a strongly admissible pattern, so $\mathrm{ad}(p) \geq 2$. We proceed by induction on the number of variables n of the pattern p. If $n = 1$ then p is equivalent to the zero pattern, so the Arf pattern induces p. Now, for the inductive step, suppose that the Arf pattern induces every pattern of admissibility degree at least 2 with at most $n - 1$ variables. Since p' induces p, it is enough to prove that the Arf pattern induces every pattern of admissibility degree 2 with n variables. So assume $\mathrm{ad}(p) = 2$. Suppose that S admits the Arf pattern. Let $s_1, \ldots, s_n \in S$ with $s_1 \geq \cdots \geq s_n$. From Lemma 1 we have

$$p(s_1, \ldots, s_n) = s_1 + \sum_{i=1}^{t-1} (b_i - 1)(s_i - s_{i+1}) + T_p(s_{t+1}, \ldots, s_n),$$

note that $b_t - 1 = 0$ and $t > 1$. From Lemma 3 the Arf pattern induces the pattern $x_1 + (b_1 - 1)(x_2 - x_3)$, so $s_1' = s_1 + (b_1 - 1)(s_1 - s_2) \in S$ with $s_1' \geq s_1$. Similarly, since the Arf pattern induces the pattern $x_1 + (b_2 - 1)(x_2 - x_3)$, then

$$s_2' = s_1' + (b_2 - 1)(s_2 - s_3) = s_1 + (b_1 - 1)(s_1 - s_2) + (b_2 - 1)(s_2 - s_3) \in S,$$

with $s_2' \geq s_1$. Iterating this process we obtain that

$$s = s_1 + \sum_{i=1}^{t-1} (b_i - 1)(s_i - s_{i+1}) \in S.$$

Since $t > 1$, the number of variables of the pattern $x_1 + T_p$ is less than n. By the inductive hypothesis, the Arf pattern induces $x_1 + T_p(x_{t+1}, \ldots, x_n)$, so

$$p(s_1, \ldots, s_n) = s + T_p(s_{t+1}, \ldots, s_n) \in S.$$

What we have so far is that for $k = 1, 2$, the subtraction pattern of degree k induces all patterns of admissibility degree at least k. As [6, Example 50] shows, this cannot be extended to $k \geq 3$.

Theorem 1 *A pattern $p = \sum_{i=1}^{n} a_i x_i$ is equivalent to the Arf pattern if and only if it has admissibility degree 2 and there exists $i \in \{1, \ldots, n\}$ such that $b_i = \sum_{j=1}^{i} a_j = 2$.*

Proof From Corollary 4, we can assume that $\mathrm{ad}(p) = 2$. Now from Proposition 8, the Arf pattern induces p. If there exists i such that $b_i = 2$, then

$$p(x_1, \ldots, x_n) = x_1 + \sum_{i=1}^{t} (b_i - 1)(x_i - x_{i+1}) + T_p(x_{t+1}, \ldots, x_n) \Rightarrow$$

$$\Rightarrow p(\underbrace{x_1, \ldots, x_1}_{i}, \underbrace{x_2, \ldots, x_2}_{t-i}, 0, \ldots, 0) = x_1 + (b_i - 1)(x_1 - x_2) = 2x_1 - x_2.$$

Therefore p induces the pattern $2x_1 - x_2$ which is equivalent to the Arf pattern. On the other hand, suppose that $b_i \neq 2$ for all $i \in \{1, \ldots, n\}$. Then, from Proposition 3, either $b_i = 1$ or $b_i \geq 3$. Let $q > 1$ and $S = \{q, q+1, q+3, \rightarrow\}$. From Lemma 2, S admits every pattern of admissibility degree greater or equal than 3. In particular, S admits $x_1 + T_p$. Now let $s_1, \ldots, s_n \in S$ with $s_1 \geq \cdots \geq s_n$. If for every $i \in \{1, \ldots, t\}$ either $b_i = 1$ or $s_i = s_{i+1}$, then

$$p(s_1, \ldots, s_n) = s_1 + T_p(s_{t+1}, \ldots, s_n) \in S.$$

Otherwise, there exists $i \in \{1, \ldots t\}$ such that $s_i > s_{i+1}$ and $b_i \geq 3$, then $s_1 \geq s_i > s_{i+1} \geq q \Rightarrow s_1 \geq q + 1$, and

$$p(s_1, \ldots, s_n) \geq s_1 + (b_i - 1)(s_{i+1} - s_i) \geq q + 3 = \mathrm{c}(S).$$

Clearly, S is not Arf since $2(q + 1) - q = q + 2 \notin S$, therefore p is not equivalent to the Arf pattern.

Note that Corollary 3 and Theorem 1, generalize and provide another proof of [6, Proposition 48], since if p is a Boolean pattern of admissibility degree k, then $b_k = k$.

5 Patterns on the Numerical Duplication

In this section S will be a numerical semigroup, E will be an ideal of S, $d \in S$ will be an odd integer and $p = \sum_{i=1}^{n} a_i x_i$ will be an admissible pattern. We say that the numerical duplication $S \bowtie^d E$ *admits p eventually with respect to d* if there exists $d' \in \mathbb{N}$ such that $S \bowtie^d E$ admits p for all $d \geq d'$.

Proposition 9 *If S admits p then also $\frac{S}{k}$ admits p for every $k \geq 1$.*

Proof If $\lambda_1 \geq \cdots \geq \lambda_n$ are elements of $\frac{S}{k}$, then $k\lambda_1 \geq \cdots \geq k\lambda_n$ are in S. Therefore

$$p(k\lambda_1, \ldots, k\lambda_n) = kp(\lambda_1, \ldots, \lambda_n) \in S \Rightarrow p(\lambda_1, \ldots, \lambda_n) \in \frac{S}{k}.$$

For the next result, recall that $\frac{S \bowtie^d E}{2} = S$.

Corollary 5 *If $S \bowtie^d E$ admits p then S admits p.*

Throughout we will assume that S admits the pattern p. Note that if p has admissibility degree 2, then by applying Corollary 1 to the center of p, we obtain that the set $B = \{i : b_i - 1 = 0\}$ is nonempty.

Proposition 10 *Suppose that p has admissibility degree 2 and set*

$$B = \{i : b_i - 1 = 0\}, \quad r = \min B, \quad t = \max B.$$

If $S \bowtie^d E$ admits p eventually with respect to d, then

1. for every $1 \leq i \leq t$, $\lfloor \frac{b_i}{2} \rfloor \in E - E$;
2. for every $r \leq i \leq t$, if b_i is even then $b_i/2 \geq \mathrm{c}(E) - \min(E)$.

Proof From Lemma 1, we can write p in the following manner

$$p = x_1 + \sum_{i=1}^{t-1} (b_i - 1)(x_i - x_{i+1}) + T_p(x_{t+1}, \ldots, x_n).$$

Now, assume $d \geq 2\,\mathrm{c}(S) - 2\min(E) + 1$, then we have that

$$(2\min(E) + d - 1) + 2 \cdot \mathbb{N} \subseteq 2\,\mathrm{c}(S) + 2 \cdot \mathbb{N} \subseteq 2 \cdot S \subseteq S \bowtie^d E.$$

Let $i \in \{1, \ldots, t-1\}$, $e \in E$ and fix $\lambda = 2e + d - 1$. By the assumption on d we have that $\lambda \in S \bowtie^d E$. If b_i is odd, it follows that

$$p(\underbrace{\underbrace{\lambda + 1, \ldots, \lambda + 1}_{i}, \lambda, \ldots, \lambda}_{t}, 0, \ldots, 0) = \lambda + 1 + (b_i - 1) =$$

$$= 2e + d + b_i - 1 = 2(e + (b_i - 1)/2) + d \in S \bowtie^d E,$$

hence $e + (b_i - 1)/2 \in E$, so by the arbitrary choice of $e \in E$ we have $(b_i - 1)/2 \in E - E$. On the other hand, if b_i is even, then

$$p(\underbrace{\underbrace{\lambda + 2, \ldots, \lambda + 2}_{i}, \lambda + 1, \ldots, \lambda + 1}_{r}, 0, \ldots, 0) = \lambda + 2 + (b_i - 1) =$$

$$= 2e + d + b_i = 2(e + b_i/2) + d \in S \bowtie^d E$$

hence $e + b_i/2 \in E$, so as before $b_i/2 \in E - E$. This proves that $\lfloor \frac{b_i}{2} \rfloor \in E - E$. Now let $i \in \{r, \dots, t\}$ such that b_i is even and set $\lambda = 2\min(E) + d + 1$. Let $x \in \mathbb{N}$ and set $\mu = \lambda + 2x$. Again by the assumption on d we have that $\mu, \lambda \in S \bowtie^d E$. Thus

$$p(\underbrace{\underbrace{\underbrace{\mu, \dots, \mu}_{r}, \lambda \dots, \lambda}_{i}, \lambda - 1, \dots, \lambda - 1}_{t}, 0, \dots, 0) = \mu + (b_i - 1) =$$

$$= \lambda + 2x + b_i - 1 = 2(\min(E) + b_i/2 + x) + d \in S \bowtie^d E,$$

hence $\min(E) + b_i/2 + x \in E$. By the arbitrary choice of x we have that $\min(E) + b_i/2 \geq \mathrm{c}(E) \Rightarrow b_i/2 \geq \mathrm{c}(E) - \min(E)$.

Proposition 11 *If p has admissibility degree 2 and is monic, then $S \bowtie^d E$ admits p eventually with respect to d if and only if*

1. for every $i \in \{1, \dots, t\}$

- *if b_i is odd then $(b_i - 1)/2 \in E - E$;*
- *if b_i is even then $b_i/2 \geq \mathrm{c}(E) - \min(E)$.*

2. $S \bowtie^d E$ admits $x_1 + T_p$.

Proof *Necessity* The first condition follows from Proposition 10 since we have $b_1 - 1 = 0$. Further, if we take $x_2 = x_3 = \cdots = x_t$, then

$$p(x_1, x_2 \dots, x_2, x_{t+1}, \dots, x_n) = x_1 + T_p(x_{t+1}, \dots, x_n).$$

Sufficiency From Proposition 6 it is enough to show that $S \bowtie^d E$ admits $p_i(x_1, x_2, x_3) = x_1 + (b_i - 1)(x_2 - x_3)$ for all $i \in \{2, \dots, n\}$. Let $i \in \{2, \dots, n\}$ and $\lambda_1, \lambda_2, \lambda_3 \in S \bowtie^d E$ with $\lambda_1 \geq \lambda_2 \geq \lambda_3$. If $\lambda_2 = \lambda_3$, then $p_i(\lambda_1, \lambda_2, \lambda_3) = \lambda_1 \in S \bowtie^d E$, so we can assume $\lambda_2 > \lambda_3$. Since S admits p, it admits also p_i, so if $\lambda_1 < 2\min(E) + d$ then $\lambda_1, \lambda_2, \lambda_3 \in 2 \cdot S$ and $p_i(\lambda_1, \lambda_2, \lambda_3) \in 2 \cdot S \subseteq S \bowtie^d E$. Now assume that $\lambda_1 \geq 2\min(E) + d$. If b_i is even then, $b_i \geq 2(\mathrm{c}(E) - \min(E))$ and we have

$$\begin{aligned} p_i(\lambda_1, \lambda_2, \lambda_3) &= \lambda_1 + (b_i - 1)(\lambda_2 - \lambda_3) \geq \lambda_1 + b_i - 1 \geq \\ &\geq 2\min(E) + d + 2(\mathrm{c}(E) - \min(E)) - 1 = \\ &= 2\,\mathrm{c}(E) + d - 1 = c(S \bowtie^d E), \end{aligned}$$

therefore $p_i(\lambda_1, \lambda_2, \lambda_3) \in S \bowtie^d E$. On the other hand, if b_i is odd, then $\mu = (b_i - 1)(\lambda_2 - \lambda_3) \in 2(E - E)$ since $E - E$ is a semigroup. Now if λ_1 is even, then $\lambda_1 + \mu$ is also even, so for $d \gg 0$ we have $\lambda_1 + \mu \in 2 \cdot S \subseteq S \bowtie^d E$. If $\lambda_1 = 2e + d \in 2 \cdot E + d$, then $\lambda_1 + \mu = 2(e + \mu/2) + d \in 2 \cdot E + d \subseteq S \bowtie^d E$ since $\mu \in 2(E - E)$.

Proposition 12 *If p has admissibility degree at least 3 and it is not monic (i.e. $a_1 \geq 2$), then $S \bowtie^d E$ admits p eventually with respect to d.*

Proof Let $\lambda_1, \ldots, \lambda_n \in S \bowtie^d E$ with $\lambda_1 \geq \cdots \geq \lambda_n$. Since S admits p, if $\lambda_1 < 2\min(E) + d$ then $\lambda_i \in 2 \cdot S$ for all $i \in \{1, \ldots, n\}$ and we have $p(\lambda_1, \ldots, \lambda_n) \in 2 \cdot S \subseteq S \bowtie^d E$. Now assume that $\lambda_1 \geq 2\min(E) + d$. Note that, since p has admissibility degree at least 3, p'' is admissible, so $p''(\lambda_1, \ldots, \lambda_n) \geq 0$. Now if we take $d \geq 2\,\mathrm{c}(E) - 4\min(E)$, then

$$p(\lambda_1, \ldots, \lambda_n) = 2\lambda_1 + p''(\lambda_1, \ldots, \lambda_n) \geq 4\min(E) + 2d \geq$$
$$\geq 2\,\mathrm{c}(E) + d \geq c(S \bowtie^d E),$$

hence $p(\lambda_1, \ldots, \lambda_n) \in S \bowtie^d E$.

Proposition 13 *If p is monic with admissibility degree at least 3, then $p'(S) \subseteq E - E$ if and only if $S \bowtie^d E$ admits p eventually with respect to d.*

Proof *Necessity* Let $\lambda_1, \ldots, \lambda_n \in S \bowtie^d E$ with $\lambda_1 \geq \cdots \geq \lambda_n$. First assume that $\lambda_2 < 2\min(E) + d$, so $\lambda_i = 2s_i$ with $s_i \in S$ for all $i \geq 2$. Now if $\lambda_1 \in 2 \cdot S$, then $p(\lambda_1, \ldots, \lambda_n) \in 2 \cdot S \subseteq S \bowtie^d E$. Otherwise, if $\lambda_1 = 2e + d \in 2 \cdot E + d$, then fix $g = p'(s_2, \ldots, s_n) \in p'(S) \subseteq E - E$, we have $g + e \in E$, hence

$$\begin{aligned} p(\lambda_1, \ldots, \lambda_n) &= 2e + d + p'(2s_2, \ldots, 2s_n) = \\ &= 2e + d + 2p'(s_2, \ldots, s_n) = \\ &= 2(e + g) + d \in 2 \cdot E + d \subseteq S \bowtie^d E. \end{aligned}$$

On the other hand, if $\lambda_2 \geq 2\min(E) + d$, take $d \geq 2\,\mathrm{c}(E) - 4\min(E)$. Since p has admissibility degree at least 3, p'' is admissible, so $p''(\lambda_2, \ldots, \lambda_n) \geq 0$, then

$$p(\lambda_1, \ldots, \lambda_n) = \lambda_1 + \lambda_2 + p''(\lambda_2, \ldots, \lambda_n) \geq 4\min(E) + 2d \geq$$
$$\geq 2\,\mathrm{c}(E) + d \geq c(S \bowtie^d E),$$

hence $p(\lambda_1, \ldots, \lambda_n) \in S \bowtie^d E$.

Sufficiency Let $g = p'(s_1, \ldots, s_{n-1}) \in p'(S)$, with $s_1, \ldots, s_{n-1} \in S$ and $s_1 \geq \cdots \geq s_{n-1}$. Let $e \in E$, it is enough to prove that $g+e \in E$. If $2s_1 < 2\min(E)+d \leq 2e+d$, it follows that

$$\begin{aligned} p(2e+d, 2s_1, \ldots, 2s_{n-1}) &= 2e+d+p'(2s_1, \ldots, 2s_{n-1}) = \\ &= 2e+d+2p'(s_1, \ldots, s_{n-1}) = \\ &= 2(e+g)+d \in S \bowtie^d E \Rightarrow e+g \in E. \end{aligned}$$

On the other hand, if $2s_1 \geq 2\min(E)+d$, take $d \geq 2\,\mathrm{c}(E) - 2\min(E)$, then

$$\begin{aligned} 2g &= p'(2s_1, \ldots, 2s_{n-1}) = \\ &= 2s_1 + p''(2s_1, \ldots, 2s_{n-1}) \geq 2s_1 \geq 2\min(E)+d \geq 2\,\mathrm{c}(E), \end{aligned}$$

hence $g \geq \mathrm{c}(E) \Rightarrow g \in E - E$.

Assembling Corollary 3, Proposition 11 and Proposition 13 and iterating these results on $H_p + T_p$, we are able to characterize when the numerical duplication $S \bowtie^d E$ admits a monic pattern p for $d \gg 0$.

Theorem 2 *Let p be a monic pattern, written as*

$$p(x_1, \ldots, x_n) = H_p(x_1, \ldots, x_h) + C_p(x_{h+1}, \ldots, x_t) + T_p(x_{t+1}, \ldots, x_n).$$

Then $S \bowtie^d E$ admits p eventually with respect to d if and only if one of the following cases occurs:

1. $\mathrm{ad}(p) = 1$, $S \bowtie^d E = \mathbb{N}$.
2. $\mathrm{ad}(p) = 2$, *for every* $i \in \{1, \ldots, t\}$
   - *if b_i is odd then* $(b_i - 1)/2 \in E - E$;
   - *if b_i is even then* $b_i/2 \geq \mathrm{c}(E) - \min(E)$;

   and $S \bowtie^d E$ admits $x_1 + T_p$.
3. $\mathrm{ad}(p) \geq 3$ *and* $p'(S) \subseteq E - E$.

From Propositions 10, 12 and Corollary 3, in order to extend the previous theorem to not monic patterns, we would need just a sufficient condition in the case $\mathrm{ad}(p) = 2$.

In the general case, that is when d can be small, we can extend the characterization of [3, Theorem 2.4] by combining it with Theorem 1. Nonetheless, as the following examples show, it seems complicated to find a sort of characterization for a generic pattern.

Example 3 The following tables show for which values of d the numerical duplication $S \bowtie^d E$ admits p.

$S = \langle 3, 19, 20 \rangle$
$E = 3 + S$
$p(x_1, x_2) = 3x_1 - x_2$.

d	admits p
3	✓
9	✓
15	✓
19	
21	✓
23	
25	
27	✓
29	✓

$S = \langle 5, 8, 19, 22 \rangle$
$E = 5 + S$
$p(x_1, x_2, x_3) = 4x_1 - x_2 - x_3$.

d	admits p
5	
13	✓
15	
19	✓
21	
23	
25	✓
27	✓
29	✓

6 Patterns on Rings

In this section, $(R, \mathfrak{m})$ will be a one-dimensional, Noetherian, Cohen–Macaulay, local ring, $\overline{R}$ will be the integral closure of R in its total ring of fractions $Q(R)$. An ideal I of R is *open* if it contains a regular element. We will assume that the residue field $k = R/\mathfrak{m}$ is infinite. From [13, Proposition 1.18, pag 74], the last condition assures that every open ideal I has an *I-transversal element*, namely an element $x \in I$ such that $xI^n = I^{n+1}$ for $n \gg 0$. On R we define the following *preorder* (namely a reflexive and transitive relation): let $x, y \in R$, then $x \leq_R y$ if $y/x \in \overline{R}$. Let p be the pattern

$$p(x_1, \ldots, x_n) = \sum_{i=1}^{n} a_i x_i.$$

Definition 2 The ring R *admits* the pattern p if for every $y_1, \ldots, y_n \in R$ with $y_1 \geq_R \cdots \geq_R y_n$, we have

$$y_1^{a_1} y_2^{a_2} \ldots y_n^{a_n} \in R.$$

With this definition, when the relation $\leq_R$ is a total preorder, R is an Arf ring if and only if it admits the Arf pattern. Note that R admits the trivializing pattern if and only if $R = \overline{R}$.

Remark 2 The ring R admits the pattern $n(x_1 - x_2)$, with $n \in \mathbb{N}$, if and only if for every $z \in \overline{R}$ it results $z^n \in R$. In fact, for every $z \in \overline{R} \subseteq Q(R)$, there exist $x, y \in R$ such that $z = y/x$, and by definition $y \geq_R x$.

From the previous remark we can determine when a ring R admits a pattern of admissibility degree 1 applying, mutatis mutandis, Corollary 3.

Corollary 6 *The ring R admits a pattern p of admissibility degree* 1 *if and only if it admits T_p and for every $z \in \overline{R}$ it results $z^{b_i} \in R$ for all $i \in \{1, \dots n\}$.*

Similarly, if p is monic and $\mathrm{ad}(p) = 2$, we can apply, mutatis mutandis, Proposition 6

Now we make additional assumptions on R. Following [1], let V be a discrete valuation domain with valuation $v : V \to \mathbb{N}$, and let $\mathcal{V}$ be the set of all subrings R of V such that R is a local, Noetherian, one-dimensional, analytically irreducible, residually rational, domain and its integral closure $\overline{R}$ is equal to V. Set $\mathscr{V}(p)$ be the family of rings in $\mathcal{V}$ that admit the pattern p. If p_1 and p_2 are two patterns, then it is clear that if $\mathscr{V}(p_1) \subseteq \mathscr{V}(p_2)$ then $\mathscr{S}(p_1) \subseteq \mathscr{S}(p_2)$, i.e. p_1 induces p_2. A question naturally arise.

Question 1 Is the implication $\mathscr{S}(p_1) \subseteq \mathscr{S}(p_2) \Rightarrow \mathscr{V}(p_1) \subseteq \mathscr{V}(p_2)$ true?

Now fix $R \in \mathcal{V}$, note that, since $\overline{R}$ is a valuation ring, $\leq_R$ is a total preorder. Further, $x \leq_R y$ if and only if $v(x) \leq v(y)$. In this setting, the integral closure of an ideal I of R is

$$\overline{I} = I\overline{R} \cap I = \{x \in R : v(x) \geq \min v(I)\},$$

(see [12, Proposition 1.6.1, Proposition 6.8.1]). It is not difficult to prove (see for instance [14, Theorem 2.2] or [1, Theorem II.2.13]) that R is an Arf ring if and only if $I^2 = xI$ for every integrally closed ideal $I \subseteq R$ and some $x \in I$ of minimum value. Actually, the inclusion $xI \subseteq I^2$ is always true, so what we actually prove is that $I^2 \subseteq xI$. We can generalize this idea to any subtraction pattern of degree k.

Proposition 14 *The ring R admits the subtraction pattern of degree k if and only if $I^k \subseteq xI$ for every integrally closed ideal $I \subseteq R$ and some $x \in I$ of minimum value.*

Proof *Necessity* Let $i_1, i_2, \dots, i_k \in I$, since $\leq_R$ is a total preorder, we can assume that $i_1 \geq_R \cdots \geq_R i_k$. If $x \in I$ is an element of minimum value, then $i_k \geq_R x$. By hypothesis $i_1 i_2 \dots i_k x^{-1} \in I$, hence $i_1 i_2 \dots i_k \in xI$.

Sufficiency Let $y_1, \dots, y_{k+1} \in R$ with $y_1 \geq_R \cdots \geq_R y_{k+1}$. Set I to be the integral closure of Ry_{k+1}. Since $v(y_1) \geq v(y_2) \geq \cdots \geq v(y_{k+1})$ and I is integrally closed, then $y_i \in I$ for all $i \in \{1, \dots, k\}$. By hypothesis $y_1 \dots y_k \in I^k \subseteq y_{k+1}I$, then $y_1 \dots y_k y_{k+1}^{-1} \in I \subseteq R$.

In [1, Theorem II.2.13] it was proved that R is Arf if and only if $v(R)$ is Arf and the multiplicity sequence of R and $v(R)$ coincides.

Question 2 For an arbitrary pattern p are there any characterization similar to the previous one?

Acknowledgements I would like to thank Marco D'Anna for his constant support, Maria Bras-Amóros for useful conversations and email exchanges, and Nicola Maugeri for indicating some good references.

References

1. Barucci, V., Dobbs, D.E., Fontana, M.: Maximality properties in numerical semigroups and applications to one-dimensional analytically irreducible local domains, vol. 598. American Mathematical Society, Providence (1997)
2. Barucci, V., D'Anna, M., Strazzanti, F.: A family of quotients of the Rees algebra. Commun. Algebra **43**(1), 130–142 (2015)
3. Borzì, A.: A characterization of the Arf property for quadratic quotients of the Rees algebra. J. Algebra Appl. (2018). https://doi.org/10.1142/S0219498820501273
4. Bras-Amorós, M.: Improvements to evaluation codes and new characterizations of Arf semigroups. In: International Symposium on Applied Algebra, Algebraic Algorithms, and Error-Correcting Codes, pp. 204–215. Springer, Berlin (2003)
5. Bras-Amorós, M.: On numerical semigroups and the redundancy of improved codes correcting generic errors. Des. Codes Crypt. **53**(2), 111 (2009)
6. Bras-Amorós, M., García-Sánchez, P.A.: Patterns on numerical semigroups. Linear Algebra Appl. **414**(2–3), 652–669 (2006)
7. Bras-Amorós, M., García-Sánchez, P.A., Vico-Oton, A.: Nonhomogeneous patterns on numerical semigroups. Int. J. Algebra Comput. **23**(6), 1469–1483 (2013)
8. D'Anna, M., Fontana, M.: An amalgamated duplication of a ring along an ideal: the basic properties. J. Algebra Appl. **6**(3), 443–459 (2007)
9. D'Anna, M., Strazzanti, F.: The numerical duplication of a numerical semigroup. In: Semigroup Forum. vol. 87, pp. 149–160. Springer, Berlin (2013)
10. Delgado, M., García-Sánchez, P.A., Morais, J.: NumericalSgps. A package for numerical semigroups, version 1.0.1 (2015). http://www.gap-system.org/Packages/numericalsgps.html
11. García-Sánchez, P.A., Heredia, B.A., Karakaş, H., Rosales, J.C.: Parametrizing Arf numerical semigroups. J. Algebra Appl. **16**(11), 1750209 (2017)
12. Huneke, C., Swanson, I.: Integral Closure of Ideals, Rings, and Modules, vol. 13. Cambridge University Press, Cambridge (2006)
13. Kiyek, K., Vicente, J.L.: Resolution of Curve and Surface Singularities in Characteristic Zero, vol. 4. Springer, Berlin (2012)
14. Lipman, J.: Stable ideals and Arf rings. Am. J. Math. **93**(3), 649–685 (1971)
15. Oneto, A., Strazzanti, F., Tamone, G.: One-dimensional Gorenstein local rings with decreasing Hilbert function. J. Algebra **489**, 91–114 (2017)
16. Rosales, J.C., García-Sánchez, P.A.: Numerical Semigroups, vol. 20. Springer, Berlin (2009)
17. Rosales, J.C., García-Sánchez, P.A., García-García, J.I., Branco, M.B.: Arf numerical semigroups. J. Algebra **276**(1), 3–12 (2004)
18. Rossi, M.E.: Hilbert functions of Cohen–Macaulay local rings. In: Commutative Algebra and Its Connections to Geometry: Pan-American Advanced Studies Institute, vol. 555, p. 173. Universidade Federal de Pernambuco, Olinda (2011)
19. Stokes, K.: Patterns of ideals of numerical semigroups. In: Semigroup Forum, vol. 93, pp. 180–200. Springer, Berlin (2016)
20. Stokes, K., Bras-Amorós, M.: Linear, non-homogeneous, symmetric patterns and prime power generators in numerical semigroups associated to combinatorial configurations. In: Semigroup Forum, vol. 88, pp. 11–20. Springer, Berlin (2014)
21. Sun, G., Zhao, Z.: Generalizing strong admissibility of patterns of numerical semigroups. Int. J. Algebra Comput. **27**(1), 107–119 (2017)
22. The GAP Group: Gap–groups, algorithms, and programming, version 4.7.9 (2015). http://www.gap-system.org

Primality in Semigroup Rings

Brahim Boulayat and Said El Baghdadi

Abstract Following P.M. Cohn, an element x in an integral domain A is primal if whenever $x \mid a_1a_2$ with $a_1, a_2 \in A$, x can be written as $x = x_1x_2$ such that $x_i \mid a_i$, $i = 1, 2$, and x is completely primal if every factor of x is primal. A ring in which every element is (completely) primal is called a pre-Schreier domain and an integrally closed pre-Schreier domain is called a Schreier domain. In this paper, we study (completely) primal elements and shed more light on the Schreier property in semigroup rings.

Keywords Monoid · Primal · Completely primal · Schreier domain · Pre-Schreier domain

2020 Mathematics Subject Classification 13A15, 13F05, 13B30, 13C11, 13F20, 13G05

1 Introduction

Let A be an integral domain. Following P.M. Cohn [7], an element $x \in A$ is primal if whenever $x \mid a_1a_2$ with $a_1, a_2 \in A$, x can be written as $x = x_1x_2$ such that $x_i \mid a_i$, $i = 1, 2$, and x is completely primal if every factor of x is primal. A ring in which every element is (completely) primal is called a pre-Schreier domain [16] and an integrally closed pre-Schreier domain is called a Schreier domain [7]. The Schreier property generalizes the GCD property.

The primality of an element in a domain depends only on the multiplicative semigroup of nonzero elements of that domain. This led several authors to study the primality in the more general context of semigroups. Let S be a commutative multiplicative cancellative monoid. For $s, t \in S$, $s \mid t$ if $t = sr$ for some $r \in S$. An

---

B. Boulayat · S. El Baghdadi (✉)
Department of Mathematics, Beni Mellal University, Faculté des Sciences et Techniques, Beni Mellal, Morocco

V. Barucci et al. (eds.), *Numerical Semigroups*, Springer INdAM Series 40,
https://doi.org/10.1007/978-3-030-40822-0_3

element $s \in S$ is primal if for $t_1, t_2 \in S$, $s \mid t_1t_2$ implies $s = s_1s_2$ where $s_1, s_2 \in S$ and $s_i \mid t_i$, $i = 1, 2$. Completely primal elements and the (pre-)Schreier property for semigroups are defined similarly.

In a polynomial ring in the indeterminate X over a ring A, the fact that the powers X^n, $n \in \mathbb{N}$, are primal, i.e., $X^n \mid fg$ for some $f, g \in A[X]$, then $f = X^r f_1$ and $g = X^{n-r} g_1$ for some $r \in \mathbb{N}$, $f_1, g_1 \in A[X]$, is crucial when working with polynomials. This raises the question of whether this result can be extended to powers X^α, $\alpha \in S$, S a semigroup. Note that in this case X^α is not necessarily a power of a prime element like in the polynomial rings. On the other hand, an interesting work on the Schreier property for semigroup rings was made by Matsuda [13] and Brookfield and Rush [6]. In [6], the authors showed that a semigroup ring is pre-Schreier if and only if it is Schreier.

The aim of this paper is to deepen and shed new light on primality in semigroup rings. In Sect. 1, we write some well known results on primal elements and Schreier property, in domains and ordered groups, in the language of monoids. In Sect. 2 we study primality in the more general context of graded domains. In [6] it was shown that in graded domains the Schreier property can be reduced to the study of the primality of the homogeneous elements. In this section we characterize the (completely) primality of an homogeneous element in terms of its (completely) primality in the multiplicative semigroup of nonzero homogeneous elements. In the integrally closed case we get an equivalence between these two primalities. As an application, in Sect. 3 we characterize primal elements in semigroup rings. In particular, we investigate the primality of the powers X^α in a semigroup ring and recover the case of polynomial rings.

2 Primal Elements in Monoids

Throughout this section a monoid means (multiplicative) commutative cancellative unitary semigroup. Let S be a monoid. If $T \subseteq S$ is a multiplicatively closed subset of S, then we get the fraction monoid $S_T := \{s/t, s \in S, t \in T\}$. If $T = S$, we have the quotient group of S, $G =< S >$.

The aim of this section is to translate and adapt the proofs of some well known results on primality and the Schreier property in domains and partially ordered groups, by using the language of monoids. These results on monoids are needed in the next sections in the case of graded domains and semigroup rings.

Let $s, t \in S$. We say that s divides t, denoted $s \mid t$, if $t = sr$ for some $r \in S$. We make use of the preoder on S: $s \leq t$ if $s \mid t$. An element $s \in S$ is primal if for $t_1, t_2 \in S$, $s \leq t_1t_2$ implies $s = s_1s_2$, where $s_1, s_2 \in S$ and $s_i \leq t_i$, $i = 1, 2$, and s is completely primal if every factor of s is primal. As for domains, a monoid in which every element is (completely) primal is called a pre-Schreier monoid and an integrally closed pre-Schreier monoid is called a Schreier monoid. Note that in the case of a domain A, the monoid in question is the multiplicative monoid $A \setminus \{0\}$, and in the case of an ordered group G, it is the positive cone G^+.

In [3], the authors believe that completely primal elements are the building blocks of the Schreier property. In what follows we give some characterizations of completely primal elements in monoids. For $x_1, \ldots, x_n \in S$, let $U(x_1, \ldots, x_n) = \{g \in S | g \geq x_1, \ldots, x_n\}$. A nonempty subset $U \subseteq S$ is lower directed if for $s_1, s_2 \in U$, there exists $s \in U$ with $s \leq s_1, s_2$. The following lemma is well known in ordered groups [3, Theorem 2.1].

Lemma 2.1 *Let S be a monoid. An element s of S is completely primal if and only if for each $x \in S$, the set $U(s, x)$ is lower directed. Moreover, if $\{s_1, s_2, \ldots, s_n\}$ is a set of completely primal elements of S, then $U(s_1, s_2, \ldots, s_n)$ is lower directed.*

Proof The proof of the first part is similar to [3, Theorem 2.1, (1)⇔(2)]. For the second part, note that the case $n = 1$ is clear and $n = 2$ follows from the first part. Suppose that $U(s_1, s_2, \ldots, s_{n-1})$ is lower directed. Let $r_1, r_2 \in U(s_1, s_2, \ldots, s_n)$. Then $r_1, r_2 \in U(s_1, s_2, \ldots, s_{n-1})$ and by induction there exists $t \in U(s_1, s_2, \ldots, s_{n-1})$ such that $t \leq r_1, r_2$. But then $r_1, r_2 \in U(t, s_n)$ and since s_n is completely primal there is $s \in U(t, s_n)$ such that $s \leq r_1, r_2$. Hence $s \in U(t, s_n) \subseteq U(s_1, s_2, \ldots, s_n)$. □

The following key characterization of completely primal elements in monoids was proven in [4, Lemma 4.6] for domains. Here we give a short proof in the case of monoids.

Proposition 2.2 *Let S be a monoid. An element s of S is completely primal if and only if $s \leq r_i t_j$, $r_i, t_j \in S$, for $i = 1, \ldots, m$ and $j = 1, \ldots, n$ implies that $s = s_1 s_2$, where $s_1 \leq r_i$ for each i and $s_2 \leq t_j$ for each j.*

Proof Let s be a completely primal element and $s \leq r_i t_j$ for $i = 1, \ldots, m$ and $j = 1, \ldots, n$. Then for each $i = 1, \ldots, m$, $s = r_{ij} t_{ji}$, where $r_{ij} \leq r_i$ and $t_{ji} \leq t_j$ for $j = 1, \ldots, n$. Since for each i, r_{ij} is completely primal (a factor of s), and $s, r_i \in U(r_{i1}, \ldots, r_{in})$, there exists $d_i \in S$ such that $r_{ij} \leq d_i \leq r_i, s$ for every $j = 1, \ldots, n$. Now, $s, t_1, \ldots, t_n \in U(s/d_1, \ldots, s/d_n)$, then there exists $t \in S$ such that $s/d_1, \ldots, s/d_n \leq t \leq s, t_1, \ldots, t_n$. Let $r \in S$ such that $s = rt$. One can easily check that $r \leq r_i$ for $i = 1, \ldots, m$.

For the converse, let $s \in S$ satisfying the condition as in the proposition, and let $x \in S$. We show that $U(s, x)$ is lower directed. Let $r_1, r_2 \in U(s, x)$. For $i = 1, 2$, write $r_i = x t_i$, so $s \leq x t_i$. By our hypothesis $s = s_1 s_2$ such that $s_1 \leq x$ and $s_2 \leq t_i$ for $i = 1, 2$. But $d = x s_2 \in U(s, x)$ and $d \leq r_1, r_2$. Thus $U(s, x)$ is lower directed and by the previous lemma s is completely primal. □

To sum up, we get the following characterization of pre-Schreier monoids, see [16, Theorem 1.1].

Corollary 2.3 *Let S be a monoid. The following are equivalent.*

(i) *S is a pre-Schreier monoid;*
(ii) *For all $s, t, x, y \in S$ with $s, t \mid x, y$ there exists $r \in S$ such that $s, t \mid r \mid x, y$;*

(iii) *For all* $s_1, \ldots, s_m \in S$ *and* $t_1, \ldots, t_n \in S$ *such that* $s_i \mid t_j$*, for each* $i = 1, \ldots, m$ *and* $j = 1, \ldots, n$*, then there exists* $r \in S$ *such that* $s_i \mid r \mid t_j$ *for each* i, j*;*

(iv) *For all* $r_1, \ldots, r_m \in S$ *and* $t_1, \ldots, t_n \in S$ *such that* $s \mid r_i t_j$*, for each* $i = 1, \ldots, m$ *and* $j = 1, \ldots, n$*, then* $s = s_1 s_2$ *for some* $s_1, s_2 \in S$ *such that* $s_1 \mid r_i$ *and* $s_2 \mid t_j$ *for each* i, j*.*

We end this section by translating to monoids the well known Nagata type theorem for Schreier domains due to Cohn [7, Theorem 2.6]. Our proof is slightly different from that in [7] for we use the characterization of completely primals in Propostion 2.2.

Let S be a monoid and T a multiplicative subset of S. The set T is called divisor-closed if T is saturated.

Proposition 2.4 *Let* S *be a monoid and* T *a multiplicative set of* S*.*

(1) *If* S *is pre-Schreier, then* S_T *is pre-Schreier.*
(2) *Assume that* T *is a divisor-closed subset of* S *such that every element of* T *is primal in* S*. If the monoid* S_T *is pre-Schreier, then* S *is pre-Schreier.*

Proof

(1) Similar to domains [16, Corollary 1.3].
(2) Assume that S_T is pre-Schreier and let $s, x_1, x_2 \in S$ such that $s \leq x_1 x_2$ in S. So $s \leq x_1 x_2$ in S_T. Since S_T is pre-Schreier, s is completely primal in S_T. Then $s = (s_1 t_1^{-1})(s_2 t_2^{-1})$ for some $s_1, s_2 \in S$ and $t_1, t_2 \in T$ such that $s_1 t_1^{-1} \leq x_1$ and $s_2 t_2^{-1} \leq x_2$ in S_T. So $x_1 = (s_1 t_1^{-1})(s_1' r_1^{-1})$ and $x_2 = (s_2 t_2^{-1})(s_2' r_2^{-1})$ for some $s_1', s_2' \in S$ and $r_1, r_2 \in T$. We put $r = t_1 r_1 t_2 r_2$, then r is an element of T which satisfies:

$$rs = (s_1 r_2)(s_2 r_1)$$
$$r x_1 = (s_1 r_2)(s_1' t_2)$$
$$r x_2 = (s_2 r_1)(s_2' t_1)$$
$$r((x_1 x_2)/s) = (s_1' t_2)(s_2' t_1)$$

So $r \leq$ to the elements in the set product $\{s_1 r_2, s_2' t_1\}\{s_2 r_1, s_1' t_2\}$. As r is completely primal in S and by Proposition 2.2, there exist $u, v \in S$ such that $r = uv$ with $u \leq s_1 r_2, s_2' t_1$ and $v \leq s_2 r_1, s_1' t_2$. Then $s = (s_1 r_2 u^{-1})(s_2 r_1 v^{-1})$ with $s_1 r_2 u^{-1} \leq x_1$ and $s_2 r_1 v^{-1} \leq x_2$, hence s is primal in S. Consequently, S is pre-Schreier. □

3 Primal Elements in a Graded Domain

Throughout, a monoid means a torsionless grading monoid, that is, a (additive) commutative cancellative torsion-free semigroup. In this section, we study primality in a graded integral domain $R = \oplus_{\alpha\in\Gamma} R_\alpha$, graded by a torsionless grading monoid Γ. We denote by H the multiplicative set (monoid) of nonzero homogeneous elements of R. If $S \subseteq H$ is a multiplicative set of R, that is, a submonoid of H, the ring of fractions R_S is graded by some fraction monoid of Γ with the nonzero homogeneous elements are of the form h/s, where $h \in H$ and $s \in S$. In particular, $\mathcal{H}(R) = R_H$ is a $<\Gamma>$-graded domain, called the homogeneous quotient field of R. Note that $\mathcal{H}(R)$ is a completely integrally closed GCD domain [2, Proposition 2.1]. Let $x \in R$. Then $x = x_1 + \cdots + x_m$ with the x_i's are homogeneous and $deg(x_i) < deg(x_j)$ for $i < j$. A fractional ideal I of R is homogeneous if $uI \subseteq R$ is a homogeneous ideal of R for some $u \in H$. Clearly, a homogeneous fractional ideal is a submodule of $\mathcal{H}(R)$. Let $x \in \mathcal{H}(R)$, $x = x_1 + \cdots + x_m$ with $deg(x_i) < deg(x_j)$ for $i < j$. The content of x is the R-submodule of $\mathcal{H}(R)$, $C(x) = (x_1, \ldots, x_m)$. Note that a fractional ideal $I \subseteq \mathcal{H}(R)$ of R is homogeneous if and only if $C(x) \subseteq I$ for every $x \in I$. The content satisfies the Dedekind–Mertens lemma for graded domains [15]. That is, for $x, y \in \mathcal{H}(R)$, there is a positive integer n so that $C(x)^n C(xy) = C(x)^{n+1} C(y)$. For more details, see [1].

We say that an element $x \in H$ is gr-primal [6] if whenever $x \mid y_1 y_2$ with $y_1, y_2 \in H$, then $x = x_1 x_2$, $x_1, x_2 \in H$, where $x_i \mid y_i$, $i = 1, 2$, and x is completely gr-primal if every homogeneous factor of x is gr-primal. These two definitions are equivalent, respectively, to x primal and completely primal in the multiplicative monoid H. The graded domain R is called gr-pre-Schreier if every element of H is (completely) gr-primal. In [6], the authors introduced gr-pre-Schreier domains and characterized graded pre-Schreier domains in terms of the gr-pre-Schreier property. In the integrally closed case, they showed that the Schreier property is equivalent to the gr-pre-Schreier property.

For an integral domain A with quotient field K and fractional ideals I, J, define $[I : J] = \{x \in K, xJ \subseteq I\}$, $I^{-1} = [A : I]$ and $I : J = [I : J] \cap A$. A homogeneous (fractional) ideal I of the graded domain R is called H-locally cyclic if every finite subset of homogeneous elements of I is contained in a (homogeneous) principal sub-ideal of I. We start this section with some characterization of gr-pre-Schreier domains.

Proposition 3.1 *Let $R = \oplus_{\alpha\in\Gamma} R_\alpha$ be a graded domain. The following statements are equivalent.*

(i) *R is a gr-pre-Schreier domain.*
(ii) *H is a pre-Schreier monoid.*
(iii) *For every nonzero homogeneous element $u \in \mathcal{H}(R)$, $(1, u)^{-1}$ is H-locally cyclic.*
(iv) *For every nonzero $x \in \mathcal{H}(R)$, $C(x)^{-1}$ is H-locally cyclic.*

Proof (i)⇔(ii) is obvious. For (ii) ⇔(iii), note that for $a, b \in H$, we have $(a, b)^{-1} = (ab)^{-1}(aR \cap bR)$, and for a homogeneous element $u \in \mathcal{H}(R)$, $u = a/b$ for some $a, b \in H$. Then apply Corollary 2.3 (ii) in H. For (iii)⇔(iv), note that $C(x)^{-1}$ is a finite intersection of homogeneous principal fractional ideals. □

Also, we get the following Nagata type theorem for gr-pre-Schreier domains analogue to that of Schreier property due to P.M. Cohn [7, Theorem 2.6].

Proposition 3.2 *Let $R = \oplus_{\alpha \in \Gamma} R_\alpha$ be a graded domain and $S \subseteq H$ a multiplicative set of R. Then*

(i) *If R is a gr-pre-Schreier domain, then R_S is a gr-pre-Schreier domain.*
(ii) *If S is generated by completely gr-primal elements and (S divisor-closed) R_S is a gr-pre-Schreier domain, then R is a gr-pre-Schreier domain.*

Proof Apply Proposition 2.4 to the quotient monoid H_S. □

Example 3.3

(1) Let $R = A[X]$ be the polynomial ring over a ring A. One can easily see that $A[X]$ is gr-pre-Schreier if and only if A is pre-Schreier. By [6, Theorem 3.2], $A[X]$ is pre-Schreier if and only if it is Schreier, if and only if A is Schreier.
(2) Let $A \subseteq B$ be an extension of integral domains and set $R = A + XB[X]$. Primality and the Schreier property for $A + XB[X]$ domains were studied in [8, 9]. We claim that $R = A + XB[X]$ is gr-pre-Schreier if and only if A is pre-Schreier and $B = A_S$, where $S = U(B) \cap A$, $U(B)$ denotes the set of invertible elements of B. Suppose that R is gr-pre-Schreier. Clearly A is pre-Schreier. On the other hand, by using the primality of X and the fact that $X \mid (bX)^2$, $b \in B$, it was shown in [8, Remark 1.1] that $B = A_S$, where $S = U(B) \cap A$. Conversely, we use Proposition 3.2. The quotient ring $R_S = A_S[X]$ is gr-pre-Schreier since A, and hence A_S, is pre-Schreier. The elements of S are gr-primal in $R = A + XA_S[X]$. Indeed, let $a \in S$ and $h_1, h_2 \in H$ such that $a \mid h_1h_2$. Since A is pre-Schreier, the case where $h_1, h_2 \in A$ is clear. Assume that $h_2 = bX^n$ for some $b \in A_S$ and $n \neq 0$. Then $a \mid h_2$ in R, and write $a = 1 \times a$.
By [9, Theorem 2.7 and Corollary 2.9], R is a pre-Schreier (resp., Schreier) domain if and only if A is a pre-Schreier (resp., Schreier) domain, $B = A_S$, where $S = U(B) \cap A$, and A_S is a Schreier domain.

Inspired by the work in [6], in the following we study (completely) primal elements in a graded domain in terms of (completely) gr-primality.

Let $h \in H$; we say that h is degree gr-primal if $h \mid x_iy_j$, $x_i, y_j \in H$, for $i = 1, \ldots, m$ and $j = 1, \ldots, n$, with $deg(x_k) < deg(x_l)$ and $deg(y_k) < deg(y_l)$ for all $k < l$, then $h = h_1h_2$ such that $h_1 \mid x_i$ for each i and $h_2 \mid y_j$ for each j. The degree gr-primality is a weak form of the completely gr-primality in H.

Theorem 3.4 *Let $R = \oplus_{\alpha \in \Gamma} R_\alpha$ be a graded domain and $h \in H$. Then*

(i) *h is primal in R if and only if h is degree gr-primal and $(h) : (x)$ is homogeneous for each $x \in R$.*

(ii) *h is completely primal in R if and only if h is completely gr-primal and $(h) : (x)$ is homogeneous for each $x \in R$.*

Proof

(i) For the "only if" condition, assume that $h \mid x_i y_j, x_i, y_j \in H$, for $i = 1, \dots, m$ and $j = 1, \dots, n$, with $deg(x_k) < deg(x_l)$ and $deg(y_k) < deg(y_l)$ for all $k < l$, Then $h \mid xy$ in R, where $x = x_1 + \cdots + x_m$ and $y = y_1 + \cdots + y_n$. By the primality $h = h_1 h_2$, $h_1, h_2 \in H$, with $h_1 \mid x$ and $h_2 \mid y$. Clearly, $h_1 \mid x_i$ and $h_2 \mid y_j$ for each i, j. To see that $(h) : (x)$ is homogeneous, let $y \in (h) : (x)$. Then $h \mid xy$. Now, $h = h_1 h_2$, $h_1, h_2 \in H$, with $h_1 \mid x$ and $h_2 \mid y$. It follows that $C(y) \subseteq (h) : (x)$.

For the "if" condition, let $x = x_1 + \cdots + x_m$ and $y = y_1 + \cdots + y_n$ be two nonzero elements of R, with $deg(x_k) < deg(x_l)$ and $deg(y_k) < deg(y_l)$ for all $k < l$, such that $h \mid xy$. Now, $y \in (h) : (x)$, a homogeneous ideal, then $h \mid x_i y_j$ for $i = 1, \dots, m$ and $j = 1, \dots, n$. On the other hand, h is degree gr-primal implies that $h = h_1 h_2$, where $h_1 \mid x_i$ and $h_2 \mid y_j$ for each i, j. Then $h = h_1 h_2$ with $h_1 \mid x$ and $h_2 \mid y$, so h is primal in R.

(ii) For the "only if" condition, clearly, if h is completely primal in R it is completely gr-primal. The remainder is similar to (i). For the "if" condition, by the same argument as in the proof of (i), h is primal in R. To prove that h is completely primal in R, let k be a factor of h. Necessarily, $k \in H$. Then k is completely gr-primal and $h = kk'$ for some $k' \in H$. Let $x \in R$ and $y \in (k) : (x)$, with $y = y_1 + \cdots + y_n$ and $deg(y_i) < deg(y_j)$ for all $i < j$. Then $k'y \in (kk') : (x) = (h) : (x)$. Since $(h) : (x)$ is homogeneous, then, for each i, $k'y_i \in (h) : (x)$, so $y_i \in (k) : (x)$. Thus $(k) : (x)$ is homogeneous. Hence, like h, k is primal in R. Therefore, h is completely primal in R. □

Example 3.5 We give an example of a degree gr-primal element which is not completely gr-primal. Let $R = \mathbb{Z} + X\mathbb{R}[X]$. By [8, Example 1.7(ii)], X^2 is primal in R, but X is not primal in R, so X^2 is not completely primal. By Theorem 3.4, X^2 is degree gr-primal but not completely gr-primal.

Let $R = \oplus_{\alpha \in \Gamma} R_\alpha$ be a graded domain, $h \in H$, and let R_h be the quotient ring of R with respect to the multiplicative set generated by h. Note that R_h is a graded subring of $\mathcal{H}(R)$. We say that R is R_h-almost normal if every homogeneous element $x \in R_h$ of nonzero degree which is integral over R is actually in R. Note that R is R_h-integrally closed, that is, R is integrally closed in R_h, if R is integrally closed in R_h with respect to the homogeneous elements of R_h. Thus R is R_h-integrally closed if and only if R is R_h-almost normal and R_0 is integrally closed in $(R_h)_0$. Almost normality defined in [1] is a globalization of R_h-almost normality, $h \in H$. Thus, R is almost normal if and only if R is R_h-almost normal for every $h \in H$. A similar statement is true for the integrally closed case.

Recall that an extension of domains $A \subseteq B$ is inert if whenever $bb' \in A$ for some $b, b' \in B$, then $b = au$ and $b' = a'u^{-1}$ for some $a, a' \in A$ and u a unit of B.

Proposition 3.6 *Let $R = \oplus_{\alpha\in\Gamma} R_\alpha$ be a graded domain and $h \in H$. Consider the following statements.*

(i) *R is R_h-integrally closed.*
(ii) *$(h) : (x)$ is homogeneous for each $x \in R_h$.*
(iii) *R is R_h-almost normal.*

Then (i)⇒(ii)⇒(iii). Moreover, if R contains a (homogeneous) unit of nonzero degree the three conditions are equivalent, and if $R_0 \subseteq R$ is inert, then (ii)⇔(iii).

Proof The proof is inspired from [1].

(i)⇒(ii). Let $x \in R_h$ and $y \in R$ such that $C(xy) \in (h)$. Then $C(x)^n C(xy) \subseteq hC(x)^n$ implies $C(x)^{n+1}C(y) \subseteq hC(x)^n$, for some integer n in the Dedekind Mertens lemma. Thus $\frac{1}{h}C(x)C(y) \subseteq [C(x)^n : C(x)^n] \cap R_h = R$, since R is R_h-integrally closed. Hence $C(x)C(y) \in (h)$. Therefore, $(h) : (x)$ is homogeneous.

(ii)⇒(iii). Let $x = a/h^k \in R_h$, $a \in H$, a homogeneous element of nonzero degree which is integral over R. Let $f(Y) = Y^n + a_{n-1}Y^{n-1} + \cdots + a_0$ with coefficients in R such that $f(x) = 0$. Since x is homogeneous, we may assume that we have an equation of the form $x^n + a_{n-1}x^{n-1} + \cdots + a_0 = 0$ with the a_i's homogeneous and $deg(a_i) = (n-i)deg(x)$. Then $f(Y) = (Y - x)g(Y)$ with $g(Y) = Y^{n-1} + b_{n-2}Y^{n-2} + \cdots + b_0$. We may assume that the elements $b_i \in R_h$ are homogeneous of distinct nonzero degree. From $f(1) = (1 - x)g(1)$, it follows that $(1 - x)g(1) \subseteq R$. Now, $(h^k - a)(g(1)/h^{k-1}) \subseteq hR$ implies $h^k - a \subseteq (h) : (g(1)/h^{k-1})$, which is homogeneous. Since $1/h^{k-1} \in C(g(1)/h^{k-1})$, it follows that $(1/h^{k-1})(h^k - a) \in hR$. So $1 - x \in R$. Hence $x \in R$.

For the moreover statements, assume that R contains a (homogeneous) unit u of nonzero degree. If $x \in R_h$ is a homogeneous element of zero degree which is integral over R, then $ux \in R_h$ is a homogeneous element of nonzero degree which is integral over R. If R is R_h-almost normal, then $ux \in R$. Hence $x \in R$. This proves that (iii)⇒(i). For the last statement, we proceed as in [1, Theorem 3.7 (2)]. □

Corollary 3.7 *Let $R = \oplus_{\alpha\in\Gamma} R_\alpha$ be a graded domain. Assume that R is integrally closed or $R_0 \subseteq R$ is inert and R is almost normal. Then*

(1) *A homogeneous element is primal in R if and only if it is degree gr-primal.*
(2) *A homogeneous element is completely primal in R if and only if it is completely gr-primal.*

Proof This follows from Theorem 3.4 and Proposition 3.6. □

Remark 3.8

(1) In [12, Section 3], the author gave un example which show that R may be an almost normal graded domain, that is, R is R_h-almost normal for every $h \in H$, but there exist $h \in H$ and $x \in R$ such that $(h) : (x)$ is not homogeneous.
(2) Let $h \in H$. In Theorem 3.4, we can check that h is primal (resp., completely primal) if and only if h is degree (resp., completely) gr-primal and $(h) : (x)$ is homogeneous for every $x \in R_h$.

Example 3.9

(1) Let A be an integral domain with quotient field K. Let $R = A[X]$, a polynomial ring. Note that the extension $A \subseteq A[X]$ is inert. If every element of A is primal in $A[X]$, then, by Cohn's Nagata type theorem for Schreier domains, $A[X]$ is Schreier since $K[X] = A[X]_S$, where $S = A \setminus \{0\}$, is Schreier (UFD). The above results shed more light on the primality of elements of A in $A[X]$. Let $0 \neq a \in A$. Clearly, a is degree gr-primal if and only if a is completely gr-primal, if and only if a is completely primal in A. Thus a is (completely) primal in $A[X]$ if and only if a is completely primal in A and A is integrally closed in A_a. For more details, see the next section.
(2) For an extension of integral domains $A \subseteq B$, consider the pullback $R = A + XB[X]$. Since the extension $A \subseteq R$ is inert, then by Theorem 3.4 and Proposition 3.6, $h = aX^n \in H$ is primal (resp., completely primal) in R if and only if h is degree (resp., completely) gr-primal and B is integrally closed in B_a (Here $R_h = B_a[X, X^{-1}]$ if $n \geq 1$, and $R_h = A_a + XB_a[X]$ if $n = 0$.)

As a corollary of Theorem 3.4, Proposition 3.6, and Cohn's Nagata type theorem for Schreier domains, we reobtain the characterization of the Schreier property in graded domains.

Corollary 3.10 *[6, Theorem 2.2] Let $R = \bigoplus_{\alpha \in \Gamma} R_\alpha$ be a graded domain. Then the following statements are equivalent.*

(i) *R is Schreier.*
(ii) *R is pre-Schreier and R_0 is integrally closed in $(R_H)_0$.*
(iii) *R is gr-pre-Schreier and integrally closed.*

4 Primal Elements in Semigroup Rings

As an application of the previous sections, we study the primality in semigroup rings. Throughout this section, Γ denotes a nonzero torsionless commutative cancellative monoid (written additively) with quotient group G, and A is an integral domain with quotient field K. Let $A[\Gamma]$ be the semigroup ring of Γ over A. Then $A[\Gamma]$ is a Γ-graded integral domain and each nonzero element $f \in A[\Gamma]$ can be written uniquely as $f = a_1X^{s_1} + \cdots + a_nX^{s_n}$, where $0 \neq a_i \in A$ and $s_i \in \Gamma$ with $s_1 < \ldots < s_n$. Note that here, $H = \{aX^\alpha, 0 \neq a \in A, \alpha \in \Gamma\}$ and $A[\Gamma]_H = K[G]$. For more on semigroup rings, see [11].

Proposition 4.1 *Let $A[\Gamma]$ be the semigroup ring of Γ over A, and consider an element of the form aX^α where $0 \neq a \in A$ and $\alpha \in \Gamma$. The followings statements are equivalent.*

(i) *aX^α is primal in $A[\Gamma]$.*
(ii) *a and X^α are both primal in $A[\Gamma]$.*

Proof *(i)* $\Rightarrow$ *(ii)*. Suppose that aX^{α} is primal in $A[\Gamma]$. Let $f, g \in A[\Gamma]$ such that $a \mid fg$, then $aX^{\alpha} \mid f(gX^{\alpha})$. Since aX^{α} is primal $aX^{\alpha} = a_1X^{\alpha_1}a_2X^{\alpha_2}$ where $a_1X^{\alpha_1} \mid f$ and $a_2X^{\alpha_2} \mid gX^{\alpha}$, so $a_1 \mid f$ and $a_2 \mid g$. Thus $a = a_1a_2$ such that $a_1 \mid f$ and $a_2 \mid g$, so a is primal in $A[\Gamma]$.

To prove that X^{α} is primal in $A[\Gamma]$, let $f, g \in A[\Gamma]$ such that $X^{\alpha} \mid fg$, then $aX^{\alpha} \mid (af)g$. Thus $aX^{\alpha} = a_1X^{\alpha_1}a_2X^{\alpha_2}$, where $a_1X^{\alpha_1} \mid af$ and $a_2X^{\alpha_2} \mid g$. Hence $X^{\alpha} = X^{\alpha_1}X^{\alpha_2}$ with $X^{\alpha_1} \mid f$ and $X^{\alpha_2} \mid g$.

(ii) $\Rightarrow$ *(i)*. Assume that a and X^{α} are both primal in $A[\Gamma]$ and let $f, g \in A[\Gamma]$ such that $aX^{\alpha} \mid fg$. Then $a \mid fg$ and $X^{\alpha} \mid fg$. Since a and X^{α} are primal in $A[\Gamma]$, we have $a = a_1a_2$ such that $a_1 \mid f$ and $a_2 \mid g$ for some $a_1, a_2 \in A$; and $X^{\alpha} = X^{\alpha_1}X^{\alpha_2}$ such that $X^{\alpha_1} \mid f$ and $X^{\alpha_2} \mid g$ for some $\alpha_1, \alpha_2 \in \Gamma$. Hence $aX^{\alpha} = a_1X^{\alpha_1}a_2X^{\alpha_2}$, where $a_1X^{\alpha_1} \mid f$ and $a_2X^{\alpha_2} \mid g$, so aX^{α} is primal in $A[\Gamma]$. □

For a semigroup ring $A[\Gamma]$, let $h = aX^{\alpha} \in H$. Then $A[\Gamma]_h = A_a[\Gamma_{\alpha}]$, where Γ_{α} is the quotient monoid with respect to the additive set generated by α. Note that $A[\Gamma]$ is integrally closed in $A_a[\Gamma_{\alpha}]$ if and only if A is integrally closed in A_a and Γ is integrally closed in Γ_{α}.

Proposition 4.2 *Let $A[\Gamma]$ be the semigroup ring of Γ over A and $h = aX^{\alpha} \in H$. The following statements are equivalent.*

(i) *$A[\Gamma]$ is $A_a[\Gamma_{\alpha}]$-integrally closed.*
(ii) *$(h) : (f)$ is homogeneous for each $f \in A_a[\Gamma_{\alpha}]$.*
(iii) *$A[\Gamma]$ is $A_a[\Gamma_{\alpha}]$-almost normal.*

Proof By Proposition 3.6, it remains to show that (iii)$\Rightarrow$(i). Let $\lambda \in A_a$ be integral over $A[\Gamma]$. Take $0 \neq \gamma \in \Gamma$. Then $\lambda X^{\gamma} \in A_a[\Gamma_{\alpha}]$ is a homogeneous element of nonzero degree which is integral over $A[\Gamma]$. So $\lambda X^{\gamma} \in A[\Gamma]$, hence $\lambda \in A$. Now, by the $A_a[\Gamma_{\alpha}]$-almost normality, $A[\Gamma]$ is $A_a[\Gamma_{\alpha}]$-integrally closed. □

The following lemmas characterize degree (resp., completely) gr-primality in semigroup rings.

Lemma 4.3 *Let $A[\Gamma]$ be the semigroup ring of Γ over A and $0 \neq a \in A$. The following statements are equivalent.*

(i) *a is completely gr-primal.*
(ii) *a is degree gr-primal.*
(iii) *a is completely primal in A.*

Proof (i)$\Rightarrow$(ii). This is clear.

(ii)$\Rightarrow$(iii). Suppose that $a \mid b_ic_j$ in A for $i = 1, \ldots, m$ and $j = 1, \ldots, n$. Let $0 \neq \alpha \in \Gamma$; set $\beta_i = i\alpha$ and $\gamma_j = j\alpha$ for $i = 1, \ldots, m$ and $j = 1, \ldots, n$. Then $a \mid (b_iX^{\beta_i})(c_jX^{\gamma_j})$ in $A[\Gamma]$, for $i = 1, \ldots, m$ and $j = 1, \ldots, n$. By (ii), there exist $a_1, a_2 \in A$ such that $a = a_1a_2$ where $a_1 \mid b_i$ for each i and $a_2 \mid c_j$ for each j. Hence a is completely primal in A (cf. Proposition 2.2).

(iii)$\Rightarrow$(i). Assume that $a \mid b_iX^{\beta_i}c_jX^{\gamma_j}$ in $A[\Gamma]$ for $i = 1, \ldots, m$ and $j = 1, \ldots, n$.

Then $a \mid b_i c_j$ in A for each i, j. So $a = a_1 a_2$, where $a_1 \mid b_i$ for each i and $a_2 \mid c_j$ for each j. Thus $a = a_1 a_2$ such that $a_1 \mid b_i X^{\beta_i}$ for each i and $a_2 \mid c_j X^{\gamma_j}$ for each j. This proves that a is completely gr-primal. □

Lemma 4.4 *Let $A[\Gamma]$ be the semigroup ring of Γ over A and $\alpha \in \Gamma$. The following statements are equivalent.*

(i) *X^α is completely gr-primal.*
(ii) *X^α is degree gr-primal.*
(iii) *α is completely primal in Γ.*

Proof (i)⇒(ii). This is clear.

(ii)⇒(iii). Suppose that $\alpha \mid \beta_i + \gamma_j$ in Γ for $i = 1, \ldots, m$ and $j = 1, \ldots, n$. We may assume that $\beta_1 < \cdots < \beta_m$ and $\gamma_1 < \cdots < \gamma_n$. Then $X^\alpha \mid X^{\beta_i} X^{\gamma_j}$ for $i = 1, \ldots, m$ and $j = 1, \ldots, n$. By (ii), there exist $\alpha_1, \alpha_2 \in \Gamma$ such that $\alpha = \alpha_1 + \alpha_2$, where $\alpha_1 \mid \beta_i$ for each i and $\alpha_2 \mid \gamma_j$ for each j. Hence α is completely primal in Γ.

(iii)⇒(i). Assume that $X^\alpha \mid b_i X^{\beta_i} c_j X^{\gamma_j}$ in $A[\Gamma]$ for $i = 1, \ldots, m$ and $j = 1, \ldots, n$. Then $\alpha \mid \beta_i + \gamma_j$ in Γ for each i, j. So $\alpha = \alpha_1 + \alpha_2$, where $\alpha_1 \mid \beta_i$ for each i and $\alpha_2 \mid \gamma_j$ for each j. Thus $X^\alpha = X^{\alpha_1} X^{\alpha_2}$ such that $X^{\alpha_1} \mid b_i X^{\beta_i}$ for each i and $X^{\alpha_2} \mid c_j X^{\gamma_j}$ for each j. This proves (i). □

Next, we state our main result of this section.

Theorem 4.5 *Let $A[\Gamma]$ be the semigroup ring of Γ over A, and let $0 \neq a \in A$ and $\alpha \in \Gamma$. Then*

(i) *a is (completely) primal in $A[\Gamma]$ if and only if a is completely primal in A and A is integrally closed in A_a.*
(ii) *X^α is (completely) primal in $A[\Gamma]$ if and only if α is completely primal in Γ and Γ is integrally closed in Γ_α.*

Proof This follows from Theorem 3.4, Remark 3.8 (2), Proposition 4.2, and Lemmas 4.3 and 4.4. □

From Theorem 4.5 and Corollary 3.7, we get:

Corollary 4.6 *Let $A[\Gamma]$ be the semigroup ring of Γ over A, and let $0 \neq a \in A$ and $\alpha \in \Gamma$. Then*

(i) *Assume that A is integrally closed. Then a is (completely) primal in $A[\Gamma]$ if and only if a is completely primal in A.*
(ii) *Assume that Γ is integrally closed. Then X^α is (completely) primal in $A[\Gamma]$ if and only if α is completely primal in Γ.*

Corollary 4.7 *[6, Theorem 3.2] Let $A[\Gamma]$ be the semigroup ring of Γ over A. The following statements are equivalent.*

(i) *$A[\Gamma]$ is pre–Schreier.*
(ii) *$A[\Gamma]$ is Schreier.*
(iii) *A and Γ are Schreier.*

Proof For (i)⇒(ii)⇒(iii) use Proposition 4.1 and Theorem 4.5, and remark that A (resp., Γ) is integrally closed if and only if A (resp., Γ) is integrally closed in A_a (resp., Γ_α) for each $0 \neq a \in A$ (resp., $0 \neq \alpha \in \Gamma$). For (iii)⇒(i) we need the Cohn's Nagata type theorem for Schreier domains. □

In the case of polynomial rings, we recover some results established in [5, Proposition 6] and [4, Lemma 4.7]. Note that in a polynomial ring the powers of X are primary, so they are primal. Thus, by Proposition 4.1, a nonzero homogeneous element of the form aX^n, $a \in A$, is primal in $A[X]$ if and only if a is primal in $A[X]$.

Corollary 4.8 *Let A be an integral domain and X an indeterminate. Then*

(i) *a is (completely) primal in $A[X]$ if and only if a is completely primal in A and A is integrally closed in A_a.*
(ii) *$A[X]$ is Schreier if and only if $A[X]$ is pre-Schreier, if and only if A is Schreier.*

Acknowledgement The authors are grateful to the referee for comments that helped to improve the exposition.

References

1. Anderson, D.D., Anderson, D.F.: Divisorial ideals and invertible ideals in a graded integral domain. J. Algebra **76**(2), 549–569 (1982)
2. Anderson, D.D., Anderson, D.F.: Divisibility properties of graded domains. Can. J. Math. **34**(1), 196–215 (1982)
3. Anderson, D.D., Zafrullah, M.: P.M. Cohn's completely primal elements. In: Anderson, D.F., Dobbs, D. (eds.) Zero-dimensional Commutative Rings. Lecture Notes in Pure and Applied Mathematics, vol. 171, pp. 115–123. Dekker, New York (1995)
4. Anderson, D.D., Zafrullah, M.: The Schreier property and Gauss' lemma. Bolletino U.M.I. (8)10-B **2007**, 43–62 (2007)
5. Anderson, D.D., Dumitrescu, T., Zafrullah, M.: Quasi-Schreier domains II. Commun. Algebra **35**, 2096–2104 (2007)
6. Brookfield, G., Rush, D.E.: When Graded domains are Schreier or pre-Schreier. J. Pure Appl. Algebra **195**, 225–230 (2005)
7. Cohn, P.M.: Bezout rings and their subrings. Proc. Camb. Philos. Soc. **64**, 251–264 (1968)
8. Dumitrescu, T., Al Salihi, S.I.: A note on composite domains $A + XB[X]$ and $A + XB[[X]]$. Math. Rep. **2**(52), 175–182 (2000)
9. Dumitrescu, T., Al Salihi, S.I., Radu, N., Shah, T.: Some factorization properties of composite domains $A + XB[X]$ and $A + XB[[X]]$. Commun. algebra **28**(3), 1125–1139 (2000)
10. Gilmer, R.: Multiplicative Ideal Theory. Marcel Dekker, New York (1972)
11. Gilmer, R.: Commutative Semigroup Rings. The University of Chicago Press, Chicago and London (1984)
12. Matsuda, R.: On the content condition of a graded integral domain. Comment. Math. Univ. St. Paul. **33**, 79–86 (1984)
13. Matsuda, R.: Note on Schreier semigroup rings. Math. J. Okayama Univ. **39**, 41–44 (1997)
14. McAdam, S., Rush, D.E.: Schreier rings. Bull. Lond. Math. Soc. **10**, 77–80 (1978)
15. Northcott, D.G.: A generalization of a theorem on the content of polynomials. Proc. Camb. Philos. Soc. **55**, 282–288 (1959)
16. Zafrullah, M.: On a property of pre–Schreier domains. Commun. Algebra **15**(9), 1895–1920 (1987)

Conjecture of Wilf: A Survey

Manuel Delgado

Abstract This paper intends to survey the vast literature devoted to a problem posed by Wilf in 1978 which, despite the attention it attracted, remains unsolved. As it frequently happens with combinatorial problems, many researchers who got involved in the search for a solution thought at some point that a solution would be just around the corner, but in the present case that corner has never been reached.

By writing this paper I intend to give the reader a broad approach on the problem and, when possible, connections between the various available results. With the hope of gathering some more information than just using set inclusion, at the end of the paper a slightly different way of comparing results is developed.

Keywords Numerical semigroup · Wilf semigroup · Wilf's conjecture

1 Introduction

At the beginning, my personal motivation was to build a list of references, each with a summary of the results therein related to Wilf's conjecture. This would have helped me by not having to dive into a collection of papers each time I needed a result. Then I thought that making the list public could also be a contribution to Wilf's conjecture. This process ended up in the writing of this paper, which is in

---

The author was partially supported by CMUP (UID/MAT/00144/2019), which is funded by FCT (Portugal) with national (MCTES) and European structural funds (FEDER), under the partnership agreement PT2020, and also by the Spanish project MTM2017-84890-P. Furthermore, the author acknowledges a sabbatical from the FCT: SFRH/BSAB/142918/2018.

M. Delgado (✉)
CMUP, Departamento de Matemática, Faculdade de Ciências, Universidade do Porto, Porto, Portugal
e-mail: mdelgado@fc.up.pt

V. Barucci et al. (eds.), *Numerical Semigroups*, Springer INdAM Series 40,
https://doi.org/10.1007/978-3-030-40822-0_4

some sense *yet another survey*. Another, because most papers fully dedicated to the conjecture provide good literature reviews. Although not aiming to be complete, these could be taken as surveys.

As it frequently happens with easy to state combinatorial problems, while working on them one thinks that a solution is at reach. This is certainly the case of the problem posed by Wilf, but nevertheless no one has found the aimed solution so far. Taking into account the number of published papers on the theme, one can infer that much time has globally been dedicated to the problem. This may lead people to classify the problem in the category of dangerous problems, in the sense that one risks to spend too much time struggling with it and have to give up without getting a solution. Fortunately, partial results may be of some interest.

The plan of the paper follows.

This introductory section contains most of the terminology and notation to be used along the paper. There are not many differences to what is commonly used. This section contains also what I consider a convenient way to visualize numerical semigroups. Although almost all further images appear only in the last section, we provided sufficient information to produce images of semigroups appearing in the remaining parts of the text.

Some problems posed by Wilf are described in the second section, which can be seen as a kind of motivation for the paper.

The third section is the real survey. It contains a large introductory part and then the statements of results grouped into several subsections.

In the final section we introduce the notion of *quasi-generalization* (roughly speaking, a set quasi-generalizes another if it contains all its elements, except possibly a finite number of them). It allows to draw a lattice involving some important properties that give rise to semigroups satisfying Wilf's conjecture.

1.1 Terminology and Notation

Most of the notation and terminology used appears in the book by Rosales and Garcìa-Sánchez [31]. Results referred as "well known" can be found in the same reference.

Let S be a numerical semigroup. Recall that a numerical semigroup S is a subset of $\mathbb{N}$ (the set of nonnegative integers) such that $0 \in S$, S is closed under addition and the complement $\mathbb{N} \setminus S$ is finite (possibly empty). Throughout the paper, when the letter S appears and nothing else is said, it should be understood as being a numerical semigroup.

The *minimal generators* of S are also known as *primitive elements* of S. The set of primitive elements of S is denoted $\mathrm{P}(S)$. It is well known to be finite. When there is no possible confusion on which is the semigroup at hand, the notation is often simplified and we write P instead of $\mathrm{P}(S)$. This kind of simplification in the notation is made for all the other combinatorial invariants introduced along the paper.

The *multiplicity* of S is the least positive integer of S and is denoted $\mathrm{m}(S)$, or simply m. The *Frobenius number* of S is the largest integer that does not belong to S, and is denoted $\mathrm{F}(S)$. The *conductor* of S is simply $\mathrm{F}(S)+1$. Wilf's notation will be used for the conductor: $\chi(S)$, or simply χ. Note that $\chi(S)$ is the smallest integer in S from which all the larger integers belong to S. Let $\mathrm{q}(S) = \lceil \chi(S)/\mathrm{m}(S) \rceil$ be the smallest integer greater than or equal to $\chi(S)/\mathrm{m}(S)$. This number is called the *depth* of S and is frequently denoted just by q. It is worth to keep in mind that $\chi(S) \le \mathrm{m}(S)\,\mathrm{q}(S)$.

The set of *left elements* of S consists of the elements of S that are smaller than $\chi(S)$. It is denoted $\mathrm{L}(S)$ (or simply L). A positive integer that does not belong to S is said to be a *gap* of S (*omitting value* in Wilf's terminology). The cardinality of the set of gaps is said to be the *genus* of S and, following Wilf, is denoted by $\Omega(S)$, or simply by Ω.

If $x \in S$, then $\mathrm{F}(S) - x \notin S$. Thus, the following well known remark holds.

Remark 1.1 Let S be a numerical semigroup. Then $\Omega(S) \ge \chi(S)/2$.

As usual, $|X|$ denotes the cardinality of a set X. It is immediate that $\Omega(S) + |\mathrm{L}(S)| = \chi(S)$. From the above remark it follows that $\chi(S) \ge 2\,|\mathrm{L}(S)|$.

The number of primitives of S is called the *embedding dimension* of S. As it is just the cardinality of $\mathrm{P}(S)$, it can be denoted $|\mathrm{P}(S)|$, but in this paper I will mainly use the notation $\mathrm{d}(S)$, or simply d; d stands for *dimension* (a short for embedding dimension).

An integer x is said to be a *pseudo-Frobenius number* of S if $x \notin S$ and $x+s \in S$, for all $s \in S \setminus \{0\}$. The cardinality of the set of pseudo-Frobenius numbers of S is said to be the *type* of S and is denoted by $\iota(S)$. The notion of type has been an important ingredient in the discovery of various families of numerical semigroups satisfying Wilf's conjecture, due to Proposition 3.1 below. Another important tool, which is used in a crucial (and frequently rather technical) way in the proofs of some results presented in this survey is the *Apéry set (with respect to the multiplicity)*: $\mathrm{Ap}(S, \mathrm{m}) = \{s \in S \mid s - \mathrm{m} \notin S\}$.

Let X be a set of positive integers. The notation $\langle X \rangle_t$ is used to represent the smallest numerical semigroup that contains X and all the integers greater than or equal to t.

For a numerical semigroup S, the interval of integers starting in $\chi(S)$ and having $\mathrm{m}(S)$ elements is called the *threshold interval* of S (following a suggestion of Eliahou).

1.2 A Convenient Way to Visualize Numerical Semigroups

The pictures in this paper were produced using the GAP [21] package IntPic [9], while the computations have been carried out using the GAP package numericalsgps [11].

Fig. 1 Pictorial representation of the numerical semigroup ⟨5, 13, 21, 22, 24⟩

20	21	22	23	24
15	16	17	18	19
10	11	12	13	14
5	6	7	8	9
0	1	2	3	4

Let S be a numerical semigroup. The set of nonnegative integers up to $\chi + \mathrm{m} - 1$ clearly contains L and it is easy to see that it contains P as well. It is helpful to dispose the mentioned integers into a table and to highlight those that, in some sense, are special.

Several figures will be presented to give pictorial views of numerical semigroups. Each of them consists of a rectangular $(\mathrm{q}+1) \times \mathrm{m}$-table and the entries corresponding to elements of the semigroup are highlighted in some way. Some gaps can also be emphasized. The entries in uppermost row are those of the threshold interval.

Example 1.1 Figure 1 is a pictorial representation of the numerical semigroup $\langle 5, 13\rangle_{20} = \langle 5, 13, 21, 22, 24\rangle$. The elements of the semigroup are highlighted and, among them, the primitive elements and the conductor are emphasized. When an element is highlighted for more than one reason, gradient colours are used.

Observe that there is at most one primitive per column. This happens exactly when the semigroup is of maximal embedding dimension.

Note that all the integers in a given column are congruent modulo m. In particular, an element belongs to the Apéry set relative to m if and only if it is the lowest emphasized element in some column (provided that no gaps (for instance the pseudo-Frobenius numbers) are highlighted).

For the benefit of the reader, I explain the way I produced Fig. 1, including the GAP code used. To start, GAP is taught what my numerical semigroup is (see the manual of numericalsgps for details).

```
ns := NumericalSemigroup(5,13,21,22,24);
```

Then one can use the following commands to produce the TikZ code for the picture shown (which can be included in a LaTeX document):

GAP-code 1
```
#cls is given just to make a change to the default colors
cls := [ "blue","-red","red!70", "black!40" ];
P := MinimalGenerators(ns);
m := Multiplicity(ns);
c := Conductor(ns);
q := CeilingOfRational(c/m);
rho := q*m-c;
list := [-rho .. c+m-1];
ti := [c..c+m-1];
importants := Union(SmallElements(ns),ti);
options := rec(colors := cls,highlights:=[[c],importants,P]);
tkz := IP_TikzArrayOfIntegers(list,m,options);;
Print(tkz);
```

38	39	40	41	42	43	44	45	46	47	48	49
26	27	28	29	30	31	32	33	34	35	36	37
14	15	16	17	18	19	20	21	22	23	24	25
2	3	4	5	6	7	8	9	10	11	12	13
-10	-9	-8	-7	-6	-5	-4	-3	-2	-1	0	1

Fig. 2 Pictorial representation of $\langle 12, 19, 20, 22, 23, 26, 27, 28, 29\rangle$, with the pseudo-Frobenius numbers highlighted

The function `IP_TikzArrayOfIntegers` (which produces the TikZ code from the information previously computed using numericalsgps) is part of the intpic package. The manual of the package can be consulted for details and examples. In particular, the manual contains a complete example showing a possible way to include the picture (or its TikZ code) in a LaTeX document.

Executing the following command, the created picture should pop up. As this command depends on some other software, namely the operating system, some extra work on the configuration may be needed.

```
IP_Splash(tkz);
```

If everything goes well, the figure (in pdf format) can be saved and included in the LaTeX document in some standard way.

Example 1.2 Figure 2 is just another example. The following GAP session shows some important data: a numerical semigroup and its pseudo-Frobenius numbers. These are highlighted in the figure, in addition to elements of the semigroup, as in Example 1.1.

```
gap> ns := NumericalSemigroup(12, 19, 20, 22, 23, 26, 27, 28, 29);;
gap> Conductor(ns);
38
gap> pf := PseudoFrobenius(ns);
[ 16, 30, 33, 37 ]
```

The picture can be obtained with just small changes from GAP-code 1. Besides redefining the numerical semigroup, it suffices to replace the line beginning with options by the following two lines of code:

```
pf := PseudoFrobenius(ns);;
options := rec(colors := cls,highlights:=[[c],importants,P,pf]);;
```

In order to obtain an image just showing the shape, the options can be changed as follows:

GAP-code 2
```
# options to produce the shape
options := rec(
highlights:=[[],[],[c],importants,[],[],[],[],P],
cell_width := "6",colsep:="0",rowsep:="0",inner_sep:="2",
shape_only:=" ",line_width:="0",line_color:="black!20");;
```

2 Two Problems Posed by Wilf

This section starts with a few words on Wilf's paper [36], including a transcript of the two problems that Wilf left open. Then, a little about each problem is said.

2.1 *Wilf's Paper*

Wilf's concern was: to present an algorithm, which, given a numerical semigroup S and a finite generating set for S, finds the conductor of S, decides whether a given integer is representable (in terms of the elements of the generating set), finds a representation of an element of S, and determines the number of omitted values of S.

These are problems that many researchers interested in combinatorial problems related to numerical semigroups are nowadays still concerned with. In a somewhat more modern language, one would say that Wilf was concerned with *the Frobenius problem* (see [30]), the membership problem, factorization problems (see [22]) and the problem of determining the genus. These problems continue to be (are at the base of) active fields of research.

The circle of lights algorithm explicitly given in [36] determines both the conductor and the genus of a numerical semigroup provided that a finite generating set is at hand. Wilf also suggests a few changes to the circle of lights algorithm in order test membership and also to find a factorization. He also observed that to test membership, a suggestion of Brauer [6] should be incorporated: it involves the use of the Apéry set (relative to the multiplicity). Another observation that I would like to make is that the (space and time) complexity is explicitly given, which is a relevant contribution to the overall quality of Wilf's paper. It is extremely agreeable to read and this without doubt contributes to the success of the problems stated in it.

At the end of Wilf's article one finds the following two problems. It should be understood that the positive integer k represents the embedding dimension of some numerical semigroup.

Problem 2.1 **([36])** Wilf asked:

(a) Is it true that for a fixed k the fraction Ω/χ of omitted values is at most $1-(1/k)$ with equality only for the generators $k, k+1, \ldots, 2k-1$?
(b) Let $f(n)$ be the number of semigroups whose conductor is n. What is the order of magnitude of $f(n)$ for $n \to \infty$?

The first problem consists in fact of two problems. They can be stated explicitly as follows:

Problem 2.2 Wilf's problem (a) splits into two problems.

(a.i) Is it true that for a fixed k the fraction Ω/χ of omitted values is at most $1 - (1/k)$?

(a.ii) Is it true that for a fixed k the fraction Ω/χ of omitted values is $1-(1/k)$ only for the generators $k, k+1, \ldots, 2k-1$?

Problem $(a.i)$ is nowadays known as *Wilf's conjecture*. Sylvester's result (which is mentioned in the first page of Wilf's paper) gives counter examples to Problem $(a.ii)$. Apparently Wilf forgot about them. Nowadays there are other counter examples known, but a characterization of those semigroups for which the equality holds is an open problem (see Sect. 2.3).

2.2 Problem (a.i): Wilf's Conjecture

To a numerical semigroup S one can associate the following number denoted $\mathrm{W}(S)$ and called the *Wilf number of* S:

$$\mathrm{W}(S) = |\mathrm{P}(S)||\mathrm{L}(S)| - \chi(S). \tag{1}$$

A numerical semigroup is said to be a *Wilf semigroup* if and only if its Wilf number is nonnegative. *Wilf's conjecture* can be stated as follows:

Conjecture 2.1 (Wilf [36]) Every numerical semigroup is a Wilf semigroup.

It is a simple exercise to verify that Conjecture 2.1 is precisely Problem 2.2 $(a.i)$.

There is another number that can be associated to a numerical semigroup, just as Wilf number is, and which has revealed great importance in recent research (as the reader will be able to confirm, in particular when reading Sect. 3.8). Let S be a numerical semigroup and let $D = \mathrm{P}(S) \cap \{\chi, \ldots, \chi + \mathrm{m} - 1\}$ be the set of non primitives in the threshold interval. Eliahou [16] associated to S the number $\mathrm{E}(S)$ that appears in Eq. (2) below and used the notation $\mathrm{W}_0(S)$ to represent it. I prefer the notation $\mathrm{E}(S)$, and use the terminology *Eliahou number of* S:

$$\mathrm{E}(S) = |\mathrm{P} \cap \mathrm{L}||\mathrm{L}| - \mathrm{q}|D| + \mathrm{q}\,\mathrm{m} - \chi\,. \tag{2}$$

Eliahou [16, Pg. 2112] observed that there are numerical semigroups with negative Eliahou number and stated the following problem which is still open.

Problem 2.3 Give a characterization of the class of numerical semigroups whose Eliahou number is negative.

2.3 Problem (a.ii): Another Open Problem

A very nice result of Sylvester [34] gives a formula for the Frobenius number of a numerical semigroup of embedding dimension 2. A closed formula (of a certain type) for the Frobenius number of a numerical semigroup of higher embedding dimension is not at reach (see [8] or [30, Cor. 2.2.2]). Sylvester's results can be written as follows (see [31]): if $S = \langle a, b\rangle$ is a numerical semigroup of embedding dimension 2, then $\mathrm{F}(S) = ab - a - b$, and $\Omega(S) = \chi(S)/2$. From this, it is immediate that for a numerical semigroup S of embedding dimension 2, $\mathrm{W}(S) = 0$. The fact that numerical semigroups of the form $\langle \mathrm{m}, k\,\mathrm{m}+1, \dots k\,\mathrm{m}+\mathrm{m}-1\rangle$ (which are of maximal embedding dimension and generated by some generalized arithmetic sequences) have Wilf number equal to 0 is straightforward (see [19, 28]).

Whether these are the only numerical semigroups for which Wilf number is 0 is a slight modification of Problem 2.2($a.ii$) and is open. I rephrase the question stated by Moscariello and Sammartano.

Problem 2.4 ([28, Question 8]) Let $S = \langle \mathrm{m}, g_2, \dots, g_{\mathrm{d}}\rangle$ be a numerical semigroup with multiplicity m and embedding dimension d. Is it true that if $\mathrm{W}(S) = 0$, then $\mathrm{d}(S) = 2$ or $\mathrm{d}(S) = \mathrm{m}(S)$ and there exists an integer $k \geq 1$ such that $g_i = k\,\mathrm{m}+(i-1)$, for $i \in \{2, \dots, \mathrm{d}\}$?

Moscariello and Sammartano observed that in order to answer affirmatively this question it suffices to prove that for a semigroup with Wilf number equal to 0, either its embedding dimension is 2 or it has maximal embedding dimension.

They also observed that no numerical semigroup of genus up to 35 provides a negative answer to the question.

Kaplan [23, Prop. 26] has shown that Problem 2.4 has a positive answer in the case of numerical semigroups whose multiplicity is at least half of the conductor. The same holds for numerical semigroups of depth 3 (see a remark by Sammartano in [16, Rem. 6.6]), thus concluding that there are no counter examples among the semigroups satisfying $\chi \leq 3\,\mathrm{m}$.

2.4 *Problem (b): Counting Numerical Semigroups*

Wilf's Problem 2.1(b) can be viewed as a problem about counting numerical semigroups by conductor. Backelin [1] addressed this problem. A slight modification consists on counting by genus. Great attention has been given to this problem after Bras-Amorós [5] proposed some conjectures on the theme. Some of these conjectures were solved by Zhai [37], while others remain open. For an excellent survey (which in particular contains references for counting by conductor and has an outline of Zhai's proofs), see Kaplan [24].

Denote respectively by $N(g)$ and $t(g)$ the number of numerical semigroups of genus g and the number of numerical semigroups of genus g satisfying $\chi(S) \leq 3\,\mathrm{m}(S)$.

In the paper where he proved some of the conjectures of Bras-Amorós (one of them being that the sequence $\big(N(g)\big)$ behaves like the Fibonacci sequence), Zhai also proved that the proportion of numerical semigroups such that $\chi(S) \leq 3\,\mathrm{m}(S)$ tends to 1 as g tends to infinity, as conjectured by Zhao [38].

Proposition 2.1 ([37]) *With the notation introduced, the following holds:*

$$\lim_{g\to\infty} \frac{t(g)}{N(g)} = 1.$$

I will leave here a question that can be stated in a similar way to Zhao's conjecture. It will be better appreciated when reading Sect. 3.6 (and confronting with Sect. 3.8).

Denote by $p(g)$ the number of numerical semigroups of genus g satisfying $\mathrm{d}(S) \geq \mathrm{m}(S)/3$.

Question 2.1 Does $\lim_{g\to\infty} \frac{p(g)}{N(g)}$ exist?

3 Some Classes of Wilf Semigroups

As already observed, it follows from a result of Sylvester that semigroups of embedding dimension 2 have Wilf number equal to 0. In particular, semigroups of embedding dimension 2 are Wilf semigroups. Many other classes are known to consist of Wilf semigroups. This section gives an account of a large number of them.

The theme is rather popular and it is frequent to check a family of numerical semigroups against Wilf's conjecture, whenever that new family of numerical semigroups is investigated for some reason. It may well happen that some results are not referred to in this paper. This is far from meaning that I do not consider the ideas involved important. In a few cases this may be a matter of choice, but most probably it simply means that the results are not part of my very restricted knowledge. For that, I humbly express my apologies both to the authors and the readers.

The results are split into several subsections, according to a criterion that seems difficult to explain. It finally just aims at putting results together so that they can be compared with ease.

Most of the families considered are described through at least two combinatorial invariants such as embedding dimension, the multiplicity or the conductor. Exceptions (besides finite sets) are families that are completely described by using only one invariant among the embedding dimension, the multiplicity or the number of left elements.

Inside each subsection several results are mentioned (through precise numbered statements or just in the text) and there are cases in which the most general one is stated as a theorem. In a few cases there are results mentioned in more than one subsection.

In the first subsection, there is an emphasis on a particular ingredient used along the proofs of the various results. The ingredient is an inequality involving the type. As illustrations of the results that can be found there, we shall see that semigroups with embedding dimension up to three, almost symmetric numerical semigroups and those semigroups generated by generalized arithmetic sequences are Wilf semigroups.

The second subsection is about families of semigroups (somehow explicitly) given by some sets of generators. The examples therein share the particularity that all the members have negative Eliahou number.

The third subsection refers to numerical semigroups with nonnegative Eliahou number.

The fourth subsection is dedicated to constructions that are somehow natural. In fact, only one such construction is given here: dilations of numerical semigroups. This subsection could certainly be filled with other constructions. My choice just reflects the feeling that possible generalizations could be worth exploring.

Then there is a subsection devoted to numerical semigroups of small multiplicity.

The sixth subsection is about results in which the main attention is given to semigroups with large embedding dimension, when compared to the multiplicity.

Next there appears a subsection containing a result involving numerical semigroups with big multiplicities and possibly small embedding dimensions.

The eighth subsection is similar to the sixth, but now the results have an emphasis on semigroups with large multiplicity, when compared to the conductor.

In ninth subsection there is a result taking into account an invariant not previously considered (at least in a fundamental way, to the best of my knowledge). It is the second smallest primitive, sometimes called the *ratio*.

The final subsection is concerned with families of numerical semigroups that can be described using only one combinatorial invariant.

3.1 The Type as an Important Ingredient

The following proposition, due to Fröberg, Gottlieb and Haeggkvist, is at the base of some results on Wilf's conjecture. It implies that semigroups whose type is smaller than its embedding dimension are Wilf.

Proposition 3.1 ([19, Theorem 20]) *Let S be a numerical semigroup. Then* $\chi(S) \leq (\mathrm{t}(S)+1)\,|\mathrm{L}(S)|$.

By proving that the type of a numerical semigroup of embedding dimension 3 is either 1 or 2 ([19, Th. 11]) and using the fact that $\chi \geq 2\,|\mathrm{L}|$ referred in Sect. 1.1, they obtained that numerical semigroups of embedding dimension 3 are Wilf, a result that Dobbs and Matthews [14, Cor. 2.6] reproved using a different approach. Using Sylvester's result for embedding dimension 2 and the fact that $\mathbb{N}$ is Wilf, the same authors obtained the following result.

Theorem 3.1 ([19, Th. 20], [14, Th. 2.11]) *Numerical semigroups of embedding dimension smaller than* 4 *are Wilf.*

Since there is no upper bound for the type of numerical semigroups of embedding dimension bigger than 3 (see [19, pg. 75], for an example due to Backelin),

Proposition 3.1 can not be used to obtain other general results (but can, and has been applied successfully to particular families of semigroups).

A numerical semigroup S is said to be *irreducible* if it cannot be expressed as the intersection of two numerical semigroups properly containing it. S is said to be *symmetric* if it is irreducible and $\mathrm{F}(S)$ is odd and it is said to be *pseudo-symmetric* if it is irreducible and $\mathrm{F}(S)$ is even. One could take the following as definition (see [31, Cor 4.5]): S is *symmetric* if and only if $\Omega(S) = \chi(S)/2$, while S is *pseudo-symmetric* if and only if $\Omega(S) = (\chi(S)+1)/2$.

It can be proved as a simple exercise that irreducible numerical semigroups are Wilf. A more involved proof could be to observe that the type of this class of semigroups does not exceed 2.

Proposition 3.2 ([14, Prop. 2.2]) *Irreducible numerical semigroups are Wilf.*

The above result was generalized by Marco La Valle in [2, Th. 5.5]. Before stating this generalization, a further definition is needed. A numerical semigroup is said to be *almost symmetric* if its genus is the arithmetic mean of its Frobenius number and its type (see [3]). It is a class of semigroups that includes the symmetric and the pseudo-symmetric ones.

Proposition 3.3 ([2, Th. 5.5]) *Almost symmetric numerical semigroups are Wilf.*

As a consequence of Theorem 3.1 Dobbs and Matthews derived an interesting corollary:

Corollary 3.1 ([14, Cor 2.7]) *If S is a numerical semigroup with $\chi \leq 4\,|\mathrm{L}|$, then S is Wilf.*

Also making use of Proposition 3.1, Kunz [25] obtained the following result (for p and q coprime). See also Kunz and Waldi [26] for some other generalizations.

Proposition 3.4 ([26, Cor. 3.1]) *Let S be a numerical semigroup with $\mathrm{d}(S) \geq 3$. Let p and q be two distinct primitives of S. If $g+h \in (p+S)\cup(q+S)$, for any (non necessarily distinct) primitives g and h of S, then $\mathrm{t}(S) \leq \mathrm{d}(S)-1$. In particular, S is Wilf.*

A semigroup generated by a generalized arithmetic sequence is a semigroup of the form $S = \langle m, hm+d, hm+2d, \ldots, hm+\ell d\rangle$, where m, d, h, ℓ are positive integers such that $m \geq 2$, $\gcd(m,d) = 1$ and $\ell \leq m-2$. Note that m and d being coprime ensures that S is a numerical semigroup. For a picture made up from a semigroup generated by a generalized arithmetic sequence with $m = 20, d = 9, h = 2, \ell = 8$, see Fig. 3. By using a result of Matthews [27, Cor. 3.4] that computes the type of a numerical semigroup generated by a generalized arithmetic sequence, Sammartano observed the following:

Proposition 3.5 ([32, Prop. 20]) *Numerical semigroups generated by generalized arithmetic sequences are Wilf.*

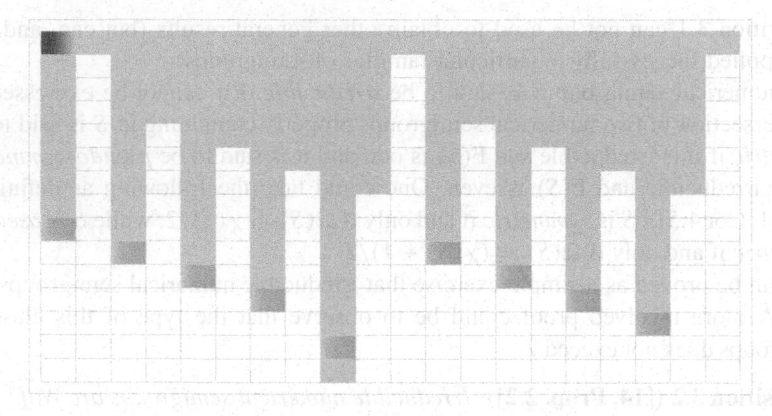

Fig. 3 Shape of the semigroup $\langle 20, 49, 58, 67, 76, 85, 94, 103, 112\rangle$

3.2 Semigroups Given by Sets of Generators

Let G be an abelian group. Let $A \subseteq G$ be a nonempty finite subset and let h be a positive integer. (For the purpose of this paper the reader may think of the group as being a cyclic group $\mathbb{Z}/m$ and take $h = 3$; for more details, see [35, Chap. 4].)

The set A is said to be a B_h *set* if, for all $a_1, \ldots, a_h, b_1, \ldots, b_h \in A$, the equality

$$a_1 + \cdots + a_h = b_1 + \cdots + b_h$$

holds if and only if $(a_1, \ldots, a_h)$ is a permutation of $(b_1, \ldots, b_h)$.

Let $m, a, b, n \in \mathbb{N}_{>0}$ be such that $n \geq 3$ and

$$(3m + 1)/2 \leq a < b \leq (5m - 1)/3.$$

Let $A \subseteq \{a, \ldots, b\}$ be such that $|A| = n - 1$ and A induces a B_3 set in $\mathbb{Z}/m$. That such a set exists follows from [17, Proposition 3.1]. Finally, let

$$S = \langle\{m\} \cup A\rangle_{4m}.$$

Eliahou and Fromentin proved the following result:

Proposition 3.6 ([17, Th. 4.1]) *Let $S = \langle\{m\}\cup A\rangle_{4m}$ be a semigroup as constructed above. Then $\mathrm{W}(S) \geq 9$ and, in particular, S is a Wilf semigroup.*

Let p be an even positive integer, let $\mu = \mu(p) = \frac{p^2}{4} + 2p + 2$ and let $\gamma = \gamma(p) = 2\mu(p) - \left(\frac{p}{2} + 4\right)$. The following holds:

Proposition 3.7 ([10, Prop.6]) *Let $S = S(p) = \langle \mu, \gamma, \gamma + 1\rangle_{p\mu}$. Then $\mathrm{W}(S) > 0$ and, in particular, S is a Wilf semigroup.*

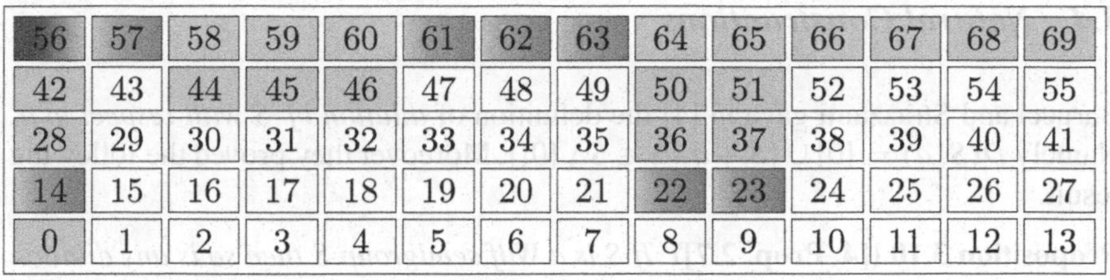

56	57	58	59	60	61	62	63	64	65	66	67	68	69
42	43	44	45	46	47	48	49	50	51	52	53	54	55
28	29	30	31	32	33	34	35	36	37	38	39	40	41
14	15	16	17	18	19	20	21	22	23	24	25	26	27
0	1	2	3	4	5	6	7	8	9	10	11	12	13

Fig. 4 Pictorial representation of $S(4) = \langle 14, 22, 23 \rangle_{56}$

Example 3.1 Figure 4 is a pictorial representation of the numerical semigroup $S(4)$. Note that $S(4)$ is of the form $\langle \{m\} \cup A \rangle_{4m}$, with $m = 14$ and $A = \{22, 23\}$.

To end this subsection I would like to make the following observations:

Remark 3.1 The numerical semigroups $S = \langle \{m\} \cup A \rangle_{4m}$ and $S(p) = \langle \mu, \gamma, \gamma + 1 \rangle_{p\mu}$ defined above have (possibly large) negative Eliahou numbers (see [10, 17]). The proof that they are Wilf involves explicit counting.

Remark 3.2 The semigroups $\langle \{m\} \cup A \rangle_{4m}$ have depth 4, while, since the conductor of $S(p)$ is $p\mu$, there is no bound for the depths of the semigroups $S(p)$.

Remark 3.3 Several other families obtained using similar constructions to the one in Proposition 3.7 can be found in the same paper. In particular, for any given integer n, an infinite family of numerical semigroups with Eliahou number equal to n is obtained. All these families consist entirely of Wilf semigroups.

3.3 Semigroups with Nonnegative Eliahou Numbers

Recall that the Eliahou number of a numerical semigroup was introduced in page 45. The following result, which states that semigroups with nonnegative Eliahou number are Wilf, appears in [16, Prop. 3.11] (see also [17, Cor. 2.3]).

Proposition 3.8 *Let S be a numerical semigroup with* $\mathrm{E}(S) \geq 0$. *Then* $\mathrm{W}(S) \geq 0$.

This result is similar to Proposition 3.1 in the sense that in order to prove that a numerical semigroup is Wilf it suffices to prove that it has nonnegative Eliahou number. The main consequences are referred to in Sect. 3.8.

While waiting for those consequences, let me refer a result obtained by Eliahou and Marín-Aragón. As they observed, the number 12 that appears in the statement is the best that can be obtained in this way: Example 3.1 gives a counter example for $|\mathrm{L}| = 13$.

Proposition 3.9 ([18]) *If S is a numerical semigroup with* $|\mathrm{L}| \leq 12$, *then S has nonnegative Eliahou number and therefore is Wilf.*

3.4 Natural Constructions

Barucci and Strazzanti gave in [4] the definition of *dilation of S with respect to a*. Namely $D(S,a) = \{0\} \cup \{S+a \mid s \in S \setminus \{0\}\}$. Moreover they proved the following result.

Proposition 3.10 ([4, Prop. 2.7]) *If S is a Wilf semigroup S then so is any dilation of S.*

Thus, for each Wilf semigroup S, the class $\{D(S,a) \mid a \in S\}$ is an infinite family of Wilf semigroups.

O'Neill and Pelayo [29] defined *shifted* numerical semigroup (they used the (less common for some historical reason, but more accurate) terminology "monoid") as follows. Let $r_1, \ldots, r_k$ be positive integers such that $r_1 < \cdots < r_k$, and let $d = \gcd(r_1, \ldots, r_k)$. Let $n > r_k$ be an integer and assume that $\gcd(n, d) = 1$. Under these conditions, $M_n = \langle n, n+r_1, \ldots, n+r_k \rangle$ is a numerical semigroup. It is called a *shifted numerical semigroup* (with respect to the shift parameter n). Note that it is the monoid obtained by shifting by n each generator of $S = \langle r_1, \ldots, r_k \rangle$. As one of the applications of the characterization obtained for the Apèry set of M_n in terms of the Apèry set of the base semigroup S, they obtained the following result:

Proposition 3.11 ([29, Cor. 4.6]) *If $n > r_k^2$, then the shifted numerical semigroup M_n is Wilf.*

3.5 Semigroups with Small Multiplicity

Sammartano [32, Cor. 19] proved that semigroups of multiplicity not greater than 8 are Wilf. Eliahou [15] announced the same kind of result but for multiplicity 12 (with a similar proof; see the comment just after Theorem 3.2). In the meantime, Dhany [13, Cor. 4.10] had obtained the result for multiplicity 9.

A big breakthrough was obtained by Bruns, García-Sánchez, O'Neil and Wilburne, who achieved multiplicity 17. (The first version of this paper referred "multiplicity 16", personally communicated by Pedro García-Sánchez.) Their proof involves computational methods, combined with geometrical ones, such as the use of Kunz polytopes.

Proposition 3.12 ([7]) *Let S be a numerical semigroup with* $\mathrm{m} \leq 17$. *Then S is Wilf.*

3.6 Semigroups with Large Embedding Dimension (Compared to the Multiplicity)

As stated in Theorem 3.1, numerical semigroups with very small (≤3) embedding dimension are Wilf. The same happens for those with large embedding dimension (when compared to the multiplicity). There are several proofs of the fact that semigroups of maximal embedding dimension (i.e., with embedding dimension equal to the multiplicity). The first one (to the best of my knowledge) is due to Dobbs and Matthews [14, Cor. 2.4]. Sammartano [32, Th. 18], by means of a rather technical proof involving Apéry sets and the counting in intervals of length m of elements in the semigroup, proved that numerical semigroups with embedding dimension at least half the multiplicity are Wilf. With refined arguments, Dhayni (see also her thesis [12, Th. 2.3.13]) generalized Sammartano's result.

Proposition 3.13 ([13, Th. 4.12]) *Let S be a numerical semigroup with* $\left(2+\frac{1}{q}\right)\mathrm{d} \geq \mathrm{m}$. *Then S is Wilf.*

Eliahou obtained the following impressive generalization by using a graph theoretical approach. The concept of matching (set of independent edges) in a certain graph associated to the Apéry set is used.

Theorem 3.2 ([15]) *Let S be a numerical semigroup with* $3\,\mathrm{d}(S) \geq \mathrm{m}(S)$. *Then S is Wilf.*

3.6.1 Some Comments

1. The previous result implies that semigroups of multiplicity up to 12 are Wilf, as already mentioned. We rephrase an argument due to Sammartano: by Theorem 3.1, a non Wilf semigroup satisfies $\mathrm{d}(S) \geq 4$. As it must also satisfy $3\,\mathrm{d}(S) < \mathrm{m}(S)$, it follows that if S is non Wilf, then $\mathrm{m}(S) > 12$.
2. A positive answer to Question 2.1 would lead to possibly interesting consequences. For instance, if the limit were $1/2$, one could conclude that asymptotically, as Ω grows, half of the numerical semigroups satisfy $3\,\mathrm{d} \geq \mathrm{m}$, and consequently are Wilf.

Two other simple consequences of Theorem 3.2 follow. The first is a generalization of a result of Dhany [13, Th. 4.9], who proved that semigroups satisfying $\mathrm{d} \geq \mathrm{m} - 5$ are Wilf. At this stage, this remark does not give anything new, but gives a very explicit result. (The first version of this paper contained a weaker result, with the same proof. This is due to the fact that at the time the first version of the paper was written the number appearing in Proposition 3.12 was 16 and now is 17. The statement in the first version was "If $\mathrm{d}(S) \geq \mathrm{m}(S) - 10$, then S is Wilf".)

Remark 3.4 If $\mathrm{d}(S) \geq \mathrm{m}(S) - 12$, then S is Wilf.

Proof If $\mathrm{m}(S) \leq 17$, then Proposition 3.12 can be used. If $\mathrm{m}(S) > 17$, then $\mathrm{m}(S)/3 \geq 6$. But in this case $\mathrm{m}(S) - 12 \geq \mathrm{m}(S)/3$. In fact, this last inequality is equivalent to $2\,\mathrm{m}(S) \geq 36$, which holds by hypothesis. □

Another consequence (it suffices Sammartano's result to get it), was observed by Eliahou.

Proposition 3.14 ([16, Prop. 7.6]) *Let S be a numerical semigroup with* $\gcd(\mathrm{L} \cap \mathrm{P}) \geq 2$. *Then S is Wilf.*

3.7 Semigroups with Big Multiplicity (and Possibly Small Embedding Dimension)

Moscariello and Sammartano proved that for every fixed value of $\lceil \mathrm{m}/\mathrm{d} \rceil$ the conjecture holds for all values of m which are sufficiently large and are not divisible by a finite set of primes. Recall from previous subsection that the cases $\lceil \mathrm{m}/\mathrm{d} \rceil \leq 3$ have been solved.

Proposition 3.15 ([28, Th. 1]) *Let S be a numerical semigroup. Let* $\rho = \left\lceil \frac{\mathrm{m}(S)}{\mathrm{d}(S)} \right\rceil$ *and let* ϕ *be the product of prime factors of* ρ. *If* $\rho > 3$, $\mathrm{m}(S) \geq \frac{\rho(3\rho^2-\rho-4)(3\rho^2-\rho-2)}{8(\rho-2)}$ *and* $\gcd(\mathrm{m}(S), \phi) = 1$, *then S is Wilf.*

The multiplicities of the semigroups that arise from Proposition 3.15 are large, as the following GAP session suggests (by showing that for $\rho = 4$ the smallest multiplicity is 1680).

```
gap> mult := r -> (r*(3*r^2-r-4)*(3*r^2-r-2))/8*(r-2);
function( r ) ... end
gap> mult(4);
1680
```

3.8 Semigroups with Large Multiplicity (Compared to the Conductor)

It was proved by Kaplan [23, Th. 24] that numerical semigroups with conductor not greater than twice the multiplicity are Wilf. This result was generalized by Eliahou. One of the ingredients he used is a theorem of Macaulay on the growth of Hilbert functions of standard graded algebras. In fact, he proved the following:

Proposition 3.16 ([16, Th. 6.4]) *Let S be a numerical semigroup with* $\chi(S) \leq 3\,\mathrm{m}(S)$. *Then S has nonnegative Eliahou number.*

This result combined with Proposition 3.8 leads to the following major fact. (Its importance can be better appreciated by seeing the comments that follow the statement.)

Theorem 3.3 ([16, Cor. 6.5]) *Let S be a numerical semigroup with $\chi(S) \leq 3\,\mathrm{m}(S)$. Then S is Wilf.*

3.8.1 Some Comments

1. Denote by $e(g)$ the number of numerical semigroups of genus g having positive Eliahou number. Combining Proposition 3.16 with Zhai's Proposition 2.1 one sees that $\lim_{g\to\infty} \frac{e(g)}{N(g)} = 1$. Thus, in the sense given by this limit, one can say that asymptotically, as the genus grows, all numerical semigroups have nonnegative Eliahou number. Consequently, asymptotically, as the genus grows, all numerical semigroups are Wilf.
2. I observe that, despite this asymptotic result concerning Eliahou numbers, there are infinitely many numerical semigroups with negative Eliahou number (see [10, 17]). All the examples given in the mentioned papers are Wilf semigroups (some of them appear in Sect. 3.2).

3.9 *Considering Unusual Invariants*

The second smallest primitive is sometimes called the *ratio* (see [31, Exercise 2.12]).

Spirito proved that if the ratio is large and the multiplicity is bounded by a quadratic function of the embedding dimension, then S is Wilf. He also proved various related statements. As an illustration, I choose one that is rather explicit:

Proposition 3.17 ([33, Prop 4.6]) *Let S be a numerical semigroup with ratio r and embedding dimension* $\mathrm{d} \geq 10$. *If*

$$r > \frac{\chi(S) + \mathrm{m}(S)}{3} \text{ and } \mathrm{m}(S) \leq \frac{8}{25}\mathrm{d}^2 + \frac{1}{5}\mathrm{d} - \frac{5}{4} \tag{3}$$

then S is a Wilf semigroup.

Remark 3.5 It is straightforward to check that if $d \leq 9$, then $8/25d^2 + 1/5d - 5/4 \leq 3d$. The following GAP session may help to quickly convince the reader:

```
gap> f := d -> 8/25*d^2 + 1/5*d - 5/4;
gap> Int(f(9));
26
```

One concludes by using Theorem 3.2 that the restriction $d \geq 10$ can be removed from the statement of Proposition 3.17. Moreover, when $d < 10$ there is no need to impose any restriction on the ratio.

3.10 Families Described Through One Invariant

Families of semigroups described by limiting the multiplicity of its members were already considered in Sect. 3.5. Proposition 3.12 could have been stated in this subsection, as well as Proposition 3.9 which refers to the number of left elements.

Dobbs and Matthews [14, Th. 2.11] proved that numerical semigroups with $|L| \leq 4$ are Wilf semigroups. As a corollary they obtained that semigroups with $\chi \leq 21$ are Wilf. Eliahou [16, Prop. 7.4] observed that numerical semigroups with less than 7 left elements are Wilf. These results have been largely superseded.

Recall that Proposition 3.9 gives a similar result, but the restriction on the number of left elements was weakened: numerical semigroups with $|L| \leq 12$ are Wilf.

Let S be a non Wilf semigroup. By Proposition 3.12, $m(S) \geq 18$. Using Proposition 3.16, which guarantees that non Wilf semigroups satisfy $\chi(S) > 3\,m(S)$, one gets that $\chi(S) > 54$. This proves that semigroups with conductor smaller than 55 are Wilf.

Fromentin and Hivert, through exhaustive computation, have shown that there are no non Wilf semigroups with genus smaller than 61. The previous published record, genus 51, had been obtained by Bras-Amorós [5].

Theorem 3.4 ([20]) *Every numerical semigroups of genus up to* 60 *is Wilf.*

Since the genus of a numerical semigroup is not smaller than its conductor plus one, the following consequence, which supersedes the above results concerning the conductor, is immediate.

Corollary 3.2 *Semigroups whose conductor does not exceed* 61 *are Wilf.*

(I am currently developing techniques to replace in Theorem 3.4 the integer 60 by a larger one. It will probably be part of an experimental preprint of mine which is in an advanced phase of preparation and is provisionally entitled "Wilf's conjecture on numerical semigroups holds for small genus".)

4 Quasi-Generalization

Denote by $\mathfrak{S}$ the class of all numerical semigroups. Let $\mathfrak{W} = \{S \in \mathfrak{S} \mid W(S) \geq 0\}$ and let $\mathfrak{E} = \{S \in \mathfrak{S} \mid E(S) \geq 0\}$.

Whether $\mathfrak{S} = \mathfrak{W}$ is presently not known (Wilf's conjecture says that the equality holds, but it is still a conjecture). That $\mathfrak{E} \subseteq \mathfrak{W}$ follows from Proposition 3.8, and up to genus 60 there are exactly 5 numerical semigroups not in $\mathfrak{E}$, thus showing that the inclusion is strict (the examples were obtained by Fromentin and appear in [16, pgs 2112,2113]). The following is a consequence of Remark 3.1.

Fact 4.1 $\mathfrak{W} \setminus \mathfrak{E}$ is infinite.

Let $\mathcal{P}$ be a property (about numerical semigroups). For instance, "d $\geq$ 3" is such a property. Let $\mathfrak{P} = \{S \in \mathfrak{S} \mid S \models \mathcal{P}\}$ be the class of numerical semigroups satisfying $\mathcal{P}$. With this notation, most results in the previous sections can be written in the following form: "If $S \in \mathfrak{P}$, then $S \in \mathfrak{W}$.", or "If S satisfies $\mathcal{P}$, then S is Wilf".

I invite the reader to think of all the results as if they had been written in this form. Some properties cannot be as nicely written as in the above example ("d $\geq$ 3"). However, for instance, the statement "We say that S satisfies property $\mathcal{P}$ if and only if S is of the form $S(p)$, with p an even positive integer." allows to write Proposition 3.7 in the above form.

By doing so, one can associate a property to each result and conversely. Although I do not intend to explicitly give names to all the properties corresponding to the results stated, there are some exceptions:

- $\mathcal{D}_3$ stands for the property "d $\geq$ 3", which is associated to Theorem 3.1. The corresponding class of semigroups is $\mathfrak{D}_3$.

Similarly, one has the correspondences:

- $\mathcal{D}$—"3 d $\geq$ m"—Theorem 3.2—$\mathfrak{D}$;
- $\mathcal{M}$—"$\chi \leq 3$ m"—Theorem 3.3—$\mathfrak{M}$;
- $\mathcal{G}_{60}$—"$\Omega \leq 60$"—Theorem 3.4—$\mathfrak{G}_{60}$.

It seems reasonable to add other exceptions: $\mathcal{S}, \mathcal{W}, \mathcal{E}$ are the properties about numerical semigroups associated, respectively, to $\mathfrak{S}, \mathfrak{W}, \mathfrak{E}$. (Note that $\mathcal{S}$ is trivial: it is satisfied by all numerical semigroups.)

Fact 4.2 All the classes $\mathfrak{D}_3, \mathfrak{D}, \mathfrak{M}, \mathfrak{G}_{60}$ are strictly contained in $\mathfrak{W}$. Furthermore, for every $\mathfrak{P} \in \{\mathfrak{D}_3, \mathfrak{D}, \mathfrak{M}, \mathfrak{G}_{60}\}$, $\mathfrak{W} \setminus \mathfrak{P}$ is infinite.

Proof By Proposition 3.16, $\mathfrak{M} \subseteq \mathfrak{E}$. Thus $\mathfrak{W} \setminus \mathfrak{E} \subseteq \mathfrak{W} \setminus \mathfrak{M}$. Since, by Fact 4.1, $\mathfrak{W} \setminus \mathfrak{E}$ is infinite, it follows that $\mathfrak{W} \setminus \mathfrak{M}$ is infinite. The reader will have no difficulties in giving examples showing that also $\mathfrak{W} \setminus \mathfrak{D}_3$, $\mathfrak{W} \setminus \mathfrak{D}$ and $\mathfrak{W} \setminus \mathfrak{G}_{60}$ are infinite. □

In what follows I will define a relation on properties (about numerical semigroups). It can be used to, in some sense, compare classes of numerical semigroups, or even results taking into account the above correspondences. Note that I do not want to make any judgement on the results and even less on their proofs. It may well happen that the ideas involved in the proof of a given result will in the future have a greater impact than the ideas involved in a proof of one of its generalizations.

A property $\mathcal{P}$ is said to be a *generalization* of a property $\mathcal{Q}$ if all the semigroups satisfying $\mathcal{Q}$ also satisfy $\mathcal{P}$, that is, $\mathfrak{Q} \subseteq \mathfrak{P}$ (or $\mathfrak{Q} \setminus \mathfrak{P}$ is empty). It is clear that a result that proves a generalization is better, but this can not be said in a definitive way when the arguments in the proofs are different. Except in some obvious cases (such as happens in several subsections of Sect. 3), just comparing through set inclusion is not of great help.

A property $\mathcal{P}$ is a *quasi-generalization* of a property $\mathcal{Q}$ if all but finitely many numerical semigroups satisfying $\mathcal{Q}$ also satisfy $\mathcal{P}$, that is, $\mathfrak{Q} \setminus \mathfrak{P}$ is either empty or finite. It is straightforward to check that quasi-generalization is reflexive and transitive (a partial quasiorder) in the set of properties on numerical semigroups. The notation $\mathcal{Q} \prec \mathcal{P}$ is used for "$\mathcal{P}$ is a quasi-generalization of $\mathcal{Q}$". In symbols: $\mathcal{Q} \prec \mathcal{P}$ if and only if $|\mathfrak{Q} \setminus \mathfrak{P}| < \infty$.

I am far from saying that properties that are quasi-generalized by others are not important (even without taking the proofs into account). According to this definition, any property defining an infinite class of numerical semigroups quasi-generalizes all the properties defining finite classes. For instance, the property $\mathcal{G}_{60}$ is quasi-generalized by the properties associated to the results stated in previous section that define infinite classes. But none of these results generalizes $\mathcal{G}_{60}$ (as the reader can easily check), which, from my point of view, makes it a property of high interest.

I encourage anyone who finds a new property (such that all numerical semigroups in the class of semigroups satisfying that property are Wilf) to compare it with other properties for quasi-generalization. Observing that it is not known any quasi-generalization $\mathcal{P}$ of the property under consideration such that $\mathfrak{P} \subseteq \mathfrak{W}$ probably will count in favour of the results obtained.

My aim now is to compare, under quasi-generalization, the properties for which a name was given: $\mathcal{S}, \mathcal{W}, \mathcal{E}, \mathcal{D}_3, \mathcal{D}, \mathcal{M}$, and $\mathcal{G}_{60}$. They do not form a chain, as it follows from next result. The impatient reader may already take a look at the lattice depicted in Fig. 7.

Proposition 4.1 *$\mathcal{M}$ and $\mathcal{D}$ are not comparable under quasi-generalization.*

Proof For a given $m > 1$, $S = \langle m \rangle_{mk} = \langle m, km+1, \dots, km+m-1 \rangle$ is a semigroup of maximal embedding dimension, thus satisfies $\mathcal{D}$ and, for $k > 3$, S does not satisfy $\mathcal{M}$. As there are infinitely many such semigroups, it follows that $\mathfrak{D} \setminus \mathfrak{M}$ is infinite and therefore $\mathcal{D} \not\prec \mathcal{M}$.

It remains to prove that $\mathcal{M} \not\prec \mathcal{D}$, which amounts to show that there are infinitely many semigroups in $\mathfrak{M}$ with small embedding dimension.

The proof of this fact begins with a trivial observation. As usual, for sets of integers A and B, $A + B$ denotes the set $\{a + b \mid a \in A, b \in B\}$. Let $X = \{0, 1, 2, 3\} \cup \{7k \mid k \in \mathbb{N}\}$.

Claim The set $X + X + X$ consists of all nonnegative integers. □

Proof of the Claim Clearly $X+X+X \subseteq \mathbb{N}$. It is also clear that $\{0, \dots, 6\} \subseteq X+X$. By the Euclidean algorithm every integer can be written in the form $7k + \rho$, with k an integer and $\rho \in \{0, \dots, 6\}$. Consequently, any nonnegative integer belongs to $X + X + X$, which proves the claim.

Let m be a positive integer and let $Y = \{m\} + X$. Consider the semigroup $S = \langle Y \rangle$. Example 4.1 helps to visualize it for two possible values of m.

Since, by the above Claim, $Y + Y + Y = \{3m\} + X + X + X = \{3m\} + \mathbb{N}$ holds, it follows that $\chi(S) \leq 3\,\mathrm{m}(S)$. Thus S satisfies $\mathcal{M}$.

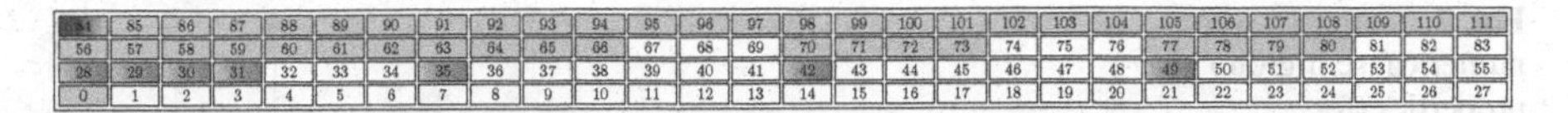

Fig. 5 Pictorial representation of the semigroup obtained with $m = 28$

Fig. 6 Shape of the numerical semigroup obtained with $m = 80$

It is straightforward to check that $\mathrm{P}(S) = \{m, m+1, m+2, m+3\} \cup \{7k+m \mid 0 < k \leq \lceil \frac{m}{7} \rceil - 1\}$. Thus the embedding dimension of S is $4 + \lceil \frac{m}{7} \rceil - 1 = 3 + \lceil \frac{m}{7} \rceil$. Note that $3 + \frac{m}{7} \leq 3 + \lceil \frac{m}{7}$, and that $3 + \frac{m}{7} < \frac{m}{3}$ if and only if $7m - 3m > 3 \times 7 \times 3$, that is $4m > 63$. Thus one concludes that, for $m > 15$, S does not satisfy $\mathcal{D}$. □

Example 4.1 Let $Y = \{m, m+1, m+2, m+3\} \cup \{7k+m \mid 0 < k \leq \lceil \frac{m}{7} \rceil\}$ be the set introduced in the proof of Proposition 4.1. Consider the semigroup $S = \langle Y \rangle$. Figures 5 and 6 give a pictorial representation for the cases $m = 28$ and $m = 80$ (in the latter case only the shape is drawn).

The following GAP code can be used to give the semigroups.

```
m := 28;;
small_gens  :=  [m,m+1,m+2,m+3];
other_gens  :=  List([1..Int(m/7)],  k -> 7*k+m);
ns := NumericalSemigroup(Union(small_gens,other_gens));;
```

Then an adaptation of the GAP-code 1 can be used to get the TikZ code. In order to obtain an image just showing the shape, one can use the options in GAP-code 2.

Most of the indicated relations in the lattice represented in Fig. 7 have been treated along the text in this section. That $\mathcal{D}$ and $\mathcal{M}$ are not comparable under quasi-generalization is shown in Proposition 4.1. Thus, the following has been proved:

Proposition 4.2 *With the notation introduced, one has the lattice in Fig. 7.*

Other comparisons could be made. As an example, fix a Wilf semigroup S (possibly with small multiplicity when compared to the conductor). It is easy to check that $\mathrm{m}(D(S,a)) = \mathrm{m}(S) + a$ and that $\chi(D(S,a)) = \chi(S) + a$. Thus, $\frac{\mathrm{m}(D(S,a))}{\chi(D(S,a))} = \frac{\mathrm{m}(S)+a}{\chi(S)+a}$ tends to 1 when a tends to infinity. In particular, from a certain point on, the quotient $\frac{\mathrm{m}}{\chi}$ is greater than $1/3$ and so, from that point on, all the semigroups satisfy $\mathcal{M}$. Therefore, $\mathcal{M}$ quasi-generalizes the property corresponding to Proposition 3.10, for any fixed S.

Denote by $\mathcal{P}_4$ the property associated to Proposition 3.15 with $\rho = 4$. As there are infinitely many semigroups satisfying $\mathcal{D}$ whose multiplicity is even, we get that $\mathcal{D} \not\nearrow \mathcal{P}_4$. On the other hand, it is straightforward to check that there are infinitely many semigroups satisfying $\mathcal{P}_4$ and with small embedding dimension (less than $\mathrm{m}/3$). Thus $\mathcal{D} \not\swarrow \mathcal{P}_4$, and we conclude that $\mathcal{D}$ and $\mathcal{P}_4$ are not comparable under quasi-generalization.

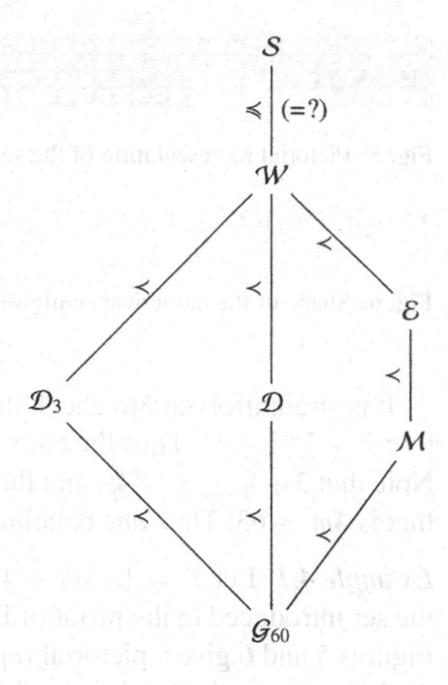

Fig. 7 Lattice of some numerical semigroup properties (for quasi-generalization)

Let $d \geq 10$ be an integer and denote by $\mathfrak{S}_{(d,\chi)}$ the class of numerical semigroups with embedding dimension d and conductor χ. Denote by $r = r(S)$ the ratio (second smallest primitive) of a numerical semigroup S. For fixed d and χ, consider the set

$$\mathfrak{R}_{(d,\chi)} = \left\{ S \in \mathfrak{S}_{(d,\chi)} \mid m \leq \frac{8\,d^2}{25} + \frac{d}{5} - \frac{5}{4} \text{ and } r > \left\lfloor \frac{\chi + m}{3} \right\rfloor \right\}.$$

Since no primitive of a numerical semigroup exceeds $\chi + m - 1$, the ratio of a semigroup of embedding dimension d must be at most $\chi + m - d + 1$.

Note that the class of numerical semigroups satisfying Eq. (3) in Proposition 3.17 is:

$$\mathfrak{R}_d = \bigcup_{\chi \geq m} \mathfrak{R}_{(d,\chi)}.$$

Consider now the class $\mathfrak{R}_d \setminus (\mathfrak{M} \cup \mathfrak{D})$. In set notation it may be written as follows:

$$\left\{ S \in \mathfrak{S} \mid 3\,d < m \leq \frac{8}{25}\,d^2 + \frac{1}{5}\,d - \frac{5}{4},\ \chi > 3\,m \text{ and } \left\lfloor \frac{\chi + m}{3} \right\rfloor < r \leq \chi + m - d + 1 \right\}.$$

A natural question is whether this class is finite, that is, does the disjunction of the properties $\mathcal{M}$ and $\mathcal{D}$ quasi-generalize the property associated to Proposition 3.17 (with d fixed)? Apparently there is no bound for the conductor, so one would be temped to answer "yes". But, observing that a big conductor will force a big ratio and that, on the other hand, a small embedding dimension and a big ratio leads to a huge conductor, one sees the question may be challenging.

Acknowledgements I would like to thank my colleges ate the FCUP's Mathematics department who made possible for me to benefit of a sabbatical year. The hospitality found in the Instituto de Matemáticas de la Universidad de Granada (IEMath-GR) was amazing, which made the writing of this paper a lot easier. Many thanks to IEMath-GR and especially to Pedro García-Sanchez who made possible this stay in Granada to happen. I want also to thank Claude Marion whose interest shown on the topic highly contributed for my decision on writing this survey. His comments greatly contributed to improve the paper. For the many very pertinent comments, I am also indebted to the anonymous referee.

References

1. Backelin, J.: On the number of semigroups of natural numbers. Math. Scand. **66**(2), 197–215 (1990). https://doi.org/10.7146/math.scand.a-12304
2. Barucci, V.: On propinquity of numerical semigroups and one-dimensional local Cohen Macaulay rings. In: Commutative Algebra and Its Applications, pp. 49–60. Walter de Gruyter, Berlin (2009)
3. Barucci, V., Fröberg, R.: One-dimensional almost Gorenstein rings. J. Algebra **188**(2), 418–442 (1997). https://doi.org/10.1006/jabr.1996.6837
4. Barucci, V., Strazzanti, F.: Dilatations of numerical semigroups. Semigroup Forum **98**(2), 251–260 (2019). https://doi.org/10.1007/s00233-018-9922-9
5. Bras-Amorós, M.: Fibonacci-like behavior of the number of numerical semigroups of a given genus. Semigroup Forum **76**(2), 379–384 (2008). https://doi.org/10.1007/s00233-007-9014-8
6. Brauer, A.: On a problem of partitions. Am. J. Math. **64**, 299–312 (1942). https://doi.org/10.2307/2371684
7. Bruns, W., Garcia-Sanchez, P., O'Neill, C., Wilburne, D.: Wilf's conjecture in fixed multiplicity (2019). arXiv:1903.04342
8. Curtis, F.: On formulas for the Frobenius number of a numerical semigroup. Math. Scand. **67**(2), 190–192 (1990). https://doi.org/10.7146/math.scand.a-12330
9. Delgado, M.: IntPic—a GAP package for drawing integers (2017). Version 0.2.3. http://www.gap-system.org/Packages/intpic.html.
10. Delgado, M.: On a question of Eliahou and a conjecture of Wilf. Math. Z. **288**(1–2), 595–627 (2018). https://doi.org/10.1007/s00209-017-1902-3
11. Delgado, M., García-Sánchez, P.A., Morais, J.: Numericalsgps—a GAP package on numerical semigroups (2018). Version number 1.1.7. http://www.gap-system.org/Packages/numericalsgps.html
12. Dhayni, M.: Problems in numerical semigroups. Ph.D. Thesis, Université d'Angers, 2017. https://www.theses.fr/2017ANGE0041.pdf.
13. Dhayni, M.: Wilf's conjecture for numerical semigroups. Palest. J. Math. **7**(2), 385–396 (2018)
14. Dobbs, D.E., Matthews, G.L.: On a question of Wilf concerning numerical semigroups. In: Focus on Commutative Rings Research, pp. 193–202. Nova Science Publishers, New York (2006)
15. Eliahou, S.: A Graph-Theoretic Approach to Wilf's Conjecture. Slides presented at the Meeting of the Catalan, Spanish, Swedish Math Societies, 2017. http://www.ugr.es/~semigrupos/Umea-2017/
16. Eliahou, S.: Wilf's conjecture and Macaulay's theorem. J. Eur. Math. Soc. **20**(9), 2105–2129 (2018). https://doi.org/10.4171/JEMS/807
17. Eliahou, S., Fromentin, J.: Near-misses in Wilf's conjecture. Semigroup Forum **98**(2), 285–298 (2019). https://doi.org/10.1007/s00233-018-9926-5
18. Eliahou, S., Marín-Aragón, D.: Personal communication by Shalom Eliahou (2019)
19. Fröberg, R., Gottlieb, C., Häggkvist, R.: On numerical semigroups. Semigroup Forum **35**(1), 63–83 (1987). https://doi.org/10.1007/BF02573091

20. Fromentin, J., and Hivert, F.: Exploring the tree of numerical semigroups. Math. Comput. **85**(301), 2553–2568 (2016). https://doi.org/10.1090/mcom/3075
21. The GAP Group: GAP—groups, algorithms, and programming, version 4.9.1 (2018). https://www.gap-system.org
22. Geroldinger, A., Halter-Koch, F: Non-unique factorizations: a survey. In: Multiplicative Ideal Theory in Commutative Algebra, pp. 207–226. Springer, New York (2006). https://doi.org/10.1007/978-0-387-36717-0_13
23. Kaplan, N.: Counting numerical semigroups by genus and some cases of a question of Wilf. J. Pure Appl. Algebra **216**(5), 1016–1032 (2012). https://doi.org/10.1016/j.jpaa.2011.10.038
24. Kaplan, N.: Counting numerical semigroups. Am. Math. Mon. **124**(9), 862–875 (2017). https://doi.org/10.4169/amer.math.monthly.124.9.862
25. Kunz, E.: On the type of certain numerical semigroups and a question of Wilf. Semigroup Forum **93**(1), 205–210 (2016). https://doi.org/10.1007/s00233-015-9755-8
26. Kunz, E., Waldi, R.: On the deviation and the type of certain local Cohen-Macaulay rings and numerical semigroups. J. Algebra **478**, 397–409 (2017). https://doi.org/10.1016/j.jalgebra.2017.01.041
27. Matthews, G.L.: On integers nonrepresentable by a generalized arithmetic progression. Integers **5**(2), A12 (2005)
28. Moscariello, A., Sammartano, A.: On a conjecture by Wilf about the Frobenius number. Math. Z. **280**(1–2), 47–53 (2015). https://doi.org/10.1007/s00209-015-1412-0
29. O'Neill, C., Pelayo, R.: Apéry sets of shifted numerical monoids. Adv. Appl. Math. **97**, 27–35 (2018). https://doi.org/10.1016/j.aam.2018.01.005
30. Ramírez-Alfonsín, J.L.: The Diophantine Frobenius Problem. Oxford Lecture Series in Mathematics and Its Applications, vol. 30. Oxford University Press, Oxford. (2005). https://doi.org/10.1093/acprof:oso/9780198568209.001.0001
31. Rosales, J.C., García Sánchez, P.A.: Numerical Semigroups. Developments in Mathematics, vol. 20. Springer, New York (2009). https://doi.org/10.1007/978-1-4419-0160-6
32. Sammartano, A.: Numerical semigroups with large embedding dimension satisfy Wilf's conjecture. Semigroup Forum **85**(3), 439–447 (2012). https://doi.org/10.1007/s00233-011-9370-2
33. Spirito, D.: Wilf's conjecture for numerical semigroups with large second generator (2017). arXiv:1710.09245
34. Sylvester, J.J.: Mathematical questions with their solutions. Educ. Times **41**, 21 (1884). Solution by W.J. Curran Sharp
35. Tao, T., Vu, V.: Additive Combinatorics. Cambridge Studies in Advanced Mathematics, vol. 105. Cambridge University Press, Cambridge (2006). https://doi.org/10.1017/CBO9780511755149
36. Wilf, H.S.: A circle-of-lights algorithm for the "money-changing problem". Am. Math. Mon. **85**(7), 562–565 (1978). https://doi.org/10.2307/2320864
37. Zhai, A.: Fibonacci-like growth of numerical semigroups of a given genus. Semigroup Forum **86**(3), 634–662 (2013). https://doi.org/10.1007/s00233-012-9456-5
38. Zhao, Y.: Constructing numerical semigroups of a given genus. Semigroup Forum **80**, 242–254 (2010). https://doi.org/10.1007/s00233-009-9190-9

Gapsets of Small Multiplicity

Shalom Eliahou and Jean Fromentin

Abstract A *gapset* is the complement of a numerical semigroup in $\mathbb{N}$. In this paper, we characterize all gapsets of multiplicity $m \leq 4$. As a corollary, we provide a new simpler proof that the number of gapsets of genus g and fixed multiplicity $m \leq 4$ is a nondecreasing function of g.

Keywords Numerical semigroups · Genus · Kunz coordinates · Gapset filtrations

1 Introduction

Denote $\mathbb{N} = \{0, 1, 2, 3, \ldots\}$ and $\mathbb{N}_+ = \mathbb{N} \setminus \{0\} = \{1, 2, 3, \ldots\}$. For $a, b \in \mathbb{Z}$, let $[a, b] = \{z \in \mathbb{Z} \mid a \leq z \leq b\}$ and $[a, \infty[= \{z \in \mathbb{Z} \mid a \leq z\}$ denote the integer intervals they span. A *numerical semigroup* is a subset $S \subseteq \mathbb{N}$ containing 0, stable under addition and with finite complement in $\mathbb{N}$. Equivalently, it is a subset $S \subseteq \mathbb{N}$ of the form $S = \langle a_1, \ldots, a_n \rangle = \mathbb{N}a_1 + \cdots + \mathbb{N}a_n$ for some globally coprime positive integers $a_1, \ldots, a_n$.

For a numerical semigroup $S \subseteq \mathbb{N}$, its *gaps* are the elements of $\mathbb{N} \setminus S$, its *genus* is $g = |\mathbb{N} \setminus S|$, its *multiplicity* is $m = \min S \setminus \{0\}$, its *Frobenius number* is $f = \max \mathbb{Z} \setminus S$, its *conductor* is $c = f + 1$, and its *embedding dimension*, usually denoted e, is the least number of generators of S, i.e. the least n such that $S = \langle a_1, \ldots, a_n \rangle$. Note that the conductor c of S satisfies $c + \mathbb{N} \subseteq S$, and is minimal with respect to this property since $c - 1 = f \notin S$.

Given $g \geq 0$, the number n_g of numerical semigroups of genus g is finite, as easily seen. The values of n_g for $g = 0, \ldots, 15$ are as follows:

$$1, 1, 2, 4, 7, 12, 23, 39, 67, 118, 204, 343, 592, 1001, 1693, 2857.$$

S. Eliahou (✉) · J. Fromentin
Université du Littoral Côte d'Opale, Calais, France
e-mail: eliahou@univ-littoral.fr; fromentin@math.cnrs.fr

V. Barucci et al. (eds.), *Numerical Semigroups*, Springer INdAM Series 40,
https://doi.org/10.1007/978-3-030-40822-0_5

In 2006, Maria Bras-Amorós made some remarkable conjectures concerning the growth of n_g. In particular, she conjectured [1] that

$$n_g \geq n_{g-1} + n_{g-2} \tag{1}$$

for all $g \geq 2$. This conjecture is widely open. Indeed, even the weaker inequality

$$n_g \geq n_{g-1} \tag{2}$$

whose validity has been settled by Alex Zhai [9] for all sufficiently large g, remains to be proved for all $g \geq 1$. See also [2, 4, 7] for closely related information on the numbers n_g.

Still in [9], Zhai showed that 'most' numerical semigroups S satisfy $c \leq 3m$, where c and m are the conductor and multiplicity of S, respectively. For a more precise statement, let us denote by n'_g the number of numerical semigroups of genus g satisfying $c \leq 3m$. The values of n'_g for $g = 0, \ldots, 15$ are as follows:

$$1, 1, 2, 4, 6, 11, 20, 33, 57, 99, 168, 287, 487, 824, 1395, 2351.$$

Zhai showed then that $\lim_{g\to\infty} n'_g/n_g = 1$, as had been earlier conjectured by Yufei Zhao [10]. In that sense, numerical semigroups satisfying $c \leq 3m$ may be considered as *generic*.

Recently, the strong conjecture (1) has been established for generic numerical semigroups. Here is the precise statement, first announced at the IMNS 2018 conference in Cortona (https://www.ugr.es/~imns2010/2018/).

Theorem 1 ([3], Theorem 6.4) *The inequalities*

$$n'_{g-1} + n'_{g-2} + n'_{g-3} \geq n'_g \geq n'_{g-1} + n'_{g-2},$$

hold for all $g \geq 3$.

The proof of this result essentially rests on the notion of *gapset filtrations*, a new flexible framework to investigate numerical semigroups introduced in [3]. More details are given in Sect. 2 since, here also, gapsets filtrations are at the core of the present results.

Remark Let $g \geq 0, m \geq 1$ be two integers. We denote by $\Gamma_{g,m}$ the finite set of all numerical semigroups of genus g and multiplicity m, and by $n_{g,m} = |\Gamma_{g,m}|$ its cardinality.

Since, for a numerical semigroup S of multiplicity m and genus g, the integers $1, \ldots, m-1$ belong to the complement $\mathbb{N} \setminus S$, the relation $g \geq m-1$ holds. Thus $n_{g,m} = 0$ for $m \geq g+2$, and so we have

$$n_g = \sum_{m=1}^{g+1} n_{g,m}.$$

The first values of $n_{g,m}$ for $g \geq 0$ and small fixed m are given below.

g	0	1	2	3	4	5	6	7	8	9	10	11	12	13	14 ...
$m = 1$	1	0	0	0	0	0	0	0	0	0	0	0	0	0	0 ...
$m = 2$	0	1	1	1	1	1	1	1	1	1	1	1	1	1	1 ...
$m = 3$	0	0	1	2	2	2	3	3	3	4	4	4	5	5	5 ...
$m = 4$	0	0	0	1	3	4	6	7	9	11	13	15	18	20	23 ...
$m = 5$	0	0	0	0	1	4	7	10	13	16	22	24	32	35	43 ...
$m = 6$	0	0	0	0	0	1	5	11	17	27	37	49	66	85	106 ...

For instance, the unique numerical semigroup of multiplicity 1 is $\mathbb{N}$. Nathan Kaplan proposed the following conjecture in [6], a refinement of the conjectured inequality (2).

Conjecture 1 Let $m \geq 2$. Then

$$n_{g,m} \geq n_{g-1,m} \tag{3}$$

for all $g \geq 1$.

On the other hand, still for $m \geq 2$ fixed, there is no hope a stronger inequality such as (1) may hold for the $n_{g,m}$, as the reader can check by looking at the rows of the above table.

Conjecture 1 is trivial for $m = 2$ since $n_{g,2} = 1$ for all $g \geq 1$, and has been settled for $m = 3, 4, 5$ in 2018 by Pedro A. García-Sánchez, Daniel Marín-Aragón and Aureliano M. Robles-Pérez [5]. For that, they used a linear integer software to count the number of integral points of the associated Kunz polytope. With it, they first achieved formulas for $n_{g,m}$ for $m = 3, 4, 5$, and then proved them to be increasing using a computer algebra system. The conjecture remains open for $m \geq 6$.

Our purpose in this paper is to give new proofs of Conjecture 1 for $m = 3$ and $m = 4$ by constructing explicit injections

$$\Gamma_{g,3} \rightarrow \Gamma_{g+1,3} \quad \text{and} \quad \Gamma_{g,4} \rightarrow \Gamma_{g+1,4}$$

for $g \geq 0$, thereby establishing the desired inequalities $n_{g+1,3} \geq n_{g,3}$ and $n_{g+1,4} \geq n_{g,4}$. Thus, our proofs are computer-free and do not rest on counting formulas for $n_{g,3}$ and $n_{g,4}$. These injections were first announced in [3].

2 Gapset Filtrations

The content of this section is mostly taken from [3].

Definition 1 Let $n \in \mathbb{N}_+$. An *additive decomposition* of n is any expression of the form $n = a + b$ with $a, b \in \mathbb{N}_+$. We refer to the positive integers a, b as the *summands* of this decomposition.

Definition 2 A *gapset* is a finite set $G \subset \mathbb{N}_+$ satisfying the following property: for all $z \in G$, if $z = x + y$ with $x, y \in \mathbb{N}_+$, then $x \in G$ or $y \in G$. That is, for any additive decomposition of $z \in G$, at least one of its summands belongs to G.

Notice the similarity of this definition with that of a prime ideal P in a ring R, where for any $z \in P$, any decomposition $z = xy$ with $x, y \in R$ implies $x \in P$ or $y \in P$.

Remark 1 It follows from the definition that a gapset G is nothing else than the set of gaps of a numerical semigroup S, where $S = \mathbb{N} \setminus G$.

Definition 3 We naturally extend the definitions of *multiplicity*, *Frobenius number*, *conductor* and *genus* of a gapset G as being those of the corresponding numerical semigroup $S = \mathbb{N} \setminus G$, respectively.

More directly, for a gapset G, these notions may be described as follows:

- the multiplicity of G is the smallest integer $m \geq 1$ such that $m \notin G$;
- the Frobenius number of G is $\max(G)$ if $G \neq \emptyset$, and -1 otherwise;
- the conductor of G is $1 + \max(G)$ if $G \neq \emptyset$, and 0 otherwise;
- the genus of G is $g(G) = \text{card}(G)$.

Example 1 The set $G = \{1, 2, 3, 4, 6, 7, 11\}$ is a gapset. For instance, for each additive decomposition of 11, namely

$$1 + 10, \quad 2 + 9, \quad 3 + 8, \quad 4 + 7, \quad 5 + 6,$$

at least one of the two summands belongs to G. Let $S = \mathbb{N} \setminus G = \{0, 5, 8, 9, 10\} \cup [12, +\infty[$. Then $S = \langle 5, 8, 9, 12 \rangle$ as easily seen, whence S is indeed a numerical semigroup. The multiplicity, conductor, Frobenius number, genus and embedding dimension of G and S are $m = 5$, $c = 12$, $f = 11$, $g = 7$ and $e = 4$, respectively.

2.1 The Canonical Partition

Lemma 1 *Let G be a gapset of multiplicity m. Then*

$$[1, m - 1] \subseteq G,$$

$$G \cap m\mathbb{N} = \emptyset.$$

Proof By definition of the multiplicity, G contains $[1, m - 1]$ but not m. Let $a \geq 2$ be an integer. The formula $am = m + (a - 1)m$ and induction on a imply that $am \notin G$. □

This motivates the following notation and definition.

Remark Let G be a gapset of multiplicity m. We denote $G_0 = [1, m-1]$ and, more generally,

$$G_i = G \cap [im+1, (i+1)m-1] \quad \text{for all } i \geq 0. \tag{4}$$

Definition 4 Let G be a gapset of multiplicity m and conductor c. The *depth* of G is the integer $q = \lceil c/m \rceil$.

Proposition 1 *Let G be a gapset of multiplicity m and depth q. Let G_i be defined as in* (4)*. Then*

$$G = G_0 \sqcup G_1 \sqcup \cdots \sqcup G_{q-1} \tag{5}$$

and $G_{q-1} \neq \emptyset$. Moreover $G_{i+1} \subseteq m + G_i$ for all $i \geq 0$.

Proof As $G \cap m\mathbb{N} = \emptyset$, it follows that G is the disjoint union of the G_i for $i \geq 0$. Let c be the conductor of G. Then $G \subseteq [1, c-1]$. Since $(q-1)m < c \leq qm$ by definition of q, it follows that $G_i = \emptyset$ for $i \geq q$, whence (5). Let $f = c-1$. Since $f \in G$, $(q-1)m \leq f < qm$ and $f \not\equiv 0 \bmod m$, it follows that $f \in G_{q-1}$.

It remains to show that $G_{i+1} \subseteq m + G_i$ for all $i \geq 0$. Let $x \in G_{i+1}$. Since $G_{i+1} \subseteq [(i+1)m+1, (i+2)m-1]$, we have

$$x - m \in [im+1, (i+1)m-1].$$

Now $x - m \in G$ since $x = m + (x-m)$ and $m \notin G$. So $x - m \in G_i$. □

Definition 5 Let G be a gapset. The *canonical partition* of G is the partition $G = G_0 \sqcup G_1 \sqcup \cdots \sqcup G_{q-1}$ given by Proposition 1.

Remark 2 The multiplicity m, genus g and depth q of a gapset G may be read off from its canonical partition $G = \sqcup_i G_i$ as follows:

$$m = \max(G_0) + 1,$$
$$g = \sum_i |G_i|,$$
$$q = \text{the number of parts of the partition.}$$

2.2 *Gapset Filtrations*

Let $G \subset \mathbb{N}_+$ be a gapset. Let $G = G_0 \sqcup G_1 \sqcup \cdots \sqcup G_{q-1}$ be its canonical partition. For all $0 \leq i \leq q-1$, denote

$$F_i = -im + G_i. \tag{6}$$

Then $F_{i+1} \subseteq F_i$ for all i, as follows from the inclusion $G_{i+1} \subseteq m + G_i$ stated in Proposition 1. This gives rise to the following definition.

Definition 6 Let $G \subset \mathbb{N}_+$ be a gapset of multiplicity m and depth q. The *gapset filtration* associated to G is the finite sequence

$$(F_0, F_1, \ldots, F_{q-1}) = (G_0, -m + G_1, \ldots, -(q-1)m + G_{q-1}),$$

i.e. with F_i defined as in (6) for all i. Thus, as seen above, we have

$$F_0 = [1, m-1] \supseteq F_1 \supseteq \cdots \supseteq F_{q-1}. \tag{7}$$

We define the *multiplicity*, *Frobenius number*, *conductor* and *genus* of a gapset filtration $F = (F_0, \ldots, F_{q-1})$ from those of the corresponding gapset G, namely:

- the multiplicity of F is $1 + \max(F_0)$ if $F_0 \neq \emptyset$ and 0 otherwise;
- the Frobenius number of F is $(q-1)m + \max(F_{q-1})$ if $F_0 \neq \emptyset$ and -1 otherwise;
- the conductor of F is $1 + (q-1)m + \max(F_{q-1})$ if $F_0 \neq \emptyset$ and 0 otherwise;
- the genus of F is $\operatorname{card}(F_0) + \cdots + \operatorname{card}(F_{q-1})$.

Example 2 Consider the gapset $G = \{1, 2, 3, 4, 6, 7, 11\}$ of Example 1. Its multiplicity is $m = 5$, and its canonical partition is given by $G_0 = \{1, 2, 3, 4\}$, $G_1 = \{6, 7\}$ and $G_2 = \{11\}$. Thus, its associated filtration is

$$F = (\{1, 2, 3, 4\}, \{1, 2\}, \{1\}).$$

Definition 7 For integers $g \geq 1, m \geq 1$, we denote by $\mathcal{F}(g, m)$ the set of all gapset filtrations of genus g and multiplicity m.

Note that any given gapset filtration $F = (F_0, \ldots, F_{q-1})$ corresponds to a *unique* gapset G, since (6) is equivalent to

$$G_i = im + F_i. \tag{8}$$

In particular, *there is a straightforward bijection between gapsets G and gapset filtrations F*, which naturally preserves the multiplicity, Frobenius number, conductor and genus. Here is a direct consequence.

Proposition 2 *For any integers $g \geq 1, m \geq 1$, we have*

$$n_{g,m} = |\mathcal{F}(g, m)|.$$

Proof Straightforward from the above discussion. □

This result allows us to study properties of the sequence $g \mapsto n_{g,m}$ in the setting of gapset filtrations of multiplicity m. In particular, in order to establish its growth, it suffices to exhibit injections from $\mathcal{F}(g, m)$ to $\mathcal{F}(g+1, m)$. This is what we achieve in subsequent sections for $m = 3$ and $m = 4$.

We start with the separate case $m = 3$, which can be treated in a straightforward way and which points to a general strategy for larger values of m. Then, following those clues, we introduce some general tools, and we end up applying them to the case $m = 4$.

3 The Case $m = 3$

Any filtration $(F_0, \dots, F_t)$ such that

$$\{1, 2\} = F_0 \supseteq F_1 \supseteq \cdots \supseteq F_t \neq \emptyset$$

is of one of the two possible forms below, with the terms on the left standing as a compact notation:

$$(12)^r(1)^s = (\underbrace{\{1, 2\}, \dots, \{1, 2\}}_{r}, \underbrace{\{1\}, \dots, \{1\}}_{s}),$$

$$(12)^r(2)^s = (\underbrace{\{1, 2\}, \dots, \{1, 2\}}_{r}, \underbrace{\{2\}, \dots, \{2\}}_{s}),$$

both with $r \geq 1$ since $F_0 = \{1, 2\}$, and $s \geq 0$. We now characterize those filtrations which are gapset filtrations of multiplicity 3.

Theorem 2 *The gapset filtrations of multiplicity $m = 3$ are exactly the following ones:*

$$(12)^r(2)^s \quad \textit{with } 0 \leq s \leq r,$$

$$(12)^r(1)^s \quad \textit{with } 0 \leq s \leq r + 1,$$

both with $r \geq 1$.

Note that $g = 2r + s$ in both cases, since the genus of a gapset filtration $F = (F_0, \dots, F_{q-1})$ is given by the sum of the $|F_i|$.

Proof We start with the second case.

Case $F = (12)^r(2)^s$ Then

$$F_0 = \cdots = F_{r-1} = \{1, 2\},$$

$$F_r = \cdots = F_{r+s-1} = \{2\}.$$

Using (8) with $m = 3$, namely $G_i = 3i + F_i$ for all i, let

$$G = G_0 \cup \cdots \cup G_{r+s-1} \tag{9}$$

be the corresponding finite set. By construction, F is a gapset filtration if and only if G is a gapset. So, when is it the case that G is a gapset? We now proceed to answer this question.

Step 1 The set G given by (9) has the following properties:

$$\begin{aligned} 3\mathbb{N} \cap G &= \emptyset \\ 3i+1 \in G &\iff i \leq r-1 \\ 3i+2 \in G &\iff i \leq r+s-1. \end{aligned}$$

Indeed, this directly follows from the definition $G_i = 3i + F_i$ and (9).

Step 2 For $i \in \mathbb{N}$, any additive decomposition $3i + 1 = a + b$ is of the form

$$(a, b) = (3x + 1, 3(i - x)) \text{ or } (3y + 2, 3(i - 1 - y) + 2)$$

for some integers $0 \leq x \leq i - 1$ or $0 \leq y \leq i - 1$. Similarly, any additive decomposition $3i + 2 = a + b$ is of the form

$$(a, b) = (3x + 2, 3(i - x)) \text{ or } (3y + 1, 3(i - y) + 1)$$

for some integers $0 \leq x \leq i - 1$ or $0 \leq y \leq i$.

Step 3 Let $3i + 1 \in G$, i.e. with $i \leq r - 1$ according to Step 1. We now show that for any additive decomposition $3i + 1 = a + b$, either a or b belongs to G. Using Step 1, if $(a, b) = (3x + 1, 3(i - x))$, then $a \in G$ since $x \leq i - 1$ and we are done. Similarly, if $(a, b) = (3y + 2, 3(i - 1 - y) + 2)$, then $a \in G$ since $y \leq i \leq r - 1 \leq r + s - 1$ and we are done again.

Step 4 Let $3i + 2 \in G$, i.e. with $i \leq r + s - 1$. Let $3i + 2 = a + b$ be any additive decomposition. If $(a, b) = (3x+2, 3(i-x))$, then $a \in G$ since $x \leq i-1$ and we are done. Assume now $(a, b) = (3y + 1, 3(i - y) + 1)$ with $0 \leq y \leq i$. Then $a, b \notin G$ if and only if $y, i - y \geq r$. This is only possible if $i \geq 2r$ and, since $i \leq r + s - 1$ by hypothesis, the latter is equivalent to $s - 1 \geq r$. In particular, if $s \leq r$, then either a or b belongs to G. In summary, we have

$$(12)^r(2)^s \text{ is a gapset filtration} \iff G \text{ is a gapset} \iff s \leq r,$$

as desired.

Case $F = (12)^r(1)^s$ The arguments are similar to those of the previous case. Here, to start with, we have

$$F_0 = \cdots = F_{r-1} = \{1, 2\},$$
$$F_r = \cdots = F_{r+s-1} = \{1\}.$$

The corresponding set G defined by $G_i = 3i + F_i$ for all i and (8) has the following properties:

$$3\mathbb{N} \cap G \quad = \emptyset$$
$$3i + 1 \in G \iff i \leq r + s - 1$$
$$3i + 2 \in G \iff i \leq r - 1.$$

Analogously to Step 3 above, it is easy to see that for any additive decomposition $a + b = 3i + 2$ where $3i + 2 \in G$, then either a or b belongs to G.

On the other hand, let $3i + 1 \in G$. Then, analogously to Step 4 above, we find that there exists an additive decomposition $3i + 1 = a + b$ with $a, b \notin G$ if and only if $s \geq r + 2$. The details, using Step 2 and the above properties of G, are straightforward and left to the reader. Therefore, G is a gapset if and only if $s \leq r + 1$, as claimed. This concludes the proof of the proposition. □

Here is a straightforward consequence of the above characterization and the main result of this section.

Corollary 1 *For all* $g \geq 0$*, there is a natural injection*

$$\mathcal{F}(g, 3) \longrightarrow \mathcal{F}(g + 1, 3).$$

In particular, we have $n_{g+1,3} \geq n_{g,3}$ *for all* $g \geq 0$.

Proof Since $\mathcal{F}(g, 3) = \emptyset$ for $g \leq 1$, the statement holds in this case. Assume now $g \geq 2$. For $F = (F_0, \ldots, F_{q-1}) \in \mathcal{F}(g, 3)$, let us denote by $f_1(F)$ the insertion of a 1 in F at the unique possible position to get a new nonincreasing sequence of subsets of $[1, 2]$. That is, for $r, s \geq 1$, we define

$$(12)^r \overset{f_1}{\longmapsto} (12)^r(1)$$
$$(12)^r(1)^s \overset{f_1}{\longmapsto} (12)^r(1)^{s+1}$$
$$(12)^r(2)^s \overset{f_1}{\longmapsto} (12)^{r+1}(2)^{s-1}.$$

When is it the case that $f_1(F)$ is still a *gapset* filtration, of course automatically of genus $g + 1$? In other words, when do we have that $f_1(F)$ belongs $\mathcal{F}(g + 1, 3)$? Theorem 2 easily provides the following answer.

- If $F = (12)^r(2)^s \in \mathcal{F}(g, 3)$, then $f_1(F) \in \mathcal{F}(g+1, 3)$ for all r, s.
- If $F = (12)^r(1)^s \in \mathcal{F}(g, 3)$, then $f_1(F) \in \mathcal{F}(g+1, 3)$ if and only if $s \leq r$.

Recall that $g = 2r + s$ in both cases. In particular, the only case where $F \in \mathcal{F}(g, 3)$ but $f_1(F) \notin \mathcal{F}(g+1, 3)$ is for $F = (12)^r(1)^s$ with $s = r+1$, i.e. for $F = (12)^r(1)^{r+1} \in \mathcal{F}(g, 3)$ where $g = 3r+1$.

Consequently, f_1 provides a well-defined map

$$f_1 \colon \mathcal{F}(g, 3) \longrightarrow \mathcal{F}(g+1, 3),$$

obviously injective by construction, whenever $g \not\equiv 1 \bmod 3$.

Similarly, for $F \in \mathcal{F}(g, 3)$, denote by $f_2(F)$ the insertion of a 2 in F where it makes sense. That is, for $r, s \geq 1$, define

$$(12)^r \stackrel{f_2}{\longmapsto} (12)^r(2)$$

$$(12)^r(1)^s \stackrel{f_2}{\longmapsto} (12)^{r+1}(1)^{s-1}$$

$$(12)^r(2)^s \stackrel{f_2}{\longmapsto} (12)^r(2)^{s+1}.$$

By Theorem 2 again, we have

- If $F = (12)^r(2)^s \in \mathcal{F}(g, 3)$, then $f_2(F) \in \mathcal{F}(g+1, 3)$ if and only if $s \leq r-1$.
- If $F = (12)^r(1)^s \in \mathcal{F}(g, 3)$, then $f_2(F) \in \mathcal{F}(g+1, 3)$ for all $r, s \geq 1$.

In particular, the only case where $F \in \mathcal{F}(g, 3)$ but $f_2(F) \notin \mathcal{F}(g+1, 3)$ is for $F = (12)^r(2)^r \in \mathcal{F}(g, 3)$ with $g = 3r$. Therefore, f_2 provides a well-defined injective map

$$f_2 \colon \mathcal{F}(g, 3) \longrightarrow \mathcal{F}(g+1, 3)$$

whenever $g \not\equiv 0 \bmod 3$.

Summarizing, we end up with a well-defined injective map

$$f \colon \mathcal{F}(g, 3) \longrightarrow \mathcal{F}(g+1, 3)$$

defined by $f = f_1$ if $g \equiv 0, 2 \bmod 3$, and $f = f_2$ otherwise. □

4 Some More General Tools

In order to facilitate discussing gapsets and gapset filtrations, and gather more tools to treat more cases, it is useful to consider somewhat more general subsets of $\mathbb{N}_+$.

4.1 On m-Extensions and m-Filtrations

Definition 8 Let $m \in \mathbb{N}_+$. An *m-extension* is a finite set $A \subset \mathbb{N}_+$ containing $[1, m-1]$ and admitting a partition

$$A = A_0 \sqcup A_1 \sqcup \cdots \sqcup A_t \tag{10}$$

for some $t \geq 0$, where $A_0 = [1, m-1]$ and $A_{i+1} \subseteq m + A_i$ for all $i \geq 0$.

In particular, an m-extension A satisfies $A \cap m\mathbb{N} = \emptyset$. Moreover, the above conditions on the A_i imply

$$A_i = A \cap [im+1, (i+1)m-1] \tag{11}$$

for all $i \geq 0$, whence the A_i are *uniquely determined* by A.

Remark 3 Every gapset of multiplicity m is an m-extension. This follows from Proposition 1.

Closely linked is the notion of *m-filtration*.

Definition 9 Let $m \in \mathbb{N}_+$. An *m-filtration* is a finite sequence

$$F = (F_0, F_1, \ldots, F_t)$$

of nonincreasing subsets of $\mathbb{N}_+$ such that

$$F_0 = [1, m-1] \supseteq F_1 \supseteq \cdots \supseteq F_t.$$

The *genus* g of F is defined as $g = \sum_{i=0}^{t} |F_i|$.

For $m \in \mathbb{N}_+$, there is a straightforward bijection between m-extensions and m-partitions.

Proposition 3 *Let $A = A_0 \sqcup A_1 \sqcup \cdots \sqcup A_t$ be an m-extension. Set $F_i = -im + A_i$ for all i. Then $(F_0, F_1, \ldots, F_t)$ is an m-filtration. Conversely, let $(F_0, F_1, \ldots, F_t)$ be an m-filtration. Set $A_i = im + F_i$ for all i, and let*

$$A = \bigsqcup_{i=0}^{t} A_i = \bigsqcup_{i=0}^{t} (im + F_i).$$

Then A is an m-extension.

Proof We have $F_i = -im + A_i$ if and only if $A_i = im + F_i$. □

Remark If A is an m-extension, we denote by $F = \varphi(A)$ the m-filtration associated to it by Proposition 3. Conversely, if F is an m-filtration, we denote by $A = \tau(F)$ its associated m-extension.

By Proposition 3, *the maps φ and τ are inverse to each other.*

4.2 Gapset Filtrations Revisited

Definition 10 Let $G \subset \mathbb{N}_+$ be a gapset of multiplicity m. The *gapset filtration* associated to G is the m-filtration $F = \varphi(G)$.

By Remark 3, every gapset G of multiplicity m is an m-extension, whence $\varphi(G)$ is well-defined.

Concretely, let G be a gapset of multiplicity m and depth q. As in (4), let $G_i = G \cap [im + 1, (i + 1)m - 1]$ for all $i \geq 0$, so that $G_0 = [1, m - 1]$ and

$$G = G_0 \sqcup \cdots \sqcup G_{q-1}.$$

The associated m-filtration $F = \varphi(G)$ is then given by $F = (F_0, \dots, F_{q-1})$ where $F_i = -im + G_i$ for all $i \geq 0$.

It follows from Remark 2 and the equality $|F_i| = |G_i|$ for all i, that the genus of F is equal to $|F_0| + \cdots + |F_{q-1}|$ and that its depth is equal to the number of nonzero F_i.

4.3 A Compact Representation

In this section, we use permutations of $[1, m - 1]$ and exponent vectors to represent m-filtrations in a compact way. We denote by $\mathfrak{S}_{m-1}$ the symmetric group on $[1, m - 1]$.

Proposition 4 *Let $F = (F_0, \dots, F_t)$ be an m-filtration. Then there exists a permutation $\sigma \in \mathfrak{S}_{m-1}$ and exponents $e_0, \dots, e_{m-2} \in \mathbb{N}$ such that*

$$F = (\underbrace{F'_0, \dots, F'_0}_{e_0}, \underbrace{F'_1, \dots, F'_1}_{e_1}, \dots, \underbrace{F'_{m-2}, \dots, F'_{m-2}}_{e_{m-2}}),$$

where $F'_0 = [1, m - 1]$ and $F'_i = F'_{i-1} \setminus \{\sigma(i)\}$ for $1 \leq i \leq m - 2$. In particular, we have $|F'_i| = m - 1 - i$ for all $0 \leq i \leq m - 2$.

Proof By hypothesis, we have

$$[1, m - 1] = F_0 \supseteq F_1 \supseteq \cdots \supseteq F_t.$$

Equalities may occur in this chain. Removing repetitions, let

$$[1, m - 1] = H_0 \supsetneq H_1 \supsetneq \cdots \supsetneq H_s$$

denote the underlying descending chain, i.e. with

$$\{F_0, F_1, \dots, F_t\} = \{H_0, H_1, \dots, H_s\}$$

and $H_i \neq H_j$ for all $i \neq j$. Each H_i comes with some repetition frequency $\mu_i \geq 1$ in $\{F_0, F_1, \dots, F_t\}$. Thus, we have

$$F = (\underbrace{H_0, \dots, H_0}_{\mu_0}, \underbrace{H_1, \dots, H_1}_{\mu_1}, \dots, \underbrace{H_s, \dots, H_s}_{\mu_s}).$$

Now, between each consecutive pair $H_{i-1} \supsetneq H_i$, we insert some maximal descending chain of subsets $H'_{i,j}$, i.e.

$$H_{i-1} = H'_{i,0} \supsetneq H'_{i,1} \supsetneq \cdots \supsetneq H'_{i,k_i} = H_i,$$

where $k_i = |H_{i-1}| - |H_i|$. Thus $|H'_{i,j}| = |H'_{i-1}| - j$ for all $0 \leq j \leq k_i$.

We end up with a maximal descending chain of subsets

$$F' = [1, m-1] = F'_0 \supsetneq F'_1 \supsetneq \cdots \supsetneq F'_{m-2},$$

where each term has one less element than the preceding one, i.e. where $|F'_j| = |F'_{j-1}| - 1$ for all $1 \leq j \leq m-2$. By construction, we have

$$\{F_0, F_1, \dots, F_t\} = \{H_0, H_1, \dots, H_s\} \subseteq \{F'_0, F'_1, \dots, F'_{m-2}\},$$

and each F'_i arises with some frequency $e_i \geq 0$ in $\{F_0, F_1, \dots, F_t\}$. Thus

$$F = (\underbrace{F'_0, \dots, F'_0}_{e_0}, \underbrace{F'_1, \dots, F'_1}_{e_1}, \dots, \underbrace{F'_{m-2}, \dots, F'_{m-2}}_{e_{m-2}}).$$

Finally, since each F'_i is obtained by removing one distinct element from F'_{i-1} for $1 \leq i \leq m-2$, there is a permutation σ of $[1, m-1]$ such that

$$F'_i = F'_{i-1} \setminus \{\sigma(i)\}$$

for $1 \leq i \leq m-2$. □

Remark Given $\sigma \in \mathfrak{S}_{m-1}$ and $e = (e_0, \dots, e_{m-2}) \in \mathbb{N}^{m-1}$ such that $e_0 \geq 1$, we denote by $F(\sigma, e)$ the m-filtration

$$F = (\underbrace{F'_0, \dots, F'_0}_{e_0}, \underbrace{F'_1, \dots, F'_1}_{e_1}, \dots, \underbrace{F'_{m-2}, \dots, F'_{m-2}}_{e_{m-2}})$$

where $F'_i = F'_{i-1} \setminus \{\sigma(i)\}$ for $1 \leq i \leq m-2$.

Example 3 Consider the 5-filtration $F = (\{1, 2, 3, 4\}, \{1, 2\}, \{1\})$ of Example 2. Let $\sigma = (3, 4, 2, 1) \in \mathfrak{S}_4$ and $e = (1, 0, 1, 1)$. Then $F = F(\sigma, e)$ as readily checked. Note that we also have $F = F(\sigma', e)$ where $\sigma' = (4, 3, 2, 1)$.

One important question is: when is the m-filtration $F = F(\sigma, e)$ a *gapset* filtration? The next section provides an answer.

4.4 Complementing an m-Extension

Remark Let $F = F(\sigma, e)$ be an m-filtration, where $\sigma \in \mathfrak{S}_{m-1}$ and $e = (e_0, \ldots, e_{m-2}) \in \mathbb{N}^{m-1}$ with $e_0 \geq 1$. We denote by $G = G(\sigma, e)$ the corresponding m-extension, i.e. $G = \tau(F)$ using Notation 4.1.

Here is how to determine the set complement in $\mathbb{N}$ of the m-extension $G = G(\sigma, e)$.

Proposition 5 *Let $F = F(\sigma, e)$ be an m-filtration, where $\sigma \in \mathfrak{S}_{m-1}$ and $e = (e_0, \ldots, e_{m-2}) \in \mathbb{N}^{m-1}$ with $e_0 \geq 1$. Let $G = G(\sigma, e)$ be the corresponding m-extension, i.e. $G = \tau(F)$. Then*

$$\mathbb{N} \setminus G = \bigsqcup_{i=0}^{m-1} \sigma(i) + m(e_0 + \cdots + e_{i-1} + \mathbb{N}), \tag{12}$$

with the conventions $\sigma(0) = 0$ and $e_0 + \cdots + e_{i-1} = 0$ for $i = 0$.

Proof For $0 \leq i \leq m-1$, denote $F_i = [1, m-1] \setminus \{\sigma(0), \ldots, \sigma(i)\}$. Thus $F_0 = [1, m-1]$, $F_1 = [1, m-1] \setminus \{\sigma(1)\}$, $F_2 = [1, m-1] \setminus \{\sigma(1), \sigma(2)\}$ and so on. By definition of $F = F(\sigma, e)$, we have

$$F = (\underbrace{F_0, \ldots, F_0}_{e_0}, \underbrace{F_1, \ldots, F_1}_{e_1}, \ldots, \underbrace{F_{m-2}, \ldots, F_{m-2}}_{e_{m-2}}).$$

Let $G = \tau(F)$. For $k \in [0, m-1]$, set $G^{(k)} = \{x \in G \mid x \equiv k \bmod m\}$. Then $G = \bigsqcup_{k=0}^{m-1} G^{(k)}$. Since G is an m-extension, we have $G \cap m\mathbb{N} = \emptyset$, i.e. $G^{(0)} = \emptyset$. We now proceed to determine $G^{(k)}$ for $k \geq 1$. Since σ is a permutation of $[1, m-1]$, there exists $i \in [1, m-1]$ such that $k = \sigma(i)$. We claim that

$$G^{(k)} = G^{(\sigma(i))} = \sigma(i) + m[0, e_0 + \cdots + e_{i-1} - 1]. \tag{13}$$

Indeed by construction, for all $r \geq 0$ we have

$$\sigma(i) \in F_r \Leftrightarrow r \leq i - 1. \tag{14}$$

Now, by definition of the map τ, we have

$$G = \bigsqcup_{l=0}^{m-2} \Big(\bigsqcup_{j=e_0+\cdots+e_{l-1}}^{e_0+\cdots+e_l-1} (jm + F_l) \Big). \tag{15}$$

It follows from (14) and (15) that

$$\sigma(i) + jm \in G \Leftrightarrow j < e_0 + \cdots + e_{i-1}$$

for all $j \geq 0$. This proves (13). Taking the complement in $\mathbb{N}$, it follows that

$$\sigma(i) + jm \in \mathbb{N} \setminus G \Leftrightarrow j \geq e_0 + \cdots + e_{i-1}.$$

This proves (12). □

Remark Given $\sigma \in \mathfrak{S}_{m-1}$ and $e = (e_0, \ldots, e_{m-2}) \in \mathbb{N}^{m-1}$ with $e_0 \geq 1$, we denote

$$S(\sigma, e) = \bigsqcup_{i=0}^{m-1} \sigma(i) + m(e_0 + \cdots + e_{i-1} + \mathbb{N}).$$

Thus, the above proposition amounts to the statement

$$\mathbb{N} = G(\sigma, e) \sqcup S(\sigma, e)$$

for all $\sigma \in \mathfrak{S}_{m-1}$ and $e = (e_0, \ldots, e_{m-2}) \in \mathbb{N}^{m-1}$ with $e_0 \geq 1$.

This yields the following way to construct all gapsets of given multiplicity $m \geq 3$.

Proposition 6 *Let $m \geq 2$. Every numerical semigroup S of multiplicity m is of the form $S = S(\sigma, e)$ for some $\sigma \in \mathfrak{S}_{m-1}$ and $e = (e_0, \ldots, e_{m-2}) \in \mathbb{N}^m$ with $e_0 \geq 1$.*

Proof Let S be a numerical semigroup of multiplicity m. Let $G = \mathbb{N} \setminus S$ and $F = \varphi(G)$ be the associated gapset and gapset filtration, respectively. Then F is an m-filtration, whence by Proposition 4, it is of the form $F = F(\sigma, e)$ for some σ and e of the desired type. Then $G = \tau(F) = G(\sigma, e)$, whence $S = \mathbb{N} \setminus G = S(\sigma, e)$. □

We now determine the conditions under which a set of the form $S(\sigma, e)$ is a numerical semigroup.

Theorem 3 *Let $m \geq 3$. Let $\sigma \in \mathfrak{S}_{m-1}$ and $e = (e_0, \ldots, e_{m-2}) \in \mathbb{N}^m$ with $e_0 \geq 1$. Then $S(\sigma, e)$ is a numerical semigroup if and only if for all $1 \leq i, j, k \leq m-1$ with $i \leq j < k$, we have*

$$e_j + \cdots + e_{k-1} \leq \begin{cases} e_0 + \cdots + e_{i-1} & \text{if } \sigma(i) + \sigma(j) = \sigma(k), \\ e_0 + \cdots + e_{i-1} + 1 & \text{if } \sigma(i) + \sigma(j) = \sigma(k) + m. \end{cases}$$

Proof Denote $S_0 = \mathbb{N}$ and $S_i = \sigma(i) + m(e_0 + \cdots + e_{i-1} + \mathbb{N})$ for $1 \leq i \leq m-1$. Let $S' = S(\sigma, e)$. Then

$$S' = \bigsqcup_{i=0}^{m-1} S_i$$

by definition. We have $0 \in S_0 \subset S'$. The complement of S' in $\mathbb{N}$ is finite, since $\mathbb{N} \setminus S(\sigma, e) = G(\sigma, e)$. It remains to prove that S' is stable under addition if and only if the stated inequalities are satisfied.

Let i, j be integers such that $0 \leq i \leq j \leq m-1$. If $i = 0$ then $S_i + S_j = S_j + m\mathbb{N} = S_j$. We now assume $i \neq 0$. There are three cases.

- **Case** $\sigma(i) + \sigma(j) \leq m - 1$. There exists $k \in [1, m-1]$ satisfying $\sigma(k) = \sigma(i) + \sigma(j)$. Then

$$\begin{aligned} S_i + S_j &= \sigma(i) + m(e_0 + \cdots + e_{i-1} + \mathbb{N}) + \sigma(j) + m(e_0 + \cdots + e_{j-1} + \mathbb{N}) \\ &= \sigma(k) + m(e_0 + \cdots + e_{i-1} + e_0 + \cdots + e_{j-1} + \mathbb{N}). \end{aligned}$$

Therefore $S_i + S_j$ is contained in S' if and only if it is contained in S_k, and this occurs if and only if

$$e_0 + \cdots + e_{k-1} \leq e_0 + \cdots + e_{i-1} + e_0 + \cdots + e_{j-1}.$$

This condition is plainly satisfied if $k < j$, and is equivalent to

$$e_j + \cdots + e_{k-1} \leq e_0 + \cdots + e_{i-1}$$

if $k > j$.

- **Case** $\sigma(i) + \sigma(j) \geq m + 1$. There exists $k \in [1, m-1]$ satisfying $\sigma(k) + m = \sigma(i) + \sigma(j)$. Then

$$\begin{aligned} S_i + S_j &= \sigma(i) + m(e_0 + \cdots + e_{i-1} + \mathbb{N}) + \sigma(j) + m(e_0 + \cdots + e_{j-1} + \mathbb{N}) \\ &= \sigma(k) + m(e_0 + \cdots + e_{i-1} + e_0 + \cdots + e_{j-1} + 1 + \mathbb{N}). \end{aligned}$$

Again, $S_i + S_j$ is contained in S' if and only if it is contained in S_k, and this occurs if and only if

$$e_0 + \cdots + e_{k-1} \leq e_0 + \cdots + e_{i-1} + e_0 + \cdots + e_{j-1} + 1.$$

This is plainly satisfied if $j < k$, and is equivalent to

$$e_j + \cdots + e_{k-1} \leq e_0 + \cdots + e_{i-1} + 1$$

otherwise.

- **Case** $\sigma(i) + \sigma(j) = m$. Then $S_i + S_j \subseteq m\mathbb{N} = S_0 \subset S'$. □

Remark 4 For a gapset filtration $F = F(\sigma, e)$ of multiplicity m, there is a strong connection between its exponent vector $e \in \mathbb{N}^{m-1}$ and the Kunz coordinates of the associated numerical semigroup $S(\sigma, e)$.

Indeed, let S be a numerical semigroup of multiplicity m. Recall that the Apéry set of S is $\mathrm{Ap}(S) = \{x \in S \mid x - m \notin S\}$. By Lemma 1.4 of [8], we have $\mathrm{Ap}(S) = \{0 = w(0), w(1), \ldots, w(m-1)\}$ where $w(i)$ is the smallest element of S which is congruent to i modulo m. Hence for $i \in [0, m-1]$ there exist $k_i \in \mathbb{N}$ such that $w(i) = i + mk_i$. The integers $k_1, \ldots, k_{m-1}$ are the *Kunz coordinates* of S. From (12), we obtain that the smallest element of $S(\sigma, e)$ which is congruent to $\sigma(i)$ modulo m is $\sigma(i) + m(e_0 + \cdots + e_{i-1})$. Hence for all $i \in [1, m-1]$, we have

$$k_{\sigma(i)} = e_0 + \cdots + e_{i-1}.$$

4.5 The Insertion Maps f_i

Let $m \geq 3$ and let $F = (F_0, \ldots, F_t)$ be an m-filtration, i.e. with

$$[1, m-1] = F_0 \supseteq F_1 \supseteq \cdots \supseteq F_t.$$

Let $g = \sum_{j=0}^{t} |F_j|$ be the genus of F. Given $i \in [1, m-1]$, we wish to insert i in F so as to end up with an m-filtration of genus $g+1$. There is only one way to do this, namely to insert i in the first F_j for which $i \notin F_j$. More formally, we define $f_i(F)$ as follows:

- If $i \in F_s \setminus F_{s+1}$ for some $s \leq t-1$, then $f_i(F) = (F'_0, \ldots, F'_t)$ where

$$F'_j = \begin{cases} F_j & \text{if } j \neq s+1, \\ F_{s+1} \sqcup \{i\} & \text{if } j = s+1. \end{cases}$$

- If $i \in F_t$, then $f_i(F) = (F_0, \ldots, F_t, F_{t+1})$ where $F_{t+1} = \{i\}$.

By construction, for all $i \in [1, m-1]$, we have that $f_i(F)$ is an m-filtration of genus $g+1$.

One delicate question is the following. If F is a *gapset* filtration of multiplicity m, for which $i \in [1, m-1]$ does it hold that $f_i(F)$ remains a gapset filtration? This question was successfully addressed in Sect. 3 for $m = 3$.

5 The Case $m = 4$

We now use the above tools to characterize all gapset filtrations of multiplicity $m = 4$ and to derive a counting-free proof of the inequality $n_{g+1,4} \geq n_{g,4}$ for all $g \geq 0$.

Let F be a gapset filtration of multiplicity $m = 4$. By Proposition 4, there exists $\sigma \in \mathfrak{S}_3$ and $e = (a, b, c) \in \mathbb{N}^3$ with $a \geq 1$ such that $F = F(\sigma, e)$. Moreover, Theorem 3 gives the exact conditions for $S(\sigma, e)$ to be a numerical semigroup, i.e.

for $F(\sigma, e)$ to be a gapset filtration. This yields the following characterization, where the six elements of $\mathfrak{S}_3$ are displayed in window notation.

Theorem 4 *The gapset filtrations of multiplicity* $m = 4$ *are exactly the filtrations* $F = F(\sigma, e)$ *given in the table below, with* $\sigma \in \mathfrak{S}_3$ *and* $e = (a, b, c) \in \mathbb{N}^3$ *such that* $a \geq 1$ *and subject to the stated conditions:*

$\sigma \in \mathfrak{S}_3$	$F = F(\sigma, e)$	conditions on a, b, c
$(1, 2, 3)$	$(123)^a(23)^b(3)^c$	$b \leq a,\ c \leq a$
$(1, 3, 2)$	$(123)^a(23)^b(2)^c$	$b + c \leq a$
$(2, 1, 3)$	$(123)^a(13)^b(3)^c$	$c \leq a$
$(2, 3, 1)$	$(123)^a(13)^b(1)^c$	$c \leq a + 1$
$(3, 1, 2)$	$(123)^a(12)^b(2)^c$	$b + c \leq a + 1,\ c \leq a + b$
$(3, 2, 1)$	$(123)^a(12)^b(1)^c$	$b \leq a + 1,\ c \leq a + 1$

(16)

Proof Consider for instance the case $\sigma = (1, 3, 2)$. We have $\sigma(1) + \sigma(1) = \sigma(3)$ and $\sigma(2) + \sigma(2) = \sigma(3) + m$. Hence, by Theorem 3, the conditions on $e = (a, b, c)$ for $S = S(\sigma, e)$ to be a numerical semigroup, i.e. for $F = F(\sigma, e)$ to be a gapset filtration, are exactly $b + c \leq a$ and $c \leq a + b + 1$. Since the latter condition is implied by the former, it may be ignored. We end up with the sole condition $b + c \leq a$, as stated in the table. The proof in the five other cases is again a straightforward application of Theorem 3 and is left to the reader. □

Corollary 2 *For all* $g \geq 0$*, there is an explicit injection*

$$\mathcal{F}(g, 4) \longrightarrow \mathcal{F}(g + 1, 4).$$

In particular, we have $n_{g+1,4} \geq n_{g,4}$ *for all* $g \geq 0$.

Proof The statement is trivial for $g \leq 2$ since $\mathcal{F}(g, 4) = \emptyset$ in this case. Assume now $g \geq 3$. Let $F \in \mathcal{F}(g, 4)$ be a gapset filtration of genus g. Write $F = F(\sigma, e)$ for some $\sigma \in \mathfrak{S}_3$ and $e = (a, b, c) \in \mathbb{N}^3$ with $a \geq 1$. For $i = 1, 3$, consider the 4-filtrations $F' = f_1(F)$ and $F'' = f_3(F)$ of genus $g + 1$ obtained by the insertion maps f_1 and f_3, respectively. Then $F' = F(\sigma, e')$ where

$$e' = \begin{cases} (a + 1, b - 1, c) & \text{if } \sigma \in \{(1, 2, 3), (1, 3, 2)\}, \\ (a, b + 1, c - 1) & \text{if } \sigma \in \{(2, 1, 3), (3, 1, 2)\}, \\ (a, b, c + 1) & \text{if } \sigma \in \{(2, 3, 1), (3, 2, 1)\}. \end{cases}$$

It follows from (16) that F' fails to be a gapset filtration, i.e. $F' \notin \mathcal{F}(g + 1, 4)$, if and only if $\sigma = (2, 3, 1)$ or $(3, 2, 1)$ and $e = (a, b, a + 1)$. This corresponds to F being one of

$$(123)^a(13)^b(1)^{a+1} \text{ or } (123)^a(23)^b(1)^{a+1}.$$

Here $g = 3a + 2b + a + 1 = 4a + 2b + 1$, whence g is odd. In particular, if $g \not\equiv 1 \bmod 2$, then F' is always a gapset filtration. We conclude that, whenever g is *even*, then f_1 yields a well-defined injection

$$\mathcal{F}(g, 4) \longrightarrow \mathcal{F}(g+1, 4).$$

Let us now turn to $F'' = f_3(F)$. Then $F'' = F(\sigma, e'')$ where e'' is easily described by a table similar to (16). Omitting details, it follows that F'' fails to be a gapset filtration if and only if F is one of

$$(123)^a(13)^b(3)^a \text{ or } (123)^a(23)^b(3)^a.$$

In this case we have $g = 3a + 2b + a = 4a + 2b$, which is even. We conclude that whenever g is *odd*, then f_3 yields a well-defined injection

$$\mathcal{F}(g, 4) \longrightarrow \mathcal{F}(g+1, 4).$$

This conclude the proof of the corollary. □

5.1 Concluding Remark

We have shown that for $m = 3$ and 4, an injection $\mathcal{F}(g, m) \longrightarrow \mathcal{F}(g+1, m)$ is provided by one of the insertion maps f_i, where $i \in [1, m-1]$ depends on the class of g modulo 3 and 2, respectively.

Unfortunately, for any given $m \geq 5$, this is no longer true in general. That is, one should not expect that for each $g \geq 1$, an injection $\mathcal{F}(g, m) \longrightarrow \mathcal{F}(g+1, m)$ will be provided by just one of the insertion maps f_i. Constructing such injections for all m, g remains open at the time of writing.

References

1. Bras-Amorós, M.: Fibonacci-like behavior of the number of numerical semigroups of a given genus. Semigroup Forum **76**, 379–384 (2008)
2. Bras-Amorós, M.: Bounds on the number of numerical semigroups of a given genus. J. Pure Appl. Algebra **213**, 997–1001 (2009)
3. Eliahou, S., Fromentin, J.: Gapsets and numerical semigroups. J. Comput. Theor. A **169**, 105129 (2020). arXiv:1811.10295
4. Fromentin, J., Hivert, F.: Exploring the tree of numerical semigroups. Math. Comp. **85**, 2553–2568 (2016)
5. García-Sánchez, P.A., Marín-Aragón, D., Robles-Pérez, A.M.: The tree of numerical semigroups with low multiplicity. arXiv:1803.06879 [math.CO] (2018)
6. Kaplan, N.; Counting numerical semigroups by genus and some cases of a question of Wilf. J. Pure Appl. Algebra **216**, 1016–1032 (2012)

7. Kaplan, N.: Counting numerical semigroups. Am. Math. Mon. **124**, 862–875 (2017)
8. Rosales, J.C., García-Sánchez, P.A., García-García, J.I., Jiménez Madrid, J.A.: The oversemigroups of a numerical semigroup. Semigroup Forum **67**, 145–158 (2003)
9. Zhai, A.: Fibonacci-like growth of numerical semigroups of a given genus. Semigroup Forum **86**, 634–662 (2013)
10. Zhao, Y.: Constructing numerical semigroups of a given genus. Semigroup Forum **80**, 242–254 (2010). http://dx.doi.org/10.1007/s00233-009-9190-9

Generic Toric Ideals and Row-Factorization Matrices in Numerical Semigroups

Kazufumi Eto

Abstract In this paper, we give conditions in which the defining ideal of the semigroup ring associated with a numerical semigroup is a generic toric ideal. As an application, we prove that the defining ideals of almost Gorenstein monomial curves are not generic, if their embedding dimension is greater than three.

Keywords Numerical semigroup · Symmetric semigroup · Almost symmetric semigroup · Generic lattice ideal

1 Preliminaries

1.1 Numerical Semigroups

Let $\mathbb{Z}$ be the ring of integers and $\mathbb{N}_0$ the set of non negative integers. For $s > 1$ and $n_1, \ldots, n_s \in \mathbb{N}_0$, we write $[1, s] = \{1, \ldots, s\}$ and

$$\langle n_1, \ldots, n_s \rangle = \left\{ \sum_{i=1}^{s} a_i n_i : a_i \in \mathbb{N}_0 \text{ for each } i \right\},$$

which is called the semigroup generated by $n_1, \ldots, n_s$ of embedding dimension s. We always assume the minimality of $n_1, \ldots, n_s$, that is $n_i - n_j \notin \langle n_1, \ldots, n_s \rangle$ for each $i \neq j$. Let $S = \langle n_1, \ldots, n_s \rangle$. We write $T_i(S) = \langle n_1, \ldots, \check{n}_i, \ldots, n_s \rangle$ for $i \in [1, s]$. If $\gcd(n_1, \ldots, n_s) = 1$, then we say that S is numerical. Note that S is a numerical semigroup if and only if $\mathbb{N}_0 \setminus S$ is finite. We define an order $\leq_S$ in S as

$$d_1 \leq_S d_2 \iff d_2 - d_1 \in S$$

K. Eto (✉)
Department of Mathematics, Nippon Institute of Technology, Saitama, Japan
e-mail: etou@nit.ac.jp

V. Barucci et al. (eds.), *Numerical Semigroups*, Springer INdAM Series 40,
https://doi.org/10.1007/978-3-030-40822-0_6

Assume that $S = \langle n_1, \ldots, n_s \rangle$ is numerical. We call the greatest integer in $\mathbb{Z} \setminus S$ the Frobenius number of S, denoted by $\mathrm{F}(S)$. For $f \in \mathbb{Z}$, if $f \notin S$ and $f + n_i \in S$ for each i, then it is called a pseudo-Frobenius number of S. By definition, the Frobenius number is a pseudo-Frobenius number. Put $\mathrm{PF}(S)$ the set of the pseudo-Frobenius numbers. The number of elements of the set $\mathrm{PF}(S)$ is called the type of S. If the type of S is one, we say that S is symmetric. If $\mathrm{F}(S) - f \in \mathrm{PF}(S)$ for each $f \in \mathrm{PF}(S)$ except $\mathrm{F}(S)$, then we say that S is almost symmetric. For $d \in S$, we define the Apèry set for d in S (cf [7]) as

$$\mathrm{Ap}(S, d) = \{x \in S : x - d \notin S\}.$$

Lemma 1 *Let $d_1, d_2 \in S$.*

(1) $d_1 \in \mathrm{Ap}(S, d_2)$ *if and only if* $d_1 - d_2 \notin S$.
(2) *If* $d_1 \leq_S d_2$, *then* $\mathrm{Ap}(S, d_1) = \{x \in S : x + (d_2 - d_1) \in \mathrm{Ap}(S, d_2)\}$.

Proof (1) follows from the definition of the Apèry set. In (2), $x \in \mathrm{Ap}(S, d_1)$ if and only if $x - d_1 = x + (d_2 - d_1) - d_2 \notin S$ and it is equivalent to $x + (d_2 - d_1) \in \mathrm{Ap}(S, d_2)$. □

For $f \in \mathbb{Z} \setminus S$, we define RF-matrices (row-factorization matrices) as follows (cf. [2, 5]): For each i, there is a unique negative number a_{ii} satisfying $f - a_{ii}n_i \in \mathrm{Ap}(S, n_i)$. Then there are $a_{ij} \geq 0$ for $j \neq i$ with $f - a_{ii}n_i = \sum_{j \neq i} a_{ij}n_j$. And we consider the matrix $(a_{ij})_{i,j}$ called an RF-matrix for f in S. Note that it is not necessarily unique. However, we denote it by $\mathrm{RF}(f)$ in abbreviation, and it indicates one of them.

1.2 The Fibers of Elements in Numerical Semigroups

In this subsection, we always assume that $S = \langle n_1, \ldots, n_s \rangle$ is a numerical semigroup. For $a = (a_i)_{i \in [1,s]} \in \mathbb{Z}^s = \bigoplus_{i=1}^s \mathbb{Z}e_i$, we define the support of a as $\operatorname{supp} a = \{i : a_i \neq 0\}$. We also define the degree map with respect to S, sending $a \in \mathbb{Z}^s$ to $\deg_S a = \sum_i a_i n_i \in \mathbb{Z}$. For $d \in \mathbb{Z}$, we write the fiber of d as

$$V_d(S) = \deg_S^{-1}(d) \cap \mathbb{N}_0^s \quad \text{and put} \quad \operatorname{supp}_S d = \bigcup_{a \in V_d(S)} \operatorname{supp} a$$

(cf. [1]). Note $d \in S$ if and only if $V_d(S) \neq \emptyset$. The following are clear.

Lemma 2 *Let $d_1, d_2 \in S$.*

(1) $d_1 \in \mathrm{Ap}(S, n_l)$ *if and only if* $l \notin \operatorname{supp}_S d_1$.
(2) *If* $d_1 \leq_S d_2$, *then* $|V_{d_1}(S)| \leq |V_{d_2}(S)|$.

For $\Lambda \subset [1, s]$, we denote $\sum_{i\in\Lambda} n_i$ by n_Λ and define

$$\mathrm{Ap}(S, \Lambda) = \begin{cases} \emptyset & \text{if } \Lambda = \emptyset, \\ \mathrm{Ap}(S, n_\Lambda) \setminus \bigcup_{l\in\Lambda} \mathrm{Ap}(S, n_{\Lambda-\{l\}}) & \text{otherwise.} \end{cases}$$

If $\Lambda = \{l\}$, then $n_\Lambda = n_l$ and $\mathrm{Ap}(S, \Lambda) = \mathrm{Ap}(S, n_l)$.

We also denote $S \setminus \bigcup_{l=1}^{s} \mathrm{Ap}(S, n_l)$ by $\mathrm{NAP}(S)$, and the set of minimal numbers in $\mathrm{Ap}(S, \Lambda)$ by $\underline{\mathrm{Ap}}(S, \Lambda)$. We put $\mathrm{AP}_j(S) = \bigcup_{|\Lambda|=j+1} \underline{\mathrm{Ap}}(S, \Lambda)$ for $j \geq 0$ and $\mathrm{AP}(S) = \bigcup_{j\geq 0} \mathrm{AP}_j(S)$. Note that $\mathrm{AP}_0(S) = \{0\}$ for any S.

Example 1 If $S = \langle 3, 4, 5\rangle$, then $\mathrm{AP}_1(S) = \{8, 9, 10\}$ and $\mathrm{AP}_2(S) = \{13, 14\}$. If $S = \langle 4, 5, 6\rangle$, then $\mathrm{AP}_1(S) = \{10, 12\}$ and $\mathrm{AP}_2(S) = \{22\}$.

Lemma 3 *Let $d \in S$. Then $d \in \mathrm{NAP}(S)$ if and only if $\mathrm{supp}_S\, d = [1, s]$.*

Proof For $d \in S$, $\mathrm{supp}_S\, d = [1, s]$ if and only if $d \notin \mathrm{Ap}(S, n_l)$ for each l by Lemma 2(1). The assertion follows from this. □

Lemma 4 *Let $\Lambda \subset [1, s]$ be a non empty subset and $d \in S$. Then the following are equivalent:*

(1) $d \in \mathrm{Ap}(S, \Lambda)$,
(2) $d - n_\Lambda \notin S$, *and* $d - n_{\Lambda\setminus\{l\}} \in S$ *for each* l.
(3) $\Lambda \not\subset \mathrm{supp}\, a$ *for each* $a \in V_d(S)$, *and there is* $a_l \in V_d(S)$ *satisfying* $\mathrm{supp}\, a_l \cap \Lambda = \Lambda \setminus \{l\}$ *for each* l.

If the conditions are satisfied, then $|V_d(S)| \geq |\Lambda|$ and $d - n_{\Lambda\setminus\Lambda'} \in \mathrm{Ap}(S, \Lambda')$ for $\Lambda' \subsetneq \Lambda$.

Proof (1) ⇔ (2) follows from Lemma 1(1). We have $d - n_\Lambda \notin S$ if and only if $\Lambda \not\subset \mathrm{supp}\, a$ for each $a \in V_d(S)$, and $d - n_{\Lambda\setminus\{l\}} \in S$ if and only if there is $a_l \in V_d(S)$ satisfying $\Lambda \setminus \{l\} \subset \mathrm{supp}\, a_l$. This proves (2) ⇔ (3). Finally, (3) implies $|V_d(S)| \geq |\Lambda|$.

We prove $d - n_{\Lambda\setminus\Lambda'} \in \mathrm{Ap}(S, \Lambda')$ for $\Lambda' \subsetneq \Lambda$. By (2), $d - n_{\Lambda\setminus\Lambda'} \in S$. By Lemma 1(2), this implies $d - n_{\Lambda\setminus\Lambda'} \in \mathrm{Ap}(S, n_{\Lambda'})$, since $d \in \mathrm{Ap}(S, n_\Lambda)$ and $d = d - n_{\Lambda\setminus\Lambda'} + (n_\Lambda - n_{\Lambda'})$. Suppose that there is $\Lambda'' \subsetneq \Lambda'$ with $d - n_{\Lambda\setminus\Lambda'} \in \mathrm{Ap}(S, n_{\Lambda''})$. Again by Lemma 1(2), we have $d \in \mathrm{Ap}(S, n_{\Lambda\setminus\Lambda'} + n_{\Lambda''})$, a contradiction. Hence, $d - n_{\Lambda\setminus\Lambda'} \notin \mathrm{Ap}(S, n_{\Lambda''})$ and $d - n_{\Lambda\setminus\Lambda'} \in \mathrm{Ap}(S, \Lambda')$. □

Proposition 1 $\underline{\mathrm{Ap}}(S, [1, s]) = \mathrm{Ap}(S, [1, s]) = \{f + n_{[1,s]} : f \in \mathrm{PF}(S)\}$.

Proof Put $\Lambda = [1, s]$ and let $d \in \mathrm{Ap}(S, \Lambda)$. By Lemma 4, we have $d - n_l \notin \mathrm{Ap}(S, \Lambda)$ for each l. Hence d is minimal w.r.t. $\leq_S$ and $d \in \underline{\mathrm{Ap}}(S, \Lambda)$.

If $f \in \mathrm{PF}(S)$, then $f + n_\Lambda \in \mathrm{Ap}(S, n_\Lambda)$, by Lemma 1(1). Since $f + n_l \in S$ and $f + n_\Lambda = (f + n_l) + n_{\Lambda\setminus\{l\}}$ for each l, we have $f + n_\Lambda \notin \mathrm{Ap}(S, n_{\Lambda\setminus\{l\}})$. This implies $f + n_\Lambda \in \mathrm{Ap}(S, \Lambda)$. Further, if $d \in \mathrm{Ap}(S, \Lambda)$, then $d - n_\Lambda \notin S$ and there is $f \in \mathrm{PF}(S)$ with $f - (d - n_\Lambda) \in S$. Thus $d \leq_S f + n_\Lambda$ and $d = f + n_\Lambda$ since $\underline{\mathrm{Ap}}(S, \Lambda) = \mathrm{Ap}(S, \Lambda)$. This completes the proof. □

For $j \geq 0$, we put

$$O_j(S) = \{d \in \mathbb{Z} : |V_d(S)| = j\}.$$

We note $d \in O_0(S)$ if and only if $d \notin S$, and $d \in O_1(S)$ if and only if d has a unique factorization in S. We denote the set of the minimal elements in $O_j(S)$ (resp. $S \setminus \bigcup_{l \leq j} O_l(S)$) with respect to the order $\leq_S$ by $\underline{O}_j(S)$ (resp. $\widetilde{O}_j(S)$) for $j \geq 0$. Note $\underline{O}_{j+1}(S) \subset \widetilde{O}_j(S)$ for $j \geq 0$.

Proposition 2

(1) *Let $d \in \widetilde{O}_1(S)$ and $a_1, a_2 \in V_d(S)$ with $a_1 \neq a_2$. Then* $\operatorname{supp} a_1 \cap \operatorname{supp} a_2 = \emptyset$.
(2) $O_1(S) \cap T_l(S) \subset \mathrm{Ap}(S, n_l)$ *for each l.*
(3) $\widetilde{O}_1(S) \subset \mathrm{AP}_1(S) \subset \bigcup_l (T_l(S) \setminus \mathrm{Ap}(S, n_l)) \subset S \setminus O_1(S)$.

Proof

(1) If $l \in \operatorname{supp} a_1 \cap \operatorname{supp} a_2$, then $a_1 - e_l = a_2 - e_l \in V_{d-n_l}(S)$ and $d - n_l \notin O_1(S)$, a contradiction. Thus $\operatorname{supp} a_1 \cap \operatorname{supp} a_2 = \emptyset$.
(2) Let $d \in T_l(S) \setminus \mathrm{Ap}(S, n_l)$. Since $d \in T_l(S)$, there is $a \in V_d(S)$ with $l \notin \operatorname{supp} a$. Since $d \notin \mathrm{Ap}(S, n_l)$, there is $b \in V_d(S)$ with $l \in \operatorname{supp} b$ by Lemma 2(1). Hence $|V_d(S)| > 1$ and $d \notin O_1(S)$. Therefore $T_l(S) \setminus \mathrm{Ap}(S, n_l) \subset S \setminus O_1(S)$ and $O_1(S) \cap T_l(S) \subset \mathrm{Ap}(S, n_l)$.
(3) Let $d \in \widetilde{O}_1(S)$ and $a_1, a_2 \in V_d(S)$ with $a_1 \neq a_2$. By (1), we have $\operatorname{supp} a_1 \cap \operatorname{supp} a_2 = \emptyset$. We choose $l_j \in \operatorname{supp} a_j$ for $j = 1, 2$. Then $d - n_{l_j} \in O_1(S)$ for $j = 1, 2$ and $d - (n_{l_1} + n_{l_2}) \notin S$, by the choice of l_1, l_2. By Lemma 4, $d \in \mathrm{Ap}(S, \{l_1, l_2\}) \subset \mathrm{AP}_1(S)$ and $\widetilde{O}_1(S) \subset \mathrm{AP}_1(S)$.
Let $d \in \mathrm{AP}_1(S)$. Then there are $l_1 \neq l_2 \in [1, s]$ with $d \in \mathrm{Ap}(S, \{l_1, l_2\})$. And there are $a_1, a_2 \in V_d(S)$ satisfying $l_j \in \operatorname{supp} a_j$, since $d \notin \mathrm{Ap}(S, n_{l_j})$ for $j = 1, 2$. Then $l_2 \notin \operatorname{supp} a_1$ since $d \in \mathrm{Ap}(S, n_{l_1} + n_{l_2})$. This implies $d \in T_{l_2} \setminus \mathrm{Ap}(S, n_{l_2})$. This completes the proof. □

Example 2 Let $S = \langle 7, 8, 10 \rangle$. Then $25 = 7 + 8 + 10 \in O_1(S) \setminus \mathrm{Ap}(S, n_l)$ for each l. Hence, in general, we have $O_1(S) \not\subset \bigcup_l \mathrm{Ap}(S, n_l)$ and $\mathrm{NAP}(S) \cap O_1(S) \neq \emptyset$.

Example 3 Let $S = \langle 23, 37, 48 \rangle$. Then

$$9 \cdot 23 = 3 \cdot 37 + 2 \cdot 48, \quad 7 \cdot 37 = 5 \cdot 23 + 3 \cdot 48, \quad 5 \cdot 48 = 4 \cdot 23 + 4 \cdot 37$$

are minimal relations. Put $d = 4 \cdot 23 + 5 \cdot 37 + 6 \cdot 48$. Then

$$V_d(S) = \{(4, 5, 6), (8, 9, 1), (13, 2, 4)\}$$

and $d \notin O_1(S)$. This implies

$$\bigcup_i (T_i(S) \setminus \mathrm{Ap}(S, n_i)) \subsetneq S \setminus O_1(S).$$

Proposition 3 *The following are equivalent:*

(1) $\mathrm{RF}(f)$ *is unique for each* $f \in \mathrm{PF}(S)$,
(2) $\mathrm{Ap}(S, n_l) \subset O_1(S)$, *that is* $\mathrm{Ap}(S, n_l) = O_1(S) \cap T_l(S)$ *for each* l,
(3) $S \setminus O_1(S) \subset \mathrm{NAP}(S)$.

Proof (1) $\Leftrightarrow$ (2) Assume that $\mathrm{RF}(f)$ is unique for each $f \in \mathrm{PF}(S)$. Then $f + n_l \in O_1(S)$ for each l. Fix l and let $d \in \mathrm{Ap}(S, n_l)$. By Lemma 1(1), $d - n_l \notin S$ and there is $f \in \mathrm{PF}(S)$ with $d - n_l \leq_S f$, thus $d \leq_S f + n_l$. Since $f + n_l \in O_1(S)$, we have $d \in O_1(S)$ and $\mathrm{Ap}(S, n_l) \subset O_1(S)$. By Proposition 2(2), we have $O_1(S) \cap T_l(S) \subset \mathrm{Ap}(S, n_l)$ and $\mathrm{Ap}(S, n_l) = O_1(S) \cap T_l(S)$. Conversely, assume $\mathrm{Ap}(S, n_l) \subset O_1(S)$ for each l. If $f \in \mathrm{PF}(S)$, then $f + n_l \in \mathrm{Ap}(S, n_l) \subset O_1(S)$, hence $\mathrm{RF}(f)$ is unique.

(2) $\Leftrightarrow$ (3). The assertion follows from the definition of $\mathrm{NAP}(S) = S \setminus \bigcup_l \mathrm{Ap}(S, n_l)$. □

1.3 Semigroup Rings

Let k be a field and $S = \langle n_1, \ldots, n_s \rangle$ a numerical semigroup. Then we define the semigroup ring of S as

$$k[S] = k[t^d]_{d \in S}.$$

There is a canonical surjection form the polynomial ring $k[X] = k[X_1, \ldots, X_s]$ to $k[S]$ sending $X^a = \prod_i X_i^{a_i}$ to $t^{\deg_S a}$ where $a = (a_i) \in \mathbb{N}_0^s$. And we denote its kernel by $I(S)$, called the defining ideal of $k[S]$. Then $I(S)$ is a binomial ideal generated by $X^a - X^b$ satisfying $\deg_S a = \deg_S b$. Indeed, $I(S)$ is a prime ideal, called a toric ideal.

Let $g = X^a - X^b$ be a binomial. If g is contained in any minimal binomial generating system of $I(S)$, we say that g is indispensable. It follows that g is indispensable if and only if its degree is contained in $\underline{O}_2(S)$. We also say that a monomial X^a is indispensable, if any minimal binomial generating system of $I(S)$ contains a binomial of the form $X^a - X^b$ for some $b \in \mathbb{N}_0^s$ (cf. [1]).

Example 4 Let $S = \langle 4, 5, 6 \rangle$. Then $X_1^4 - X_3^2$ and $X_2^2 - X_1X_3$ are indispensable, since $\underline{O}_2(S) = \{10, 12\}$. Hence $I(S)$ is minimally generated by indispensable binomials.

Example 5 Let $S = \langle 6, 10, 15 \rangle$. Then the monomials X_1^5, X_2^3 and X_3^2 are indispensable, since $\widetilde{O}_1(S) = \{30\}$. We also have $\underline{O}_2(S) \subset O_2(S) = \emptyset \subsetneq \widetilde{O}_1(S)$. Note that $I(S)$ is not minimally generated by indispensable binomials. On the other hand,

$$\mathrm{RF}(\mathrm{F}(S)) = \mathrm{RF}(29) = \begin{pmatrix} -1 & 2 & 1 \\ 4 & -1 & 1 \\ 4 & 2 & -1 \end{pmatrix}$$

is unique. Note $\mathrm{AP}_1(S) = \{30\}$ and $|V_{30}(S)| = 3 > 2$.

2 Generic Toric Ideals

2.1 Main Results

Let $g = X^a - X^b$ be a binomial. We write $\operatorname{supp} g = \operatorname{supp} a \cup \operatorname{supp} b$. If $\operatorname{supp} g = [1, s]$, we say that g has full support. If a toric ideal I has a minimal generating system consisting of binomials with full support, then I is called generic.

Theorem 1 *The following are equivalent:*

(1) *$I(S)$ is generic,*
(2) $\underline{O}_2(S) = \widetilde{O}_1(S) = \mathrm{AP}_1(S) \subset \mathrm{NAP}(S)$,
(3) *For each $f \in \mathrm{PF}(S)$, $\mathrm{RF}(f) = (m_{ij})$ is unique and $m_{ij} \neq m_{i'j}$ if $i \neq i'$.*

If the above conditions are satisfied, $\widetilde{O}_1(S)$ is the set of degrees of the binomials contained in a minimal generating system of $I(S)$.

Proof (1) $\Rightarrow$ (2) Assume that $I(S)$ is generic. Then each binomial of the minimal generating system is indispensable (cf [1, 6]). This implies $\widetilde{O}_1(S) = \underline{O}_2(S)$. If $d \in \widetilde{O}_1(S)$, we have $\operatorname{supp}_S d = [1, s]$ since $I(S)$ is generic. Hence $\widetilde{O}_1(S) \subset \mathrm{NAP}(S)$ by Lemma 3. Let $d \in \mathrm{AP}_1(S)$. Then there are $l_1 \neq l_2 \in [1, s]$ satisfying $d \in \underline{\mathrm{Ap}}(S, \{l_1, l_2\})$. By Proposition 2(3), we have $d \in S \setminus O_1(S)$ and there is $d' \in \widetilde{O}_1(S)$ with $d' \leq_S d$. Since $d' \leq_S d$, d' is contained in $\mathrm{Ap}(S, n_{l_1} + n_{l_2})$. We also have $\operatorname{supp}_S d' = [1, s]$ and $d' \notin \mathrm{Ap}(S, n_{l_j})$ for $j = 1, 2$. Hence $d' \in \mathrm{Ap}(S, \{l_1, l_2\})$ and $d = d' \in \widetilde{O}_1(S)$. Again by Proposition 2(3), we conclude $\mathrm{AP}_1(S) = \widetilde{O}_1(S)$.

(2) $\Rightarrow$ (3) Let $f \in \mathrm{PF}(S)$. By Proposition 3, $\mathrm{RF}(f) = (m_{ij})$ is unique. We choose $i \neq i' \in [1, s]$ and put $a_1 = \sum_{j=1}^{s} \max\{m_{ij} - m_{i'j}, 0\} e_j$, $a_2 = \sum_{j=1}^{s} \max\{m_{i'j} - m_{ij}, 0\} e_j$ and $d = \deg_S a_1 = \deg_S a_2$. Then $d \in \mathrm{AP}_1(S)$. Since $d \in \underline{O}_2(S) \subset \mathrm{NAP}(S)$, we have $V_d(S) = \{a_1, a_2\}$ and $\operatorname{supp} a_1 \cup \operatorname{supp} a_2 = [1, s]$. This implies $m_{ij} \neq m_{i'j}$ for each j.

(3) $\Rightarrow$ (1) Put $\mathrm{PF}(S) = \{f_1, \ldots, f_u\}$ and $\mathrm{RF}(f_l) = (m^l_{ij})$ for each l. And put $a_{li} = \sum_j \max\{m^l_{1j} - m^l_{ij}, 0\} e_j$ and $b_{li} = \sum_j \max\{m^l_{ij} - m^l_{1j}, 0\} e_j$ for each i, l. Then $\deg_S a_{li} = \deg_S b_{li}$ and put $g_{li} = X^{a_{li}} - X^{b_{li}} \in I(S)$ for each i, l and $J = (g_{li})_{i,l}$ a binomial ideal in $k[X]$. By the assumption, g_{li} is a binomial with full support for each i, l and $J + (X_1) = (X^{a_{li}})_{i,l} + (X_1)$. For each i, l, we claim $d - n_k \in \mathrm{Ap}(S, n_1)$ for $k \in \operatorname{supp} a_{li}$, where $d = \deg_S a_{li} = \deg_S b_{li}$. Otherwise, there is $d' \in S$ with $d - n_k = d' + n_1$ and $d - n_1 = d' + n_k \in S$. Note $d - n_1 \leq_S f_l + n_i$ and $d - n_1 \in O_1(S)$. Since $d = \deg_S b_{li}$ and $1 \in \operatorname{supp} b_{li}$, we have $k \in \operatorname{supp}(b_{li} - e_1) \subset \operatorname{supp} b_{li}$. This contradicts to $k \in \operatorname{supp} a_{li}$, since $\operatorname{supp} a_{li} \cap \operatorname{supp} b_{li} = \emptyset$. Hence $d - n_k \in \mathrm{Ap}(S, n_1)$ for each $k \in \operatorname{supp} a_{li}$. This implies $\dim_k k[X]/J + (X_1) = |\mathrm{Ap}(S, n_1)| = n_1 = \dim_k k[X]/I(S) + (X_1)$. From $J \subset I(S)$, we have $I(S) = J$ (cf [4]) and it is generic. Note that a subset of $\{g_{li}\}$ forms a minimal generating system of $I(S)$. □

For a numerical semigroup S, we say that the semigroup ring $k[S]$ is almost Gorenstein (resp. Gorenstein), if S is almost symmetric (resp. symmetric) (cf [3]).

Note that $k[S]$ is Gorenstein if and only if $k[S]$ is almost Gorenstein of type one, where the type of semigroup rings means that of semigroups.

Theorem 2 *If $s > 3$ and if $k[S]$ is almost Gorenstein, then $I(S)$ is not generic. In case of $s = 3$, $I(S)$ is always generic, if $k[S]$ is almost Gorenstein of type greater than one. (note that $I(S)$ is not generic, if $s = 3$ and if $k[S]$ is Gorenstein).*

Proof Assume $s = 3$. If $k[S]$ is Gorenstein, then $I(S)$ is a complete intersection (cf. [4]) and not generic. If the type of $k[S]$ is grater than one, then $I(S)$ is an almost complete intersection (cf. [4]) and generic.

Assume $s > 3$. If $k[S]$ is Gorenstein and if there is a unique RF-matrix for the Frobenius number F(S) of S, then $I(S)$ is a complete intersection by Eto [2, Proposition 3.5], thus not generic. If the RF-matrix of F(S) is not unique, then $I(S)$ is not generic by Theorem 1.

Suppose that $k[S]$ is almost Gorenstein of type greater than one and that $I(S)$ is generic. Note that S is almost symmetric. Then, for each $f \in \mathrm{PF}(S)$, $\mathrm{RF}(f) = (m_{ij})$ is unique and $m_{ij} \neq m_{i'j}$ for $i \neq i'$. Since the type of S is greater than one, we choose $f_1, f_2 \in \mathrm{PF}(S)$ with $f_1+f_2 = \mathrm{F}(S)$. Let $a \in V_{f_1+n_i}(S)$. Suppose $|\operatorname{supp} a| > 1$ and choose $l \neq l' \in \operatorname{supp} a$. Then $f_1+n_i-n_l$ (resp. $f_1+n_i-e_{l'}$) is contained in S and $l \notin \operatorname{supp}_S(f_2+n_l)$ (resp. $l' \notin \operatorname{supp}_S(f_2+n'_l)$). And $(a-e_l)+b \neq (a-e_{l'})+b'$ for $b \in V_{f_2+n_l}(S)$ and $b' \in V_{f_2+n_{l'}}(S)$. Thus

$$\mathrm{F}(S) + n_i = (f_1 + n_i) + f_2 \notin O_1(S),$$

a contradiction. Hence $|\operatorname{supp} a| = 1$ for each i. This implies that there are $s - 2$ zeros in each row of $\mathrm{RF}(f) = (m_{ij})$ for $f \in \mathrm{PF}(S)$ with $f \neq \mathrm{F}(S)$ and that there exists j with $m_{ij} = m_{i'j}$ for $i \neq i'$ since $s > 3$. This contradicts that $I(S)$ is generic by Theorem 1. □

2.2 Basic Fibers

For $d \in S$, the fiber $V_d(S)$ is called basic if

$$\bigcap_{a \in V_d(S)} \operatorname{supp} a = \emptyset \quad \text{and} \quad \bigcap_{\substack{a \in V_d(S) \\ a \neq b}} \operatorname{supp} a \neq \emptyset \text{ for } \forall b \in V_d(S).$$

Note that $V_d(S)$ is basic if and only if $\gcd\{x^a : a \in V_d(S)\} = 1$, that is the general common divisor of all x^a with $a \in V_d(S)$ is one, and $\gcd\{x^a : a \in V_d(S), a \neq b\} \neq 1$ for each $b \in V_d(S)$. Hence this definition agrees with that in [6].

Proposition 4 *Let $d \in S$. Then, $V_d(S)$ is a basic fiber if and only if there is $\Lambda \subset [1, s]$ with $|\Lambda| = |V_d(S)|$ and $d \in \underline{\mathrm{Ap}}(S, \Lambda)$. Equivalently, $d \in \mathrm{AP}_{|V_d(S)|-1}(S)$.*

Proof Assume that $V_d(S) = \{b_j\}$ is a basic fiber. By definition, for each l, there is $i_l \in [1, s]$ with $i_l \in \operatorname{supp} b_j$ if $j \neq l$. Note $i_l \notin \operatorname{supp} b_l$ and $\operatorname{supp} b_l \cap \Lambda = \Lambda \setminus \{i_l\}$ for each l. Put $\Lambda = \{i_l\}_l$. Note $|\Lambda| = |V_d(S)|$. Then $d - n_\Lambda \notin S$. Otherwise, there is $b_l \in V_d(S)$ with $\operatorname{supp} b_l \supset \Lambda$, and this contradicts to $d - n_\Lambda \notin S$. Since $\operatorname{supp} b_l \cap \Lambda = \Lambda \setminus \{i_l\}$ for each l, we obtain $d - n_{\Lambda\setminus\{l\}} \in S$ for each l. By Lemma 4, $d \in \operatorname{Ap}(S, \Lambda)$. Suppose $d - n_i \in \operatorname{Ap}(S, \Lambda)$. By Lemma 4, we have $|V_{d-n_i}(S)| = |V_d(S)| = |\Lambda|$. This implies $\bigcap \operatorname{supp} b_j \ni i$, since we may write $b_j = b'_j + e_i$ where $b'_j \in V_{d-n_i}(S)$ for each j. By assumption, we conclude $d - n_i \notin \operatorname{Ap}(S, \Lambda)$ for each i and $d \in \overline{\operatorname{Ap}}(S, \Lambda)$.

Conversely, assume $d \in \overline{\operatorname{Ap}}(S, \Lambda)$ and $|\Lambda| = |V_d(S)|$. Since $d \in \overline{\operatorname{Ap}}(S, \Lambda)$, we have $\bigcap_{a\in V_d(S)} \operatorname{supp} a = \emptyset$. Since $|V_d(S)| = |\Lambda|$, we may assume $V_d(S) = \{a_l\}_{l\in\Lambda}$ and $\operatorname{supp} a_l \cap \Lambda = \Lambda \setminus \{l\}$ for each l, by Lemma 4. Then $\bigcap_{j\neq l} \operatorname{supp} a_j \ni l$. Therefore $V_d(S)$ is a basic fiber. □

Example 6 Let $S = \langle 4, 5, 6\rangle$. Then

$$V_{22}(S) = \{(1, 0, 3), (4, 0, 1), (3, 2, 0), (0, 2, 2)\}.$$

Thus $|V_{22}(S)| = 4 > 3$. By Proposition 4, $V_{22}(S)$ is not a basic fiber. On the other hand, we have $22 \in \operatorname{AP}_2(S)$.

Theorem 3 *Assume that $I(S)$ is generic. For $\Lambda \subset [1, s]$, $\overline{\operatorname{Ap}}(S, \Lambda) \subset O_{|\Lambda|}(S)$. For $d \in \overline{\operatorname{Ap}}(S, \Lambda)$, we have*

$$V_d(S) = \left\{ \sum_{j=1}^{s} m_{lj} e_j + \sum_{j\in\Lambda} e_j : l \in \Lambda \right\},$$

where $\operatorname{RF}(d - n_\Lambda) = (m_{ij})_{i,j}$. *(Note* $\operatorname{RF}(d - n_\Lambda)$ *is unique in this case.)*

Proof Let $\Lambda \subset [1, s]$. We prove $|V_d(S)| = |\Lambda|$ for $d \in \overline{\operatorname{Ap}}(S, \Lambda)$. Since $d - n_\Lambda \notin S$, there is $f \in \operatorname{PF}(S)$ satisfying $d \leq_S f + n_\Lambda$. Put $d' = f + n_\Lambda$. Since $|\Lambda| \leq |V_d(S)| \leq |V_{d'}(S)|$, if $|V_{d'}(S)| = |\Lambda|$, we obtain $|V_d(S)| = |\Lambda|$. Hence we may assume $d = f + n_\Lambda$.

By Theorem 1, $\operatorname{RF}(f) = (m_{ij})$ is unique. Let $a = \sum_j a_j e_j \in V_d(S)$. Then there is $l \in \Lambda \setminus \operatorname{supp} a$. Put $b = \sum_j b_j e_j = \sum_j m_{lj} e_j + \sum_{j\in\Lambda} e_j$. Then $l \notin \operatorname{supp} b$. Suppose $a \neq b$. Put $a' = \sum_j \max\{a_j - b_j, 0\} e_j$, $b' = \sum_j \max\{b_j - a_j, 0\} e_j$, and $d' = \deg_S a' = \deg_S b'$. Then $a' \neq b'$ and $|V_{d'}(S)| > 1$. Thus $d' \in \widetilde{O}_1(S) \subset \operatorname{NAP}(S)$ and $l \notin \operatorname{supp} a' \cup \operatorname{supp} b'$. This contradicts to $\operatorname{supp}_S d' = [1, s]$. Hence $a = b$ and $|V_d(S)| = |\Lambda|$. □

Example 7 We write $V_d(S)$ as a matrix (m_{ij}), that is $V_d(S) = \{\sum_j m_{ij} e_j\}$. If $d = f + n_{[1,s]}$ where $f \in \operatorname{PF}(S)$, we have $V_d(S) = \operatorname{RF}(f) + U$ where U

is the matrix whose entries are 1. For example, let $S = \langle 20, 24, 25, 31\rangle$. Then $\mathrm{PF}(S) = \{61, 66, 67, 77, 78, 83\}$ and

$$V_{161}(S) = \begin{pmatrix} 0 & 1 & 3 & 2 \\ 4 & 0 & 2 & 1 \\ 1 & 2 & 0 & 3 \\ 2 & 4 & 1 & 0 \end{pmatrix}, V_{166}(S) = \begin{pmatrix} 0 & 2 & 1 & 3 \\ 3 & 0 & 3 & 1 \\ 4 & 1 & 0 & 2 \\ 1 & 4 & 2 & 0 \end{pmatrix}, V_{167}(S) = \begin{pmatrix} 0 & 1 & 2 & 3 \\ 4 & 0 & 1 & 2 \\ 2 & 4 & 0 & 1 \\ 1 & 3 & 3 & 0 \end{pmatrix},$$

$$V_{177}(S) = \begin{pmatrix} 0 & 4 & 2 & 1 \\ 2 & 0 & 3 & 2 \\ 3 & 1 & 0 & 3 \\ 4 & 3 & 1 & 0 \end{pmatrix}, V_{178}(S) = \begin{pmatrix} 0 & 3 & 3 & 1 \\ 3 & 0 & 1 & 3 \\ 1 & 4 & 0 & 2 \\ 4 & 2 & 2 & 0 \end{pmatrix}, V_{183}(S) = \begin{pmatrix} 0 & 4 & 1 & 2 \\ 2 & 0 & 2 & 3 \\ 4 & 3 & 0 & 1 \\ 3 & 2 & 3 & 0 \end{pmatrix}.$$

We denote the i-th Betti number of $k[S]$ as $k[X]$-module with degree j, by $\beta_{i,j}(k[S])$. From [6, Corollary 3.4] and Theorem 3, we obtain

Theorem 4 (cf. [6, Corollary 3.4]) *Assume that $I(S)$ is generic. Then every element in $\mathrm{AP}(S)$ defines a basic fiber. Therefore $\beta_{i,j}(k[S]) > 0$, equivalently $\beta_{i,j}(k[S]) = 1$ if and only if $j \in \mathrm{AP}_i(S)$.*

Acknowledgements The author is grateful to Professor Kei-ichi Watanabe, who informed me of a question related to the main theorem in this paper.

References

1. Charalambous, H., Katsabekis, A., Thoma, A.: Minimal systems of binomial generators and the indispensable complex of a toric ideal. Proc. Am. Math. Soc. **135**, 3443–3451 (2007)
2. Eto, K.: Almost Gorenstein monomial curves in affine four space. J. Algebra **488**, 362–387 (2017)
3. Herzog, J., Watanabe, K.-i.: Almost symmetric numerical semigroups. Semigroup Forums **98**, 589–630 (2019)
4. Kunz, E.: Introduction to Commutative Algebra and Algebraic Geometry. Birkhäuser, Boston (1985)
5. Moscariello, A.: On the type of an almost Gorenstein monomial curve. J. Algebra **456**, 266–277 (2016)
6. Peeva, I., Sturmfels, B.: Generic lattice ideals. J. Am. Math. Soc. **11**, 363–373 (1998)
7. Rosales, J.C., García-Sánchez, P.A.: Numerical Semigroups. Springer, New York (2009)

Symmetric (Not Complete Intersection) Semigroups Generated by Six Elements

Leonid G. Fel

Abstract We consider symmetric (not complete intersection) numerical semigroups S_6, generated by a set of six positive integers $\{d_1, \ldots, d_6\}$, $\gcd(d_1, \ldots, d_6) = 1$, and derive inequalities for degrees of syzygies of such semigroups and find the lower bound for their Frobenius numbers. We show that this bound may be strengthened if S_6 satisfies the Watanabe lemma.

Keywords Symmetric (not complete intersection) semigroups · Betti's numbers · Frobenius number

1 Introduction

Among a vast number of numerical semigroups of different types, symmetric numerical semigroups $\mathsf{S}_m = \langle d_1, \ldots, d_m \rangle$ of embedding dimension (*edim*) m, generated by positive integers $\{d_1, \ldots, d_m\}$, are of particular interest in commutative algebra. The whole set of symmetric semigroups may be decomposed in two parts depending on the associated with S_m graded Gorenstein semigroup rings $k[\mathsf{S}_m]$, whose defining ideals I_{S_m} are generated either by $m-1$ or more than m elements[1] when $m \geq 4$. The former and latter semigroups are refered to as complete intersection (CI) and symmetric (not CI) semigroups. They both possess the duality properties of Betti's numbers $\beta_k(\mathsf{S}_m)$ and of syzygies degrees in the Hilbert series of $k[\mathsf{S}_m]$ with the Cohen-Macaulay type 1.

Unlike the CI semigroups, a study of their not CI counterparts is far from full completion. Bresinsky [3, 4] studied symmetric (not CI) semigroups with small *edim*

[1]By Kunz theorems [11, 12], if a cardinality $\#I_{\mathsf{S}_m}$ is equal m, then a semigroup S_m cannot be symmetric and, therefore, $k[\mathsf{S}_m]$ cannot be Gorenstein. The defining ideals I_{S_2} and I_{S_3} are always generated by 1 and 2 elements, respectively.

L. G. Fel (✉)
Department of Civil Engineering, Technion–Israel Institute of Technology, Haifa, Israel
e-mail: lfel@technion.ac.il

V. Barucci et al. (eds.), *Numerical Semigroups*, Springer INdAM Series 40,
https://doi.org/10.1007/978-3-030-40822-0_7

and found $\beta_1(\mathsf{S}_4) = 5$ and an upper bound for $\beta_1(\mathsf{S}_5)$ when d_i are constraint by linear relation. However, the other characteristics of S_m are left unknown even for small $m \geq 4$.

Recently, we have derived $m-1$ polynomial identities [7] for degrees of syzygies in non-symmetric semigroups S_m. They became a source of various relations for semigroups of a different nature. In the case of CI, the whole set of $m-1$ identities was reduced up to one identity for degrees of the 1st syzygy [7]. In [6] and [8], we have studied two symmetric (not CI) semigroups with $m = 4$ and $m = 5$, respectively, and found that the number of identities in both cases is reduced up to two. New lower bounds for the Frobenius numbers in these semigroups were given in [6–8]. In the present paper we apply the approach of polynomial identities to study the next case of symmetric (not CI) semigroups with $m = 6$, which have two independent Betti's numbers β_1, β_2.

The paper is organized in five sections. In Sect. 2 we derive polynomial identities for symmetric (not CI) semigroups S_6 and show that only three identities are independent. We prove Stanley's conjecture 4b [13] on the unimodal sequence of Betti's numbers in the Gorenstein rings $k[\mathsf{S}_m]$ if $m = 6$. Combining polynomial identities with Cauchy-Schwarz's inequality, we find the final inequality (22) for symmetric polynomials X_k built of the 1st syzygy degrees and arrive at the lower bound (23) for Frobenius number $F(\mathsf{S}_6)$ which provides a sufficient condition to satisfy inequality (22). In Sect. 3 we improve the lower bound (37) for $F(\mathsf{S}_6)$ by providing the necessary condition to satisfy inequality (22). In Sect. 4 we compare the lower bounds (38) for $F(\mathsf{S}_6)$ in CI, symmetric (not CI) and non-symmetric semigroups, generated by six elements, and find the upper bound (41) for the difference $\beta_2-\beta_1$, while its lower bound (7) follows in proof of Stanley's conjecture 4b. The upper bound (41) coexists with an absence of the upper bounds for Betti's numbers β_1, β_2. In Sect. 5 we consider the symmetric (not CI) semigroups S_6 satisfying Watanabe's Lemma [14] and find the lower bound of their Frobenius numbers, which is stronger than its counterparts for symmetric (not CI) semigroups S_6 not satisfying Watanabe's Lemma.

2 Symmetric (Not CI) Semigroups Generated by Six Integers and Polynomial Identities

Consider a symmetric numerical semigroup S_6, which is not CI and generated by six positive integers. Its Hilbert series $H\left(\mathsf{S}_6;t\right)$ with independent Betti's numbers β_1, β_2 reads:

$$H\left(\mathsf{S}_6;t\right) = \frac{Q_6(t)}{\prod_{i=1}^{6}\left(1-t^{d_i}\right)}, \tag{1}$$

$$Q_6(t) = 1 - \sum_{j=1}^{\beta_1} t^{x_j} + \sum_{j=1}^{\beta_2} t^{y_j} - \sum_{j=1}^{\beta_2} t^{g-y_j} + \sum_{j=1}^{\beta_1} t^{g-x_j} - t^g,$$

$$x_j, y_j, g \in \mathbb{Z}_>, \qquad 2d_1 \leq x_j, y_j < g.$$

The Frobenius number $F(\mathsf{S}_6)$ of numerical semigroup S_6 is related to the largest degree g as follows:

$$F(\mathsf{S}_6)=g-\sigma_1,\qquad \sigma_1=d_1+\ldots+d_6.$$

There are two constraints more,

$$a)\quad \beta_1\ge 7\quad\text{and}\quad b)\quad d_1\ge 7. \tag{2}$$

The inequality (2a) holds since S_6 is neither CI ($\beta_1=5$) nor almost CI ($\beta_1=6$) according to [12], and the inequality (2b) is necessary since a semigroup $\langle m,d_2,\ldots,d_m\rangle$ is never symmetric [5].

Polynomial identities for degrees of syzygies for numerical semigroups were derived in [7, Thm 1]. In the case of a symmetric (not CI) semigroup S_6, they read:

$$\sum_{j=1}^{\beta_1}x_j^r-\sum_{j=1}^{\beta_2}y_j^r+\sum_{j=1}^{\beta_2}(g-y_j)^r-\sum_{j=1}^{\beta_1}(g-x_j)^r+g^r=0,\quad r\le 4, \tag{3}$$

$$\sum_{j=1}^{\beta_1}x_j^5-\sum_{j=1}^{\beta_2}y_j^5+\sum_{j=1}^{\beta_2}(g-y_j)^5-\sum_{j=1}^{\beta_1}(g-x_j)^5+g^5=120\pi_6,\quad \pi_6=\prod_{j=1}^{6}d_j.$$

Only three of five identities in (3) are not trivial, these are for $r=1,3,5$:

$$\mathcal{B}_6 g+\sum_{j=1}^{\beta_1}x_j=\sum_{j=1}^{\beta_2}y_j,\qquad \mathcal{B}_6=\frac{\beta_2-\beta_1+1}{2}, \tag{4}$$

$$\mathcal{B}_6 g^3+\sum_{j=1}^{\beta_1}x_j^2\,(3g-2x_j)=\sum_{j=1}^{\beta_2}y_j^2\,(3g-2y_j), \tag{5}$$

$$\mathcal{B}_6 g^5+\sum_{j=1}^{\beta_1}x_j^3\left(10g^2-15gx_j+6x_j^2\right)-360\pi_6=\sum_{j=1}^{\beta_2}y_j^3\left(10g^2-15gy_j+6y_j^2\right), \tag{6}$$

where $\mathcal{B}_6$ is defined according to the expression for an arbitrary symmetric semigroup S_m in [7], Formulas (5.7, 5.9). The sign of $\mathcal{B}_6$ is strongly related to the famous Stanley Conjecture 4b [13] on the unimodal sequence of Betti's numbers in the 1-dim local Gorenstein semigroup rings $k[\mathsf{S}_m]$. We give its simple proof in the case $edim=6$.

Lemma 1 *Let a symmetric (not CI) semigroup* S_6 *be given with the Hilbert series* $H\left(\mathsf{S}_6;z\right)$ *in accordance with (1). Then*

$$\beta_2\ge\beta_1+1. \tag{7}$$

Proof According to identity (4) and the constraints on degrees x_j of the 1st syzygies (1) we have,

$$\sum_{j=1}^{\beta_2} y_j < B_6 g + \beta_1 g = \frac{\beta_2 + \beta_1 + 1}{2} g. \tag{8}$$

On the other hand, there holds another constraint on degrees y_j of the 2nd syzygies,

$$\sum_{j=1}^{\beta_2} y_j < \beta_2 g. \tag{9}$$

Inequality (9) holds always, while inequality (8) is not valid for every set $\{x_1, \ldots, x_{\beta_1}\}$, but only when (4) holds. In order to make the both inequalities consistent, we have to find a relation between β_1 and β_2 where both inequalities (8) and (9) are satisfied, even if (8) is stronger than (9). To satisfy these inequalities, it is enough to require $(\beta_2 + \beta_1 + 1)/2 \leq \beta_2$, that leads to (7). □

Another constraint for Betti's numbers β_j follows from the general inequality for the sum of β_j in the case of non-symmetric semigroups [5], Formula (1.9),

$$\sum_{j=0}^{m-1} \beta_j \leq d_1 2^{m-1} - 2(m-1), \qquad \beta_0 = 1. \tag{10}$$

Applying the duality relation for Betti's numbers, $\beta_j = \beta_{m-j-1}$, $\beta_{m-1} = 1$, in symmetric semigroups S_6 to inequality (10) and combining it with Lemma 1, we obtain

$$\beta_1 < 2(4d_1 - 1). \tag{11}$$

To study polynomial identities (4, 5, 6) and their consequences, start with estimation for two real functions $R_1(z)$, $R_2(z)$, presented in Fig. 1,

$$R_1(z) \geq A_* R_2(z), \quad 0 \leq z \leq 1, \qquad \text{where} \tag{12}$$
$$R_1(z) = z^2\sqrt{10 - 15z + 6z^2}, \qquad R_2(z) = z^2(3 - 2z), \quad A_* = 0.9682.$$

The constant A_* is chosen by requirement of the existence of such a coordinate $z_* \in [0, 1]$ providing
two equalities,

$$R_1(z_*) = A_* R_2(z_*), \qquad R_1'(z_*) = A_* R_2'(z_*), \quad z_* \simeq 0.8333,$$

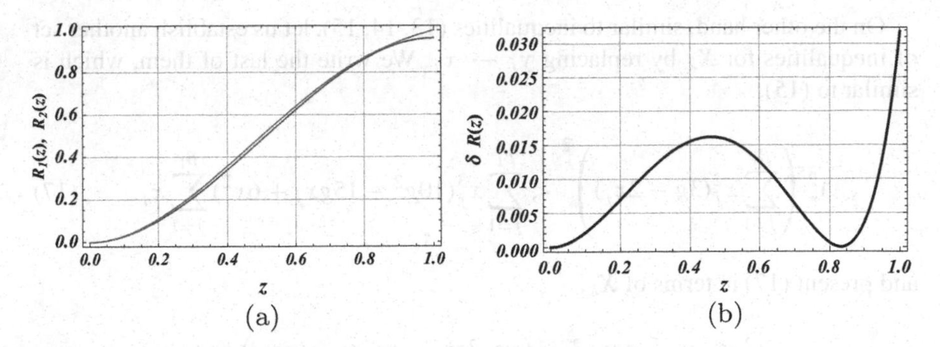

Fig. 1 Plot of the functions (**a**) $R_1(z)$ *in red color*, $A_*R_2(z)$ *in blue color* and a discrepancy (**b**) $\delta R(z) = R_1(z) - A_*R_2(z)$ in the range $z \in [0, 1]$ (colour figure online)

where $R_j'(z_*) = dR_j(z)/dz$ at $z = z_*$. Substituting $z = y_j/g$, $0 < z < 1$, into inequality (12) and making summation over $1 \le j \le \beta_2$, we get

$$A_* \sum_{j=1}^{\beta_2} y_j^2(3g - 2y_j) < \sum_{j=1}^{\beta_2} y_j^2\sqrt{10g^2 - 15gy_j + 6y_j^2}. \tag{13}$$

Applying the Cauchy-Schwarz inequality $\left(\sum_{j=1}^{N} a_j b_j\right)^2 \le \left(\sum_{j=1}^{N} a_j^2\right)\left(\sum_{j=1}^{N} b_j^2\right)$ to the right-hand side of inequality (13), we obtain

$$\left(\sum_{j=1}^{\beta_2} y_j^{3/2}\sqrt{10g^2 - 15gy_j + 6y_j^2}\ y_j^{1/2}\right)^2 \le \sum_{j=1}^{\beta_2} y_j^3(10g^2 - 15gy_j + 6y_j^2)\sum_{j=1}^{\beta_2} y_j. \tag{14}$$

Combining (13) and (14), we arrive at the inequality

$$A_*^2\left(\sum_{j=1}^{\beta_2} y_j^2(3g - 2y_j)\right)^2 < \sum_{j=1}^{\beta_2} y_j^3(10g^2 - 15gy_j + 6y_j^2)\sum_{j=1}^{\beta_2} y_j. \tag{15}$$

Denote by X_k the k-th power symmetric polynomial $X_k(x_1, \ldots, x_{\beta_1}) = \sum_{j=1}^{\beta_1} x_j^k$, where $x_j < g$, and substitute identities (4, 5, 6) into inequality (15),

$$A_*^2\left(\mathcal{B}_6 g^3 + 3gX_2 - 2X_3\right)^2 < \left(\mathcal{B}_6 g^5 - 360\pi_6 + 10g^2X_3 - 15gX_4 + 6X_5\right)\left(\mathcal{B}_6 g + X_1\right). \tag{16}$$

On the other hand, similar to inequalities (13, 14, 15), let us establish another set of inequalities for X_k by replacing $y_j \to x_j$. We write the last of them, which is similar to (15),

$$A_*^2 \left(\sum_{j=1}^{\beta_1} x_j^2 (3g - 2x_j) \right)^2 < \sum_{j=1}^{\beta_1} x_j^3 (10g^2 - 15gx_j + 6x_j^2) \sum_{j=1}^{\beta_1} x_j, \tag{17}$$

and present (17) in terms of X_k,

$$A_*^2 (3gX_2 - 2X_3)^2 < \left(10g^2 X_3 - 15gX_4 + 6X_5\right) X_1. \tag{18}$$

Represent the both inequalities (16) and (18) as follows:

$$360\pi_6 - \mathcal{B}_6 g^5 + A_*^2 \frac{\left(\mathcal{B}_6 g^3 + 3gX_2 - 2X_3\right)^2}{\mathcal{B}_6 g + X_1} < 10g^2 X_3 - 15gX_4 + 6X_5, \tag{19}$$

$$A_*^2 \frac{(3gX_2 - 2X_3)^2}{X_1} < 10g^2 X_3 - 15gX_4 + 6X_5. \tag{20}$$

Inequality (20) holds always, while inequality (19) is not valid for every set $\{x_1, \ldots, x_{\beta_1}, g\}$. In order to make both inequalities consistent, we have to find a range of g where both inequalities (19) and (20) are satisfied. To satisfy these inequalities, it is enough to require that inequality (20) implies inequality (19), i.e.,

$$\frac{360\pi_6 - \mathcal{B}_6 g^5}{A_*^2} + \frac{\left(\mathcal{B}_6 g^3 + 3gX_2 - 2X_3\right)^2}{\mathcal{B}_6 g + X_1} < \frac{(3gX_2 - 2X_3)^2}{X_1}. \tag{21}$$

Simplifying the above expressions, we present the last inequality (21) as follows:

$$CX_1(X_1 + \mathcal{B}_6 g) < \left(3gX_2 - 2X_3 - g^2 X_1\right)^2, \qquad C = \frac{360\pi_6 - \alpha \mathcal{B}_6 g^5}{A_*^2 \mathcal{B}_6 g}, \tag{22}$$

where $\alpha = 1 - A_*^2 \simeq 0.06259$ and $\mathcal{B}_6 \geq 1$ due to Lemma 1. Inequality (22) holds always if its left-hand side is negative, i.e., $C < 0$, that results in the following constraint,

$$g > q_6, \qquad q_6 = \sqrt[5]{\frac{360}{\alpha\, \mathcal{B}_6}} \sqrt[5]{\pi_6}, \quad \text{where} \quad \sqrt[5]{\frac{360}{\alpha}} \simeq 5.649. \tag{23}$$

The lower bound q_6 in (23) provides a sufficient condition to satisfy the inequality (22). In fact, a necessary condition has to produce another bound $g_6 < q_6$.

3 The Lower Bound for the Frobenius Numbers of Semigroups $\mathbf{S_6}$

An actual lower bound of g precedes the bound, given in (23), since the inequality (22) may be satisfied for a sufficiently small $C > 0$. To find it, we introduce another kind of symmetric polynomials $\mathcal{X}_k$:

$$\mathcal{X}_k = \sum_{i_1<i_2<\ldots<i_k}^{\beta_1} x_{i_1} x_{i_2} \ldots x_{i_k},$$

$$\mathcal{X}_0 = 1, \quad \mathcal{X}_1 = \sum_{i=1}^{\beta_1} x_i, \quad \mathcal{X}_2 = \sum_{i<j}^{\beta_1} x_i x_j, \quad \mathcal{X}_3 = \sum_{i<j<r}^{\beta_1} x_i x_j x_r, \quad \ldots, \quad \mathcal{X}_{\beta_1} = \prod_{i=1}^{\beta_1} x_i,$$

which are related to polynomials X_k by the Newton recursion identities,

$$m\mathcal{X}_m = \sum_{k=1}^{m} (-1)^{k-1} X_k \mathcal{X}_{m-k}, \quad \text{i.e.},$$

$$X_1 = \mathcal{X}_1, \quad X_2 = \mathcal{X}_1^2 - 2\mathcal{X}_2, \quad X_3 = \mathcal{X}_1^3 - 3\mathcal{X}_2\mathcal{X}_1 + 3\mathcal{X}_3, \quad \ldots \quad (24)$$

Recall the Newton-Maclaurin inequalities [9] for polynomials $\mathcal{X}_k$,

$$\frac{\mathcal{X}_1}{\beta_1} \geq \left(\frac{\mathcal{X}_2}{\binom{\beta_1}{2}}\right)^{\frac{1}{2}} \geq \left(\frac{\mathcal{X}_3}{\binom{\beta_1}{3}}\right)^{\frac{1}{3}} \geq \ldots \geq \sqrt[\beta_1]{\mathcal{X}_{\beta_1}}. \quad (25)$$

Consider the inequality (22) in the following form

$$CX_1(X_1 + \mathcal{B}_6 g) < 9g^2X_2^2 + 4X_3^2 + g^4X_1^2 + 4g^2X_1X_3 - 12gX_2X_3 - 6g^3X_1X_2, \quad (26)$$

and substitute Newton's identities (24) into (26),

$$\mathcal{X}_1 P(\mathcal{X}_1, \mathcal{X}_2, \mathcal{X}_3) < \mathcal{X}_1 Q_1(\mathcal{X}_1) + \mathcal{X}_2 Q_2(\mathcal{X}_1, \mathcal{X}_2) + \mathcal{X}_3 Q_3(\mathcal{X}_1, \mathcal{X}_2, \mathcal{X}_3), \quad (27)$$

where

$$\begin{aligned}
P(\mathcal{X}_1, \mathcal{X}_2, \mathcal{X}_3) &= 4g^2\mathcal{X}_1\mathcal{X}_2 + 2\mathcal{X}_1^3\mathcal{X}_2 + 6\mathcal{X}_2\mathcal{X}_3 + 6g\mathcal{X}_2^2 + 3g\mathcal{X}_1\mathcal{X}_3, \\
Q_1(\mathcal{X}_1) &= \frac{1}{3}\mathcal{X}_1^5 - g\mathcal{X}_1^4 + \frac{13}{12}g^2\mathcal{X}_1^3 - \frac{1}{2}g^3\mathcal{X}_1^2 + \frac{g^4 - C}{12}\mathcal{X}_1 - \frac{C}{12}\mathcal{B}_6 g, \\
Q_2(\mathcal{X}_1, \mathcal{X}_2) &= 3g^2\mathcal{X}_2 + 3\mathcal{X}_2\mathcal{X}_1^2 + 5g\mathcal{X}_1^3 + g^3\mathcal{X}_1, \\
Q_3(\mathcal{X}_1, \mathcal{X}_2, \mathcal{X}_3) &= 3\mathcal{X}_3 + 2\mathcal{X}_1^3 + g^2\mathcal{X}_1 + 6g\mathcal{X}_2.
\end{aligned}$$

Applying inequalities (25) to $Q_2(\mathcal{X}_1,\mathcal{X}_2)$ and $Q_3(\mathcal{X}_1,\mathcal{X}_2,\mathcal{X}_3)$, we obtain

$$Q_2(\mathcal{X}_1,\mathcal{X}_2) < \mathcal{X}_1 Q_{21}(\mathcal{X}_1), \quad Q_{21}(\mathcal{X}_1) = 3\frac{\mathcal{X}_1}{\beta_1^2}\binom{\beta_1}{2}\left(g^2+\mathcal{X}_1^2\right) + 5g\mathcal{X}_1^2 + g^3,$$

$$Q_3(\mathcal{X}_1,\mathcal{X}_2,\mathcal{X}_3) < \mathcal{X}_1 Q_{31}(\mathcal{X}_1), \quad Q_{31}(\mathcal{X}_1) = 3\frac{\mathcal{X}_1^2}{\beta_1^3}\binom{\beta_1}{3} + 2\mathcal{X}_1^2 + g^2 + 6g\frac{\mathcal{X}_1}{\beta_1^2}\binom{\beta_1}{2}. \quad (28)$$

Substituting inequalities (28) into (27) and applying again (25), we obtain

$$P(\mathcal{X}_1,\mathcal{X}_2,\mathcal{X}_3) < Q_1(\mathcal{X}_1) + \frac{\mathcal{X}_1^2}{\beta_1^2}\binom{\beta_1}{2}Q_{21}(\mathcal{X}_1) + \frac{\mathcal{X}_1^3}{\beta_1^3}\binom{\beta_1}{3}Q_{31}(\mathcal{X}_1). \quad (29)$$

Represent the right-hand side of inequality (29) as a polynomial $E(\mathcal{X}_1)$ of the 5th order in $\mathcal{X}_1$,

$$E(\mathcal{X}_1) = \sum_{k=0}^{5} E_k g^{5-k}\mathcal{X}_1^k, \qquad \text{where} \qquad (30)$$

$$E_0 = -\frac{B_6 C g^{-4}}{12}, \qquad E_1 = \frac{1-Cg^{-4}}{12}, \qquad E_2 = \frac{1}{\beta_1^2}\binom{\beta_1}{2} - \frac{1}{2} = -\frac{1}{2\beta_1},$$

$$E_3 = \frac{3}{\beta_1^4}\binom{\beta_1}{2}^2 + \frac{1}{\beta_1^3}\binom{\beta_1}{3} + \frac{13}{12}, \qquad E_4 = \frac{5}{\beta_1^2}\binom{\beta_1}{2} + \frac{6}{\beta_1^5}\binom{\beta_1}{2}\binom{\beta_1}{3} - 1,$$

$$E_5 = \frac{3}{\beta_1^4}\binom{\beta_1}{2}^2 + \frac{3}{\beta_1^6}\binom{\beta_1}{3}^2 + \frac{2}{\beta_1^3}\binom{\beta_1}{3} + \frac{1}{3}.$$

Thus, the inequality (22) reads:

$$P(\mathcal{X}_1,\mathcal{X}_2,\mathcal{X}_3) < E(\mathcal{X}_1). \qquad (31)$$

On the other hand, applying (25) to the polynomial $P(\mathcal{X}_1,\mathcal{X}_2,\mathcal{X}_3)$, we have another inequality,

$$P(\mathcal{X}_1,\mathcal{X}_2,\mathcal{X}_3) < J(\mathcal{X}_1), \qquad J(\mathcal{X}_1) = \sum_{k=3}^{5} J_k g^{5-k}\mathcal{X}_1^k, \qquad (32)$$

$$J_5 = \frac{2}{\beta_1^2}\binom{\beta_1}{2}\left[1 + \frac{3}{\beta_1^3}\binom{\beta_1}{3}\right], \qquad J_4 = \frac{6}{\beta_1^4}\binom{\beta_1}{2}^2 + \frac{3}{\beta_1^3}\binom{\beta_1}{3}, \qquad J_3 = \frac{4}{\beta_1^2}\binom{\beta_1}{2}.$$

Inequality (32) holds always, while inequality (31) is not valid for every set $\{x_1,\ldots,x_{\beta_1},g\}$. In order to make both inequalities consistent, we have to find a range for g where both inequalities (31) and (32) are satisfied. To satisfy both inequalities, it is enough to require that (32) implies (31),

$$E(\mathcal{X}_1) > J(\mathcal{X}_1), \qquad \text{or}$$

$$(E_5 - J_5)\mathcal{X}_1^5 + (E_4 - J_4)g\mathcal{X}_1^4 + (E_3 - J_3)g^2\mathcal{X}_1^3 + E_2 g^3\mathcal{X}_1^2 + E_1 g^4\mathcal{X}_1 + E_0 g^5 > 0, \quad (33)$$

where

$$E_5 - J_5 = \frac{3}{\beta_1^4}\left(\frac{\beta_1}{2}\right)^2 + \frac{3}{\beta_1^6}\left(\frac{\beta_1}{3}\right)^2 + \frac{2}{\beta_1^3}\left(\frac{\beta_1}{3}\right) + \frac{1}{3} - \frac{2}{\beta_1^2}\left(\frac{\beta_1}{2}\right)\left[1 + \frac{3}{\beta_1^3}\left(\frac{\beta_1}{3}\right)\right] = \frac{1}{3\beta_1^4},$$

$$E_4 - J_4 = \frac{5}{\beta_1^2}\left(\frac{\beta_1}{2}\right) + \frac{6}{\beta_1^5}\left(\frac{\beta_1}{2}\right)\left(\frac{\beta_1}{3}\right) - 1 - \frac{6}{\beta_1^4}\left(\frac{\beta_1}{2}\right)^2 - \frac{3}{\beta_1^3}\left(\frac{\beta_1}{3}\right) = -\frac{1}{\beta_1^3},$$

$$E_3 - J_3 = \frac{3}{\beta_1^4}\left(\frac{\beta_1}{2}\right)^2 + \frac{1}{\beta_1^3}\left(\frac{\beta_1}{3}\right) + \frac{13}{12} - \frac{4}{\beta_1^2}\left(\frac{\beta_1}{2}\right) = \frac{13}{12\beta_1^2}. \quad (34)$$

Substituting expressions $E_k - J_k$, $k = 3, 4, 5$ from (34) and E_0, E_1, E_2 from (30) into (33), we obtain

$$\frac{C}{g^4} < G(b,u), \qquad G(b,u) = \frac{u}{u+b}(1-u)^2(1-2u)^2, \quad u = \frac{\mathcal{X}_1}{\beta_1 g}, \quad b = \frac{\mathcal{B}_6}{\beta_1}. \quad (35)$$

The function $G(b,u)$ is continuous (see Fig. 2) and attains its global maximal value $G(b,u_m)$ at $u_m(b) \in (0, 1/2)$, where $u_m = u_m(b)$ is a smaller positive root of cubic equation,

$$8u_m^3 + 2(5b-3)u_m^2 - 9bu_m + b = 0,$$

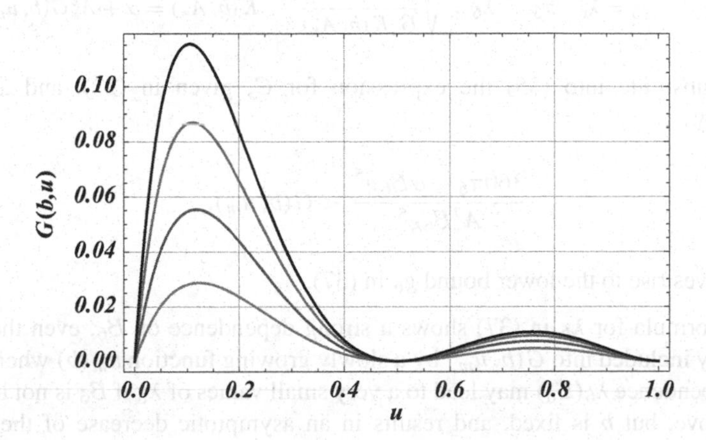

Fig. 2 Plot of the functions $G(b,u)$ with different b: (*in brown*) $b = 1.75$, $u_m = 0.125$; (*in blue*) $b = 0.85$, $u_m = 0.117$; (*in red*) $b = 0.5$, $u_m = 0.112$; (*in black*) $b = 0.35$, $u_m = 0.107$ (colour figure online)

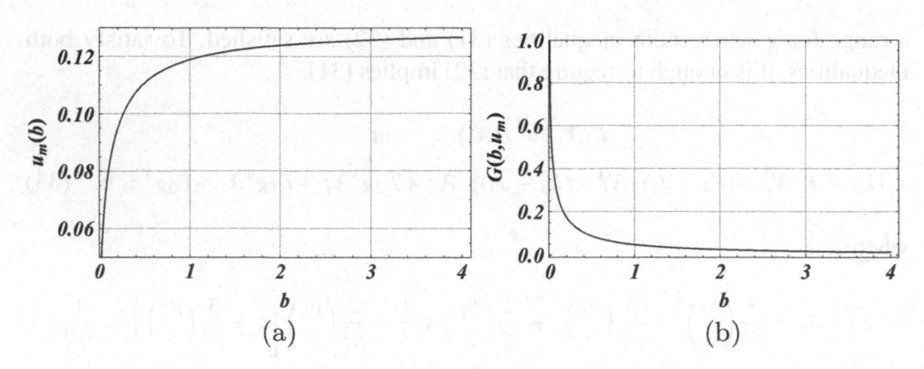

Fig. 3 Plot of the functions (**a**) $u_m(b)$ and (**b**) $G(b, u_m)$ in a wide range of b

with asymptotic behavior of $u_m(b)$ and $G(b, u_m)$ (see Fig. 3),

$$u_m(b) \stackrel{b\to 0}{\longrightarrow} \sqrt{\frac{b}{6}}, \tag{36}$$

$$u_m(b) \stackrel{b\to\infty}{\longrightarrow} v_1 - \frac{v_2}{b}, \quad v_1 = \frac{9-\sqrt{41}}{20} \simeq 0.1298, \quad v_2 = \frac{7\sqrt{41}-3}{500\sqrt{41}} \simeq 0.013,$$

$$G(b, u_m) \stackrel{b\to 0}{\longrightarrow} 1, \qquad G(b, u_m) \stackrel{b\to\infty}{\longrightarrow} \frac{w}{b}, \qquad w = \frac{411+41\sqrt{41}}{12500} \simeq 0.05388.$$

Theorem 1 *Let a symmetric (not CI) semigroup* S_6 *be given with its Hilbert series* $H\left(\mathsf{S}_6; z\right)$ *in accordance with (1). Then the following inequality holds:*

$$g > g_6, \qquad g_6 = \lambda_6 \sqrt[5]{\pi_6}, \quad \lambda_6 = \sqrt[5]{\frac{360}{\mathcal{B}_6 K(b, A_*)}}, \qquad K(b, A_*) = \alpha + A_*^2 G(b, u_m). \tag{37}$$

Proof Substitute into (35) the expression for C, given in (22), and arrive at inequality

$$\frac{360\pi_6 - \alpha\mathcal{B}_6 g^5}{A_*^2 \mathcal{B}_6 g^5} < G(b, u_m),$$

which gives rise to the lower bound g_6 in (37). □

The formula for λ_6 in (37) shows a strong dependence on $\mathcal{B}_6$, even the last is implicitly included into $G(b, u_m)$ by a slowly growing function $u_m(b)$ when $b > 1$. Such dependence $\lambda_6(\mathcal{B}_6)$ may lead to a very small values of λ_6 if $\mathcal{B}_6$ is not bounded from above, but b is fixed, and results in an asymptotic decrease of the bound, $g_6 \stackrel{\mathcal{B}_6\to\infty}{\longrightarrow} 0$. The last limit poses a question: does formula (37) for g_6 contradict the known lower bound [10] for the Frobenius number in the case of 6-generated numerical semigroups of arbitrary nature, i.e., not assuming symmetry. If the answer is affirmative then it arises another question: what should be required in order to avoid such contradiction. We address both questions in the next section in a slightly different form: are there any constraints on Betti's numbers.

4 Are There Any Constraints on Betti's Numbers of Symmetric (Not CI) Semigroups S_6?

Denote by $\widetilde{g_6}$ and $\overline{g_6}$ the lower bounds of the largest degree of syzygies for non-symmetric [10] and symmetric CI [7] semigroups generated by six integers, respectively. Compare g_6 with $\widetilde{g_6}$ and $\overline{g_6}$ and note that the following double inequality hold:

$$\widetilde{g_6} < g_6 < \overline{g_6}, \qquad \widetilde{g_6} = \sqrt[5]{120}\sqrt[5]{\pi_6}, \quad \overline{g_6} = 5\sqrt[5]{\pi_6}. \tag{38}$$

Substituting the expression for g_6 from (37) into (38), we obtain

$$\frac{72}{625}\frac{1}{K(b, A_*)} < \mathcal{B}_6 < \frac{3}{K(b, A_*)}, \qquad \frac{72}{625} = 0.1152, \tag{39}$$
$$K(b, A_*) \stackrel{b \to 0}{\longrightarrow} 1, \qquad K(b, A_*) \stackrel{b \to \infty}{\longrightarrow} \alpha,$$

where the two limits follow by (36, 37). The double inequality (39) determines the upper and lower bounds for varying $\mathcal{B}_6$ in the plane $(b, \mathcal{B}_6)$ as monotonic functions (see Fig. 4a) with asymptotic behavior,

$$\begin{aligned} Upp.\ bound: &\quad \mathcal{B}_6 \stackrel{b \to 0}{\longrightarrow} 3, \quad \mathcal{B}_6 \stackrel{b \to \infty}{\longrightarrow} 47.92; \\ Low.\ bound: &\quad \mathcal{B}_6 \stackrel{b \to 0}{\longrightarrow} 0.1152, \quad \mathcal{B}_6 \stackrel{b \to \infty}{\longrightarrow} 1.84. \end{aligned} \tag{40}$$

According to Lemma 1, the lower bound in (39, 40) may be chosen as $\mathcal{B}_6 = 1$.

Find the constraints on Betti's numbers. In order to find the constraints on Betti's numbers, inequality (39) has to be replaced by

$$1 < \beta_2 - \beta_1 < \frac{6}{K(b, A_*)} - 1, \qquad \beta_2 - \beta_1 \stackrel{\beta_1 \to 7}{\longrightarrow} 83.8, \quad \beta_2 - \beta_1 \stackrel{\beta_1 \to \infty}{\longrightarrow} 5, \tag{41}$$

and the plot in Fig. 4a has to be transformed by rescaling the coordinates $(b, \mathcal{B}_6)$ with inversion, $b \to \beta_1 = \mathcal{B}_6/b$, and shift, $\mathcal{B}_6 \to \beta_2 - \beta_1 = 2\mathcal{B}_6 - 1$ (see Fig. 4b). Following Sect. 2, the constraints (39) have to be supplemented by another double inequality $7 \le \beta_1 < 2(4d_1 - 1)$ in accordance with (2a) and (11).

The double inequality (41) manifests a phenomenon, which does not exist in symmetric (not CI) semigroups S_m, generated by four [6] and five [8] integers, where inequalities $\widetilde{g_m} < g_m < \overline{g_m}$, are always satisfied and independent of Betti's

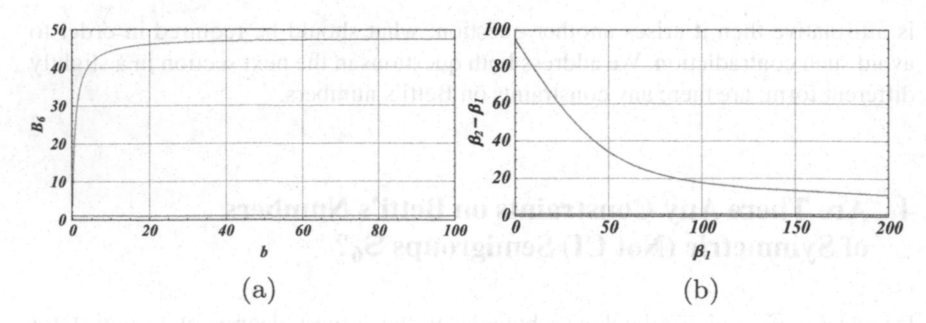

Fig. 4 Plots of the lower (*blue*) and upper (*red*) bounds in the planes (**a**) $(b, \mathcal{B}_6)$ and (**b**) $(\beta_1, \beta_2 - \beta_1)$. The lower bounds are given by the values (**a**) $\mathcal{B}_6 = 1$ and (**b**) $\beta_2 - \beta_1 = 1$ (colour figure online)

numbers ($\beta_1 = 5$ for S_4 and $\beta_1 = \beta$ for S_5):

$$\widetilde{g_m} < \lambda_m \sqrt[m-1]{\pi_m} < \overline{g_m}, \tag{42}$$

$$\lambda_4 = \sqrt[3]{25}, \qquad \widetilde{g_m} = \sqrt[m-1]{(m-1)!}\,\sqrt[m-1]{\pi_m},$$
$$\lambda_5 = \sqrt[4]{192(\beta-1)/\beta}, \qquad \overline{g_m} = (m-1)\sqrt[m-1]{\pi_m}.$$

Note, that constraints (41) do not contradict Bresinsky's theorem [2] on the arbitrarily large finite value of β_1 for generic semigroup S_m, $m \geq 4$. Below, we put forward some considerations about validity of (41) for Betti's numbers β_1, β_2 of symmetric (not CI) semigroup S_6.

The double inequality (41) has arisen by comparison of g_6 with two other bounds $\widetilde{g_6}$ and $\overline{g_6}$ and, strictly speaking, a validity of (41) is dependent on how small is a discrepancy $\delta R(z)$ in Fig. 1. If $\delta R(z)$ is not small enough and neglecting it in (12) is far too crude approximation, then there may exist symmetric (not CI) semigroups S_6 with Betti's numbers β_1, β_2, where (41) is not true. Such violation should indicate a necessity to improve the lower bound g_6 in (37) to restore the relationship $\widetilde{g_6} < g_6 < \overline{g_6}$. Note, that such improvement is very hard to provide even by replacing $A_* \to A$ in inequality (13), where $A_* < A < 1$, and still preserving (13) with a new A. Such replacement leads again to (37) with $K(b, A)$ instead $K(b, A_*)$, i.e., the constraints on β_1, β_2 still exist, even the area of admissible Betti's numbers becomes wider.

However, if there are no such symmetric (not CI) semigroups S_6, for which the double inequality (41) does not hold, then there arises a much more deep question: why do the constraints on Betti's numbers exist. This problem is strongly related to the structure of minimal relations of the first and second syzygies in the minimal free resolution for the 1–dim Gorenstein (not CI) semigroup ring $k[S_6]$ and has to be addressed in a separate paper.

5 Symmetric (Not CI) Semigroups S_6 with the W and W^2 Properties

In [8], we introduced a notion of the W property for the m-generated symmetric (not CI) semigroups S_m satisfying Watanabe's Lemma [14]. We recall this Lemma together with the definition of the W property and two other statements relevant in this section.

Lemma 2 ([14]) *Let a semigroup* $S_{m-1} = \langle \delta_1, \ldots, \delta_{m-1} \rangle$ *be given and* $a \in \mathbb{Z}$, $a > 1$, *such that* $\gcd(a, d_m) = 1$, $d_m \in S_{m-1} \setminus \{\delta_1, \ldots, \delta_{m-1}\}$. *Consider a semigroup* $S_m = \langle a\delta_1, \ldots, a\delta_{m-1}, d_m \rangle$ *and denote it by* $S_m = \langle aS_{m-1}, d_m \rangle$. *Then* S_m *is symmetric if and only if* S_{m-1} *is symmetric, and* S_m *is symmetric CI if and only if* S_{m-1} *is symmetric CI.*

Corollary 1 ([8]) *Let a semigroup* $S_{m-1} = \langle \delta_1, \ldots, \delta_{m-1} \rangle$ *be given and* $a \in \mathbb{Z}$, $a > 1$, *such that* $\gcd(a, d_m) = 1$, $d_m \in S_{m-1} \setminus \{\delta_1, \ldots, \delta_{m-1}\}$. *Consider a semigroup* $S_m = \langle aS_{m-1}, d_m \rangle$. *Then* S_m *is symmetric (not CI) if and only if* S_{m-1} *is symmetric (not CI).*

Definition 1 ([8]) A symmetric (not CI) semigroup S_m has *the property* W if there exists another symmetric (not CI) semigroup S_{m-1} giving rise to S_m by the construction, described in Corollary 1.

Theorem 2 ([8]) *A minimal* edim *of symmetric (not CI) semigroup* S_m *with the property* W *is* $m = 5$.

In this section we study the symmetric (not CI) semigroups S_6 satisfying Watanabe's Lemma [14]. To distinguish such semigroups from the rest of symmetric (not CI) semigroups S_6 without the property W we denote them by W_6.

Lemma 3 *Let two symmetric (not CI) semigroups* $W_6 = \langle aS_5, d_6 \rangle$ *and* $S_5 = \langle q_1, \ldots, q_5 \rangle$ *be given and* $\gcd(a, d_6) = 1$, $d_6 \in S_5 \setminus \{q_1, \ldots, q_5\}$. *Let the lower bound* F_{6w} *of the Frobenius number* $F(W_6)$ *be represented as,* $F_{6w} = g_{6w} - \left(a \sum_{j=1}^{5} q_j + d_6\right)$. *Then*

$$g_{6w} = a\left(\lambda_5 \sqrt[4]{\pi_5(q)} + d_6\right), \qquad \pi_5(q) = \prod_{j=1}^{5} q_j. \tag{43}$$

where λ_5 *is defined in (42).*

Proof Consider a symmetric (not CI) numerical semigroup S_5 generated by five integers (without the W property), and apply the recent result [8] on the lower bound F_5 of its Frobenius number, $F(S_5)$,

$$F(S_5) \geq F_5, \qquad F_5 = h_5 - \sum_{j=1}^{5} q_j, \qquad h_5 = \lambda_5 \sqrt[4]{\pi_5(q)}. \tag{44}$$

The following relationship between the Frobenius numbers $F(\mathsf{W}_6)$ and $F(\mathsf{S}_5)$ was derived in [1]:

$$F(\mathsf{W}_6) = aF(\mathsf{S}_5) + (a-1)d_6. \tag{45}$$

Substituting $F(\mathsf{W}_6) = g - \left(a\sum_{j=1}^{5} q_j + d_6\right)$ and the representation (44) for $F(\mathsf{S}_5)$ into (45), we obtain

$$g - a\sum_{j=1}^{5} q_j - d_6 = ah_5 - a\sum_{j=1}^{5} q_j + (a-1)d_6 \quad \text{and hence} \quad g = a(h_5 + d_6). \tag{46}$$

Comparing the last equality in (46) with the lower bound of h_5 in (44), we arrive at (43). □

Following Corollary 1, let us apply the construction of a symmetric (not CI) semigroup S_m with the W property to a symmetric (not CI) semigroup S_{m-1}, which already has such property.

Definition 2 A symmetric (not CI) semigroup S_m has *the property* W^2 if there exist two symmetric (not CI) semigroup $\mathsf{S}_{m-1} = \langle q_1, \ldots, q_{m-1}\rangle$ and $\mathsf{S}_{m-2} = \langle p_1, \ldots, p_{m-2}\rangle$ giving rise to S_m by the construction, described in Corollary 1,

$$\mathsf{S}_m = \langle a_1\mathsf{S}_{m-1}, d_m\rangle, \qquad d_m \in \mathsf{S}_{m-1} \setminus \{q_1, \ldots, q_{m-1}\}, \qquad \gcd(a_1, d_m) = 1,$$
$$\mathsf{S}_{m-1} = \langle a_2\mathsf{S}_{m-2}, q_{m-1}\rangle, \quad q_{m-1} \in \mathsf{S}_{m-2} \setminus \{p_1, \ldots, p_{m-2}\}, \quad \gcd(a_2, q_{m-1}) = 1.$$

Theorem 3 *A minimal* edim *of symmetric (not CI) semigroup* S_m *with the property* W^2 *is* $m = 6$.

Proof This statement follows if we combine Definition 2 and Theorem 2. □

In this section we denote the symmetric (not CI) semigroups S_6 with the property W^2 by W_6^2.

Lemma 4 *Let three symmetric (not CI) semigroups* $\mathsf{W}_6^2 = \langle a_1\mathsf{W}_5, d_6\rangle$, $\mathsf{W}_5 = \langle a_2\mathsf{S}_4, q_5\rangle$, *and* $\mathsf{S}_4 = \langle p_1, \ldots, p_4\rangle$, *where* $q_j = a_2 p_j$, $1 \le j \le 4$, *be given in such a way that*

$$d_6 \in \mathsf{W}_5\backslash\{q_1, \ldots, q_5\}, \quad q_5 \in \mathsf{S}_4\backslash\{p_1, \ldots, p_4\}, \quad \gcd(a_1, d_6) = \gcd(a_2, q_5) = 1.$$

Let the lower bound F_{6w} *of the Frobenius number* $F(\mathsf{W}_6^2)$ *be represented as,* $F_{6w^2} = g_{6w^2} - \left(a_1 a_2 \sum_{j=1}^{4} p_j + a_1 q_5 + d_6\right)$. *Then*

$$g_{6w^2} = a_1\left[a_2\left(\lambda_4\sqrt[3]{\pi_4(p)} + q_5\right) + d_6\right], \qquad \pi_4(p) = \prod_{j=1}^{4} p_j, \tag{47}$$

where λ_4 is defined in (42).

Proof By Lemma 2 in [8], the lower bound F_{5w} of its Frobenius number $F(\mathsf{W}_5)$ of the symmetric (not CI) semigroup W_5 reads:

$$F_{5w} = g_{5w} - \left(a_2 \sum_{j=1}^{4} p_j + q_5 \right), \qquad g_{5w} = a_2 \left(\lambda_4 \sqrt[3]{\pi_4(p)} + q_5 \right). \quad (48)$$

Consider a symmetric (not CI) semigroup W_6^2, generated by six integers, and make use of a relationship between the Frobenius numbers $F(\mathsf{W}_6^2)$ and $F(\mathsf{W}_5)$ derived in [1]:

$$F(\mathsf{W}_6^2) = a_1 F(\mathsf{W}_5) + (a_1 - 1) d_6. \quad (49)$$

Substituting $F(\mathsf{W}_6^2) = g_{6w^2} - \left(a_1 a_2 \sum_{j=1}^{4} p_j + a_1 q_5 + d_6 \right)$ and the representation (48) for $F(\mathsf{S}_5)$ into (49), we obtain

$$g_{6w^2} - a_1 a_2 \sum_{j=1}^{4} p_j - a_1 q_5 - d_6 = a_1 \left[g_{5w} - \left(a_2 \sum_{j=1}^{4} p_j + q_5 \right) \right] + (a_1 - 1) d_6. \quad (50)$$

Simplifying the last equality (50), we arrive at (47). □

Among the subsets $\{\mathsf{W}_6^2\}$, $\{\mathsf{W}_6\}$ and the entire set $\{\mathsf{S}_6\}$ of symmetric (not CI) semigroups, generated by six integers, the following containment holds:

$$\{\mathsf{W}_6^2\} \subset \{\mathsf{W}_6\} \subset \{\mathsf{S}_6\}.$$

Below we present twelve symmetric (not CI) semigroup generated by six integers: V_1, V_2, V_3, V_4—without the W property, V_5, V_6, V_7, V_8—with the W property, and $V_9, V_{10}, V_{11}, V_{12}$—with the W^2 property.

$$V_1 = \langle 7, 9, 11, 12, 13, 15 \rangle, \quad V_5 = \langle 12, 20, 28, 30, 38, 41 \rangle, \quad V_9 = \langle 30, 33, 36, 37, 42, 48 \rangle,$$
$$V_2 = \langle 7, 9, 10, 11, 12, 13 \rangle, \quad V_6 = \langle 12, 20, 28, 38, 46, 47 \rangle, \quad V_{10} = \langle 42, 45, 48, 54, 59, 78 \rangle,$$
$$V_3 = \langle 12, 13, 14, 15, 17, 19 \rangle, \quad V_7 = \langle 14, 24, 26, 36, 46, 49 \rangle, \quad V_{11} = \langle 40, 42, 48, 54, 71, 78 \rangle,$$
$$V_4 = \langle 12, 13, 14, 15, 18, 19 \rangle, \quad V_8 = \langle 38, 46, 58, 62, 74, 79 \rangle, \quad V_{12} = \langle 46, 48, 75, 78, 90, 102 \rangle.$$

We give a comparative Table 1 for the largest degree g of syzygies and its lower bounds g_6, g_{6w}, g_{6w^2} and $\widetilde{g}_6$, calculated by formula (38).

For symmetric (not CI) semigroups W_6^2, presented in Table 1, the following inequalities hold:

$$g > g_{6w^2} > g_{6w} > g_6 > \widetilde{g}_6. \quad (51)$$

Table 1 The largest degree g of syzygies for symmetric (not CI) semigroups S_6 with different Betti's numbers β_1, β_2 and its lower bounds $g_6, g_{6w}, g_{6w^2}, \widetilde{g_6}$

	–				W property				W^2 property			
S_6	V_1	V_2	V_3	V_4	V_5	V_6	V_7	V_8	V_9	V_{10}	V_{11}	V_{12}
β_1	13	14	10	10	8	9	10	14	7	7	7	7
β_2	31	35	19	22	19	18	23	37	16	16	16	16
$\mathcal{B}_6$	9.5	11	5	6.5	6	5	7	12	5	5	5	5
g	84	77	125	126	256	292	302	638	387	603	598	816
g_{6w^2}	–	–	–	–	–	–	–	–	385.6	595.3	590.3	811.2
g_{6w}	–	–	–	–	240.4	271.2	286	609.2	359.8	554.8	548	746.9
g_6	55	49.6	88	86.5	173.3	196	196.6	395.4	274.4	420.8	426.6	586.6
$\widetilde{g_6}$	45.5	42	66.2	66.9	130.4	146	153.4	338	199.9	306.5	310.7	427.2

For the rest of symmetric (not CI) semigroups W_6 and S_6 the bounds g_{6w^2} and g_{6w} are skipped in inequalities (51) depending on the existence (or absence) of the W property in these semigroups. It is easy to verify that Betti's numbers of all semigroups from Table 1 satisfy the constraints (41).

Acknowledgements The research was partly supported by the Kamea Fellowship. The author is grateful to the referee for valuable suggestions to improve the article.

References

1. Brauer, A., Shockley, J.E.: On a problem of Frobenius. J. Reine Angew. Math. **211**, 215–220 (1962)
2. Bresinsky, H.: On prime ideals with generic zero $X_i = t^{n_i}$. Proc. Am. Math. Soc. **47**, 329–332 (1975)
3. Bresinsky, H.: Symmetric semigroups of integers generated by four elements. Manuscripta Math. **17**, 205–219 (1975)
4. Bresinsky, H.: Monomial Gorenstein ideals. Manuscripta Math. **29**, 159–181 (1979)
5. Fel, L.G.: Duality relation for the Hilbert series of almost symmetric numerical semigroups. Israel J. Math. **185**, 413–444 (2011); a detailed proof of Formula (10) is given in author's preprint: *Frobenius problem for semigroups* $S(d_1, d_2, d_3)$, sect. 7.2. https://arxiv.org/pdf/math/0409331.pdf
6. Fel, L.G.: On Frobenius numbers for symmetric (not complete intersection) semigroups generated by four elements. Semigroup Forum **93**, 423–426 (2016)
7. Fel, L.G.: Restricted partition functions and identities for degrees of syzygies in numerical semigroups. Ramanujan J. **43**, 465–491 (2017)
8. Fel, L.G.: Symmetric (not complete intersection) semigroups generated by five elements. Int. Electron. J. Comb. Number Theory **18**, # A44 (2018)
9. Hardy, G.H., Littlewood, J.E., Polya, G.: Inequalities. University Press, Cambridge (1959)
10. Killingbergtrø, H.G.: Betjening av figur i Frobenius' problem. Normat (Nordisk Matematisk Tidskrift) **2**, 75–82 (2000, in Norwegian)
11. Kunz, E.: The value-semigroup of a one-dimensional Gorenstein ring. Proc. Am. Math. Soc. **25**, 748–751 (1970)

12. Kunz, E.: Almost complete intersections are not Gorenstein rings. J. Algebra **28**, 111–115 (1974)
13. Stanley, R.P.: Log-concave and unimodal sequences in algebra, combinatorics, and geometry. In: Graph Theory and its Applications: East and West. Annals of the New York Academy of Sciences, vol. 576, pp. 500–535. Academy Sciences, New York (1989)
14. Watanabe, K.: Some examples of 1–dim Gorenstein domains. Nagoya Math. J. **49**, 101–109 (1973)

Syzygies of Numerical Semigroup Rings, a Survey Through Examples

Philippe Gimenez and Hema Srinivasan

Abstract This survey presents recent results on minimal free resolutions of numerical semigroup rings. We focus on two classes of numerical semigroups where the resolution is explicitly given: Gorenstein semigroups of embedding dimension 4 that are not a complete intersection and semigroups generated by a sequence of integers in arithmetic progression. Finally, we describe how the resolution is constructed when the semigroup is obtained by gluing of two numerical semigroups of smaller embedding dimension. Along the paper, we provide several non-trivial examples to illustrate our results.

Keywords Semigroup rings · Syzygies · Gorenstein rings · Arithmetic sequence · Gluing

1 Introduction

Given a sequence of positive integers $\mathbf{a} = (a_1, \ldots a_n)$ and an arbitrary field k, consider the ring homomorphism $\phi_{\mathbf{a}} : k[x_1, \ldots, x_n] \to k[t]$ defined by $\phi_{\mathbf{a}}(x_i) = t^{a_i}$. Then $I_{\mathbf{a}} := \ker \phi_{\mathbf{a}}$ is a prime binomial ideal of height $n-1$ in $R := k[x_1, \ldots, x_n]$ and it is weighted homogeneous with the weighting $\deg x_i = a_i$ on R. It is the defining ideal of the affine monomial curve $C_{\mathbf{a}} \subset \mathbb{A}_k^n$ parametrically defined by $\mathbf{a}$ and whose coordinate ring is $k[\mathbf{a}] := \operatorname{Im} \phi_{\mathbf{a}} = k[t^{a_1}, \ldots, t^{a_n}] \simeq R/I_{\mathbf{a}}$. As $k[\mathbf{a}]$ is isomorphic to $k[d\mathbf{a}]$ for any integer $d \geq 1$, we may assume without loss of generality that $a_1, \ldots, a_n$ are relatively prime. The ring $k[\mathbf{a}]$ is also known as the semigroup

---

P. Gimenez (✉)
Instituto de Matemáticas de la Universidad de Valladolid (IMUVA), Facultad de Ciencias, Valladolid, Spain
e-mail: pgimenez@agt.uva.es

H. Srinivasan
Mathematics Department, University of Missouri, Columbia, MO, USA
e-mail: SrinivasanH@missouri.edu

V. Barucci et al. (eds.), *Numerical Semigroups*, Springer INdAM Series 40, https://doi.org/10.1007/978-3-030-40822-0_8

ring of the numerical semigroup $\langle \mathbf{a} \rangle = \langle a_1, \ldots a_n \rangle \subset \mathbb{N}$ generated by $a_1, \ldots, a_n$, and it is always a Cohen-Macaulay ring. When the sequence $\mathbf{a}$ minimally generates the semigroup $\langle \mathbf{a} \rangle$, then the semigroup ring $k[\mathbf{a}]$ has embedding dimension n which is equivalent to the condition $I_{\mathbf{a}} \subseteq \langle x_1, \ldots, x_n \rangle^2$. We also say $\mathbf{a}$ is numerically independent if it is a minimal generating set for the semigroup $\langle \mathbf{a} \rangle$.

In this survey, we are interested in the minimal graded free resolution of $k[\mathbf{a}]$ as R-module with respect to the grading on R induced by $\deg x_i = a_i$, and in particular in determining the graded Betti numbers of $k[\mathbf{a}]$ and other invariants related to its syzygies like the Hilbert series, the Cohen-Macaulay type or the Castelnuovo-Mumford regularity. The complete intersection case is under control since the Kozsul complex gives a minimal graded free resolution in this case so one only needs to give a minimal generating set of $I_{\mathbf{a}}$. When the embedding dimension is ≤ 2, things are hence trivial and for $n = 3$, one can easily describe the syzygies of $k[\mathbf{a}]$ using the results of Herzog [10]. For $n \geq 4$, the number of elements in a minimal generating set of $I_{\mathbf{a}}$ is not bounded and the description of a minimal resolution in general becomes a hard problem. The first interesting case is thus when $n = 4$ and $k[\mathbf{a}]$ is Gorenstein and not a complete intersection. In this case, Bresinsky shows in [2] that $I_{\mathbf{a}}$ is minimally generated by 5 elements and he gives a complete description of a minimal generating set of $I_{\mathbf{a}}$. This information also encodes the whole minimal graded free resolution of $k[\mathbf{a}]$ in this case as we will recall in Proposition 1. The concept of principal matrix, that we recall in Sect. 2, is useful for understanding Gorenstein monomial curves of embedding dimension 4 that are not a complete intersection. Bresinsky's construction shows that when $k[\mathbf{a}]$ is Gorenstein and not a complete intersection, then $\mathbf{a}$ has a principal matrix that satisfies a property that we call pseudo-Gorenstein (Definition 2) but not any pseudo-Gorenstein matrix provides a Gorenstein sequence (Theorem 1). We will also recall from [9] how, given a Gorenstein monomial curve, one gets two families of Gorenstein monomial curves by translation (Theorem 2). Note that in [1, Theorem 6], one can find a minimal graded free resolution of $k[\mathbf{a}]$ in another interesting case of embedding dimension 4: when $\langle \mathbf{a} \rangle$ is pseudosymmetric, i.e., almost symmetric of type 2. The case of almost symmetric numerical semigroups of embedding dimension 4 and type t is treated in [11, Sec. 6].

In Sect. 3, we focus on the case of semigroup rings defined by a sequence in arithmetic progression. This is, to our knowledge, the biggest familly of numerical semigroup rings where the minimal graded free resolution is completely described in arbitrary embedding dimension, [8]. The construction of the sygygies and the computation of the graded Betti numbers are recalled in Theorem 3. Using this result, one can describe all the patterns of minimal graded free resolutions of semigroup rings defined by arithmetic sequences in low embedding dimension (Examples 5–7).

Finally, we will show in Sect. 4 how the syzygies behave when one glues two numerical semigroups. When $\mathbf{c} = k_1\mathbf{a} \sqcup k_2\mathbf{b}$ is a gluing, a minimal graded free resolution of $k[\mathbf{c}]$ can be obtained from minimal resolutions of $k[\mathbf{a}]$ and $k[\mathbf{b}]$ (Theorem 4) and formulas for all the numerical invariants can then easily be deduced (Corollary 2).

Along this survey, we illustrate results and phenomenons with many examples where we construct syzygies and compute invariants like Betti numbers and regularity. The computer algebra system SINGULAR [3] was very useful to check our computations and build non-trivial examples.

2 Principal Matrices and Gorenstein Sequences of Length 4

Given a sequence of relatively prime positive integers $\mathbf{a} = (a_1, \ldots, a_n)$, $n \geq 2$, there is an integer s such that $x > s \Longrightarrow x \in \langle \mathbf{a} \rangle$. The smallest integer with this property is called the *Frobenius number* of $\langle \mathbf{a} \rangle$ and we will denote it by $F(\mathbf{a})$. When $n = 2$, one has a formula for $F(\mathbf{a})$ but in higher embedding dimension, there is no such a simple formula; see [12].

Lemma 1 *If a_1 and a_2 are two relatively prime positive integers, then $F(a_1, a_2) = a_1a_2 - a_1 - a_2$, i.e., if $x \geq a_1a_2 - a_2 - a_1 + 1$ then $x \in \langle a_1, a_2 \rangle$, and $a_1a_2 - a_1 - a_2 \notin \langle a_1, a_2 \rangle$.*

For each i, $1 \leq i \leq n$, there exists a multiple of a_i that belongs to the numerical semigroup generated by the rest of the elements in the sequence and we denote by $r_i > 0$ the smallest positive integer such that $r_i a_i \in \langle a_1, \ldots, a_{i-1}, a_{i+1}, \ldots, a_n \rangle$. So we have that

$$\forall i,\ 1 \leq i \leq n,\ r_i a_i = \sum_{j \neq i} r_{ij} a_j,\ r_{ij} \geq 0,\ r_i > 0\,. \tag{1}$$

Definition 1 The $n \times n$ integer matrix $D(\mathbf{a}) := (r_{ij})$ where $r_{ii} := -r_i$ is called a *principal matrix* associated to $\mathbf{a}$.

Principal matrix $D(\mathbf{a})$ is not uniquely defined. Although the diagonal entries $-r_i$ are uniquely determined, there is not a unique choice for r_{ij} in general. We have the "map" $D : \mathbb{N}^{[n]} \to T_n^*$ from the set $\mathbb{N}^{[n]}$ of sequences of n relatively prime positive integers to the set T_n^* of $n \times n$ singular matrices with negative integers on the diagonal and non-negative integers outside the diagonal. When $D(\mathbf{a})$ has rank $n-1$, the maximum possible, we can recover $\mathbf{a}$ from $D(\mathbf{a})$ by factoring out the greatest common divisor of the n maximal minors of the $n-1 \times n$ submatrix of $D(\mathbf{a})$ obtained by removing the first row. In other words, call $D^{-1} : T_n^* \to N^{[n]}$ the operation that, for $M \in T_n^*$, takes the first column of $\mathrm{adj}(M)$ and then factors out the g.c.d. and removes the signs to get an element in $\mathbb{N}^{[n]}$. Then $D^{-1}(D(\mathbf{a})) = \mathbf{a}$ if $D(\mathbf{a})$ has rank $n-1$. Now given a matrix $M \in T_n^*$, $D(D^{-1}(M)) \neq M$ in general.

Example 1 Consider the matrix $M = \begin{pmatrix} -4 & 0 & 1 & 1 \\ 1 & -5 & 4 & 0 \\ 0 & 4 & -5 & 1 \\ 3 & 1 & 0 & -2 \end{pmatrix}$. It has rank 3 and $D^{-1}(M) = (7, 11, 12, 16)$ but $D(D^{-1}(M)) \neq M$. It is easy to check, for example, that for $\mathbf{a} = (7, 11, 12, 16)$, one has that $r_2 = 3 < 5$.

As observed in [4] where Delorme characterizes sequences $\mathbf{a}$ such that $k[\mathbf{a}]$ is a complete intersection, this fact does not depend on the field k by [10, Corollary 1.13]. On the other hand, it is well-known that $k[\mathbf{a}]$ is Gorenstein if and only if the numerical semigroup $\langle a_1, \dots a_n\rangle \subset \mathbb{N}$ is symmetric, which does not depend either on the field k. We will thus say that $\mathbf{a}$ is a complete intersection (respectively Gorenstein) if the semigroup ring $k[\mathbf{a}]$ is a complete intersection (respectively Gorenstein).

In his classical paper [2], Bresinsky gives a characterization of monomial curves in $\mathbb{A}_k^4$ that are Gorenstein but not a complete intersection. As shown in [9], principal matrices turn out to play an important role in this characterization as we will recall now.

Assume that $n = 4$. If $\mathbf{a}$ is Gorenstein but is not a complete intersection, by [2, Theorems 3 and 5], there is a principal matrix $D(\mathbf{a})$ that has the following form:

$$D(\mathbf{a}) = \begin{pmatrix} -c_1 & 0 & d_{13} & d_{14} \\ d_{21} & -c_2 & 0 & d_{24} \\ d_{31} & d_{32} & -c_3 & 0 \\ 0 & d_{42} & d_{43} & -c_4 \end{pmatrix} \tag{2}$$

with $c_i \geq 2$ and $d_{ij} > 0$ for all $1 \leq i, j \leq 4$, the columns summing to zero and all the columns of the adjoint being relatively prime. The first column of the adjoint of this matrix is $-\mathbf{a}^T$ and Bresinsky's characterization also says that the first column of the adjoint of every such matrix defines a Gorenstein monomial curve (after removing the signs) provided they are relatively prime. In fact, we can do this with any column of the matrix $\mathrm{adj}(D(\mathbf{a}))$ in (2), not only the first one, that all give $\mathbf{a}$ after factoring out the g.c.d.

Definition 2 We say that a 4×4 matrix with integer entries $A = (a_{ij})$ is *pseudo-Gorenstein* if

1. the columns add up to zero;
2. entries on the diagonal are all negative;
3. the other entries are all non-negative;
4. there are exactly 4 entries that are zero: $a_{12} = a_{23} = a_{34} = a_{41} = 0$.

Remark 1 Any pseudo-Gorenstein matrix A will be of rank 3 so $\operatorname{adj}(A)$ has rank 1, i.e., its columns are all equal up to a multiple. Moreover, since the columns of A add up to zero, the 4 columns of $\operatorname{adj}(A)$ are the same. This means that $\operatorname{adj}(A) = \mathbf{a}^T \times [-1 - 1 - 1 - 1]$ for some $\mathbf{a} = (a_1, a_2, a_3, a_4)$.

Thus, by Bresinsky, any sequence in $\mathbb{N}^4$ that is Gorenstein but not a complete intersection has a principal matrix which is pseudo-Gorenstein with the four entries in the first column of the adjoint being relatively prime. We can prove the following strengthening of this criterion.

Theorem 1 *If A is a pseudo-Gorenstein 4×4 matrix, then the first column of the adjoint of A (after removing the signs) defines a Gorenstein monomial curve provided these entries are relatively prime.*

Proof Consider such a matrix A and let a_1, a_2, a_3, a_4 be the entries in the first column of the adjoint of A (after removing the signs). Since we are assuming that they are relatively prime, there exist integers $\lambda_1, \dots, \lambda_4$ such that $\lambda_1 a_1 + \cdots + \lambda_4 a_4 = 1$.

It suffices to show that the four relations in the rows of A are principal relations. We will show this for the first row and the other rows are similar. Suppose that $b_{11}a_1 = b_{12}a_2 + b_{13}a_3 + b_{14}a_4$ is a relation with $b_{11} \geq 2$ and $b_{12}, b_{13}, b_{14} \geq 0$ and let's show that $b_{11} \geq c_1$.

Since the system $\begin{bmatrix} -b_{11} & b_{12} & b_{13} & b_{14} \\ d_{21} & -c_2 & 0 & d_{24} \\ d_{31} & d_{32} & -c_3 & 0 \\ 0 & d_{42} & d_{43} & -c_4 \end{bmatrix} Y = 0$ has a non-trivial solution, namely $Y = (a_1, a_2, a_3, a_4)^T$, we see that it has determinant zero. So there exist x_i such that

$$(1, x_2, x_3, x_4) \begin{bmatrix} -b_{11} & b_{12} & b_{13} & b_{14} \\ d_{21} & -c_2 & 0 & d_{24} \\ d_{31} & d_{32} & -c_3 & 0 \\ 0 & d_{42} & d_{43} & -c_4 \end{bmatrix} = 0 . \tag{3}$$

Consider the matrix $T_4 = \begin{bmatrix} -b_{11} & b_{12} & b_{13} & b_{14} \\ d_{21} & -c_2 & 0 & d_{24} \\ d_{31} & d_{32} & -c_3 & 0 \\ \lambda_1 & \lambda_2 & \lambda_3 & \lambda_4 \end{bmatrix}$. If the determinant of T_4 is $-t_4$, then the last column of its adjoint is $-t_4(a_1, a_2, a_3, a_4)^T$. This is because $T_4(a_1, a_2, a_3, a_4)^T = (0, 0, 0, 1)^T$. Hence, looking at the element in the last row and last column of the adjoint of T_4, one gets using (3) that

$$-t_4 a_4 = \begin{vmatrix} -b_{11} & b_{12} & b_{13} \\ d_{21} & -c_2 & 0 \\ d_{31} & d_{32} & -c_3 \end{vmatrix} = \begin{vmatrix} 0 & -x_4 d_{42} & -x_4 d_{43} \\ d_{21} & -c_2 & 0 \\ d_{31} & d_{32} & -c_3 \end{vmatrix} = -x_4 a_4 .$$

Hence $t_4 = x_4$, and since t_4 is an integer, so is x_4. Now, looking at the element in the last column and first row of the adjoint of T_4, one has

$$t_4 a_1 = \begin{vmatrix} b_{12} & b_{13} & b_{14} \\ -c_2 & 0 & d_{24} \\ d_{32} & -c_3 & 0 \end{vmatrix} = b_{12}c_3d_{24} + b_{13}d_{32}d_{24} + b_{14}c_2c_3 > 0\,.$$

So, $x_4 = t_4$ is now a positive integer.

Consider the matrix $T_2 = \begin{bmatrix} -b_{11} & b_{12} & b_{13} & b_{14} \\ d_{31} & d_{32} & -c_3 & 0 \\ 0 & d_{42} & d_{43} & -c_4 \\ \lambda_1 & \lambda_2 & \lambda_3 & \lambda_4 \end{bmatrix}$ which determinant is denoted by $-t_2$. By similar calculations, we see that $x_2 = t_2$ is an integer and, focusing on the element in the last column and third row of the adjoint of T_2, one gets that

$$t_2 a_3 = \begin{vmatrix} -b_{11} & b_{12} & b_{14} \\ d_{31} & d_{32} & 0 \\ 0 & d_{42} & -c_4 \end{vmatrix} = b_{11}c_4d_{32} + b_{12}d_{31}c_4 + b_{14}d_{31}d_{42} > 0$$

so $x_2 = t_2$ is also a positive integer.

Similarly, using the matrix $T_3 = \begin{bmatrix} -b_{11} & b_{12} & b_{13} & b_{14} \\ d_{21} & -c_2 & 0 & d_{24} \\ 0 & d_{42} & d_{43} & -c_4 \\ \lambda_1 & \lambda_2 & \lambda_3 & \lambda_4 \end{bmatrix}$ of determinant $-t_3$, one gets that $x_3 = -t_3$ and hence x_3 is an integer. However, by calculating the entry in the last column and second row of the adjoint of T_3, one gets that

$$(-t_3)a_2 = \begin{vmatrix} -b_{11} & b_{13} & b_{14} \\ d_{21} & 0 & d_{24} \\ 0 & d_{43} & -c_4 \end{vmatrix} = b_{11}d_{43}d_{24} + b_{13}d_{21}c_4 + b_{14}d_{21}d_{43} > 0,$$

and hence, x_3 is again a positive integer.

So, $b_{11} = x_2d_{21} + x_3d_{31} \geq d_{21} + d_{31} = c_1$ as desired.

Since we can make any of the c_i's the first row, by rearranging the a_i's suitably, this proves that all of the rows are principal relations and this is a principal matrix. Hence $\mathbf{a} = (a_1, a_2, a_3, a_4)$ is Gorenstein by Bresinsky's criterion. □

Example 2 The matrix $A = \begin{pmatrix} -3 & 0 & 5 & 2 \\ 2 & -5 & 0 & 3 \\ 1 & 1 & -6 & 0 \\ 0 & 4 & 1 & -5 \end{pmatrix}$ is pseudo-Gorenstein and the first column of its adjoint is $(-75, -63, -23, -55)^T$. The sequence $\mathbf{a} = (75, 63, 23, 55)$ is, by our criterion, Gorenstein and not a complete intersection and A is a principal matrix for $\mathbf{a}$, i.e., $D(D^{-1}(A)) = A$.

However not all pseudo-Gorenstein matrices provide a Gorenstein sequence through this process.

Example 3 Consider the matrix $A = \begin{pmatrix} -4 & 0 & 2 & 1 \\ 3 & -7 & 0 & 4 \\ 1 & 5 & -5 & 0 \\ 0 & 2 & 3 & -5 \end{pmatrix}$. It is pseudo-Gorenstein. But the columns of adj (A) do not have relatively prime entries: all the columns of the adjoint are $-3\mathbf{a}$ for $\mathbf{a} = (25, 29, 34, 32)$. Hence $\mathbf{a}$ cannot be Gorenstein. In fact, the principal matrix of $(25, 29, 34, 32)$ can be seen to be $D(\mathbf{a}) = \begin{pmatrix} -4 & 0 & 2 & 1 \\ 2 & -4 & 1 & 1 \\ 3 & 1 & -4 & 1 \\ 0 & 2 & 3 & -5 \end{pmatrix}$ which is not pseudo-Gorenstein! One has that $I_{\mathbf{a}} \subset R = k[x_1, \ldots, x_4]$ is minimally generated by 7 elements, the semigroup ring $k[\mathbf{a}] \simeq R/I_{\mathbf{a}}$ is not Gorenstein, and its minimal free resolutions have the following shape:

$$0 \to R^4 \to R^{10} \to R^7 \to R \to k[\mathbf{a}] \to 0\,.$$

The following result [9, Thm. 4] gives two families of Gorenstein monomial curves in $\mathbb{A}_k^4$ by translation from a given Gorenstein curve.

Theorem 2 *Given any Gorenstein non-complete intersection sequence* $\mathbf{a}$ *of length* 4 *with principal pseudo-Gorenstein matrix* $D(\mathbf{a})$*, there exist two vectors* $\mathbf{u}$ *and* $\mathbf{v}$ *in* $\mathbb{N}^4$ *such that, for all* $t \geq 0$*,* $\mathbf{a} + t\mathbf{u}$ *and* $\mathbf{a} + t\mathbf{v}$ *are also Gorenstein non-complete intersection whenever the entries of the corresponding sequence* ($\mathbf{a} + t\mathbf{u}$ *for the first* *family,* $\mathbf{a}+t\mathbf{v}$ *for the second) are relatively prime. When this occurs, their principal matrices are*

$$D(\mathbf{a}+t\mathbf{u}) = D(\mathbf{a}) + t\begin{pmatrix} -1 & 0 & 1 & 0 \\ 0 & 0 & 0 & 0 \\ 1 & 0 & -1 & 0 \\ 0 & 0 & 0 & 0 \end{pmatrix} \text{ and } D(\mathbf{a}+t\mathbf{v}) = D(\mathbf{a}) + t\begin{pmatrix} 0 & 0 & 0 & 0 \\ 0 & -1 & 0 & 1 \\ 0 & 0 & 0 & 0 \\ 0 & 1 & 0 & -1 \end{pmatrix}.$$

Example 4 Let $\mathbf{a} = (34, 33, 42, 64)$. This is a Gorenstein non-complete intersection sequence with the principal matrix

$$D(\mathbf{a}) = \begin{pmatrix} -5 & 0 & 1 & 2 \\ 2 & -4 & 0 & 1 \\ 3 & 2 & -4 & 0 \\ 0 & 2 & 3 & -3 \end{pmatrix}.$$

Here, the vector $\mathbf{u}$ defined in [9] is $\mathbf{u} = (10, 9, 10, 16)$. So, if one sets $\mathbf{a}_t := \mathbf{a} + t\mathbf{u}$, it is not a relatively prime sequence if $t = 1$ but it is relatively prime sequence for $t = 2$. Denoting by A_t the principal matrix of $\mathbf{a}_t$, one has that $A_1 = \begin{pmatrix} -6 & 0 & 2 & 2 \\ 2 & -4 & 0 & 1 \\ 4 & 2 & -5 & 0 \\ 0 & 2 & 3 & -3 \end{pmatrix}$ and $A_2 = \begin{pmatrix} -7 & 0 & 3 & 2 \\ 2 & -4 & 0 & 1 \\ 5 & 2 & -6 & 0 \\ 0 & 2 & 3 & -3 \end{pmatrix}$. It is clear from the first row of A_1, that the first column of its adjoint is not relatively prime and hence it is not the principal matrix of a Gorenstein not complete intersection. However, for A_2, the sequence $\mathbf{a} + 2\mathbf{u} = (54, 51, 62, 96)$ is Gorenstein and not complete intersection for it is relatively prime.

At the end of [9], we also show that when $\mathbf{a}$ is Gorenstein and not a complete intersection, the principal matrix (2) encodes the resolution of $k[\mathbf{a}]$. This result was also obtained, independently, in [1].

Proposition 1 *If* $\mathbf{a}$ *is Gorenstein and not a complete intersection, then a minimal graded free resolution of* $k[\mathbf{a}]$ *is*

$$0 \to R \xrightarrow{\delta_3} R^5 \xrightarrow{\phi} R^5 \xrightarrow{\delta_1} R \to k[\mathbf{c}] \to 0$$

with $\phi = \begin{pmatrix} 0 & 0 & x_2^{d_{32}} & x_3^{d_{43}} & x_4^{d_{24}} \\ 0 & 0 & x_1^{d_{21}} & x_4^{d_{14}} & x_2^{d_{42}} \\ -x_2^{d_{32}} & -x_1^{d_{21}} & 0 & 0 & x_3^{d_{13}} \\ -x_3^{d_{43}} & -x_4^{d_{14}} & 0 & 0 & x_1^{d_{31}} \\ -x_4^{d_{24}} & -x_2^{d_{42}} & -x_3^{d_{13}} & -x_1^{d_{31}} & 0 \end{pmatrix}$ *and* $\delta_3 = (\delta_1)^T = \begin{pmatrix} x_1^{c_1} - x_3^{d_{13}} x_4^{d_{14}} \\ x_3^{c_3} - x_1^{d_{31}} x_2^{d_{32}} \\ x_4^{c_4} - x_2^{d_{42}} x_3^{d_{43}} \\ x_2^{c_2} - x_1^{d_{21}} x_4^{d_{24}} \\ x_1^{d_{21}} x_3^{d_{43}} - x_2^{d_{32}} x_4^{d_{14}} \end{pmatrix}.$

Fact
If A is a pseudo-Gorenstein matrix, then there is always a Gorenstein ideal associated to it: if $A = \begin{pmatrix} -c_1 & 0 & d_{13} & d_{14} \\ d_{21} & -c_2 & 0 & d_{24} \\ d_{31} & d_{32} & -c_3 & 0 \\ 0 & d_{42} & d_{43} & -c_4 \end{pmatrix}$ is as in Definition 2, consider the ideal $I(A) = \langle x_1^{c_1} - x_3^{d_{13}} x_4^{d_{14}}, x_2^{c_2} - x_1^{d_{21}} x_4^{d_{24}}, x_3^{c_3} - x_1^{d_{31}} x_2^{d_{32}}, x_4^{c_4} - x_2^{d_{42}} x_3^{d_{43}}, x_1^{d_{21}} x_3^{d_{43}} - x_2^{d_{32}} x_4^{d_{14}} \rangle$ and the sequence $\mathbf{a} = (a_1, a_2, a_3, a_4)$ given in Remark 1. Then $I(A)$ is homogeneous if we give weight a_i to x_i in $R = k[x_1, \ldots, x_4]$ and if $I(A)$ has height 3, then the resolution given in Proposition 1 is a minimal graded free resolution of $R/I(A)$ so $I(A)$ is Gorenstein. The ideal $I(A)$ is the pfaffian ideal of the 4×4 minors of the skew symmetric matrix ϕ. The thing is that the ideal $I(A)$ might not coincide with $I_{\mathbf{a}}$, one can just say that it is contained in $I_{\mathbf{a}}$. If it is equal to $I_{\mathbf{a}}$ then, of course, $I_{\mathbf{a}}$ will be Gorenstein and $A = D(\mathbf{a})$.

Remark 2 Fröberg proved in [5] that the Cohen-Macaulay type of $k[\mathbf{a}]$ coincides with the number of elements $\ell \in \mathbb{N}$ that are not in the semigroup $\langle \mathbf{a} \rangle$ and such that $\ell + s \in \langle \mathbf{a} \rangle$ for all $s \in \langle \mathbf{a} \rangle$; see also [14, Thm 10.2.10]. Of course $F(\mathbf{a})$, the Frobenious number of $\langle \mathbf{a} \rangle$ defined before Lemma 1, is always such an element so $k[\mathbf{a}]$ is Gorenstein if and only if it is the only one.

Question
When $\mathbf{a}$ is not Gorenstein, can one tell the Cohen-Macaulay type of $k[\mathbf{a}]$ from the principal matrix $D(\mathbf{a})$ or its adjoint when it is of maximal rank?

Question
Can we characterize the Goresntein sequences of length $n \geq 5$ by their principal matrices?

3 Arithmetic Sequences

In this section, we will consider *arithmetic sequences* of length $n \geq 3$, i.e., sequences of integers of the form $\mathbf{a} = (a, a + d, \ldots, a + (n - 1)d)$ for some $a, d \in \mathbb{N}$. We will assume that the elements in the sequence are relatively prime, i.e., $\gcd(a, d) = 1$, and that they minimally generate the semigroup $\langle \mathbf{a} \rangle$.

In [8], we construct a minimal graded free resolution for $k[\mathbf{a}]$ when $\mathbf{a}$ is an arithmetic sequence and derive formulae for various invariants. These form an important class of semigroup rings and, as of now, it is the only big class of semigroup rings, besides complete intersections, for which we have a minimal resolution in all embedding dimensions. In Sect. 4, we will define the concept of gluing (or decomposition) and semigroup rings defined by arithmetic sequences of length at least 4 are significant because they are not decomposable or equivalently cannot be obtained by gluing two semigroup rings of smaller embedding dimensions as we will recall in Remark 4.

Let us recall here the construction of a minimal graded free resolution of semigroup rings defined by an arithmetic sequence. The main preliminary result is a description of the ideal $I_{\mathbf{a}}$ as the sum of two determinantal ideals. Write

$$a = q(n-1) + r$$

for q, r positive integers and $r \in [1, n-1]$. As observed in [7], $q \geq 1$ because we have assumed that $\mathbf{a}$ minimally generates the semigroup $\langle \mathbf{a} \rangle$. Now consider the following two matrices,

$$A = \begin{pmatrix} x_1 & \cdots & x_{n-1} \\ x_2 & \cdots & x_n \end{pmatrix} \quad \text{and} \quad B = \begin{pmatrix} x_n^q & x_1 & \cdots & x_{n-r} \\ x_1^{q+d} & x_{r+1} & \cdots & x_n \end{pmatrix}.$$

By [7, Thm. 1.1], $I_{\mathbf{a}}$ is the sum of the two determinantal ideals on maximal minors of A and B, i.e., $I_{\mathbf{a}} = I_2(A) + I_2(B)$. Using the minimal resolution of $R/I_2(A)$ given by the Eagon-Northcott complex and then an iterated mapping cone construction that we make minimal at each step, we construct a minimal graded free resolution of $k[\mathbf{a}]$ that we recall in Theorem 3; see [8, Thms. 3.10 and 4.1].

Notations
Given two integers $m \geq t \geq 1$, denote by $\sigma(m, t)$ the collection (with repetitions) of all possible sums of t distinct non-negative integers which are all strictly smaller than m, i.e., $\sigma(m, t) = \{\sum_{0 \leq r_1 < \cdots < r_t \leq m-1} r_i\}$. Note that $\#\sigma(m, t) = \binom{m}{t}$.

Theorem 3 *Let* $\mathbf{a} = (a, a+d, \ldots, a+(n-1)d)$ *be an arithmetic sequence of length* $n \geq 3$ *with* $\gcd(a, d) = 1$, *and write* $a = q(n-1) + r$ *for* q, r *two positive integers with* $r \in [1, n-1]$. *If* $R = k[x_1, \ldots, x_n]$ *is graded according to* $\deg(x_i) = a + (i-1)d$, *the minimal graded free resolution of* $k[\mathbf{a}]$ *as an* R*-module is*

$$0 \to F_{n-1} \longrightarrow E_{n-2} \oplus F_{n-2} \longrightarrow \cdots \longrightarrow E_1 \oplus F_1 \longrightarrow R \longrightarrow k[\mathbf{a}] \to 0$$

where, for all $s \in [2, n-1]$, $E_{s-1} = \bigoplus_{i=1}^{s-1} \left(\bigoplus_{j \in \sigma(n-1,s)} R(-(sa+id+jd)) \right)$, *and*

$$F_1 = \left(\bigoplus_{i=0}^{n-1-r} R(-[a(q+d+1)+id]) \right)$$

$$F_2 = \left(\bigoplus_{i=1}^{n-1-r} \left(\bigoplus_{j=0}^{n-2} R(-[(a(q+d+2)+id+jd]) \right) \right)$$

$$F_s = \begin{cases} \left(\bigoplus_{i=s-1}^{n-r-1} \left(\bigoplus_{j \in \sigma(n-1,s-1)} R(-[a(q+d+s)+id+jd]) \right) \right) & \text{if } s \in [3, n-r], \\ \left(\bigoplus_{i=n-r}^{s-1} \left(\bigoplus_{j \in \sigma(n-1,s)} R(-[a(q+d+s+1)+id+jd]) \right) \right) & \text{if } s \in [n-r+1, n-1]. \end{cases}$$

In particular, for $n \geq 3$ *fixed, the Betti numbers of the semigroup ring associated to an arithmetic sequence* $\mathbf{a} = (a, a+d, \ldots, a+(n-1)d)$ *of length* n*, only depend on the value of* a *modulo* $n-1$*, and they are given by the following formula:*

$$\beta_j = j\binom{n-1}{j+1} + \begin{cases} (n-r+1-j)\binom{n-1}{j-1} & \text{if } 1 \leq j \leq n-r \\ (j-n+r)\binom{n-1}{j} & \text{if } n-r < j \leq n-1 \end{cases}.$$

Corollary 1 *If* $\mathbf{a} = (a, a+d, \ldots, a+(n-1)d)$ *is an arithmetic sequence of length* $n \geq 3$ *with* $\gcd(a, d) = 1$ *and* q, r *are the integers defined in Theorem 3, the value of the Castelnuovo-Mumford regularity of* $k[\mathbf{a}]$ *is*

$$reg(k[\mathbf{a}]) = \begin{cases} \frac{(n-2)(n+1)}{2}d + a(q+d) + (n-1)(a-1) & \text{if } r = 1, \\ \frac{(n-2)(n+1)}{2}d + a(q+d+1) + (n-1)(a-1) & \text{otherwise.} \end{cases}$$

Example 5 For $n = 3$, given an arithmetic sequence $\mathbf{a} = (a, a+d, a+2d)$ with $\gcd(a, d)) = 1$, one has that:

- if a is even (and d is odd and relatively prime to $a/2$), then the minimal graded free resolution of $k[\mathbf{a}]$ is $0 \to R \to R^2 \to R \to k[\mathbf{a}] \to 0$. In this case, $k[\mathbf{a}]$ is a complete intersection;
- if a is odd (and $\gcd(a, d) = 1$), then the minimal graded free resolution of $k[\mathbf{a}]$ is $0 \to R^2 \to R^3 \to R \to k[\mathbf{a}] \to 0$. In this case, $k[\mathbf{a}]$ is Hilbert-Burch.

(continued)

Example 5 (continued)
Using [10, Thms. 3.7 and 3.8], one easily gets that an arbitrary sequence of length 3 always defines a semigroup ring that is either a complete intersection or Hilbert-Burch.

Example 6 For $n = 4$, the minimal graded free resolution of the semigroup ring $k[\mathbf{a}]$ defined by an arithmetic sequence $\mathbf{a} = (a, a+d, a+2d, a+3d)$ with $\gcd(a, d) = 1$ is of one of the following types:

$$\begin{array}{l} 0 \to R^3 \longrightarrow R^8 \longrightarrow R^6 \longrightarrow R \longrightarrow k[\mathbf{a}] \to 0 \text{ if } a \equiv 1 \bmod 3 \\ 0 \to R \longrightarrow R^5 \longrightarrow R^5 \longrightarrow R \longrightarrow k[\mathbf{a}] \to 0 \text{ if } a \equiv 2 \bmod 3, \\ 0 \to R^2 \longrightarrow R^5 \longrightarrow R^4 \longrightarrow R \longrightarrow k[\mathbf{a}] \to 0 \text{ if } a \equiv 0 \bmod 3. \end{array}$$

Note that for an arbitrary sequence $\mathbf{a}$ of length 4, it is well-known that the number of minimal generators of $I_{\mathbf{a}}$ is not bounded while, when the sequence is arithmetic, it can only be 4, 5 or 6 (in particular it is never a complete intersection).

Example 7 For $n = 5$, the semigroup ring $k[\mathbf{a}]$ associated to an arithmetic sequence $\mathbf{a} = (a, a+d, a+2d, a+3d, a+4d)$ with $\gcd(a, d) = 1$ has a minimal graded free resolution of one of the following forms:

$$\begin{array}{l} 0 \to R^4 \longrightarrow R^{15} \longrightarrow R^{20} \longrightarrow R^{10} \longrightarrow R \longrightarrow k[\mathbf{a}] \to 0 \text{ if } a \equiv 1 \bmod 4, \\ 0 \to R \longrightarrow R^9 \longrightarrow R^{16} \longrightarrow R^9 \longrightarrow R \longrightarrow k[\mathbf{a}] \to 0 \text{ if } a \equiv 2 \bmod 4, \\ 0 \to R^2 \longrightarrow R^7 \longrightarrow R^{12} \longrightarrow R^8 \longrightarrow R \longrightarrow k[\mathbf{a}] \to 0 \text{ if } a \equiv 3 \bmod 4, \\ 0 \to R^3 \longrightarrow R^{11} \longrightarrow R^{14} \longrightarrow R^7 \longrightarrow R \longrightarrow k[\mathbf{a}] \to 0 \text{ if } a \equiv 0 \bmod 4. \end{array}$$

Remark 3 In particular, an arithmetic sequence $\mathbf{a} = (a, a+d, \ldots, a+(n-1)d)$ with $\gcd(a, d) = 1$ is a complete intersection if and only if $n = 3$ and a is even (and d is odd and relatively prime to $a/2$).

Example 8 The sequence $\mathbf{a} = (4, 11, 18, 25)$ defines a semigroup ring whose minimal graded free resolution is of the first type in Example 6 while the sequences $\mathbf{b} = (7, 12, 17, 22, 27)$ and $\mathbf{d} = (11, 13, 15, 17, 19)$ both define

(continued)

Example 8 (continued)
semigroup rings whose minimal graded free resolution is of the third type in Example 7. Applying Corollary 1, one gets that $\mathrm{reg}(k[\mathbf{a}]) = 76$, $\mathrm{reg}(k[\mathbf{b}]) = 118$ and $\mathrm{reg}(k[\mathbf{d}]) = 113$.

4 Gluing

Let's recall here the concept of gluing introduced by Rosales in [13]. We say that a sequence of relatively prime integers $\mathbf{c} = (c_1, \ldots, c_n)$ is a *gluing* of two relatively prime sequences $\mathbf{a}$ and $\mathbf{b}$ if the set $\mathbf{c}$ splits into two disjoint parts, $\mathbf{c} = k_1\mathbf{a} \cup k_2\mathbf{b}$, with k_1 and k_2 relatively prime and such that $k_1 \in \langle \mathbf{b} \rangle \setminus \mathbf{b}$ and $k_2 \in \langle \mathbf{a} \rangle \setminus \mathbf{a}$. When this occurs, we also say that $\mathbf{c}$ is *decomposable*, or that $\mathbf{c}$ is a *gluing* of $\mathbf{a}$ and $\mathbf{b}$, and we denote $\mathbf{c} = k_1\mathbf{a} \sqcup k_2\mathbf{b}$.

In [6], we construct a minimal graded free resolution of the semigroup ring $k[\mathbf{c}]$ in terms of that of $k[\mathbf{a}]$ and $k[\mathbf{b}]$ when $\mathbf{c}$ is a gluing of $\mathbf{a}$ and $\mathbf{b}$. We first recall some well-known facts on gluing.

Notations
If $\mathbf{c} = k_1\mathbf{a} \sqcup k_2\mathbf{b}$ is a gluing:

- The number of elements in $\mathbf{a}$ and $\mathbf{b}$, will be denoted by p and q respectively: $\mathbf{a} = (a_1, \ldots a_p)$, $\mathbf{b} = (b_1, \ldots, b_q)$, so that $\mathbf{c} = (k_1a_1, \ldots, k_1a_p, k_2b_1, \ldots, k_2b_q)$.
- Set $R_{\mathbf{a}} = k[x_1, \ldots, x_p]$, $R_{\mathbf{b}} = k[y_1, \ldots, y_q]$ and $R = R_{\mathbf{c}} = k[x_1, \ldots, x_p, y_1, \ldots, y_q]$ and consider on those rings the grading induced giving weights to the variables according to the associated sequence. Then, $k[\mathbf{a}] \simeq R_{\mathbf{a}}/I_{\mathbf{a}}$, $k[\mathbf{b}] \simeq R_{\mathbf{b}}/I_{\mathbf{b}}$ and $k[\mathbf{c}] \simeq R/I_{\mathbf{c}}$ are graded modules on $R_{\mathbf{a}}$, $R_{\mathbf{b}}$ and R respectively.

Lemma 2 *Let* $\mathbf{c}$ *be a gluing of* $\mathbf{a}$ *and* $\mathbf{b}$, $\mathbf{c} = k_1\mathbf{a} \sqcup k_2\mathbf{b}$.

1. *If* $\mathbf{a}$ *and* $\mathbf{b}$ *are numerically independent then so is* $\mathbf{c}$.
2. *Since* $k_1 \in \langle \mathbf{b} \rangle$ *and* $k_2 \in \langle \mathbf{a} \rangle$, *there exist non-negative integers* α_i, β_i *such that* $k_1 = \sum_{j=1}^{q} \beta_j b_j$ *and* $k_2 = \sum_{i=1}^{p} \alpha_i a_i$. *Then, the ideal* $I_{\mathbf{c}}$ *is minimally generated by the union of minimal generating sets of* $I_{\mathbf{a}}$ *and* $I_{\mathbf{b}}$, *and exactly one extra generator,*

$$\rho = \prod_{i=1}^{p} x_i^{\alpha_i} - \prod_{j=1}^{q} y_j^{\beta_j} \in R\,.$$

3. ρ is homogeneous of degree k_1k_2.

Now we are ready to state the theorem on resolution; see [6, Thm. 3.1].

Theorem 4 *Suppose that $\mathbf{c} = k_1\mathbf{a} \sqcup k_2\mathbf{b}$ and let F_A and F_B be minimal resolutions of $k[\mathbf{a}]$ and $k[\mathbf{b}]$ respectively. A minimal graded free resolution of the semigroup ring $k[\mathbf{c}]$ is obtained as the mapping cone of the map of complexes $\rho : F_A \otimes F_B \to F_A \otimes F_B$, where ρ is induced by multiplication by ρ. In particular, $(I_\mathbf{a}R + I_\mathbf{b}R :_R \rho) = I_\mathbf{a}R + I_\mathbf{b}R$.*

Example 9 Consider the decomposable sequence

$$\mathbf{c} = (76, 209, 342, 475, 182, 312, 442, 572, 702)$$

where $\mathbf{c} = k_1\mathbf{a} \sqcup k_2\mathbf{b}$ for $\mathbf{a} = (4, 11, 18, 25)$, $\mathbf{b} = (7, 12, 17, 22, 27)$, $k_1 = 19 = 7 + 12 \in \langle\mathbf{b}\rangle \setminus \mathbf{b}$ and $k_2 = 26 = 2 \cdot 4 + 18 \in \langle\mathbf{a}\rangle \setminus \mathbf{a}$. Set $R_\mathbf{a} = k[x_1, \ldots, x_4]$, $R_\mathbf{b} = k[y_1, \ldots, y_5]$ and $R = k[x_1, \ldots, y_5]$.

Both $\mathbf{a}$ and $\mathbf{b}$ are arithmetic sequences. They appeared in Example 8 and the minimal graded free resolutions of $k[\mathbf{a}]$ and $k[\mathbf{b}]$ are

$$0 \to R^3 \to R^8 \to R^6 \to R \to k[\mathbf{a}] \to 0\,,$$

$$0 \to R^2 \to R^7 \to R^{12} \to R^8 \to R \to k[\mathbf{b}] \to 0\,.$$

The tensor product of these two resolutions provides a minimal graded free resolution of R/J (as R-module) where $J = I_\mathbf{a}R + I_\mathbf{b}R$:

$$0 \to R^6 \to R^{37} \to R^{104} \to R^{164} \to R^{146} \to R^{68} \to R^{14} \to R \to R/J \to 0\,.$$

Note that the differentials can be easily written down if needed.

Finally, the extra minimal generator in $I_\mathbf{c}$ is $\rho = x_1^2x_3 - y_1y_2$ and the mapping cone induced by multiplication by ρ gives a minimal resolution of $k[\mathbf{c}]$ (as R-module):

$$0 \to R^6 \to R^{43} \to R^{141} \to R^{268} \to R^{310} \to R^{214} \to R^{82} \to R^{15} \to R \to k[\mathbf{c}] \to 0\,.$$

Again, the differentials are easily given by the mapping cone construction.

We obtain many corollaries from the theorem, some are new and some are recovering results already obtained by different methods that did not require the construction of the minimal resolution. Using the explicit resolution, we can read off many invariants, such as Hilbert function, Cohen-Macaulay type, Betti numbers (global or graded), etc.

Corollary 2 *Assume that* $\mathbf{c} = k_1\mathbf{a} \sqcup k_2\mathbf{b}$ *is a gluing.*

1. *The Betti numbers of* $k[\mathbf{c}]$ *are given by the following two formulas:*

$$\forall i \geq 0, \quad \beta_i(k[\mathbf{c}]) = \sum_{i'=0}^{i} \beta_{i'}(k[\mathbf{a}])[\beta_{i-i'}(k[\mathbf{b}]) + \beta_{i-i'-1}(k[\mathbf{b}])]$$

$$= \sum_{i'=0}^{i} \beta_{i'}(k[\mathbf{b}])[\beta_{i-i'}(k[\mathbf{a}]) + \beta_{i-i'-1}(k[\mathbf{a}])] .$$

2. *The Cohen Macaulay type is given by Type*$(k[\mathbf{c}])$ = *Type*$(k[\mathbf{a}]) \cdot$ *Type*$(k[\mathbf{b}])$.
3. $k[\mathbf{c}]$ *is Gorenstein, respectively a complete intersection, if and only if* $k[\mathbf{a}]$ *and* $k[\mathbf{b}]$ *are both Gorenstein, respectively complete intersections.*
4. *If neither* $k[\mathbf{a}]$ *nor* $k[\mathbf{b}]$ *is Gorenstein, then the Cohen-Macaulay type of* $k[C]$ *is not prime.*
5. *The graded Betti numbers of* $k[\mathbf{c}]$ *are given by the following two formulas*

$$\beta_{i,j}(k[\mathbf{c}]) = \sum_{i'=0}^{i} \left(\sum_{r,s/k_1r+k_2s=j} \beta_{i'r}(k[\mathbf{a}])[\beta_{i-i',s}(k[\mathbf{b}]) + \beta_{i-i'-1,s-k_1}(k[\mathbf{b}])] \right)$$

$$= \sum_{i'=0}^{i} \left(\sum_{r,s/k_1r+k_2s=j} \beta_{i'r}(k[\mathbf{b}])[\beta_{i-i',s}(k[\mathbf{a}]) + \beta_{i-i'-1,s-k_1}(k[\mathbf{a}])] \right) .$$

6. *The Castelnuovo-Mumford regularity of* $k[\mathbf{c}]$ *can be seen as*

$$reg(k[\mathbf{c}]) = k_1 reg(k[\mathbf{a}]) + k_2 reg(k[\mathbf{b}]) + (p-1)(k_1-1) + (q-1)(k_2-1) + k_1k_2 - 1 .$$

7. *The Hilbert series of* $k[\mathbf{c}]$ *is given by* $H_{\mathbf{c}}(t) = (1 - t^{k_1k_2})H_{\mathbf{a}}(t^{k_1})H_{\mathbf{b}}(t^{k_2})$.
8. *If the minimal free resolutions of* $k[\mathbf{a}]$ *and* $k[\mathbf{b}]$ *admit a DG algebra structure, then* $k[\mathbf{c}]$ *inherits the structure from those of* $k[\mathbf{a}]$ *and* $k[\mathbf{b}]$. *That is, we can explicitly construct a multiplication on the minimal resolution of* $k[\mathbf{c}]$ *if we know the multiplication on those of* $k[\mathbf{a}]$ *and* $k[\mathbf{b}]$.

Recall that if a resolution F admits a multiplication which makes it an associative, graded commutative differential graded algebra, we say it has a *differential graded algebra structure* or a *DG algebra structure.*

Example 10 For the sequence $\mathbf{c} = (76, 209, 342, 475, 182, 312, 442, 572, 702)$, one has that $\mathbf{c} = k_1\mathbf{a} \sqcup k_2\mathbf{b}$ with $A = (4, 11, 18, 25)$, $B = (7, 12, 17, 22, 27)$, $k_1 = 19$ and $k_2 = 26$. The minimal resolutions in Example 9 give the Betti numbers of $k[\mathbf{a}]$ and $k[\mathbf{b}]$:

i	0	1	2	3
$\beta_i(A)$	1	6	8	3

i	0	1	2	3	4
$\beta_i(B)$	1	8	12	7	2

Applying both formulas in Corollary 2(1), one gets that $\beta_0 = \beta_0(C) = 1$, and

$$\begin{aligned}
\beta_1 &= 1 \cdot (8+1) + 6 \cdot 1 = \mathbf{15} = 1 \cdot (6+1) + 8 \cdot 1 \\
\beta_2 &= 1 \cdot (12+8) + 6 \cdot (8+1) + 8 \cdot 1 = \mathbf{82} = 1 \cdot (8+6) + 8 \cdot (6+1) + 12 \cdot 1 \\
\beta_3 &= 1 \cdot (7+12) + 6 \cdot (12+8) + 8 \cdot (8+1) + 3 \cdot 1 = \mathbf{214} \\
&= 1 \cdot (3+8) + 8 \cdot (8+6) + 12 \cdot (6+1) + 7 \cdot 1 \\
\beta_4 &= 1 \cdot (2+7) + 6 \cdot (7+12) + 8 \cdot (12+8) + 3 \cdot (8+1) = \mathbf{310} \\
&= 1 \cdot 3 + 8 \cdot (3+8) + 12 \cdot (8+6) + 7 \cdot (6+1) + 2 \cdot 1 \\
\beta_5 &= 1 \cdot 2 + 6 \cdot (2+7) + 8 \cdot (7+12) + 3 \cdot (12+8) = \mathbf{268} \\
&= 8 \cdot 3 + 12 \cdot (3+8) + 7 \cdot (8+6) + 2 \cdot (6+1) \\
\beta_6 &= 6 \cdot 2 + 8 \cdot (2+7) + 3 \cdot (7+12) = \mathbf{141} = 12 \cdot 3 + 7 \cdot (3+8) + 2 \cdot (8+6) \\
\beta_7 &= 8 \cdot 2 + 3 \cdot (2+7) = \mathbf{43} = 7 \cdot 3 + 2 \cdot (3+8) \\
\beta_8 &= 3 \cdot 2 = \mathbf{6},
\end{aligned}$$

and the minimal free resolution of $k[\mathbf{c}]$ shows, as announced in Example 9, as

$$0 \to R^6 \to R^{43} \to R^{141} \to R^{268} \to R^{310} \to R^{214} \to R^{82} \to R^{15} \to R \to k[\mathbf{c}] \to 0.$$

One could also get the graded Betti numbers applying Corollary 2 (5). Now recall that in Example 8 we computed the values of the regularity of $k[\mathbf{a}]$ and $k[\mathbf{b}]$ using Corollary 1: $\operatorname{reg}(k[\mathbf{a}]) = 76$ and $\operatorname{reg}(k[\mathbf{b}]) = 118$. By applying the formula in Corollary 2 (6), one has that the regularity of the semigroup ring $k[\mathbf{c}]$ is

$$\begin{aligned}
\operatorname{reg}(k[\mathbf{c}]) &= k_1\operatorname{reg}(k[\mathbf{a}]) + k_2\operatorname{reg}(k[\mathbf{b}]) + (p-1)(k_1-1) + (q-1)(k_2-1) + k_1k_2 - 1 \\
&= 19 \cdot 76 + 26 \cdot 118 + 3 \cdot 18 + 4 \cdot 25 + 19 \cdot 26 - 1 = 5159.
\end{aligned}$$

Remark 4 As observed in [6, Prop. 5.2], an arithmetic sequence $(a, a+d, \ldots, a+(n-1)d)$ is glued if and only if $n = 3$ and a is even (and d is odd and relatively prime to $a/2$). According to Remark 3, these are also the only arithmetic sequences that are a complete intersection.

Acknowledgements The first author was partially supported by *Ministerio de Ciencia e Innovación* (Spain) MTM2016-78881-P and *Consejería de Educación de la Junta de Castilla y León* VA128G18.

References

1. Barucci, V., Fröberg, R., Şahin, M.: On free resolutions of some semigroup rings. J. Pure Appl. Algebra **218**, 1107–1116 (2014)
2. Bresinsky, H.: Symmetric semigroups of integers generated by 4 elements. Manuscripta Math. **17**, 205–219 (1975)
3. Decker, W., Greuel, G.-M., Pfister, G., Schönemann, H.: SINGULAR 4-1-1, a computer algebra system for polynomial computations (2018). Available at http://www.singular.uni-kl.de
4. Delorme, C.: Sous-monoïdes d'intersection complète de N. Ann. Sci. École Norm. Sup. (4) **9**, 145–154 (1976)
5. Fröberg, R.: The Frobenius number of some semigroups. Comm. Algebra **22**, 6021–6024 (1994)
6. Gimenez, P., Srinivasan, H.: The structure of the minimal free resolution of semigroup rings obtained by gluing. J. Pure Appl. Algebra **223**, 1411–1426 (2019)
7. Gimenez, P., Sengupta, I., Srinivasan, H.: Minimal free resolution for certain affine monomial curves. In: Corso, A., Polini, C. (eds.) Commutative Algebra and Its Connections to Geometry (PASI 2009). Contemporary Mathematics, vol. 555, pp. 87–95. American Mathematical Society, Providence (2011)
8. Gimenez, P., Sengupta, I., Srinivasan, H.: Minimal graded free resolutions for monomial curves defined by arithmetic sequences. J. Algebra **388**, 294–310 (2013)
9. Gimenez, P., Srinivasan, H.: A note on Gorenstein monomial curves. Bull. Braz. Math. Soc. **45**, 671–678 (2014)
10. Herzog, J.: Generators and relations of abelian semigroups and semigroup rings. Manuscripta Math. **3**, 175–193 (1970)
11. Herzog, J., Watanabe, K.: Almost symmetric numerical semigroups. Semigroup Forum **98**, 589–630 (2019)
12. Ramírez Alfonsín, J.L.: The Diophantine Frobenius problem. Oxford Lecture Series in Mathematics and its Applications, vol. 30. Oxford University Press, Oxford (2005)
13. Rosales, J.C.: On presentations of subsemigroups of $\mathbb{N}^n$. Semigroup Forum **55**, 152–159 (1997)
14. Villarreal, R.H.: Monomial Algebras. Monographs and Textbooks in Pure and Applied Mathematics, vol. 238. Marcel Dekker, New York (2001)

Irreducibility and Factorizations in Monoid Rings

Felix Gotti

Abstract For an integral domain R and a commutative cancellative monoid M, the ring consisting of all polynomial expressions with coefficients in R and exponents in M is called the monoid ring of M over R. An integral domain R is called atomic if every nonzero nonunit element can be written as a product of irreducibles. In the study of the atomicity of integral domains, the building blocks are the irreducible elements. Thus, tools to prove irreducibility are crucial to study atomicity. In the first part of this paper, we extend Gauss's Lemma and Eisenstein's Criterion from polynomial rings to monoid rings. An integral domain R is called half-factorial (or an HFD) if any two factorizations of a nonzero nonunit element of R have the same number of irreducible elements (counting repetitions). In the second part of this paper, we determine which monoid algebras with nonnegative rational exponents are Dedekind domains, Euclidean domains, PIDs, UFDs, and HFDs. As a side result, we characterize the submonoids of $(\mathbb{Q}_{\geq 0}, +)$ satisfying a dual notion of half-factoriality known as other-half-factoriality.

Keywords Monoid algebras · Gauss lemma · Eisenstein's Criterion · Puiseux algebras · Atomic domains · Other-half-factorial monoids · Puiseux monoids · Numerical semigroups

1 Introduction

Given an integral domain R and a commutative cancellative monoid M, the ring of all polynomial expressions with coefficients in R and exponents in M is known as the monoid ring of M over R (cf. group rings). Although the study of group rings dates back to the first half of the twentieth century, it was not until the 1970s that the study of monoid rings gained significant attention. A systematic treatment of

F. Gotti (✉)
Department of Mathematics, UC Berkeley, Berkeley, CA, USA
e-mail: felixgotti@berkeley.edu

V. Barucci et al. (eds.), *Numerical Semigroups*, Springer INdAM Series 40,
https://doi.org/10.1007/978-3-030-40822-0_9

ring-theoretical properties of monoid rings was initiated by R. Gilmer and T. Parker [14, 16, 17] in 1974. Since then monoid rings have received a substantial amount of consideration and have permeated through many fields under active research, including algebraic combinatorics [7], discrete geometry [8], and functional analysis [1]. During the last decades, monoid rings have also been studied from the point of view of factorization theory; see, for instance, [2, 3, 25]. Gilmer in [15] offers a comprehensive exposition on the advances of commutative semigroup ring theory until mid 1980s.

An integral domain is called atomic if every nonzero nonunit element it contains can be written as a product of irreducibles. Irreducible elements (sometimes called atoms) are the building blocks of atomicity and factorization theory. As a result, techniques to argue irreducibility are crucial in the development of factorization theory. Gauss's Lemma and Eisenstein's Criterion are two of the most elementary but effective tools to prove irreducibility in the context of polynomial rings. After reviewing some necessary terminology and background in Sect. 2, we dedicate Sect. 3 to extend Gauss's Lemma and Eisenstein's Criterion from the context of polynomial rings to that one of monoid rings.

An atomic monoid M is called half-factorial provided that for all $x \in M$, any two factorizations of x have the same number of irreducibles (counting repetitions). In addition, an integral domain is called half-factorial (or an HFD) if its multiplicative monoid is half-factorial. The concept of half-factoriality was first investigated by L. Carlitz in the context of algebraic number fields; he proved that an algebraic number field is half-factorial if and only if its class group has size at most two [9]. Other-half-factoriality, on the other hand, is a dual version of half-factoriality, and it was introduced by J. Coykendall and W. Smith in [12].

Additive monoids of rationals have a wild atomic structure [18, 20] and a complex arithmetic of factorizations [21, 22]. The monoid rings they determine have been explored in [5]. In addition, examples of such monoid rings have also shown up in the past literature, including [23, Section 1] and [4, Example 2.1] and more recently in [11, Section 5]. In the second part of this paper, which is Sect. 4, we study half-factoriality and other-half-factoriality in the context of additive monoids of rationals and the monoid algebras they induce. We also determine which of these monoid algebras are Dedekind domains, Euclidean domains, PIDs, UFDs, and HFDs.

2 Notation and Background

2.1 General Notation

Throughout this paper, we let $\mathbb{N}_0$ denote the set of all nonnegative integers, and we set $\mathbb{N} := \mathbb{N}_0 \setminus \{0\}$. If $a, b \in \mathbb{Z}$ and $a \le b$, then we let $[[a, b]]$ denote the interval of integers from a to b, i.e.,

$$[[a, b]] := \{j \in \mathbb{Z} \mid a \le j \le b\}.$$

For a subset X of $\mathbb{R}$, we set $X^\bullet := X \setminus \{0\}$. In addition, if $r \in \mathbb{R}$, we define

$$X_{>r} := \{x \in X \mid x > r\} \quad \text{and} \quad X_{\geq r} := \{x \in X \mid x \geq r\}.$$

If $q \in \mathbb{Q}_{>0}$, then we denote the unique $m, n \in \mathbb{N}$ such that $q = m/n$ and $\gcd(m, n) = 1$ by $\mathsf{n}(q)$ and $\mathsf{d}(q)$, respectively.

2.2 Monoids

Within the scope of our exposition, a *monoid* is defined to be a commutative and cancellative semigroup with an identity element. In addition, monoids here are written multiplicatively unless we specify otherwise. Let M be a monoid. We let $U(M)$ denote the set of units (i.e., invertible elements) of M. When $U(M)$ consists of only the identity element, M is said to be *reduced*. On the other hand, M is called *torsion-free* if for all $x, y \in M$ and $n \in \mathbb{N}$, the equality $x^n = y^n$ implies $x = y$. For $S \subseteq M$, we let $\langle S \rangle$ denote the submonoid of M generated by S. Further basic definitions and concepts on commutative cancellative monoids can be found in [24, Chapter 2].

If $y, z \in M$, then *y divides z in M* provided that there exists $x \in M$ such that $z = xy$; in this case we write $y \mid_M z$. Also, the elements y and z are called *associates* if $y \mid_M z$ and $z \mid_M y$; in this case we write $y \simeq z$. An element $p \in M \setminus U(M)$ is said to be *prime* when for all $x, y \in M$ with $p \mid_M xy$, either $p \mid_M x$ or $p \mid_M y$. If every element in $M \setminus U(M)$ can be written as a product of primes, then M is called *factorial*. In a factorial monoid every nonunit element can be uniquely written as a product of primes (up to permutation and associates). In addition, an element $a \in M \setminus U(M)$ is called an *atom* if for any $x, y \in M$ such that $a = xy$ either $x \in U(M)$ or $y \in U(M)$. The set of all atoms of M is denoted by $\mathcal{A}(M)$, and M is said to be *atomic* if every nonunit element of M is a product of atoms. Since every prime element is clearly an atom, every factorial monoid is atomic.

2.3 Factorizations

Let M be a monoid, and let $x \in M \setminus U(M)$. Suppose that for an index $m \in \mathbb{N}$ and atoms $a_1, \dots, a_m \in \mathcal{A}(M)$,

$$x = a_1 \cdots a_m. \tag{1}$$

Then the right-hand side of (1) (treated as a formal product of atoms) is called a *factorization* of x, and m is called the *length* of such a factorization. Two factorizations $a_1 \cdots a_m$ and $b_1 \cdots b_n$ of x are considered to be equal provided that $m = n$ and that there exists a permutation $\sigma \in S_m$ such that $b_i \simeq a_{\sigma(i)}$ for every $i \in [[1, m]]$. The set of all factorizations of x is denoted by $\mathsf{Z}_M(x)$ or, simply, by

$\mathsf{Z}(x)$. We then set

$$\mathsf{Z}(M) := \bigcup_{x \in M \setminus U(M)} \mathsf{Z}(x).$$

For $z \in \mathsf{Z}(x)$, we let $|z|$ denote the length of z.

2.4 Monoid Rings

For an integral domain R, we let $R^\times$ denote the group of units of R. We say that R is *atomic* if every nonzero nonunit element of R can be written as a product of irreducibles (which are also called atoms).

Let M be a reduced torsion-free monoid that is additively written. For an integral domain R, consider the set $R[X; M]$ comprising all maps $f : M \to R$ satisfying that

$$\{s \in M \mid f(s) \neq 0\}$$

is finite. We shall conveniently represent an element $f \in R[X; M]$ by

$$f = \sum_{s \in M} f(s) X^s = \sum_{i=1}^{n} f(s_i) X^{s_i},$$

where $s_1, \ldots, s_n$ are those elements $s \in M$ satisfying that $f(s) \neq 0$. Addition and multiplication in $R[X; M]$ are defined as for polynomials, and we call the elements of $R[X; M]$ *polynomial expressions*. Under these operations, $R[X; M]$ is a commutative ring, which is called the *monoid ring of M over R* or, simply, a *monoid ring*. Following Gilmer [15], we will write $R[M]$ instead of $R[X; M]$. Since R is an integral domain, $R[M]$ is an integral domain [15, Theorem 8.1] with set of units $R^\times$ [16, Corollary 4.2]. If F is a field, then we say that $F[M]$ is a *monoid algebra*. Now suppose that the monoid M is totally ordered. For $k \in \mathbb{N}$, we say that

$$f = \alpha_1 X^{q_1} + \cdots + \alpha_k X^{q_k} \in R[M] \setminus \{0\}$$

is written in *canonical form* if the coefficient α_i is nonzero for every $i \in [[1, k]]$ and $q_1 > \cdots > q_k$. Observe that there is only one way to write f in canonical form. We call $\deg(f) := q_1$ the *degree* of f. In addition, α_1 is called the *leading coefficient* of f, and α_k is called the *constant coefficient* of f provided that $q_k = 0$. As it is customary for polynomials, f is called a *monomial* when $k = 1$.

Suppose that $\psi : M \to M'$ is a monoid homomorphism, where M and M' are reduced torsion-free monoids. Also, let $\psi^* : R[M] \to R[M']$ be the ring homomorphism determined by the assignment $X^s \mapsto X^{\psi(s)}$. It follows from [15,

Theorem 7.2(2)] that if ψ is injective (resp., surjective), then ψ^* is injective (resp., surjective). Let us recall the following easy observation.

Remark 1 If R is an integral domain and the monoids M and M' are isomorphic, then the monoid rings $R[M]$ and $R[M']$ are also isomorphic.

3 Irreducibility Criteria for Monoid Rings

3.1 Extended Gauss's Lemma

Our primary goal in this section is to offer extended versions of Gauss's Lemma and Eisenstein's Criterion for monoid rings.

Let R be an integral domain and take $r_1, \dots, r_n \in R \setminus \{0\}$ for some $n \in \mathbb{N}$. An element $r \in R$ is called a *greatest common divisor* of $r_1, \dots, r_n$ if r divides r_i in R for every $i \in [[1, n]]$ and r is divisible by each common divisor of $r_1, \dots, r_n$. Any two greatest common divisors of $r_1, \dots, r_n$ are associates in R. We let $\text{GCD}(r_1, \dots, r_n)$ denote the set of all greatest common divisors of $r_1, \dots, r_n$.

Definition 1 An integral domain R is called a *GCD-domain* if any finite subset of $R \setminus \{0\}$ has a greatest common divisor in R.

Let M be a reduced torsion-free monoid, and let R be an integral domain. Suppose that for the polynomial expression

$$f = \alpha_1 X^{q_1} + \cdots + \alpha_k X^{q_k} \in R[M] \setminus \{0\}$$

the exponents $q_1, \dots, q_k$ are pairwise distinct. Then $\text{GCD}(\alpha_1, \dots, \alpha_k)$ is called the *content* of f and is denoted by $\mathsf{c}(f)$. If $\mathsf{c}(f) = R^\times$, then f is called *primitive*. Notice that if R is not a GCD-domain, then $\mathsf{c}(f)$ may be the empty set. It is clear that $\mathsf{c}(rf) = r\mathsf{c}(f)$ for all $r \in R \setminus \{0\}$ and $f \in R[M] \setminus \{0\}$. As for the case of polynomials, the following lemma holds.

Lemma 1 *Let M be a reduced torsion-free monoid, and let R be a GCD-domain. If f and g are elements of $R[M] \setminus \{0\}$, then $\mathsf{c}(fg) = \mathsf{c}(f)\mathsf{c}(g)$.*

Proof Since R is a GCD-domain, there exist primitive polynomial expressions f_1 and g_1 in $R[M]$ such that $f = \mathsf{c}(f) f_1$ and $g = \mathsf{c}(g) g_1$. Because M is a torsion-free monoid, it follows from [16, Proposition 4.6] that the element $f_1 g_1$ is primitive in $R[M]$. Therefore $\mathsf{c}(f_1 g_1) = R^\times$. As a consequence, we find that

$$\mathsf{c}(fg) = \mathsf{c}\big(\mathsf{c}(f) f_1 \mathsf{c}(g) g_1\big) = \mathsf{c}(f)\mathsf{c}(g)\mathsf{c}(f_1 g_1) = \mathsf{c}(f)\mathsf{c}(g),$$

as desired. □

Let F denote the field of fractions of a GCD-domain R. Gauss's Lemma states that a non-constant polynomial f with coefficients in R is irreducible in $R[X]$ if and only if it is irreducible in $F[X]$ and primitive in $R[X]$. Now we extend Gauss's Lemma to the context of monoid rings.

Theorem 1 (Extended Gauss's Lemma) *Let M be a reduced torsion-free monoid, and let R be a GCD-domain with field of fractions F. Then an element $f \in R[M] \setminus R$ is irreducible in $R[M]$ if and only if f is irreducible in $F[M]$ and primitive in $R[M]$.*

Proof For the direct implication, suppose that f is irreducible in $R[M]$. If $r \in \mathsf{c}(f)$, then there exists $g \in R[M] \setminus R$ such that $f = rg$. Because $R[M]^\times \subset R$, the element g is not a unit of $R[M]$. As f is irreducible in $R[M]$, one finds that $r \in R[M]^\times = R^\times$. So $\mathsf{c}(f) = R^\times$, which implies that f is primitive in $R[M]$. To argue that f is irreducible in $F[M]$, take $g_1, g_2 \in F[M]$ such that $f = g_1 g_2$. Since R is a GCD-domain, there exist nonzero elements $a_1, a_2, b_1, b_2 \in R$ such that both

$$h_1 := \frac{a_1}{b_1} g_1 \quad \text{and} \quad h_2 := \frac{a_2}{b_2} g_2$$

are primitive elements of $R[M]$. Clearly, $a_1 a_2 f = b_1 b_2 h_1 h_2$. This, along with Lemma 1, implies that

$$a_1 a_2 R^\times = a_1 a_2 \mathsf{c}(f) = \mathsf{c}(a_1 a_2 f) = \mathsf{c}(b_1 b_2 h_1 h_2) = b_1 b_2 \mathsf{c}(h_1)\mathsf{c}(h_2) = b_1 b_2 R^\times.$$

Then $\frac{a_1 a_2}{b_1 b_2} \in R^\times$ and, as a consequence, $\frac{a_1 a_2}{b_1 b_2} f = h_1 h_2$ is irreducible in $R[M]$. Thus, either $h_1 \in R[M]^\times = R^\times$ or $h_2 \in R[M]^\times = R^\times$. This, in turn, implies that either g_1 or g_2 belongs to $F^\times = F[M]^\times$. Hence f is irreducible in $F[M]$.

Tor argue the reverse implication, suppose that f is irreducible in $F[M]$ and primitive in $R[M]$. Then take elements g_1 and $g_2 \in R[M]$ such that $f = g_1 g_2$. Since f is irreducible in $F[M]$, either $g_1 \in F[M]^\times = F^\times$ or $g_2 \in F[M]^\times = F^\times$. This, along with the fact that $R[M] \cap F^\times = R \setminus \{0\}$, implies that either $g_1 \in \mathsf{c}(f)$ or $g_2 \in \mathsf{c}(f)$. As $\mathsf{c}(f) = R^\times = R[M]^\times$, either g_1 or g_2 belongs to $R[M]^\times$. As a result, f is irreducible in $R[M]$, which concludes the proof. □

3.2 Extended Eisenstein's Criterion

It is hardly debatable that Eisenstein's Criterion is one of the most popular and useful criteria to argue the irreducibility of certain polynomials. Now we proceed to offer an extended version of Eisenstein's Criterion for monoid rings.

Proposition 1 (Extended Eisenstein's Criterion) *Let M be a reduced totally-ordered torsion-free monoid, and let R be an integral domain. Suppose that the* *element*

$$f = \alpha_n X^{q_n} + \cdots + \alpha_1 X^{q_1} + \alpha_0 \in R[M] \setminus \{0\},$$

written in canonical form, is primitive. If there exists a prime ideal P of R satisfying the conditions

1. $\alpha_n \notin P$,
2. $\alpha_j \in P$ *for every* $j \in [[0, n-1]]$, *and*
3. $\alpha_0 \notin P^2$,

then f is irreducible in $R[M]$.

Proof We let $\bar{R}$ denote the quotient R/P and, for any $h \in R[M]$, we let $\bar{h}$ denote the image of h under the natural surjection $R[M] \to \bar{R}[M]$, i.e., $\bar{h}$ is the result of reducing the coefficients of h modulo P. To argue that f is irreducible suppose, by way of contradiction, that $f = g_1 g_2$ for some nonzero nonunit elements g_1 and g_2 of $R[M]$. As f is primitive, $g_1 \notin R$ and $g_2 \notin R$. By the condition (2) in the statement, one obtains that $\bar{g}_1 \bar{g}_2 = \bar{f} = \bar{\alpha}_n X^{q_n}$. Thus, both $\bar{g}_1$ and $\bar{g}_2$ are monomials. This, along with the fact that none of the leading coefficients of g_1 and g_2 are in P (because $\alpha_n \notin P$), implies that the constant coefficients of both g_1 and g_2 are in P. As a result, the constant coefficient α_0 of f must belong to P^2, which is a contradiction. □

Corollary 1 *Let M be a reduced totally-ordered torsion-free monoid, and let R be an integral domain containing a prime element. Then for each $q \in M^\bullet$, there exists an irreducible polynomial expression in $R[M]$ of degree q.*

Proof Let p be a prime element of R. It suffices to verify that, for any $q \in M^\bullet$, the element $f := X^q + p \in R[M]$ is irreducible. Indeed, this is an immediate consequence of Proposition 1 once we take $P := (p)$. □

In Corollary 1, the integral domain R is required to contain a prime element. This condition is not superfluous, as the next example illustrates.

Example 1 For a prime number p, consider the monoid algebra $\mathbb{F}_p[M]$, where M is the submonoid $\langle 1/p^n \mid n \in \mathbb{N} \rangle$ of $(\mathbb{Q}_{\geq 0}, +)$ and $\mathbb{F}_p$ is a finite field of characteristic p. It is clear that M is a reduced totally-ordered torsion-free monoid. Now let

$$f := \alpha_1 X^{q_1} + \cdots + \alpha_n X^{q_n}$$

be an element of $\mathbb{F}_p[M] \setminus \mathbb{F}_p$ written in canonical form. As $\mathbb{F}_p$ is a perfect field of characteristic p, the Frobenius homomorphism $x \mapsto x^p$ is surjective and, therefore, for each $i \in [[1, n]]$ there exists $\beta_i \in \mathbb{F}_p$ with $\alpha_i = \beta_i^p$. On the other hand, it is clear that $q_i/p \in M$ for every $i \in [[1, n]]$. As

$$f = \alpha_1 X^{q_1} + \cdots + \alpha_n X^{q_n} = \left(\beta_1 X^{q_1/p} + \cdots + \beta_n X^{q_n/p}\right)^p,$$

the polynomial expression f is not irreducible in $\mathbb{F}_p[M]$. Hence the monoid algebra $\mathbb{F}_p[M]$ does not contain irreducible elements. Clearly, the field $\mathbb{F}_p$ is an integral domain containing no prime elements.

4 Factorizations in Monoid Algebras

A *numerical semigroup* is a submonoid N of $(\mathbb{N}_0, +)$ whose complement is finite, i.e., $|\mathbb{N}_0 \setminus N| < \infty$. Numerical semigroups are finitely generated and, therefore, atomic. However, the only factorial numerical semigroup is $(\mathbb{N}_0, +)$. For an introduction to numerical semigroups, see [13], and for some of their many applications, see [6]. A *Puiseux monoid*, on the other hand, is an additive submonoid of $(\mathbb{Q}_{\geq 0}, +)$. Albeit Puiseux monoids are natural generalizations of numerical semigroups, the former are not necessarily finitely generated or atomic; for example, consider $\langle 1/2^n \mid n \in \mathbb{N} \rangle$. The factorization structure of Puiseux monoids have been compared with that of other well-studied atomic monoids in [19] and, more recently, in [10]. In this section, we determine the Puiseux monoids whose monoid algebras are Dedekind domains, Euclidean domains, PIDs, UFDs, or HFDs.

Definition 2 An atomic monoid M is *half-factorial* (or an *HF-monoid*) if for all $x \in M \setminus U(M)$ and $z, z' \in \mathsf{Z}(x)$, the equality $|z| = |z'|$ holds. An integral domain is *half-factorial* (or an *HFD*) if its multiplicative monoid is an HF-monoid.

Clearly, half-factoriality is a relaxed version of being a factorial monoid or a UFD. Although the concept of half-factoriality was first considered by Carlitz in his study of algebraic number fields [9], it was A. Zaks who coined the term "half-factorial domain" [26].

Definition 3 An atomic monoid M is *other-half-factorial* (or an *OHF-monoid*) if for all $x \in M \setminus U(M)$ and $z, z' \in \mathsf{Z}(x)$ the equality $|z| = |z'|$ implies that $z = z'$.

Observe that other-half-factoriality is somehow a dual version of half-factoriality. Although an integral domain is a UFD if and only if its multiplicative monoid is an OHF-monoid [12, Corollary 2.11], OHF-monoids are not always factorial or half-factorial, even in the class of Puiseux monoids.

Proposition 2 *For a nontrivial atomic Puiseux monoid M, the following conditions hold.*

1. *M is an HF-monoid if and only if M is factorial.*
2. *M is an OHF-monoid if and only if $|\mathcal{A}(M)| \leq 2$.*

Proof For the direct implication of (1), suppose that M is an HF-monoid. Since M is an atomic nontrivial Puiseux monoid, $\mathcal{A}(M)$ is not empty. Let a_1 and a_2 be two atoms of M. Then $z_1 := \mathsf{n}(a_2)\mathsf{d}(a_1)a_1$ and $z_2 := \mathsf{n}(a_1)\mathsf{d}(a_2)a_2$ are two factorizations of the element $\mathsf{n}(a_1)\mathsf{n}(a_2) \in M$. Because M is an HF-monoid,

$|z_1| = |z_2|$ and so

$$\mathsf{n}(a_2)\mathsf{d}(a_1) = \mathsf{n}(a_1)\mathsf{d}(a_2).$$

Therefore $a_1 = a_2$, and then M contains only one atom. Hence $M \cong (\mathbb{N}_0, +)$ and, as a result, M is factorial. The reverse implication of (1) is trivial.

To prove the direct implication of (2), assume that M is an OHF-monoid. If M is factorial, then $M \cong (\mathbb{N}_0, +)$, and we are done. Then suppose that M is not factorial. In this case, $|\mathcal{A}(M)| \geq 2$. Assume, by way of contradiction, that $|\mathcal{A}(M)| \geq 3$. Take $a_1, a_2, a_3 \in \mathcal{A}(M)$ satisfying that $a_1 < a_2 < a_3$. Let $d = \mathsf{d}(a_1)\mathsf{d}(a_2)\mathsf{d}(a_3)$, and set $a'_i = da_i$ for each $i \in [[1, 3]]$. Since a'_1, a'_2, and a'_3 are integers satisfying that $a'_1 < a'_2 < a'_3$, there exist $m, n \in \mathbb{N}$ such that

$$m(a'_2 - a'_1) = n(a'_3 - a'_2). \tag{2}$$

Clearly, $z_1 := ma_1 + na_3$ and $z_2 := (m + n)a_2$ are two distinct factorizations in $\mathsf{Z}(M)$ satisfying that $|z_1| = m + n = |z_2|$. In addition, after dividing both sides of the equality (2) by d, one obtains that

$$ma_1 + na_3 = (m + n)a_2,$$

which means that z_1 and z_2 are factorizations of the same element. However, this contradicts that M is an OHF-monoid. Hence $|\mathcal{A}(M)| \leq 2$, as desired. For the reverse implication of (2), suppose that $|\mathcal{A}(M)| \leq 2$. By [18, Proposition 3.2], M is isomorphic to a numerical semigroup N. As N is generated by at most two elements, either $N = (\mathbb{N}_0, +)$ or $N = \langle a, b \rangle$ for $a, b \in \mathbb{N}_{\geq 2}$ with $\gcd(a, b) = 1$. If $N = (\mathbb{N}_0, +)$, then N is factorial and, in particular, an OHF-monoid. On the other hand, if $N = \langle a, b \rangle$, then it is an OHF-monoid by [12, Example 2.13]. □

In [16, Theorem 8.4] Gilmer and Parker characterize the monoid algebras that are Dedekind domains, Euclidean domains, or PIDs. We conclude this section extending such a characterization in the case where the exponent monoids are Puiseux monoids.

Theorem 2 *For a nontrivial Puiseux monoid M and a field F, the following conditions are equivalent:*

1. *$F[M]$ is a Euclidean domain;*
2. *$F[M]$ is a PID;*
3. *$F[M]$ is a UFD;*
4. *$F[M]$ is an HFD;*
5. *$M \cong (\mathbb{N}_0, +)$;*
6. *$F[M]$ is a Dedekind domain.*

Proof It is well known that every Euclidean domain is a PID, and every PID is a UFD. Therefore condition (1) implies condition (2), and condition (2) implies condition (3). In addition, it is clear that every UFD is an HFD, and so condition (3) implies condition (4). As Puiseux monoids are torsion-free, [25, Proposition 1.4] ensures that M is an HF-monoid when $F[M]$ is an HFD. This, along with Proposition 2(1), guarantees that $M \cong (\mathbb{N}_0, +)$ provided that $F[M]$ is an HFD. Thus, condition (4) implies condition (5). Also, if condition (5) holds, then $F[M] \cong F[\mathbb{N}_0] = F[X]$ (by Remark 1) is a Euclidean domain, which is condition (1). Then we have argued that the first five conditions are equivalent.

To include (6) in the set of already-established equivalent conditions, observe that condition (2) implies condition (6) because every PID is a Dedekind domain. On the other hand, suppose that the monoid algebra $F[M]$ is a Dedekind domain. Then the fact that M is torsion-free, along with [16, Theorem 8.4], implies that $M \cong (\mathbb{N}_0, +)$. Hence condition (6) implies condition (5), which completes the proof. □

Acknowledgements While working on this paper, the author was supported by the NSF-AGEP Fellowship and the UC Dissertation Year Fellowship. The author would like to thank an anonymous referee, whose suggestions help to simplify and improve the initially-submitted version of this paper.

References

1. Amini, M.: Module amenability for semigroup algebras. Semigroup Forum **69**, 243–254 (2004)
2. Anderson, D.D., Juett, J.R.: Long length functions. J. Algebra **426**, 327–343 (2015)
3. Anderson, D.F., Scherpenisse, D.: Factorization in $K[S]$. In: Anderson, D.D. (ed.) Factorization in Integral Domains. Lecture Notes in Pure and Applied Mathematics, vol. 189, pp. 45–56. Marcel Dekker, New York (1997)
4. Anderson, D.D., Anderson, D.F., Zafrullah, M.: Factorizations in integral domains. J. Pure Appl. Algebra **69**, 1–19 (1990)
5. Anderson, D.D., Coykendall, J., Hill, L., Zafrullah, M.: Monoid domain constructions of antimatter domains. Comm. Alg. **35**, 3236–3241 (2007)
6. Assi, A., García-Sánchez, P.A.: Numerical Semigroups and Applications. RSME Springer Series. Springer, New York (2016)
7. Briales, E., Campillo, A., Marijuán, C., Pisón, P.: Combinatorics and syzygies for semigroup algebras. Coll. Math. **49**, 239–256 (1998)
8. Bruns, W., Gubeladze, J.: Semigroup algebras and discrete geometry. In: Bonavero, L., Biron, M. (eds.) Geometry of Toric Varieties. Séminaires et Congrés, vol. 6, pp. 43–127. Mathematical Society of France, Paris (2002)
9. Carlitz, L.: A characterization of algebraic number fields with class number two. Proc. Am. Math. Soc. **11**, 391–392 (1960)
10. Chapman, S.T., Gotti, F., Gotti, M.: Factorization invariants of Puiseux monoids generated by geometric sequences. Comm. Alg. **48**, 380–396 (2020)
11. Coykendall, J., Gotti, F.: On the atomicity of monoid algebras. J. Algebra **539**, 138–151 (2019)
12. Coykendall, J., Smith, W.W.: On unique factorization domains. J. Algebra **332**, 62–70 (2011)
13. García-Sánchez, P.A., Rosales, J.C.: Numerical Semigroups. Developments in Mathematics, vol. 20. Springer, New York (2009)
14. Gilmer, R.: A two-dimensional non-Noetherian factorial ring. Proc. Am. Math. Soc. **44**, 25–30 (1974)
15. Gilmer, R.: Commutative Semigroup Rings. Chicago Lectures in Mathematics. The University of Chicago Press, London (1984)

16. Gilmer, R., Parker, T.: Divisibility properties of semigroup rings. Mich. Math. J. **21**, 65–86 (1974)
17. Gilmer, R., Parker, T.: Nilpotent elements of commutative semigroup rings. Mich. Math. J. **22**, 97–108 (1975)
18. Gotti, F.: On the atomic structure of Puiseux monoids. J. Algebra Appl. **16**, 1750126 (2017)
19. Gotti, F.: Puiseux monoids and transfer homomorphisms. J. Algebra **516**, 95–114 (2018)
20. Gotti, F.: Increasing positive monoids of ordered fields are FF-monoids. J. Algebra **518**, 40–56 (2019)
21. Gotti, F.: Systems of sets of lengths of Puiseux monoids. J. Pure Appl. Algebra **223**, 1856–1868 (2019)
22. Gotti, F., O'Neill, C.: The elasticity of Puiseux monoids. J. Commut. Algebra (to appear). https://projecteuclid.org/euclid.jca/1523433696
23. Grams, A.: Atomic domains and the ascending chain condition for principal ideals. Math. Proc. Cambridge Philos. Soc. **75**, 321–329 (1974)
24. Grillet, P.A.: Commutative Semigroups. Advances in Mathematics, vol. 2. Kluwer Academic Publishers, Boston (2001)
25. Kim, H.: Factorization in monoid domains. Commun. Algebra **29**, 1853–1869 (2001)
26. Zaks, A.: Half-factorial domains. Bull. Am. Math. Soc. **82**, 721–723 (1976)

On the Molecules of Numerical Semigroups, Puiseux Monoids, and Puiseux Algebras

Felix Gotti and Marly Gotti

Abstract A *molecule* is a nonzero non-unit element of an integral domain (resp., commutative cancellative monoid) having a unique factorization into irreducibles (resp., atoms). Here we study the molecules of Puiseux monoids as well as the molecules of their corresponding semigroup algebras, which we call Puiseux algebras. We begin by presenting, in the context of numerical semigroups, some results on the possible cardinalities of the sets of molecules and the sets of reducible molecules (i.e., molecules that are not irreducibles/atoms). Then we study the molecules in the more general context of Puiseux monoids. We construct infinitely many non-isomorphic atomic Puiseux monoids all whose molecules are atoms. In addition, we characterize the molecules of Puiseux monoids generated by rationals with prime denominators. Finally, we turn to investigate the molecules of Puiseux algebras. We provide a characterization of the molecules of the Puiseux algebras corresponding to root-closed Puiseux monoids. Then we use such a characterization to find an infinite class of Puiseux algebras with infinitely many non-associated reducible molecules.

Keywords Numerical semigroups · Puiseux monoids · Monoid algebras · Atoms · Irreducibles · Atomic monoids · Atomic algebras

1 Introduction

Let M be a commutative cancellative monoid. A non-invertible element of M is called an *atom* if it cannot be expressed as a product of two non-invertible elements. If $x \in M$ can be expressed as a formal product of atoms, then such a formal

---

F. Gotti
Department of Mathematics, UC Berkeley, Berkeley, CA, USA
e-mail: felixgotti@berkeley.edu

M. Gotti (✉)
Department of Mathematics, University of Florida, Gainesville, FL, USA
e-mail: marlycormar@ufl.edu

V. Barucci et al. (eds.), *Numerical Semigroups*, Springer INdAM Series 40,
https://doi.org/10.1007/978-3-030-40822-0_10

product (up to associate and permutation) is called a *factorization* of x. If every non-invertible element of M has a factorization, then M is called *atomic*. Furthermore, the atoms and factorizations of an integral domain are the irreducible elements and the formal products of irreducible elements, respectively. All the undefined or informally-defined terms mentioned in this section will be formally introduced later on.

The elements having exactly one factorization are crucial in the study of factorization theory of commutative cancellative monoids and integral domains. Aiming to avoid repeated long descriptions, we call such elements *molecules*. Molecules were first studied in the context of algebraic number theory by W. Narkiewicz and other authors in the 1960's. For instance, in [18] and [19] Narkiewicz studied some distributional aspects of the molecules of quadratic number fields. In addition, he gave an asymptotic formula for the number of (non-associated) integer molecules of any algebraic number field [20]. In this paper, we study the molecules of submonoids of $(\mathbb{Q}_{\geq 0}, +)$, including numerical semigroups, and the molecules of their corresponding semigroup algebras.

A *numerical semigroup* is a finite-complemented submonoid of $(\mathbb{N}_0, +)$, where $\mathbb{N}_0 = \{0, 1, 2, \dots\}$. Every numerical semigroup is finitely generated by its set of atoms and, in particular, atomic. In addition, if $N \neq \mathbb{N}_0$ is a numerical semigroup, then it contains only finitely many molecules. Notice, however, that every positive integer is a molecule of $(\mathbb{N}_0, +)$. Figure 1 shows the distribution of the sets of molecules of four numerical semigroups. We begin Sect. 3 pointing out how the molecules of numerical semigroups are related to the Betti elements. Then we show that each element in the set $\{n \in \mathbb{N}_0 : n \geq 4\} \cup \{\infty\}$ (and only such elements) can be the number of molecules of a numerical semigroup. We conclude our study of molecules of numerical semigroups exploring the possible cardinalities of the sets of reducible molecules (i.e., molecules that are not atoms).

A submonoid of $(\mathbb{Q}_{\geq 0}, +)$ is called a *Puiseux monoid*. Puiseux monoids were first studied in [11] and have been systematically investigated since then (see [3] and references therein). Albeit a natural generalization of the class of numerical

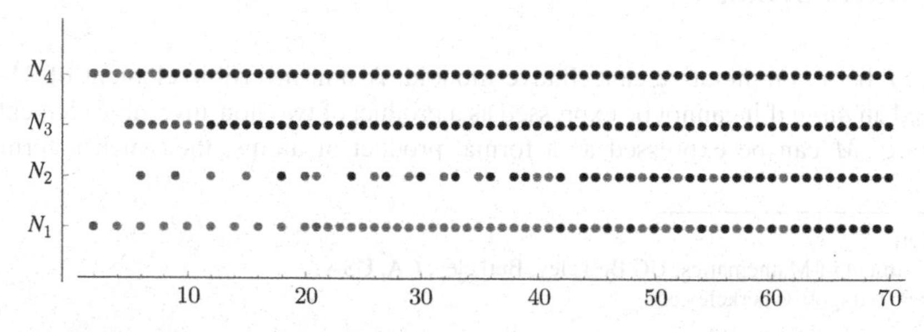

Fig. 1 The dots on the horizontal line labeled by N_i represent the nonzero elements of the numerical semigroup N_i; here we are setting $N_1 = \langle 2, 21 \rangle$, $N_2 = \langle 6, 9, 20 \rangle$, $N_3 = \langle 5, 6, 7, 8, 9 \rangle$, and $N_4 = \langle 2, 3 \rangle$. Atoms are represented in blue, molecules that are not atoms in red, and non-molecules in black (colour figure online)

semigroups, the class of Puiseux monoids contains members having infinitely many atoms and, consequently, infinitely many molecules. A Puiseux monoid is *prime reciprocal* if it can be generated by rationals of the form a/p, where p is a prime and a is a positive integer not divisible by p. In Sect. 4, we study the sets of molecules of Puiseux monoids, finding infinitely many non-isomorphic Puiseux monoids all whose molecules are atoms (in contrast to the fact that the set of molecules of a numerical semigroup always differs from its set of atoms). In addition, we construct infinitely many non-isomorphic Puiseux monoids having infinitely many molecules that are not atoms (in contrast to the fact that the set of molecules of a nontrivial numerical semigroup is always finite). We conclude Sect. 4 characterizing the sets of molecules of prime reciprocal Puiseux monoids.

The final section of this paper is dedicated to the molecules of the semigroup algebras of Puiseux monoids, which we call *Puiseux algebras*. Puiseux algebras have been studied in [1, 5, 12]. First, for a fixed field F we establish a bijection between the set molecules of a Puiseux monoid and the set of non-associated monomial molecules of its corresponding Puiseux algebra over F. Then we characterize the molecules of Puiseux algebras of root-closed Puiseux monoids. We conclude this paper using the previous characterization to exhibit a class of Puiseux algebras having infinitely many molecules that are neither monomials nor irreducibles.

2 Monoids, Atoms, and Molecules

2.1 General Notation

In this section we review the nomenclature and main concepts on commutative monoids and factorization theory we shall be using later. For a self-contained approach to the theory of commutative monoids we suggest [16] by P. A. Grillet, and for background on non-unique factorization theory of atomic monoids and integral domains the reader might want to consult [9] by A. Geroldinger and F. Halter-Koch.

We use the double-struck symbols $\mathbb{N}$ and $\mathbb{N}_0$ to denote the set of positive integers and the set of nonnegative integers, respectively, while we let $\mathbb{P}$ denote the set of primes. If $R \subseteq \mathbb{R}$ and $r \in \mathbb{R}$, then we set

$$R_{\ge r} := \{x \in R : x \ge r\}.$$

The notation $R_{>r}$ is used in a similar way. We let the symbol $\emptyset$ denote the empty set. If $q \in \mathbb{Q}_{>0}$, then the unique $a, b \in \mathbb{N}$ such that $q = a/b$ and $\gcd(a, b) = 1$ are denoted by $\mathsf{n}(q)$ and $\mathsf{d}(q)$, respectively. For $Q \subseteq \mathbb{Q}_{>0}$, we call

$$\mathsf{n}(Q) := \{\mathsf{n}(q) : q \in Q\} \quad \text{and} \quad \mathsf{d}(Q) := \{\mathsf{d}(q) : q \in Q\}$$

the *numerator set* and *denominator set* of Q, respectively. In addition, if S is a set consisting of primes and $q \in \mathbb{Q}_{>0}$, then we set

$$\mathsf{D}_S(q) := \{p \in S : p \mid \mathsf{d}(q)\} \quad \text{and} \quad \mathsf{D}_S(Q) := \cup_{q \in Q} \mathsf{D}_S(q).$$

For $p \in \mathbb{P}$, the *p-adic valuation* on $\mathbb{Q} \geq 0$ is the map defined by $\mathsf{v}_p(0) = \infty$ and $\mathsf{v}_p(q) = \mathsf{v}_p(\mathsf{n}(q)) - \mathsf{v}_p(\mathsf{d}(q))$ for $q \in \mathbb{Q} > 0$, where for $n \in \mathbb{N}$ the value $\mathsf{v}_p(n)$ is the exponent of the maximal power of p dividing n. It can be easily seen that the p-adic valuation satisfies that $\mathsf{v}_p(q_1 + \cdots + q_n) \geq \min\{\mathsf{v}_p(q_1), \ldots, \mathsf{v}_p(q_n)\}$ for every $n \in \mathbb{N}$ and $q_1, \ldots, q_n \in \mathbb{Q}_{>0}$.

2.2 Monoids

Throughout this paper, we will tacitly assume that the term *monoid* by itself always refers to a commutative and cancellative semigroup with identity. In addition, we will use additive notation by default and switch to multiplicative notation only when necessary (in which case, the notation will be clear from the context). For a monoid M, we let $M^\bullet$ denote the set $M \setminus \{0\}$. If $a, c \in M$, then we say that *a divides c in M* and write $a \mid_M c$ provided that $c = a + b$ for some $b \in M$. We write $M = \langle S \rangle$ when M is generated by a set S. The monoid M is *finitely generated* if it can be generated by a finite set; otherwise, M is said to be *non-finitely generated*. A succinct exposition of finitely generated monoids can be found in [7].

2.3 Atoms and Molecules

The set of invertible elements of M is denoted by $M^\times$, and M is said to be *reduced* if $M^\times$ contains only the identity element.

Definition 1 An element $a \in M \setminus M^\times$ is an *atom* provided that for all $u, v \in M$ the fact that $a = u + v$ implies that either $u \in M^\times$ or $v \in M^\times$. The set of atoms of M is denoted by $\mathcal{A}(M)$, and M is called *atomic* if $M = \langle \mathcal{A}(M) \rangle$.

Let M be a reduced monoid. Then the *factorization monoid* $\mathsf{Z}(M)$ of M is the free commutative monoid on $\mathcal{A}(M)$. The elements of $\mathsf{Z}(M)$, which are formal sums of atoms, are called *factorizations*. If $z = a_1 + \cdots + a_n \in \mathsf{Z}(M)$ for some $a_1, \ldots, a_n \in \mathcal{A}(M)$, then $|z| := n$ is called the *length* of z. As $\mathsf{Z}(M)$ is free on $\mathcal{A}(M)$, there is a unique monoid homomorphism from $\mathsf{Z}(M)$ to M determined by the assignment $a \mapsto a$ for all $a \in \mathcal{A}(M)$. Such a monoid homomorphism is called the *factorization homomorphism* of M and is denoted by ϕ_M (or just ϕ when there is no risk of ambiguity involved). For $x \in M$, the sets

$$\mathsf{Z}(x) := \mathsf{Z}_M(x) := \phi^{-1}(x) \subseteq \mathsf{Z}(M) \quad \text{and} \quad \mathsf{L}(x) := \mathsf{L}_M(x) := \{|z| : z \in \mathsf{Z}(x)\}$$

are called the *set of factorizations* and the *set of lengths* of x, respectively. Clearly, M is atomic if and only if $\mathsf{Z}(x) \neq \emptyset$ for all $x \in M$.

Let M_{red} denote the set of classes of M under the equivalence relation $x \sim y$ if $y = x + u$ for some $u \in M^\times$. It turns out that M_{red} is a monoid with the addition operation inherited from M. The monoid M_{red} is called the *reduced monoid* of M (clearly, M_{red} is reduced). Note that an element a belongs to $\mathcal{A}(M)$ if and only if the class of a belongs to $\mathcal{A}(M_{\text{red}})$. If M is an atomic monoid (that is not necessarily reduced), then we set $\mathsf{Z}(M) := \mathsf{Z}(M_{\text{red}})$ and, for $x \in M$, we define $\mathsf{Z}(x)$ and $\mathsf{L}(x)$ in terms of $\mathsf{Z}(M)$ as we did for the reduced case.

As one of the main purposes of this paper is to study elements with exactly one factorization in Puiseux monoids (in particular, numerical semigroups), we introduce the following definition.

Definition 2 Let M be a monoid. We say that an element $m \in M \setminus M^\times$ is a *molecule* provided that $|\mathsf{Z}(m)| = 1$. The set of all molecules of M is denoted by $\mathcal{M}(M)$.

It is clear that the set of atoms of any monoid is contained in the set of molecules. However, such an inclusion might be proper (consider, for instance, the additive monoid $\mathbb{N}_0$). In addition, for any atomic monoid M the set $\mathcal{M}(M)$ is *divisor-closed* in the sense that if $m \in \mathcal{M}(M)$ and $m' \mid_M m$ for some $m' \in M \setminus M^\times$, then $m' \in \mathcal{M}(M)$. If the condition of atomicity is dropped, then this observation is not necessarily true (see Example 3).

3 Molecules of Numerical Semigroups

In this section we study the sets of molecules of numerical semigroups, putting particular emphasis on their possible cardinalities.

Definition 3 A *numerical semigroup* is a cofinite additive submonoid of $\mathbb{N}_0$.

We let $\mathcal{N}$ denote the class consisting of all numerical semigroups (up to isomorphism). We say that $N \in \mathcal{N}$ is *nontrivial* if $\mathbb{N}_0 \setminus N$ is not empty, and we let $\mathcal{N}^\bullet$ denote the class of all nontrivial numerical semigroups. Every $N \in \mathcal{N}$ has a unique minimal set of generators A, which is finite. The cardinality of A is called the *embedding dimension* of N. Suppose that N has embedding dimension n, and let $N = \langle a_1, \dots, a_n \rangle$ (we always assume that $a_1 < \cdots < a_n$). Then $\gcd(a_1, \dots, a_n) = 1$ and $\mathcal{A}(N) = \{a_1, \dots, a_n\}$. In particular, every numerical semigroup is atomic. When N is nontrivial, the maximum of $\mathbb{N}_0 \setminus N$ is called the *Frobenius number* of N. Here we let $\mathsf{F}(N)$ denote the Frobenius number of N. See [8] for a friendly introduction to numerical semigroups.

Example 1 For $k \geq 1$, consider the numerical semigroup $N_1 = \langle 2, 21 \rangle$, whose molecules are depicted in Fig. 1. It is not hard to see that $x \in N_1^\bullet$ is a molecule if and only if every factorization of x contains at most one copy of 21. Therefore

$$\mathcal{M}(N_1) = \big\{2m + 21n : 0 \leq m < 21, n \in \{0, 1\}, \text{ and } (m, n) \neq (0, 0)\big\}.$$

In addition, if $2m + 21n = 2m' + 21n'$ for some $m, m' \in \{0, \dots, 20\}$ and $n, n' \in \{0, 1\}$, then one can readily check that $m = m'$ and $n = n'$. Hence $|\mathcal{M}(N_1)| = 41$.

3.1 Betty Elements

Let $N = \langle a_1, \dots, a_n \rangle$ be a minimally generated numerical semigroup. We always represent an element of $\mathsf{Z}(N)$ with an n-tuple $z = (c_1, \dots, c_n) \in \mathbb{N}_0^n$, where the entry c_i specifies the number of copies of a_i that appear in z. Clearly, $|z| = c_1 + \dots + c_n$. Given factorizations $z = (c_1, \dots, c_n)$ and $z' = (c'_1, \dots, c'_n)$, we define

$$\gcd(z, z') = (\min\{c_1, c'_1\}, \dots, \min\{c_n, c'_n\}).$$

The *factorization graph* of $x \in N$, denoted by $\nabla_x(N)$ (or just ∇_x when no risk of confusion exists), is the graph with vertices $\mathsf{Z}(x)$ and edges between those $z, z' \in \mathsf{Z}(x)$ satisfying that $\gcd(z, z') \neq 0$. The element x is called a *Betti element* of N provided that ∇_x is disconnected. The set of Betti elements of N is denoted by $\text{Betti}(N)$.

Example 2 Take N to be the numerical semigroup $\langle 14, 16, 18, 21, 45 \rangle$. A computation in SAGE using the `numericalsgps GAP` package reveals that N has nine Betti elements. In particular, $90 \in \text{Betti}(N)$. In Fig. 2 one can see the disconnected factorization graph of the Betti element 90 on the left and the connected factorization graph of the non-Betti element 84 on the right.

Observe that $0 \notin \text{Betti}(N)$ since $|\mathsf{Z}(0)| = 1$. It is well known that every numerical semigroup has finitely many Betti elements. Betti elements play a fundamental role in the study of uniquely-presented numerical semigroups [6] and the study of *delta sets* of BF-monoids [2]. For a more general notion of Betti element, meaning the *syzygies* of an $\mathbb{N}^n$-graded module, see [17]. In a numerical semigroup, Betti elements and molecules are closely related.

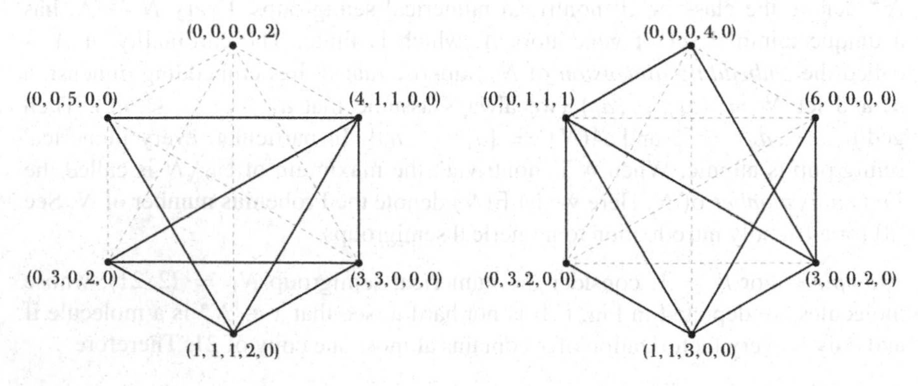

Fig. 2 The factorization graphs of $90 \in \text{Betti}(N)$ and $84 \notin \text{Betti}(N)$, where N is the numerical semigroup $\langle 14, 16, 18, 21, 45 \rangle$

Remark 1 Let N be a numerical semigroup. An element $m \in N$ is a molecule if and only if $\beta \nmid_N m$ for any $\beta \in \text{Betti}(N)$.

Proof For the direct implication, suppose that m is a molecule of N and take $\alpha \in N$ such that $\alpha \mid_N m$. As the set of molecules is closed under division, $|\mathsf{Z}(\alpha)| = 1$. This implies that ∇_α is connected and, therefore, α cannot be a Betti element. The reverse implication is just a rephrasing of [6, Lemma 1]. □

3.2 *On the Sizes of the Sets of Molecules*

Obviously, for every $n \in \mathbb{N}$ there exists a numerical semigroup having exactly n atoms. The next proposition answers the same realization question replacing the concept of an atom by that of a molecule. Recall that $\mathcal{N}^\bullet$ denotes the class of all nontrivial numerical semigroups.

Proposition 1 $\{|\mathcal{M}(N)| : N \in \mathcal{N}^\bullet\} = \mathbb{N}_{\ge 4}$.

Proof Let N be a nontrivial numerical semigroup. Then N must contain at least two atoms. Let a and b denote the two smallest atoms of N, and assume that $a < b$. Note that $2a$ and $a + b$ are distinct molecules that are not atoms. Hence $|\mathcal{M}(N)| \ge 4$. As a result, $\{|\mathcal{M}(N)| : N \in \mathcal{N}^\bullet\} \subseteq \mathbb{N}_{\ge 4} \cup \{\infty\}$. Now take $x \in \mathbb{N}$ with $x > \mathsf{F}(N) + ab$. Since $x' := x - ab > \mathsf{F}(N)$, we see that $x' \in N$ and, therefore, $\mathsf{Z}(x')$ contains at least one factorization, namely z. So we can find two distinct factorizations of x by adding to z either a copies of b or b copies of a. Then $\mathsf{F}(N) + ab$ is an upper bound for $\mathcal{M}(N)$, which means that $|\mathcal{M}(N)| \in \mathbb{N}_{\ge 4}$. Thus, $\{|\mathcal{M}(N)| : N \in \mathcal{N}^\bullet\} \subseteq \mathbb{N}_{\ge 4}$.

To argue the reverse inclusion, suppose that $n \in \mathbb{N}_{\ge 4}$, and let us find $N \in \mathcal{N}$ with $|\mathcal{M}(N)| = n$. For $n = 4$, we can take the numerical semigroup $\langle 2, 3\rangle$ (see Fig. 1). For $n > 4$, consider the numerical semigroup

$$N = \langle n-2, n-1, \dots, 2(n-2) - 1\rangle.$$

It follows immediately that $\mathcal{A}(N) = \{n-2, n-1, \dots, 2(n-2) - 1\}$. In addition, it is not hard to see that $2(n-2), 2(n-2)+1 \in \mathcal{M}(N)$ while $k \notin \mathcal{M}(N)$ for any $k > 2(n-2)+1$. Consequently, $\mathcal{M}(N) = \mathcal{A}(N) \cup \{2(n-2), 2(n-2)+1\}$, which implies that $|\mathcal{M}(N)| = n$. Therefore $\{|\mathcal{M}(N)| : N \in \mathcal{N}\} \supseteq \mathbb{N}_{\ge 4}$, which completes the proof. □

Corollary 1 *The monoid* $(\mathbb{N}_0, +)$ *is the only numerical semigroup having infinitely many molecules.*

In Proposition 1 we have fully described the set $\{|\mathcal{M}(N)| : N \in \mathcal{N}\}$. A full description of the set $\{|\mathcal{M}(N) \setminus \mathcal{A}(N)| : N \in \mathcal{N}\}$ seems to be significantly more involved. However, the next theorem offers some evidence to believe that $\{|\mathcal{M}(N) \setminus \mathcal{A}(N)| : N \in \mathcal{N}\} = \mathbb{N}_{\ge 2} \cup \{\infty\}$.

Theorem 1 *The following statements hold.*

1. $\{|\mathcal{M}(N) \setminus \mathcal{A}(N)| : N \in \mathcal{N}^\bullet\} \subseteq \mathbb{N}_{\geq 2}$.
2. $|\mathcal{M}(N)\setminus\mathcal{A}(N)| = 2$ *for infinitely many numerical semigroups* N.
3. *For each* $k \in \mathbb{N}$, *there is a numerical semigroup* N *with* $|\mathcal{M}(N)\setminus\mathcal{A}(N)| > k$.

Proof To prove (1), take $N \in \mathcal{N}^\bullet$. Then we can assume that N has embedding dimension n with $n \geq 2$. Take $a_1, \dots, a_n \in \mathbb{N}$ with $a_1 < \cdots < a_n$ such that $N = \langle a_1, \dots, a_n \rangle$. Since $a_1 < a_2 < a_j$ for every $j = 3, \dots, n$, the elements $2a_1$ and $a_1 + a_2$ are two distinct molecules of N that are not atoms. Hence $\mathcal{M}(N) \setminus \mathcal{A}(N) \subseteq \mathbb{N}_{\geq 2} \cup \{\infty\}$. On the other hand, Proposition 1 guarantees that $|\mathcal{M}(N)| < \infty$, which implies that $|\mathcal{M}(N) \setminus \mathcal{A}(N)| < \infty$. As a result, the statement (1) follows.

To verify the statement (2), one only needs to consider for every $n \in \mathbb{N}$ the numerical semigroup $N_n := \{0\} \cup \mathbb{N}_{\geq n-2}$. The minimal set of generators of N_n is the $(n-2)$-element set $\{n-2, n-1, \dots, 2(n-2)-1\}$ and, as we have already argued in the proof of Proposition 1, the set $\mathcal{M}(N_n)\setminus\mathcal{A}(N_n)$ consists precisely of two elements.

Finally, let us prove condition (3). To do this, we first argue that for any $a, b \in \mathbb{N}_{\geq 2}$ with $\gcd(a, b) = 1$ the numerical semigroup $\langle a, b \rangle$ has exactly $ab - 1$ molecules (cf. Example 1). Assume $a < b$, take $N := \langle a, b \rangle$, and set

$$\mathcal{M} = \{ma + nb : 0 \leq m < b,\ 0 \leq n < a, \text{ and } (m, n) \neq (0, 0)\}.$$

Now take $x \in N$ to be a molecule of N. As $|\mathsf{Z}(x)| = 1$, the unique factorization $z := (c_1, c_2) \in \mathsf{Z}(x)$ (with $c_1, c_2 \in \mathbb{N}_0$) satisfies that $c_1 < b$; otherwise, we could exchange b copies of the atom a by a copies of the atom b to obtain another factorization of x. A similar argument ensures that $c_2 < a$. As a consequence, $\mathcal{M}(N) \subseteq \mathcal{M}$. On the other hand, if $ma + nb = m'a + n'b$ for some $m, m', n, n' \in \mathbb{N}_0$, then $\gcd(a, b) = 1$ implies that $b \mid m - m'$ and $a \mid n - n'$. Because of this observation, the element $(b-1)a + (a-1)b$ has only the obvious factorization, namely $(b-1, a-1)$. Since $(b-1)a + (a-1)b$ is a molecule satisfying that $y \mid_N (b-1)a + (a-1)b$ for every $y \in \mathcal{M}$, the inclusion $\mathcal{M} \subseteq \mathcal{M}(N)$ holds. Hence $|\mathcal{M}(N)| = |\mathcal{M}| = ab - 1$. To argue the statement (3) now, it suffices to take $N := \langle 2, 2k+1 \rangle$. □

We conclude this section with the following conjecture.

Conjecture 1 For every $n \in \mathbb{N}_{\geq 2}$, there exists a numerical semigroup N such that $|\mathcal{M}(N)\setminus\mathcal{A}(N)| = n$.

4 Molecules of Puiseux Monoids

4.1 Molecules of Generic Puiseux Monoids

In this section we study the sets of molecules of Puiseux monoids. We will argue that there are infinitely many non-finitely generated atomic Puiseux monoids P such that $|\mathcal{M}(P) \setminus \mathcal{A}(P)| = \infty$. On the other hand, we will prove that, unlike the case of numerical semigroups, there are infinitely many non-isomorphic atomic Puiseux monoids all whose molecules are, indeed, atoms. The last part of this section is dedicated to characterize the molecules of prime reciprocal Puiseux monoids.

Definition 4 A *Puiseux monoid* is an additive submonoid of $\mathbb{Q}_{\geq 0}$.

Clearly, every numerical semigroup is naturally isomorphic to a Puiseux monoid. However, Puiseux monoids are not necessarily finitely generated or atomic, as the next example illustrates. The atomic structure of Puiseux monoids has been investigated recently [11, 14, 15]. At the end of Sect. 2 we mentioned that the set of molecules of an atomic monoid is divisor-closed. The next example indicates that this property may not hold for non-atomic monoids.

Example 3 Consider the Puiseux monoid

$$P = \left\langle \frac{2}{5}, \frac{3}{5}, \frac{1}{2^n} : n \in \mathbb{N} \right\rangle.$$

First, observe that 0 is a limit point of $P^\bullet$, and so P cannot be finitely generated. After a few easy verifications, one can see that $\mathcal{A}(P) = \{2/5, 3/5\}$. On the other hand, it is clear that $1/2 \notin \langle 2/5, 3/5 \rangle$, so P is not atomic. Observe now that $\mathsf{Z}(1)$ contains only one factorization, namely $2/5 + 3/5$. Therefore $1 \in \mathcal{M}(P)$. Since $\mathsf{Z}(1/2)$ is empty, $1/2$ is not a molecule of P. However, $1/2 \mid_P 1$. As a result, $\mathcal{M}(P)$ is not divisor-closed.

Although the additive monoid $\mathbb{N}_0$ contains only one atom, it has infinitely many molecules. The next result implies that $\mathbb{N}_0$ is basically the only atomic Puiseux monoid having finitely many atoms and infinitely many molecules.

Proposition 2 *Let P be a Puiseux monoid. Then $|\mathcal{M}(P)| \in \mathbb{N}_{\geq 2}$ if and only if $|\mathcal{A}(P)| \in \mathbb{N}_{\geq 2}$.*

Proof Suppose first that $|\mathcal{M}(P)| \in \mathbb{N}_{\geq 2}$. As every atom is a molecule, $\mathcal{A}(P)$ is finite. Furthermore, note that if $\mathcal{A}(P) = \{a\}$, then every element of the set $S = \{na : n \in \mathbb{N}\}$ would be a molecule, which is not possible as $|S| = \infty$. As a result, $|\mathcal{A}(P)| \in \mathbb{N}_{\geq 2}$. Conversely, suppose that $|\mathcal{A}(P)| \in \mathbb{N}_{\geq 2}$. Since the elements in $P \setminus \langle \mathcal{A}(P) \rangle$ have no factorizations, $\mathcal{M}(P) = \mathcal{M}(\langle \mathcal{A}(P) \rangle)$. Therefore there is no loss in assuming that P is atomic. As $1 < |\mathcal{A}(P)| < \infty$, the monoid P is isomorphic to a nontrivial numerical semigroup. The proposition now follows from the fact that nontrivial numerical semigroups have finitely many molecules. □

Corollary 2 *If P is a Puiseux monoid, then $|\mathcal{M}(P)| \neq 1$.*

The set of atoms of a numerical semigroup is always strictly contained in its set of molecules. However, there are many atomic Puiseux monoids which do not satisfy such a property. Before proceeding to formalize this observation, let us mention that if two Puiseux monoids P and P' are isomorphic, then there exists $q \in \mathbb{Q}_{>0}$ such that $P' = qP$; this is a consequence of [13, Proposition 3.2(1)].

Theorem 2 (cf. Theorem 1(1)) *There are infinitely many non-isomorphic atomic Puiseux monoids P satisfying that $\mathcal{M}(P) = \mathcal{A}(P)$.*

Proof Let $\mathcal{S} = \{S_n : n \in \mathbb{N}\}$ be a collection of infinite and pairwise-disjoint sets of primes. Now take $S = S_n$ for some arbitrary $n \in \mathbb{N}$, and label the primes in S strictly increasingly by $p_1, p_2, \dots$. Recall that $\mathsf{D}_S(r)$ denotes the set of primes in S dividing $\mathsf{d}(r)$ and that $\mathsf{D}_S(R) = \cup_{r \in R} \mathsf{D}_S(r)$ for $R \subseteq \mathbb{Q}_{>0}$. We proceed to construct a Puiseux monoid P_S satisfying that $\mathsf{D}_S(P_S) = S$.

Take $P_1 := \langle 1/p_1 \rangle$ and $P_2 := \langle P_1, 2/(p_1 p_2) \rangle$. In general, suppose that P_k is a finitely generated Puiseux monoid such that $\mathsf{D}_S(P_k) \subset S$, and let $r_1, \dots, r_{n_k}$ be all the elements in P_k which can be written as a sum of two atoms. Clearly, $n_k \ge 1$. Because $|S| = \infty$, one can take $p'_1, \dots, p'_{n_k}$ to be primes in $S \setminus \mathsf{D}_S(P_k)$ satisfying that $p'_i \nmid \mathsf{n}(r_i)$. Now consider the following finitely generated Puiseux monoid

$$P_{k+1} := \left\langle P_k \cup \left\{ \frac{r_1}{p'_1}, \dots, \frac{r_{n_k}}{p'_{n_k}} \right\} \right\rangle.$$

For every $i \in \{1, \dots, n_k\}$, there is only one element in $P_k \cup \{r_1/p'_1, \dots, r_{n_k}/p'_{n_k}\}$ whose denominator is divisible by p'_i, namely r_i/p'_i. Therefore $r_i/p'_i \in \mathcal{A}(P_{k+1})$ for $i = 1, \dots, n_k$. To check that $\mathcal{A}(P_k) \subset \mathcal{A}(P_{k+1})$, fix $a \in \mathcal{A}(P_k)$ and take

$$z := \sum_{i=1}^{m} \alpha_i a_i + \sum_{i=1}^{n_k} \beta_i \frac{r_i}{p'_i} \in \mathsf{Z}_{P_{k+1}}(a), \tag{1}$$

where $a_1, \dots, a_m$ are pairwise distinct atoms in $\mathcal{A}(P_{k+1}) \cap P_k$ and α_i, β_j are nonnegative coefficients for $i = 1, \dots, m$ and $j = 1, \dots, n_k$. In particular, $a_1, \dots, a_m \in \mathcal{A}(P_k)$. For each $i = 1, \dots, n_k$, the fact that the p'_i-adic valuation of a is nonnegative implies that $p'_i \mid \beta_i$. Hence

$$a = \sum_{i=1}^{m} \alpha_i a_i + \sum_{i=1}^{n_k} \beta'_i r_i,$$

where $\beta'_i = \beta_i / p'_i \in \mathbb{N}_0$. Since $r_i \in \mathcal{A}(P_k) + \mathcal{A}(P_k)$ and $(\beta_i/p'_i) r_i \mid_{P_k} a$ for every $i = 1, \dots, n_k$, one obtains that $\beta_1 = \dots = \beta_{n_k} = 0$. As a result, $a = \sum_{i=1}^{m} \alpha_i a_i$. Because $a \in \mathcal{A}(P_k)$, the factorization $\sum_{i=1}^{m} \alpha_i a_i$ in $\mathsf{Z}_{P_k}(a)$ must have length 1, i.e, $\sum_{i=1}^{m} \alpha_i = 1$. Thus, $\sum_{i=1}^{m} \alpha_i + \sum_{i=1}^{n_k} \beta_i = 1$, which means that z has length 1 and so $a \in \mathcal{A}(P_{k+1})$. As a result, the inclusion $\mathcal{A}(P_k) \subset \mathcal{A}(P_{k+1})$ holds. Observe that because $n_k \ge 1$, the previous containment must be strict. Now set

$$P_S = \bigcup_{k \in \mathbb{N}} P_k.$$

Let us verify that P_S is an atomic monoid satisfying that $\mathcal{A}(P_S) = \cup_{k\in\mathbb{N}}\mathcal{A}(P_k)$. Since P_k is atomic for every $k \in \mathbb{N}$, the inclusion chain $\mathcal{A}(P_1) \subset \mathcal{A}(P_2) \subset \dots$ implies that $P_1 \subset P_2 \subset \dots$. In addition, if $a_0 = a_1 + \cdots + a_m$ for $m \in \mathbb{N}$ and $a_0, a_1, \dots, a_m \in P_S$, then $a_0 = a_1 + \cdots + a_m$ will also hold in P_k for some $k \in \mathbb{N}$ large enough. This immediately implies that $\cup_{k\in\mathbb{N}}\mathcal{A}(P_k) \subseteq \mathcal{A}(P_S)$. Since the reverse inclusion follows trivially, $\mathcal{A}(P_S) = \cup_{k\in\mathbb{N}}\mathcal{A}(P_k)$. To check that P_S is atomic, take $x \in P_S^\bullet$. Then there exists $k \in \mathbb{N}$ such that $x \in P_k$ and, because P_k is atomic, $x \in \langle\mathcal{A}(P_k)\rangle \subseteq \langle\mathcal{A}(P_S)\rangle$. Hence P_S is atomic.

To check that $\mathcal{M}(P_S) = \mathcal{A}(P_S)$, suppose that m is a molecule of P_S. Take $K \in \mathbb{N}$ such that $m \in P_k$ for every $k \geq K$. Since $\mathcal{A}(P_k) \subset \mathcal{A}(P_{k+1}) \subset \dots$, we obtain that $\mathsf{Z}_{P_k}(m) \subseteq \mathsf{Z}_{P_{k+1}}(m) \subseteq \dots$. Moreover, $\cup_{k\geq K}\mathcal{A}(P_k) = \mathcal{A}(P_S)$ implies that $\cup_{k\geq K}\mathsf{Z}_{P_k}(m) = \mathsf{Z}_{P_S}(m)$. Now suppose for a contradiction that $m = \sum_{j=1}^{i} a_j$ for $i \in \mathbb{N}_{\geq 2}$, where $a_1, \dots, a_i \in \mathcal{A}(P_S)$. Take $j \in \mathbb{N}_{\geq K}$ such that $a_1, \dots, a_i \in \mathcal{A}(P_j)$. Then the way in which P_{j+1} was constructed ensures that $|\mathsf{Z}_{P_{j+1}}(a_1 + a_2)| \geq 2$ and, therefore, $|\mathsf{Z}_{P_{j+1}}(m)| \geq 2$. As $\mathsf{Z}_{P_{j+1}}(m) \subseteq \mathsf{Z}_{P_S}(m)$, it follows that $|\mathsf{Z}_{P_S}(m)| \geq 2$, which contradicts that m is a molecule. Hence $\mathcal{M}(P_S) = \mathcal{A}(P_S)$.

Finally, we argue that the monoids constructed are not isomorphic. Let S and S' be two distinct members of the collection $\mathcal{S}$ and suppose, by way of contradiction, that $\psi \colon P_S \to P_{S'}$ is a monoid isomorphism. Because homomorphisms of Puiseux monoids are given by rational multiplications, there exists $q \in \mathbb{Q}_{>0}$ such that $P_{S'} = qP_S$. In this case, all but finitely many primes in $\mathsf{D}_{\mathbb{P}}(P_S)$ belong to $\mathsf{D}_{\mathbb{P}}(P_{S'})$. Since $\mathsf{D}_{\mathbb{P}}(P_S) \cap \mathsf{D}_{\mathbb{P}}(P_{S'}) = \emptyset$ when $S \neq S'$, we get a contradiction. □

4.2 Molecules of Prime Reciprocal Monoids

For the remaining of this section, we focus our attention on the class consisting of all prime reciprocal monoids.

Definition 5 Let S be a nonempty set of primes. A Puiseux monoid P is *prime reciprocal over* S if there exists a set of positive rationals R such that $P = \langle R\rangle$, $\mathsf{d}(R) = S$, and $\mathsf{d}(r) = \mathsf{d}(r')$ implies $r = r'$ for all $r, r' \in R$.

Within the scope of this paper, the term *prime reciprocal monoid* refers to a Puiseux monoid that is prime reciprocal over some nonempty set of primes. Let us remark that if a Puiseux monoid P is prime reciprocal, then there exists a *unique* $S \subseteq \mathbb{P}$ such that P is prime reciprocal over S. It is easy to verify that every prime reciprocal Puiseux monoid is atomic.

Proposition 3 (cf. Theorem 1(1)) *There exist infinitely many non-finitely generated atomic Puiseux monoids P such that $|\mathcal{M}(P)\setminus\mathcal{A}(P)| = \infty$.*

Proof As in the proof of Theorem 2, let $\mathcal{S} = \{S_n : n \in \mathbb{N}\}$ be a collection of infinite and pairwise-disjoint subsets of $\mathbb{P}\setminus\{2\}$. For every $n \in \mathbb{N}$, let P_n be a prime reciprocal Puiseux monoid over S_n. Fix $a \in \mathcal{A}(P_n)$, and take a factorization

$$z := \sum_{i=1}^{k} \alpha_i a_i \in \mathsf{Z}(2a)$$

for some $k \in \mathbb{N}$, pairwise distinct atoms $a_1, \dots, a_k$, and $\alpha_1, \dots, \alpha_k \in \mathbb{N}_0$. Since $\mathsf{d}(a) \neq 2$, after applying the $\mathsf{d}(a)$-adic valuation on both sides of the equality $2a = \sum_{i=1}^{t} \alpha_i a_i$, one obtains that $z = 2a$. So $2a \in \mathcal{M}(P_n) \setminus \mathcal{A}(P_n)$ and, as a result, $|\mathcal{M}(P_n) \setminus \mathcal{A}(P_n)| = \infty$. Now suppose, by way of contradiction, that $P_i \cong P_j$ for some $i, j \in \mathbb{N}$ with $i \neq j$. Since isomorphisms of Puiseux monoids are given by rational multiplication, there exists $q \in \mathbb{Q}_{>0}$ such that $P_j = qP_i$. However, this implies that only finitely many primes in $\mathsf{d}(P_i)$ are not contained in $\mathsf{d}(P_j)$, which contradicts that $S_i \cap S_j = \emptyset$. Hence no two monoids in $\{P_n : n \in \mathbb{N}\}$ are isomorphic, and the proposition follows. □

Theorem 2 and Proposition 3 ensure the existence of infinitely many non-finitely generated atomic Puiseux monoids P and Q with $|\mathcal{M}(P)\setminus\mathcal{A}(P)| = 0$ and $|\mathcal{M}(Q)\setminus\mathcal{A}(Q)| = \infty$.

Conjecture 2 (cf. Conjecture 1) For every $n \in \mathbb{N}$ there exists a non-finitely generated atomic Puiseux monoid P satisfying that $|\mathcal{M}(P)\setminus\mathcal{A}(P)| = n$.

Before characterizing the molecules of prime reciprocal monoids, let us introduce the concept of maximal multiplicity. Let P be a Puiseux monoid. For $x \in P$ and $a \in \mathcal{A}(P)$ we define the *maximal multiplicity* of a in x to be

$$\mathsf{m}(a, x) := \max\{n \in \mathbb{N}_0 : na \mid_P x\}.$$

Proposition 4 *Let P be a prime reciprocal monoid. If $x \in P$ satisfies that $\mathsf{m}(a, x) < \mathsf{d}(a)$ for all $a \in \mathcal{A}(P)$, then $x \in \mathcal{M}(P)$.*

Proof Suppose, by way of contradiction, that $x \notin \mathcal{M}(P)$. Then there exist $k \in \mathbb{N}$, elements $\alpha_i, \beta_i \in \mathbb{N}_0$ (for $i = 1, \dots, k$), and pairwise distinct atoms $a_1, \dots, a_k$ such that

$$z := \sum_{i=1}^{k} \alpha_i a_i \text{ and } z' := \sum_{i=1}^{k} \beta_i a_i$$

are two distinct factorizations in $\mathsf{Z}(x)$. As $z \neq z'$, there is an index $i \in \{1, \dots, k\}$ such that $\alpha_i \neq \beta_i$. Now we can apply the $\mathsf{d}(a_i)$-adic valuation to both sides of the equality

$$\sum_{i=1}^{k} \alpha_i a_i = \sum_{i=1}^{k} \beta_i a_i$$

to verify that $\mathsf{d}(a_i) \mid \beta_i - \alpha_i$. As $\alpha_i \neq \beta_i$, we obtain that

$$\mathsf{m}(a_i, x) \geq \max\{\alpha_i, \beta_i\} \geq \mathsf{d}(a_i).$$

However, this contradicts the fact that $\mathsf{m}(a, x) < \mathsf{d}(a)$ for all $a \in \mathcal{A}(P)$. Hence $x \in \mathcal{M}(P)$. □

For $S \subseteq \mathbb{P}$, we call the monoid $E_S := \langle 1/p : p \in S \rangle$ the *elementary* prime reciprocal monoid over S; if $S = \mathbb{P}$ we say that E_S is *the* elementary prime reciprocal monoid. It was proved in [15, Section 5] that every submonoid of the elementary prime reciprocal monoid is atomic. This gives a large class of non-finitely generated atomic Puiseux monoids, which contains each prime reciprocal monoid.

Proposition 5 *Let S be an infinite set of primes, and let E_S be the elementary prime reciprocal monoid over S. For $x \in E_S$, the following conditions are equivalent:*

1. $x \in \mathcal{M}(E_S)$;
2. 1 *does not divide* x *in* E_S;
3. $\mathsf{m}(a, x) < \mathsf{d}(a)$ *for all* $a \in \mathcal{A}(E_S)$;
4. *If* $a_1, \dots, a_n \in \mathcal{A}(E_S)$ *are distinct atoms and* $\alpha_1, \dots, \alpha_n \in \mathbb{N}_0$ *satisfy that* $\sum_{j=1}^n \alpha_j a_j \in \mathsf{Z}(x)$, *then* $\alpha_j < \mathsf{d}(a_j)$ *for each* $j = 1, \dots, n$.

Proof First, let us recall that since E_S is atomic, $\mathcal{M}(E_S)$ is divisor-closed. On the other hand, note that for any two distinct atoms $a, a' \in \mathcal{A}(E_S)$, both factorizations $\mathsf{d}(a)\, a$ and $\mathsf{d}(a')\, a'$ are in $\mathsf{Z}(1)$. Therefore $1 \notin \mathcal{M}(E_S)$. Because the set of molecules of E_S is divisor-closed, $1 \nmid_{E_S} m$ for any $m \in \mathcal{M}(E_S)$; in particular, $1 \nmid_{E_S} x$. Thus, (1) implies (2). If $\mathsf{m}(a, x) \geq \mathsf{d}(a)$ for $a \in \mathcal{A}(E_S)$, then

$$x = \mathsf{m}(a, x)\, a + y = 1 + (\mathsf{m}(a, x) - \mathsf{d}(a))\, a + y$$

for some $y \in E_S$. As a result, $1 \mid_{E_S} x$, from which we can conclude that (2) implies (3). It is obvious that (3) and (4) are equivalent conditions. Finally, the fact that (3) implies (1) follows from Proposition 4. □

Corollary 3 *Let S be an infinite set of primes, and let E_S be the elementary prime reciprocal monoid over S. Then $|\mathsf{Z}(x)| = \infty$ for all $x \notin \mathcal{M}(E_S)$.*

In order to describe the set of molecules of an arbitrary prime reciprocal monoid, we need to cast its atoms into two categories.

Definition 6 Let P be a prime reciprocal monoid. We say that $a \in \mathcal{A}(P)$ is *stable* if the set $\{a' \in \mathcal{A}(P) : \mathsf{n}(a') = \mathsf{n}(a)\}$ is infinite, otherwise we say that a is *unstable*. If every atom of P is stable (resp., unstable), then we call P *stable* (resp., *unstable*).

For a prime reciprocal monoid P, we let $\mathcal{S}(P)$ denote the submonoid of P generated by the set of stable atoms. Similarly, we let $\mathcal{U}(P)$ denote the submonoid of P generated by the set of unstable atoms. Clearly, P is stable (resp., unstable) if and only if $P = \mathcal{S}(P)$ (resp., $P = \mathcal{U}(P)$). In addition, $P = \mathcal{S}(P) + \mathcal{U}(P)$, and $\mathcal{S}(P) \cap \mathcal{U}(P)$ is trivial only when either $\mathcal{S}(P)$ or $\mathcal{U}(P)$ is trivial. Clearly, if P is stable, then it cannot be finitely generated. Finally, we say that $u \in \mathcal{U}(P)$ is *absolutely unstable* provided that u is not divisible by any stable atom in P, and we let $\mathcal{U}^a(P)$ denote the set of all absolutely unstable elements of P.

Example 4 Let $\{p_n\}$ be the strictly increasing sequence with underlying set $\mathbb{P} \setminus \{2\}$, and consider the prime reciprocal monoid P defined as

$$P := \left\langle \frac{3+(-1)^n}{p_{2n-1}}, \frac{p_{2n}-1}{p_{2n}} : n \in \mathbb{N} \right\rangle.$$

Set $a_n = \frac{3+(-1)^n}{p_{2n-1}}$ and $b_n = \frac{p_{2n}-1}{p_{2n}}$. It is clear that P is an atomic monoid with $\mathcal{A}(P) = \{a_n, b_n : n \in \mathbb{N}\}$. As both sets $\{n \in \mathbb{N} : \mathsf{n}(a_n) = 2\}$ and $\{n \in \mathbb{N} : \mathsf{n}(a_n) = 4\}$ have infinite cardinality, a_n is a stable atom for every $n \in \mathbb{N}$. In addition, since $\{\mathsf{n}(b_n)\}$ is a strictly increasing sequence bounded below by $\mathsf{n}(b_1) = 4$ and $\mathsf{n}(a_n) \in \{2, 4\}$, the element b_n is an unstable atom for every $n \in \mathbb{N}_{\ge 2}$. Also, notice that $4/3 = 2a_1 \in \mathcal{S}(P)$, but $4/3 \notin \mathcal{U}(P)$ because $\mathsf{d}(4/3) = 3 \notin \mathsf{d}(\mathcal{U}(P))$. Furthermore, for every $n \in \mathbb{N}$ the element $u_n := (p_{2n} - 1)b_n \in \mathcal{U}(P)$ is not in $\mathcal{S}(P)$ because $p_{2n} = \mathsf{d}(u_n) \notin \mathsf{d}(\mathcal{S}(P))$. However, $\mathcal{S}(P) \cap \mathcal{U}(P) \neq \emptyset$ since the element $4 = 6a_1 = 5b_1$ belongs to both $\mathcal{S}(P)$ and $\mathcal{U}(P)$. Finally, we claim that $2b_n$ is absolutely unstable for every $n \in \mathbb{N}$. If this were not the case, then $2b_k \notin \mathcal{M}(P)$ for some $k \in \mathbb{N}$. By Proposition 4 there exists $a \in \mathcal{A}(P)$ such that $\mathsf{m}(a, 2b_k) \ge \mathsf{d}(a)$. In this case, one would obtain that $2b_k \ge \mathsf{m}(a, 2b_k)a \ge \mathsf{d}(a)a = \mathsf{n}(a) \ge 2$, contradicting that $b_n < 1$ for every $n \in \mathbb{N}$. Thus, $2b_n \in \mathcal{U}^a(P)$ for every $n \in \mathbb{N}$.

Proposition 6 *Let P be a prime reciprocal monoid that is stable, and let $x \in P$. Then $x \in \mathcal{M}(P)$ if and only if $\mathsf{n}(a)$ does not divide x in P for any $a \in \mathcal{A}(P)$.*

Proof For the direct implication, assume that $x \in \mathcal{M}(P)$ and suppose, by way of contradiction, that $\mathsf{n}(a) \mid_P x$ for some $a \in \mathcal{A}(P)$. Since a is a stable atom, there exist $p_1, p_2 \in \mathbb{P}$ with $p_1 \neq p_2$ such that $\gcd(p_1 p_2, \mathsf{n}(a)) = 1$ and $\mathsf{n}(a)/p_1, \mathsf{n}(a)/p_2 \in \mathcal{A}(P)$. As $\mathsf{n}(a) \mid_P x$, we can take $a_1, \dots, a_k \in \mathcal{A}(P)$ such that $x = \mathsf{n}(a) + a_1 + \cdots + a_k$. Therefore

$$p_1 \frac{\mathsf{n}(a)}{p_1} + a_1 + \cdots + a_k \text{ and } p_2 \frac{\mathsf{n}(a)}{p_2} + a_1 + \cdots + a_k$$

are two distinct factorizations in $\mathsf{Z}(x)$, contradicting that x is a molecule. Conversely, suppose that x is not a molecule. Consider two distinct factorizations $z := \sum_{i=1}^k \alpha_i a_i$ and $z' := \sum_{i=1}^k \beta_i a_i$ in $\mathsf{Z}(x)$, where $k \in \mathbb{N}$, $\alpha_i, \beta_i \in \mathbb{N}_0$, and $a_1, \dots, a_k \in \mathcal{A}(P)$ are pairwise distinct atoms. Pick an index $j \in \{1, \dots, k\}$ such that $\alpha_j \neq \beta_j$ and assume, without loss of generality, that $\alpha_j < \beta_j$. After applying the $\mathsf{d}(a_j)$-adic valuations on both sides of the equality

$$\sum_{i=1}^k \alpha_i a_i = \sum_{i=1}^k \beta_i a_i$$

one finds that the prime $\mathsf{d}(a_j)$ divides $\beta_j - \alpha_j$. Therefore $\beta_j > \mathsf{d}(a_j)$ and so

$$x = \mathsf{n}(a_j) + (\beta_j - \mathsf{d}(a_j))a_j + \sum_{i \neq j} \alpha_i a_i.$$

Hence $\mathsf{n}(a_j) \mid_P x$, which concludes the proof. □

Observe that the reverse implication of Proposition 6 does not require that $\mathcal{S}(P) = P$. However, the stability of P is required for the direct implication to hold as the following example illustrates.

Example 5 Let $\{p_n\}$ be the strictly increasing sequence with underlying set $\mathbb{P} \setminus \{2\}$, and consider the unstable prime reciprocal monoid

$$P := \left\langle \frac{1}{2}, \frac{p_n^2 - 1}{p_n} : n \in \mathbb{N} \right\rangle.$$

As the smallest two atoms of P are $1/2$ and $8/3$, it immediately follows that $m := 2(1/2) + 8/3 \notin \langle 1/2 \rangle$ must be a molecule of P. In addition, notice that $1 = \mathsf{n}(1/2)$ divides m in P.

We conclude this section characterizing the molecules of prime reciprocal monoids.

Theorem 3 *Let P be a prime reciprocal monoid. Then $x \in P$ is a molecule if and only if $x = s + u$ for some $s \in \mathcal{S}(P) \cap \mathcal{M}(P)$ and $u \in \mathcal{U}^a(P) \cap \mathcal{M}(P)$.*

Proof First, suppose that x is a molecule. As $P = \mathcal{S}(P) + \mathcal{U}(P)$, there exist $s \in \mathcal{S}(P)$ and $u \in \mathcal{U}(P)$ such that $x = s + u$. The fact that $x \in \mathcal{M}(P)$ guarantees that $s, u \in \mathcal{M}(P)$. On the other hand, since $|\mathsf{Z}(u)| = 1$ and u can be factored using only unstable atoms, u cannot be divisible by any stable atom in P. Thus, $u \in \mathcal{U}^a(P)$, and the direct implication follows.

For the reverse implication, assume that $x = s + u$, where $s \in \mathcal{S}(P) \cap \mathcal{M}(P)$ and $u \in \mathcal{U}^a(P) \cap \mathcal{M}(P)$. We first check that x can be uniquely expressed as a sum of two elements s and u contained in the sets $\mathcal{S}(P) \cap \mathcal{M}(P)$ and $\mathcal{U}^a(P) \cap \mathcal{M}(P)$, respectively. To do this, suppose that $x = s + u = s' + u'$, where $s' \in \mathcal{S}(P) \cap \mathcal{M}(P)$ and $u' \in \mathcal{U}^a(P) \cap \mathcal{M}(P)$. Take pairwise distinct stable atoms $a_1, \dots, a_k$ of P for some $k \in \mathbb{N}$ such that $z = \sum_{i=1}^k \alpha_i a_i \in \mathsf{Z}_P(s)$ and $z' = \sum_{i=1}^k \alpha'_i a_i \in \mathsf{Z}_P(s')$, where $\alpha_j, \alpha'_j \in \mathbb{N}_0$ for $j = 1, \dots, k$. Because u and u' are absolutely unstable elements, they are not divisible in P by any of the atoms a_i's. Thus, $\mathsf{d}(a_j) \nmid \mathsf{d}(u)$ and $\mathsf{d}(a_j) \nmid \mathsf{d}(u')$ for any $j \in \{1, \dots, k\}$. Now for each $j = 1, \dots, k$ we can apply the $\mathsf{d}(a_j)$-adic valuation in both sides of the equality

$$u + \sum_{i=1}^k \alpha_i a_i = u' + \sum_{i=1}^k \alpha'_i a_i$$

to conclude that the prime $\mathsf{d}(a_j)$ must divide $\alpha_j - \alpha'_j$. Therefore either $z = z'$ or there exists $j \in \{1, \dots, k\}$ such that $|\alpha_j - \alpha'_j| > \mathsf{d}(a_j)$. Suppose that $|\alpha_j - \alpha'_j| > \mathsf{d}(a_j)$ for some j, and say $\alpha_j > \alpha'_j$. As $\alpha_j > \mathsf{d}(a_j)$, one can replace $\alpha_j a_j$ by $(\alpha_j - \mathsf{d}(a_j))a_j + \mathsf{n}(a_j)$ in $s = \phi(z) = \alpha_1 a_1 + \dots + \alpha_k a_k$ to find that $\mathsf{n}(a_j)$ divides s in $\mathcal{S}(P)$, which contradicts Proposition 6. Then $z = z'$. Therefore $s' = s$ and $u' = u$.

Finally, we argue that $x \in \mathcal{M}(P)$. Write $x = \sum_{i=1}^{\ell} \gamma_i a_i + \sum_{i=1}^{\ell} \beta_i b_i$ for $\ell \in \mathbb{N}_{\ge k}$, pairwise distinct stable atoms $a_1, \dots, a_\ell$ (where $a_1, \dots, a_k$ are the atoms showing up in z), pairwise distinct unstable atoms $b_1, \dots, b_\ell$, and coefficients $\gamma_i, \beta_i \in \mathbb{N}_0$ for $i = 1, \dots, \ell$. Set $z''' := \sum_{i=1}^{\ell} \gamma_i a_i$ and $w''' = \sum_{i=1}^{\ell} \beta_i b_i$. Note that, *a priori*, $\phi(z''')$ and $\phi(w''')$ are not necessarily molecules. As in the previous paragraph, we can apply $\mathsf{d}(a_j)$-adic valuation to both sides of the equality

$$u + \sum_{i=1}^{k} \alpha_i a_i = \sum_{i=1}^{\ell} \gamma_i a_i + \sum_{i=1}^{\ell} \beta_i b_i$$

to find that $z''' = z$. Hence $\phi(z''') = s$ and $\phi(w''') = u$ are both molecules. Therefore z''' must be the unique factorization of s, while w''' must be the unique factorization of u. As a result, $x \in \mathcal{M}(P)$. □

5 Molecules of Puiseux Algebras

Let M be a monoid and let R be a commutative ring with identity. Then $R[X; M]$ denotes the ring of all functions $f : M \to R$ having finite *support*, which means that $\mathsf{Supp}(f) := \{s \in M : f(s) \neq 0\}$ is finite. We represent an element $f \in R[X; M]$ by

$$f(X) = \sum_{i=1}^{n} f(s_i) X^{s_i},$$

where $s_1, \dots, s_n$ are the elements in $\mathsf{Supp}(f)$. The ring $R[X; M]$ is called the *monoid ring* of M *over* R, and the monoid M is called the *exponent monoid* of $R[X; M]$. For a field F, we will say that $F[X; M]$ is a *monoid algebra*. As we are primarily interested in the molecules of monoid algebras of Puiseux monoids, we introduce the following definition.

Definition 7 If F is a field and P is a Puiseux monoid, then we say that $F[X; P]$ is a *Puiseux algebra*. If N is a numerical semigroup, then $F[X; N]$ is called a *numerical semigroup algebra*.

Let $F[X; P]$ be a Puiseux algebra. We write any element $f \in F[X; P] \setminus \{0\}$ in *canonical representation*, that is, $f(X) = \alpha_1 X^{q_1} + \dots + \alpha_k X^{q_k}$ with $\alpha_i \neq 0$ for

$i = 1, \dots, k$ and $q_1 > \cdots > q_k$. It is clear that any element of $F[X; P] \setminus \{0\}$ has a unique canonical representation. In this case, $\deg(f) := q_1$ is called the *degree* of f, and we obtain that the degree identity $\deg(fg) = \deg(f) + \deg(g)$ holds for all $f, g \in F[X; P] \setminus \{0\}$. As for polynomials, we say that f is a *monomial* if $k = 1$. It is not hard to verify that $F[X; P]$ is an integral domain with group of units $F^\times$, although this follows from [10, Theorem 8.1] and [10, Theorem 11.1]. Finally, note that, unless $P \cong (\mathbb{N}_0, +)$, no monomial of $F[X; P]$ can be a prime element; this is a consequence of the trivial fact that non-cyclic Puiseux monoids do not contain prime elements.

For an integral domain R, we let R_{red} denote the reduced monoid of the multiplicative monoid of R.

Definition 8 Let R be an integral domain. We call a nonzero non-unit $r \in R$ a *molecule* if $rR^\times$ is a molecule of R_{red}.

Let R be an integral domain. By simplicity, we let $\mathcal{A}(R)$, $\mathcal{M}(R)$, $\mathsf{Z}(R)$, and ϕ_R denote $\mathcal{A}(R_{\text{red}})$, $\mathcal{M}(R_{\text{red}})$, $\mathsf{Z}(R_{\text{red}})$, and $\phi_{R_{\text{red}}}$, respectively. In addition, for a nonzero non-unit $r \in R$, we let $\mathsf{Z}_R(r)$ and $\mathsf{L}_R(r)$ denote $\mathsf{Z}_{R_{\text{red}}}(rR^\times)$ and $\mathsf{L}_{R_{\text{red}}}(rR^\times)$, respectively.

Proposition 7 *Let F be a field, and let P be a Puiseux monoid. For a nonzero $\alpha \in F$, a monomial $X^q \in \mathcal{M}(F[X; P])$ if and only if $q \in \mathcal{M}(P)$.*

Proof Consider the canonical monoid monomorphism $\mu \colon P \to F[X; P] \setminus \{0\}$ given by $\mu(q) = X^q$. It follows from [4, Lemma 3.1] that an element $a \in P$ is an atom if and only if the monomial X^a is irreducible in $F[X; P]$ (or, equivalently, an atom in the reduced multiplicative monoid of $F[X; P]$). Therefore μ lifts canonically to the monomorphism $\bar{\mu} \colon \mathsf{Z}(P) \to \mathsf{Z}(F[X; P])$ determined by the assignments $a \mapsto X^a$ for each $a \in \mathcal{A}(P)$, preserving not only atoms but also factorizations of the same element. Put formally, this means that the diagram

$$\begin{array}{ccc} \mathsf{Z}(P) & \xrightarrow{\bar{\mu}} & \mathsf{Z}(F[X; P]) \\ \phi_P \big\downarrow & & \phi_{F[X;P]} \big\downarrow \\ P & \xrightarrow{\mu} & F[X; P]_{\text{red}} \end{array}$$

commutes, and the (fiber) restriction maps $\bar{\mu}_q \colon \mathsf{Z}_P(q) \to \mathsf{Z}_{F[X;P]}(X^q)$ of $\bar{\mu}$ are bijections for every $q \in P$. Hence $|\mathsf{Z}_P(q)| = 1$ if and only if $|\mathsf{Z}_{F[X;P]}(X^q)| = 1$ for all $q \in P^\bullet$, which concludes our proof. □

Corollary 4 *For each field F, there exists an atomic Puiseux monoid P whose Puiseux algebra satisfies that $|\mathcal{M}(F[X; P]) \setminus \mathcal{A}(F[X; P])| = \infty$.*

Proof It is an immediate consequence of Proposition 3 and Proposition 7. □

The *difference group* $\mathsf{gp}(M)$ of a monoid M is the abelian group (unique up to isomorphism) satisfying that any abelian group containing a homomorphic image of M will also contain a homomorphic image of $\mathsf{gp}(M)$. An element $x \in \mathsf{gp}(M)$ is called a *root element* of M if $nx \in M$ for some $n \in \mathbb{N}$. The subset $\widetilde{M}$ of $\mathsf{gp}(M)$ consisting of all root elements of M is called the *root closure* of M. If $\widetilde{M} = M$, then M is called *root-closed*. From now on, we assume that each Puiseux monoid P we mention here is root-closed. Before providing a characterization for the irreducible elements of $F[X; P]$, let us argue the following two easy lemmas.

Lemma 1 *Let P be a Puiseux monoid. Then $\mathsf{d}(P^\bullet)$ is closed under taking least common multiples.*

Proof Take $d_1, d_2 \in \mathsf{d}(P^\bullet)$ and $q_1, q_2 \in P^\bullet$ with $\mathsf{d}(q_1) = d_1$ and $\mathsf{d}(q_2) = d_2$. Now set $d = \gcd(d_1, d_2)$ and $n = \gcd(\mathsf{n}(q_1), \mathsf{n}(q_2))$. It is clear that n is the greatest common divisor of $(d_2/d)\mathsf{n}(q_1)$ and $(d_1/d)\mathsf{n}(q_2)$. So there exist $m \in \mathbb{N}$ and $c_1, c_2 \in \mathbb{N}_0$ such that

$$n\big(1 + m \operatorname{lcm}(d_1, d_2)\big) = c_1 \frac{d_2}{d}\mathsf{n}(q_1) + c_2 \frac{d_1}{d}\mathsf{n}(q_2). \tag{2}$$

Using the fact that $d \operatorname{lcm}(d_1, d_2) = d_1 d_2$, one obtains that

$$\frac{n\big(1 + m \operatorname{lcm}(d_1, d_2)\big)}{\operatorname{lcm}(d_1, d_2)} = c_1 q_1 + c_2 q_2 \in P$$

after dividing both sides of the equality (2) by $\operatorname{lcm}(d_1, d_2)$. In addition, note that $n(1 + m \operatorname{lcm}(d_1, d_2))$ and $\operatorname{lcm}(d_1, d_2)$ are relatively prime. Hence $\operatorname{lcm}(d_1, d_2) \in \mathsf{d}(P^\bullet)$, from which the lemma follows. □

Lemma 2 *Let P be a root-closed Puiseux monoid containing* 1*. Then $1/d \in P$ for all $d \in \mathsf{d}(P^\bullet)$.*

Proof Let $d \in \mathsf{d}(P^\bullet)$, and take $r \in P^\bullet$ such that $\mathsf{d}(r) = d$. As $\gcd(\mathsf{n}(r), \mathsf{d}(r)) = 1$, there exist $a, b \in \mathbb{N}_0$ such that $a\mathsf{n}(r) - b\,\mathsf{d}(r) = 1$. Therefore

$$\frac{1}{d} = \frac{a\mathsf{n}(r) - b\,\mathsf{d}(r)}{d} = ar - b \in \mathsf{gp}(P).$$

This, along with the fact that $d(1/d) = 1 \in P$, ensures that $1/d$ is a root element of P. Since P is root-closed, it must contain $1/d$, which concludes our argument. □

We are in a position now to characterize the irreducibles of $F[X; P]$.

Proposition 8 *Let F be a field, and let P be a root-closed Puiseux monoid containing* 1*. Then $f \in F[X; P] \setminus F$ is irreducible in $F[X; P]$ if and only if $f(X^m)$ is irreducible in $F[X]$ for every $m \in \mathsf{d}(P^\bullet)$ that is a common multiple of the elements of* $\mathsf{d}(\mathsf{Supp}(f))$.

Proof Suppose first that $f \in F[X; P] \setminus F$ is an irreducible element of $F[X; P]$, and let $m \in \mathsf{d}(P^\bullet)$ be a common multiple of the elements of $\mathsf{d}(\mathsf{Supp}(f))$. Then $f(X^m)$ is an element of the polynomial ring $F[X]$. Take $g, h \in F[X]$ such that

$f(X^m) = g(X)h(X)$. Since P is a root-closed and $m \in \mathsf{d}(P^\bullet)$, Lemma 2 ensures that $g(X^{1/m}), h(X^{1/m}) \in F[X; P]$. Thus, $f(X) = g(X^{1/m})h(X^{1/m})$ in $F[X; P]$. Since f is irreducible in $F[X; P]$ either $g(X^{1/m}) \in F$ or $h(X^{1/m}) \in F$, which implies that either $g \in F$ or $h \in F$. Hence $f(X^m)$ is irreducible in $F[X]$.

Conversely, suppose that $f \in F[X; P]$ satisfies that $f(X^m)$ is an irreducible polynomial in $F[X]$ for every $m \in \mathsf{d}(P^\bullet)$ that is a common multiple of the elements of the set $\mathsf{d}(\mathsf{Supp}(f))$. To argue that f is irreducible in $F[X; P]$ suppose that $f = g\,h$ for some $g, h \in F[X; P]$. Let m_0 be the least common multiple of the elements of $\mathsf{d}(\mathsf{Supp}(g)) \cup \mathsf{d}(\mathsf{Supp}(h))$. Lemma 1 guarantees that $m_0 \in \mathsf{d}(P^\bullet)$. Moreover, $f = g\,h$ implies that m_0 is a common multiple of the elements of $\mathsf{d}(\mathsf{Supp}(f))$. As a result, the equality $f(X^{m_0}) = g(X^{m_0})h(X^{m_0})$ holds in $F[X]$. Since $f(X^{m_0})$ is irreducible in $F[X]$, either $g(X^{m_0}) \in F$ or $h(X^{m_0}) \in F$ and, therefore, either $g \in F$ or $h \in F$. This implies that f is irreducible in $F[X; P]$, as desired. □

We proceed to show the main result of this section.

Theorem 4 *Let F be a field, and let P be a root-closed Puiseux monoid. Hence*

$$\mathcal{M}(F[X; P]) = \langle \mathcal{A}(F[X; P]) \rangle.$$

Proof As each molecule of $F[X; P]$ is a product of irreducible elements in $F[X; P]$, the inclusion $\mathcal{M}(F[X; P]) \subseteq \langle \mathcal{A}(F[X; P]) \rangle$ holds trivially. For the reverse inclusion, suppose that $f \in F[X; P] \setminus F$ can be written as a product of irreducible elements in $F[X; P]$. As a result, there exist $k, \ell \in \mathbb{N}$ and irreducible elements $g_1, \dots, g_k$ and $h_1, \dots, h_\ell$ in $F[X; P]$ satisfying that

$$g_1(X) \cdots g_k(X) = f(X) = h_1(X) \cdots h_\ell(X). \tag{3}$$

Let m be the least common multiple of all the elements of the set

$$\left(\bigcup_{i=1}^{k} \mathsf{d}(\mathsf{Supp}(g_i)) \right) \bigcup \left(\bigcup_{j=1}^{\ell} \mathsf{d}(\mathsf{Supp}(h_j)) \right).$$

Note that $f(X^m)$, $g_i(X^m)$ and $h_j(X^m)$ are polynomials in $F[X]$ for $i = 1, \dots, k$ and $j = 1, \dots, \ell$. Lemma 1 ensures that $m \in \mathsf{d}(P^\bullet)$. On the other hand, m is a common multiple of all the elements of $\mathsf{d}(\mathsf{Supp}(g_i))$ (or all the elements of $\mathsf{d}(\mathsf{Supp}(h_i))$). Therefore Proposition 8 guarantees that the polynomials $g_i(X^m)$ and $h_j(X^m)$ are irreducible in $F[X]$ for $i = 1, \dots, k$ and $j = 1, \dots, \ell$. After substituting X by X^m in (3) and using the fact that $F[X]$ is a UFD, one finds that $\ell = k$ and $g_i(X^m) = h_{\sigma(i)}(X^m)$ for some permutation $\sigma \in S_k$ and every $i = 1, \dots, k$. This, in turns, implies that $g_i = h_{\sigma(i)}$ for $i = 1, \dots, k$. Hence $|\mathsf{Z}_{F[X;P]}(f)| = 1$, which means that f is a molecule of $F[X; P]$. □

As we have seen before, Corollary 4 guarantees the existence of a Puiseux algebra $F[X; P]$ satisfying that $|\mathcal{M}(F[X; P]) \setminus \mathcal{A}(F[X; P])| = \infty$. Now we use

Theorem 4 to construct an infinite class of Puiseux algebras satisfying a slightly more refined condition.

Proposition 9 *For any field F, there exist infinitely many Puiseux monoids P such that the algebra $F[X; P]$ contains infinite molecules that are neither atoms nor monomials.*

Proof Let $\{p_j\}$ be the strictly increasing sequence with underlying set $\mathbb{P}$. Then for each $j \in \mathbb{N}$ consider the Puiseux monoid $P_j = \langle 1/p_j^n \mid n \in \mathbb{N} \rangle$. Fix $j \in \mathbb{N}$, and take $P := P_j$. The fact that $\mathsf{gp}(P) = P \cup -P$ immediately implies that P is a root-closed Puiseux monoid containing 1. Consider the Puiseux algebra $\mathbb{Q}[X; P]$ and the element $X + p \in \mathbb{Q}[X; P]$, where $p \in \mathbb{P}$. To argue that $X + p$ is an irreducible element in $\mathbb{Q}[X; P]$, write $X + p = g(X)\,h(X)$ for some $g, h \in \mathbb{Q}[X; P]$. Now taking m to be the maximum power of p_j in the set $\mathsf{d}(\mathsf{Supp}(g) \cup \mathsf{Supp}(h))$, one obtains that $X^m + p = g(X^m)\,h(X^m)$ in $\mathbb{Q}[X]$. Since $\mathbb{Q}[X]$ is a UFD, it follows by Eisenstein's criterion that $X^m + p$ is irreducible as a polynomial over $\mathbb{Q}$. Hence either $g(X) \in \mathbb{Q}$ or $h(X) \in \mathbb{Q}$, which implies that $X + p$ is irreducible in $\mathbb{Q}[X; P]$. Now it follows by Theorem 4 that $(X + p)^n$ is a molecule in $\mathbb{Q}[X; P]$ for every $n \in \mathbb{N}$. Clearly, the elements $(X + p)^n$ are neither atoms nor monomials.

Finally, we prove that the algebras we have defined in the previous paragraph are pairwise non-isomorphic. To do so suppose, by way of contradiction, that $\mathbb{Q}[X; P_j] \cong \mathbb{Q}[X; P_k]$ for distinct $j, k \in \mathbb{N}$. Let $\psi \colon \mathbb{Q}[X; P_j] \to \mathbb{Q}[X; P_k]$ be an algebra isomorphism. Since ψ fixes $\mathbb{Q}$, it follows that $\psi(X^q) \notin \mathbb{Q}$ for any $q \in P_j^\bullet$. This implies that $\deg(\psi(X)) \in P_k^\bullet$. As $\mathsf{d}(P_j^\bullet)$ is unbounded there exists $n \in \mathbb{N}$ such that $p_j^n > \mathsf{n}(\deg(\psi(X)))$. Observe that

$$\deg\big(\psi(X)\big) = \deg\big(\psi(X^{\frac{1}{p_j^n}})^{p_j^n}\big) = p_j^n \deg\big(\psi(X^{\frac{1}{p_j^n}})\big). \tag{4}$$

Because $\gcd(p_j, d) = 1$ for every $d \in \mathsf{d}(P_k^\bullet)$, from (4) one obtains that p_j^n divides $\mathsf{n}(\deg \psi(X))$, which contradicts that $p_j^n > \mathsf{n}(\deg(\psi(X)))$. Hence the Puiseux algebras in $\{P_j : j \in \mathbb{N}\}$ are pairwise non-isomorphic, which completes our proof. □

Acknowledgements While working on this paper, the first author was supported by the NSF-AGEP Fellowship and the UC Dissertation Year Fellowship. The authors would like to thank an anonymous referee, whose helpful suggestions help to improve the final version of this paper.

References

1. Anderson, D.D., Coykendall, J., Hill, L., Zafrullah, M.: Monoid domain constructions of antimatter domains. Comm. Alg. **35**, 3236–3241 (2007)
2. Chapman, S.T., García-Sánchez, P.A., Llena, D., Malyshev, A., Steinberg, D.: On the delta set and the Betti elements of a BF-monoid. Arab. J. Math. **1**, 53–61 (2012)

3. Chapman, S. T., Gotti, F., and Gotti, M.: Factorization invariants of Puiseux monoids generated by geometric sequences. Comm. Algebra **48**, 380–396 (2020)
4. Coykendall, J., Mammenga, B.: An embedding theorem. J. Algebra **325**, 177–185 (2011)
5. Coykendall, J., Gotti, F.: On the atomicity of monoid algebras. J. Algebra **539** 138–151 (2019)
6. García-Sánchez, P.A., Ojeda, I.: Uniquely presented finitely generated commutative monoids. Pac. J. Math. **248**, 91–105 (2010)
7. García-Sánchez, P.A., Rosales, J.C.: Finitely Generated Commutative Monoids. Nova Science Publishers Inc., New York (1999)
8. García-Sánchez, P.A., Rosales, J.C.: Numerical Semigroups. Developments in Mathematics, vol. 20. Springer, New York (2009)
9. Geroldinger, A., Halter-Koch, F.: Non-unique factorizations: a survey. In: Brewer, J.W., Glaz, S., Heinzer, W., Olberding, B. (eds.) Multiplicative Ideal Theory in Commutative Algebra, pp. 207–226. Springer, New York (2006)
10. Gilmer, R.: Commutative Semigroup Rings. Chicago Lectures in Mathematics. The University of Chicago Press, London (1984)
11. Gotti, F.: On the atomic structure of Puiseux monoids. J. Algebra Appl. **16**, 1750126 (2017)
12. Gotti, F.: Atomic and antimatter semigroup algebras with rational exponents (2018). arXiv:1801.06779v2
13. Gotti, F.: Puiseux monoids and transfer homomorphisms. J. Algebra **516**, 95–114 (2018)
14. Gotti, F.: Increasing positive monoids of ordered fields are FF-monoids. J. Algebra **518**, 40–56 (2019)
15. Gotti, F., Gotti, M.: Atomicity and boundedness of monotone Puiseux monoids. Semigroup Forum **96**, 536–552 (2018)
16. Grillet, P.A.: Commutative Semigroups. Advances in Mathematics, vol. 2. Kluwer Academic Publishers, Boston (2001)
17. Miller, E., Sturmfels, B.: Combinatorial Commutative Algebra. Graduate Texts in Mathematics, vol. 227. Springer, New York (2004)
18. Narkiewicz, W.: On natural numbers having unique factorization in a quadratic number field. Acta Arith. **12**, 1–22 (1966)
19. Narkiewicz, W.: On natural numbers having unique factorization in a quadratic number field II. Acta Arith. **13**, 123–129 (1967)
20. Narkiewicz, W.: Numbers with unique factorization in an algebraic number field. Acta Arith. **21**, 313–322 (1972)

Arf Numerical Semigroups with Multiplicity 9 and 10

Halil İbrahim Karakaş

Abstract In this work we give a new characterization of Arf numerical semigroups and use it to parametrize Arf numerical semigroups with multiplicity 9 and 10.

Keywords Numerical semigroups · Arf numerical semigroups · Multiplicity · Conductor · Frobenius number · Ratio · Major

1 Introduction

Let $\mathbb{N}$ denote the set of positive integers and $\mathbb{N}_0 = \mathbb{N} \cup \{0\}$, the set of nonnegative integers. The cardinality of a set K will be denoted by $|K|$. A subset $S \subseteq \mathbb{N}_0$ satisfying

$$\text{(i) } 0 \in S \qquad \text{(ii) } x, y \in S \Rightarrow x + y \in S \qquad \text{(iii) } |\mathbb{N}_0 \setminus S| < \infty$$

is called a *numerical semigroup*. It is well known (see, for instance, [2, 4, 7]) that the condition (iii) above is equivalent to saying that the greatest common divisor $\gcd(S)$ of elements of S is 1.

If A is a subset of $\mathbb{N}_0$, we will denote by $\langle A \rangle$ the submonoid of $\mathbb{N}_0$ generated by A. If $S = \langle A \rangle$, A is called a *set of generators* for S. If $A = \{a_1, \ldots, a_r\}$, we write $\langle A \rangle = \langle a_1, \ldots, a_r \rangle$. The monoid $\langle A \rangle$ is a numerical semigroup if and only if $\gcd(A) = 1$.

For every numerical semigroup S there exists a unique *minimal set of generators* $\{a_1, a_2, \ldots, a_e\}$ with $a_1 < a_2 < \cdots < a_e$; that is, $\{a_1, a_2, \ldots, a_e\}$ is a set of generators for S, but no proper subset of $\{a_1, a_2, \ldots, a_e\}$ generates S. The integers a_1 and e are called the *multiplicity* and the *embedding dimension* of S, and they are denoted by $m(S)$ and $e(S)$, respectively. The multiplicity $m(S)$ is the smallest

H. İ. Karakaş (✉)
Faculty of Commercial Sciences, Başkent University, Ankara, Turkey
e-mail: karakas@baskent.edu.tr

V. Barucci et al. (eds.), *Numerical Semigroups*, Springer INdAM Series 40,
https://doi.org/10.1007/978-3-030-40822-0_11

positive element of S. It is known that $e(S) \leq m(S)$ (see, for instance Chapter 1 of [7]). S is said to be a numerical semigroup of *maximal embedding dimension* if $e(S) = m(S)$.

For a numerical semigroup S, the largest integer that is not in S is called the *Frobenius number* of S and it is denoted by $\mathsf{f}(S)$; the smallest element of S for which all subsequent natural numbers belong to S is called the *conductor* of S and it is denoted by $c(S)$. Clearly, $c(S) = \mathsf{f}(S) + 1$. We have $c(\mathbb{N}_0) = 0$ and $c \geq 2$ if and only if $S \neq \mathbb{N}_0$.

If S is a numerical semigroup and $a \in S\backslash\{0\}$, the *Apéry set* of S with respect to a is the set

$$\mathrm{Ap}(S, a) = \{s \in S : s - a \notin S\}.$$

It is easy to see that $\mathrm{Ap}(S, a) = \{w(0) = 0, w(1), \ldots, w(a-1)\}$, where $w(i)$ is the least element of S such that $w(i) \equiv i \pmod{a}$. It is also easy to see that

$$S = \langle a, w(1), \ldots, w(a-1)\rangle \text{ and } \mathsf{f}(S) = \max(\mathrm{Ap}(S, a)) - a.$$

Thus

$$c = \max(\mathrm{Ap}(S, a)) - a + 1 \text{ and } \max(\mathrm{Ap}(S, a)) = c + a - 1.$$

For general concepts and facts about numerical semigroups, we refer to [7].

2 Arf Numerical Semigroups

A numerical semigroup S satisfying the additional condition

$$x, y, z \in S;\ x \geq y \geq z \Rightarrow x + y - z \in S \tag{2.1}$$

is called an *Arf numerical semigroup*.

This is the original definition of an Arf numerical semigroup given by C. Arf in [1]. We will refer to the condition (2.1) as the *Arf condition*. Fifteen conditions equivalent to the Arf condition are given in Theorem 1.3.4 of [2]. If $x, y, z \in S$; $x \geq y \geq z$ and $x \geq c$, the conductor of S, then $x + y - z \geq c$ and thus $x + y - z \in S$. Therefore, to prove that a numerical semigroup with conductor c is an Arf numerical semigroup it is enough to check the Arf condition (2.1) for small elements of S; i.e, for elements $x < c$.

The following lemma by Compillo, Farran and Munuera [3] gives a very useful condition equivalent to the Arf condition.

Lemma 2.1 ([3], Proposition 2.3) *A numerical semigroup S is an Arf numerical semigroup if and only if $2x - y \in S$ for all $x, y \in S$ with $x \geq y$.*

Combining Lemma 2.1 and our remark above about the Arf condition, we see that a numerical semigroup S is an Arf numerical semigroup if and only if $2x - y \in S$ for all $x, y \in S$ with $c > x > y$.

The following lemma of Rosales et al. [8] will be crucial for what follows.

Lemma 2.2 ([8], Lemma 11) *Let S be an Arf numerical semigroup and let s be any element of S. If $s + 1 \in S$, then $s + k \in S$ for all $k \in \mathbb{N}_0$ and thus $c \leq s$.*

It is well known (see for instance [6] or [7]) that every Arf numerical semigroup is of maximal embedding dimension, that is, $e(S) = m(S)$. Thus if S is an Arf numerical semigroup with multiplicity $m = m(S)$, then

$$\{m, w(1), \ldots, w(m-1)\}$$

is the minimal set of generators for S, where

$$\mathrm{Ap}(S, m) = \{w(0) = 0, w(1), \ldots, w(m-1)\}.$$

The number $\max(\mathrm{Ap}(S, m))$ is called the *major* of S and it is denoted by $\mathcal{M}$. The smallest minimal generator that is larger than the multiplicity of S is called the *ratio* of S and it is denoted by $\mathcal{R}$. Thus $\mathcal{M} = c + m - 1$ and $\mathcal{R} = \min(\mathrm{Ap}(S, m) \backslash \{0\})$.

From now on we stick to the above notations and we put $\mathrm{f}(S) = \mathrm{f}$, $c(S) = c$.

Let S be a numerical semigroup with multiplicity m and conductor c. Since any multiple of m is an element of S, $c \not\equiv 1 \pmod{m}$. Hence $c \equiv k \pmod{m}$, where $k \in \{0, 2, \ldots, m-1\}$. If S is an Arf numerical semigroup with multiplicity m and conductor c, the following lemma by Garcia-Sánchez et al. [5] shows that $w(1)$ and $w(m-1)$ are completely determined by c and m.

Lemma 2.3 ([5], Lemma 13) *Let S be an Arf numerical semigroup with multiplicity m and conductor c where $c \equiv k\ (\mathrm{mod} m)$, $k \in \{0, 2, \ldots, m-1\}$. Then*

$$(i)\ w(1) = \begin{cases} c + 1 & \text{if } k = 0 \ (i.e,\ c \equiv 0 \pmod{m}), \\ c - k + m + 1 & \text{if } k \neq 0 \ (i.e,\ c \not\equiv 0 \pmod{m}), \end{cases}$$

(ii) $w(m-1) = c - k + m - 1$.

Lemma 2.4 *Let S be a numerical semigroup with multiplicity $m > 2$, conductor c and $\mathrm{Ap}(S) = \{w(0) = 0, w(1), \ldots, w(m-1)\}$. Let $h, k \in \{0, 1, \ldots, m-1\}$ such that $w(h) < w(k)$. Then*

(i) $h < k \Longrightarrow w(h) \leq w(k) - k + h$ *and thus* $w(k) - k + h \in S$,
(ii) $h > k \Longrightarrow w(h) \leq w(k) - k + h - m$ *and thus* $w(k) - k + h - m \in S$.

Proof There exist $x_h, x_k \in \mathbb{N}_0$ such that $w(h) = h + x_h m$ and $w(k) = k + x_k m$.

(*i*) The assertion is clear if $h = 0$, because $k < m < w(k)$. Thus we may assume that $0 < h < k$. Then $x_h, x_k \in \mathbb{N}$ and $w(h) < w(k) \Longrightarrow x_h \leq x_k$. Hence

$$w(k) - w(h) = (k - h) + (x_k - x_h)m \geq (k - h)$$

which yields $w(h) \leq w(k) - k + h$.

(ii) We have $h > k > 0$, because $w(h) < w(k)$. Therefore $x_h < x_k$ and

$$w(k) - w(h) = (k - h) + (x_k - x_h)m \geq (k - h) + m$$

which yields $w(h) \leq w(k) - k + h - m$.

Corollary 2.1 ([5], Lemma 11) *Let S be an Arf numerical semigroup with multiplicity $m > 2$, conductor c, and $\mathrm{Ap}(S, m) = \{w(0) = 0, w(1), \ldots, w(m-1)\}$. Then*

(i) $w(k-1) < w(k) \Longrightarrow w(k) - 1 \geq c$,
(ii) $w(k-1) > w(k) \Longrightarrow w(k-1) \geq c$,
for any $k \in \{2, \ldots, m-1\}$.

Proof

(i) $w(k) - k + (k-1) = w(k-1) + 1 \in S$ by Lemma 2.4(*i*). So $w(k) - 1 \geq c$ by Lemma 2.2.
(ii) $w(k-1) - (k-1) + k - m = w(k-1) - m + 1 \in S$ by Lemma 2.4(*ii*). So $w(k-1) + 1 \in S$, and $w(k-1) \geq c$ by Lemma 2.2.

Corollary 2.1 shows that if S is an Arf numerical semigroup with multiplicity $m > 2$ and $k \in \{2, \ldots, m-1\}$, at least one of $w(k-1)$ or $w(k)$ is not less than the conductor of S.

Lemma 2.5 *Let S be a numerical semigroup with multiplicity $m > 2$ and*

$$\mathrm{Ap}(S, m) = \{w(0) = 0, w(1), \ldots, w(m-1)\}.$$

For any $h, k \in \{0, 1, \ldots, m-1\}$ with $h \neq k$, there exists a unique integer t_{kh} such that

$$t_{kh}m \leq w(k) - w(h) \leq (t_{kh} + 1)m.$$

Proof Let $h, k \in \{0, 1, \ldots, m-1\}$ such that $h \neq k$. Then there exist $x_h, x_k \in \mathbb{N}_0$ such that $w(h) = h + x_h m$, $w(k) = k + x_k m$. Thus

$$\frac{w(k) - w(h)}{m} = \frac{k - h}{m} + (x_k - x_h),$$

where $\frac{k-h}{m}$ is a rational number with $|\frac{k-h}{m}| < 1$. Let

$$t_{kh} = \begin{cases} x_k - x_h - 1 & if\ k - h < 0, \\ x_k - x_h & if\ k - h > 0. \end{cases}$$

Clearly, t_{kh} is uniquely determined and we have

$$t_{kh}m \leq w(k) - w(h) \leq (t_{kh} + 1)m.$$

Notations being as above, for any $h, k \in \{0, 1, \ldots, m-1\}$ such that $h \neq k$, there exists a unique $t_{kh} \in \mathbb{Z}$ for which $w(h) + t_{kh}m \leq w(k) \leq w(h) + (t_{kh} + 1)m$. In that case, we set

$$^*w_k(h) = w(h) + t_{kh}m\ ,\ \ w_k^*(h) = w(h) + (t_{kh} + 1)m.$$

Then $^*w_k(h)$ is called the *left h-neighbour* of $w(k)$ and $w_k^*(h)$ is called the *right h-neighbour* of $w(k)$. Thus we have

$$^*w_k(h) \leq w(k) \leq w_k^*(h)$$

for $h, k \in \{0, 1, \cdots, m-1\}$ such that $h \neq k$.

Note that $^*w_k(h)$ and $w_k^*(h)$ are both elements of S if $t_{kh} \in \mathbb{N}_0$. This is the case if $h, k \in \{0, 1, \ldots, m-1\}$ such that $w(h) < w(k)$. Because in that case

$$h < k \Longrightarrow\ ^*w_k(h) = w(k) - k + h\ ,\ \ w_k^*(h) = w(k) - k + h + m,$$

$$h > k \Longrightarrow\ ^*w_k(h) = w(k) - k + h - m\ ,\ \ w_k^*(h) = w(k) - k + h;$$

from which we conclude by Lemma 2.4 that $^*w_k(h)$ and $w_k^*(h)$ are both elements of S.

Left and right neighbours can be used to characterize Arf numerical semigroups. Namely, we have the following propostion.

Proposition 2.2 *Let S be a numerical semigroup with multiplicity $m > 2$, conductor c and* $\mathrm{Ap}(S) = \{w(0) = 0, w(1), \ldots, w(m-1)\}$. *Then S is Arf if and only if*

$$2w(k) -\ ^*w_k(h) \in S \ \textit{and}\ 2w_k^*(h) - w(k) \in S$$

for any $h, k \in \{0, 1, \cdots, m-1\}$ with $c > w(k) > w(h)$.

Proof The necessity is obvious. As for the sufficiency, we will show that $2x - y \in S$ for all $x, y \in S$ with $x > y$. For such a pair x, y, there exist $i, j \in \{0, 1, \ldots, m-1\}$ and nonnegative integers r_x, r_y such that $x = w(i) + r_x m$, $y = w(j) + r_y m$. If $x \geq c$, then $2x - y \geq c$ and therefore $2x - y \in S$. So we may assume that $x < c$. Then $w(i) < c$ and $w(j) < c$. If $i = j$, then $r_x > r_y$ and $2x - y = w(i) + (2r_x - r_y)m \in S$. Therefore we may assume that $i \neq j$. There are two cases: either $w(i) > w(j)$ or $w(i) < w(j)$.

If $w(i) > w(j)$, then

$$y = w(j) + r_y m < x = w(i) + r_x m \leq w_i^*(j) + r_x m = w(j) + (t_{ij} + 1)m + r_x m,$$

which implies $t_{ij} + 1 + r_x > r_y$ or, equivalently, $t_{ij} + r_x - r_y \geq 0$. Now we have

$$\begin{aligned} 2x - y &= 2w(i) - w(j) + (2r_x - r_y)m \\ &= 2w(i) - ({}^*w_i(j) - t_{ij}m) + (2r_x - r_y)m \\ &= 2w(i) - {}^*w_i(j) + (t_{ij} + 2r_x - r_y)m, \end{aligned}$$

where $2w(i) - {}^*w_i(j) \in S$ by hypothesis, and also $(t_{ij} + 2r_x - r_y)m \in S$ since $(t_{ij} + 2r_x - r_y) \geq 0$. Hence $2x - y \in S$ in this case.

If $w(i) < w(j)$, then

$$x = w(i) + r_x m > y = w(j) + r_y m \geq {}^*w_j(i) + r_y m = w(i) + t_{ji}m + r_y m,$$

which implies $r_x > t_{ji} + r_y$ or, equivalently, $r_x - r_y - t_{ji} - 1 \geq 0$. On the other hand,

$$\begin{aligned} 2x - y &= 2w(i) - w(j) + (2r_x - r_y)m \\ &= 2(w_j^*(i) - (t_{ji} + 1)m) - w(j) + (2r_x - r_y)m \\ &= 2w_j^*(i) - w(j) + (2r_x - r_y - 2t_{ji} - 2)m, \end{aligned}$$

where $2w_j^*(i) - w(j) \in S$ by hypothesis, and also $(2r_x - r_y - 2t_{ji})m \in S$ since $(2r_x - r_y - 2t_{ji} - 2) \geq 0$. Hence $2x - y \in S$ in this case, too.

Proposition 2.3 *Let S be an Arf numerical semigroup with multiplicity $m > 2$, conductor c and $\mathrm{Ap}(S) = \{w(0) = 0, w(1), \ldots, w(m-1)\}$. Let $k \in \{1, \ldots, m-1\}$. Then*

(i) $w(m-k) + 2k \in S$ and thus $w(k) \leq w(m-k) + 2k$.
(ii) For any positive integer q with $qk < m$, we have $w(k) + (q-1)k \in S$ and thus $w(qk) \leq w(k) + (q-1)k$.
(iii) For $k < \frac{m}{2}$, we have $w(m-2k) \leq w(m-k) + (m-k)$.

Proof

(i) $w(m-k) + k = w_{m-k}^*(0) \in S$. Since S is Arf,

$$2w_{m-k}^*(0) - w(m-k) = 2(w(m-k) + k) - w(m-k) = w(m-k) + 2k \in S.$$

Moreover, $w(m-k) + 2k \equiv k \pmod{m}$. Thus $w(k) \leq w(m-k) + 2k$.

(ii) We proceed by induction on q. The assertion is trivially true for $q = 1$. Since $w(k) - k = {}^*w_k(0) \in S$ and S is Arf, we have

$$2w(k) - {}^*w_k(0) = w(k) + k \in S.$$

Moreover, $w(k) + k \equiv 2k \pmod{m}$. Thus $w(2k) \leq w(k) + k$, proving the assertion for $q = 2$. Now let $q > 2$, $qk < m$, and assume that the assertion is true for all integers r with $1 < r < q$. Then

$$w(k) + (q-2)k,\ w(k) + (q-3)k \in S.$$

Since $w(k) + (q-2)k > w(k) + (q-3)k$ and S is Arf, we get

$$2(w(k) + (q-2)k) - (w(k) + (q-3)k) = w(k) + (q-1)k \in S.$$

This yields $w(qk) \leq w(k) + (q-1)k$, because $w(k) + (q-1)k \equiv qk (\text{mod } m)$.

(iii) Assume $k < \frac{m}{2}$. Note that ${}^*w_{m-k}(0) = w(m-k) - (m-k) \in S$ and S is Arf. So $2w(m-k) -{}^* w_{m-k}(0) = w(m-k) + (m-k) \in S$. Note also that $w(m-k)+(m-k) \equiv m-2k \ (\text{mod } m)$. Hence $w(m-2k) \leq w(m-k)+(m-k)$.

Proposition 2.4 *Let* S *be an Arf numerical semigroup with multiplicity* $m > 2$, *conductor* c *and* $\text{Ap}(S) = \{w(0) = 0, w(1), \ldots, w(m-1)\}$. *Assume that the ratio of* S *is* $\mathcal{R} = w(k)$, *where* $k \in \{1, \ldots, m-1\}$. *Let* $q \in \mathbb{N}$ *such that* $qk < m$. *Then*

(i) $w(qk) = w(k) + (q-1)k$,
(ii) $w(qk+1) \geq c+1$ *if* $qk+1 < m$,
(iii) $w(qk-1) \geq c$ *if* $qk \neq 1$.

Proof

(i) The assertion is trivially true for $q = 1$. Thus we may assume that $q > 1$. Then $w(k) < w(qk)$, because $w(k)$ is the ratio. Lemma 2.4 yields $w(k)+(q-1)k \leq w(qk)$. We also have $w(qk) \leq w(k) + (q-1)k$ by Proposition 2.3. Hence we have the equality.

(ii) Note that $qk+1 > k$. So $w(k) < w(qk+1)$, because $w(k)$ is the ratio. Applying Lemma 2.4, we get $w(k) + (q-1)k + 1 \leq w(qk+1)$. Now we use *(i)* to obtain $w(qk)+1 \leq w(qk+1)$, or equivalently, $w(qk) \leq w(qk+1) - 1$. We conclude that $w(qk+1)$ and $w(qk+1) - 1$ are both elements of S. Hence $w(qk+1) \geq c+1$ by Lemma 2.2.

(iii) $qk - 1 = k$ only if $q = 2$ and $k = 1$. In that case, the assertion is true, because $w(1) \geq c$ by Lemma 2.3. If $q \neq 2$ or $k \neq 1$, then $k \neq qk - 1$ and thus we have $w(k) < w(qk-1)$, because $w(k)$ is the ratio. Therefore $w(k) + (q-1)k - 1 \leq w(qk-1)$ by Lemma 2.4. Now we use *(i)* to obtain $w(qk) - 1 \leq w(qk-1)$, or equivalently, $w(qk) \leq w(qk-1) + 1$. It follows that $w(qk-1)$ and $w(qk-1) + 1$ both belong to S. Lemma 2.2 implies $w(qk-1) \geq c$.

Proposition 2.5 *Let* S *be an Arf numerical semigroup with multiplicity* $m > 2$, *conductor* c *and* $\text{Ap}(S) = \{w(0) = 0, w(1), \ldots, w(m-1)\}$. *Assume that the ratio of* S *is* $\mathcal{R} = w(m-k)$, *where* $k \in \{1, \ldots, m-1\}$. *Then for any* $q \in \mathbb{N}$

(i) $q \geq 2,\ qk < m \Longrightarrow w((q-1)k) = w(m-k) + qk$ *and* $w(m-2k) = w(m-k) + (m-k)$,
(ii) $m < qk < m+k \Longrightarrow w((q-1)k) \in \{w(m-k) + qk - m,\ w(m-k) + qk\}$.

Proof

(*i*) Since $w(m-k)$ is the ratio, we have $w(m-k) < w((q-1)k)$ and $w(m-k) < w(m-2k)$. Therefore

$$w(m-k) + qk \leq w((q-1)k) \text{ and } w(m-k) + (m-k) \leq w(m-2k)$$

by Lemma 2.4. On the other hand,

$$w((q-1)k) \leq w(k) + (q-2)k \leq w(m-k) + 2k + (q-2)k = w(m-k) + qk,$$

and

$$w(m-2k) \leq w(m-k) + (m-k)$$

by Proposition 2.3. Hence

$$w((q-1)k) = w(m-k) + qk \text{ and } w(m-2k) = w(m-k) + (m-k).$$

(*ii*) Since $w(m-k)$ is the ratio, we have $w(m-k) < w((q-1)k)$. Therefore $w(m-k) + qk - m \leq w((q-1)k)$ by Lemma 2.4. On the other hand,

$$w((q-1)k) \leq w(k) + (q-2)k \leq w(m-k) + 2k + (q-2)k = w(m-k) + qk$$

by Proposition 2.3. Hence $w((q-1)k) \in \{w(m-k) + qk - m, w(m-k) + qk\}$.

In [5], Arf numerical semigroups with multiplicity up to seven and given conductor are described parametrically. A similar description is given in [9] for Arf numerical semigroups with multiplicity eight and given conductor. In what follows, we use the above characterizations to describe Arf numerical semigroups with multiplicity nine and ten.

The ratio $\mathcal{R}$ will play an important part in our discussions. Recall that $\mathcal{R}$ is the least element larger than the multiplicity in the minimal set of generators of the numerical semigroup under consideration. It is easily seen that

$$\mathcal{R} \leq c + 1 \text{ if } c \equiv 0 \pmod{m} \text{ and } \mathcal{R} \leq c \text{ if } c \not\equiv 0 \pmod{m}.$$

Each proposition in the next two sections will give the list of all Arf numerical semigroups with multiplicity $m = 9$ or $m = 10$ and conductor c for a congruence class of $c \pmod{m}$. Each semigroup in each list is easily seen to be Arf by applying, for instance, Proposition 2.2. Therefore in the proof of each proposition we only verify that the list there contains all Arf numerical semigroups with the given multiplicity and conductor.

3 Arf Numerical Semigroups with Multiplicity 9

Let S be an Arf numerical semigroup with multiplicity 9 and conductor c. Then $c \equiv 0,\ 2,\ 3,\ 4,\ 5,\ 6,\ 7$ or $8 \pmod 9$.

The following proposition describes all Arf numerical semigroups S with multiplicity 9 and conductor c if $c \equiv 0 \pmod 9$, $c > 9$.

Proposition 3.1 *Let S be a numerical semigroup with multiplicity 9 and conductor c, where $c > 9$ and $c \equiv 0 \pmod 9$. Then S is an Arf semigroup if and only if S is one of the following:*

$\langle 9, c+1, c+2, c+3, c+4, c+5, c+6, c+7, c+8\rangle$; *or*
$\langle 9, 9u+3, 9u+6, c+1, c+2, c+4, c+5, c+7, c+8\rangle$, $1 \le u \le \frac{c-9}{9}$; *or*
$\langle 9, c-4, c-2, c+1, c+2, c+3, c+4, c+6, c+8\rangle$, *or*
$\langle 9, c-4, c+1, c+2, c+3, c+4, c+6, c+7, c+8\rangle$; *or*
$\langle 9, 9v+6, 9v+12, c+1, c+2, c+4, c+5, c+7, c+8\rangle$, $1 \le v \le \frac{c-9}{9}$; *or*
$\langle 9, c-2, c+1, c+2, c+3, c+4, c+5, c+6, c+8\rangle$.

Proof Let S be an Arf numerical semigroup with multiplicity 9 and conductor $c \equiv 0$ *(mod 9)*, where $c > 9$. Then $w(1) = c+1$ and $w(8) = c+8 = \mathcal{M}$ by Lemma 2.3. Using Proposition 2.3(*ii*), $c+8 = w(8) \le w(2)+6 \Longrightarrow w(2) = c+2$. Similarly, $w(4) = c+4$. Now we have $c+2 = w(2) \le w(7)+4 \Longrightarrow w(7) \ge c-2$, and $c+4 = w(4) \le w(5)+8 \Longrightarrow w(5) \ge c-4$ by Proposition 2.3(*i*). It follows that

$$\mathcal{R} \in \{w(1) = c+1,\ w(3),\ w(5) = c-4,\ w(6),\ w(7) = c-2\}.$$

If $\mathcal{R} = w(1) = c+1$, then

$$S = \langle 9, c+1, c+2, c+3, c+4, c+5, c+6, c+7, c+8\rangle.$$

If $\mathcal{R} = w(3)$, then $w(3) = 9u+3$ for some $u \in \{1, \dots, \frac{c}{9}-1\}$. In that case, $w(5) = c+5$, $w(7) = c+7$ and $w(6) = 9u+6$ by Proposition 2.4. Hence

$$S = \langle 9, 9u+3, 9u+6, c+1, c+2, c+4, c+5, c+7, c+8\rangle,\ 1 \le u \le \frac{c-9}{9}.$$

If $\mathcal{R} = w(5) = c-4$, then $w(6) = c+6$ by Proposition 2.4 and therefore $w(3) = c+3$ by Proposition 2.3. Moreover, since $c-4 = w(5) < w(7)$, we have $w(7) = c-2$ or $w(7) = c+7$. Hence

$$S = \langle 9, c-4, c-2, c+1, c+2, c+3, c+4, c+6, c+8\rangle, \text{ or}$$
$$S = \langle 9, c-4, c+1, c+2, c+3, c+4, c+6, c+7, c+8\rangle.$$

If $\mathcal{R} = w(6)$, then $w(6) = 9v+6$ for some $v \in \{1, \dots, \frac{c}{9}-1\}$. We have $w(5) = c+5$, $w(7) = c+7$ by Proposition 2.4; and $w(3) = 9v+12$ by Proposition 2.5(i).

Hence

$$S = \langle 9, 9v+6, 9v+12, c+1, c+2, c+4, c+5, c+7, c+8\rangle, \ 1 \le u \le \frac{c-9}{9}.$$

If $\mathcal{R} = w(7) = c-2$, then $w(6) = c+6$ by Proposition 2.4. Since $c-2$ is the ratio, $c-2 < w(3)$ and $c-2 < w(5)$, which imply $w(3) = c+3$ and $w(5) = c+5$. Thus

$$S = \langle 9, c-2, c+1, c+2, c+3, c+4, c+5, c+6, c+8\rangle.$$

Proposition 3.2 *Let S be a numerical semigroup with multiplicity* 9 *and conductor c, where $c > 11$ and $c \equiv 2 \pmod 9$. Then S is an Arf semigroup if and only if S is one of the following:*

$\langle 9, c, c+1, c+2, c+3, c+4, c+5, c+6, c+8\rangle$; *or*
$\langle 9, 9u+3, 9u+6, c, c+2, c+3, c+5, c+6, c+8\rangle$, $1 \le u \le \frac{c-11}{9}$; *or*
$\langle 9, 9v+6, 9v+12, c, c+2, c+3, c+5, c+6, c+8\rangle$, $1 \le v \le \frac{c-11}{9}$; *or*
$\langle 9, c-4, c, c+1, c+2, c+3, c+4, c+6, c+8\rangle$.

Proof Let S be an Arf numerical semigroup with multiplicity 9 and conductor $c \equiv 2$ *(mod 9)*, where $c > 11$. Then $w(1) = c+8 = \mathcal{M}$ and $w(8) = c+6$ by Lemma 2.3. Using Proposition 2.3, we get $w(2) = c$, $w(4) = c+2$, $w(5) = c+3$ and $w(7) \ge c-4$. It follows that

$$\mathcal{R} \in \{w(2) = c, w(3), w(6), w(7) = c-4\}.$$

The rest of the proof is similar to the proof of Proposition 3.1 and we omit it.

Proposition 3.3 *Let S be a numerical semigroup with multiplicity* 9 *and conductor c, where $c > 12$ and $c \equiv 3 \pmod 9$. Then S is an Arf semigroup if and only if S is one of the following:*

$\langle 9, 9u+3, 9u+6, c+1, c+2, c+4, c+5, c+7, c+8\rangle$, $1 \le u \le \frac{c-3}{9}$; *or*
$\langle 9, 9v+6, 9v+12, c+1, c+2, c+4, c+5, c+7, c+8\rangle$, $1 \le v \le \frac{c-12}{9}$.

Proof Let S be an Arf numerical semigroup with multiplicity 9 and conductor $c \equiv 3$ *(mod 9)*, where $c > 12$. Then $w(1) = c+7$, $w(8) = c+5$ by Lemma 2.3. We also have $\mathcal{M} = c+8 = w(2)$. Applying Proposition 2.3, we get $w(4) = c+1$, $w(5) = c+2$ and $w(7) = c+4$. Therefore either $\mathcal{R} = w(3)$ or $\mathcal{R} = w(6)$ and the assertions follow.

We have an unexpected result for Arf numerical semigroups with multiplicity 9 if the conductor c is congruent to 4 (mod 9).

Proposition 3.4 *Let S be a numerical semigroup with multiplicity* 9 *and conductor c, where c* > 13 *and c* ≡ 4 (mod 9). *Then S is an Arf semigroup if and only if S is one of the following:*

$\langle 9, c, c+1, c+2, c+3, c+4, c+6, c+7, c+8\rangle$; *or*
$\langle 9, c-2, c, c+1, c+2, c+3, c+4, c+6, c+8\rangle$; *or*
$\langle 9, c-6, c-2, c, c+1, c+2, c+4, c+6, c+8\rangle$.

Proof Let S be an Arf numerical semigroup with multiplicity 9 and conductor $c \equiv 4$ *(mod 9)*, where $c > 13$. Then we have $w(1) = c+6$, $w(8) = c+4$ by Lemma 2.3; and $\mathcal{M} = c+8 = w(3)$. Using Proposition 2.3, we get $w(2) \geq c-2$, $w(4) = c$, $w(5) = c+1$, $w(6) = c+2$ and $w(7) \geq c-6$. Thus

$$\mathcal{R} \in \{w(7) = c-6, w(2) = c-2, w(4) = c\}.$$

Furthermore, if $\mathcal{R} = w(7) = c-6$, then $w(2) = w(7)+4 = c-2$ by Proposition 2.3 and the assertions follow.

Proposition 3.5 *Let S be a numerical semigroup with multiplicity* 9 *and conductor c, where c* > 14 *and c* ≡ 5 (mod 9). *Then S is an Arf semigroup if and only if S is one of the following:*

$\langle 9, c, c+1, c+2, c+3, c+5, c+6, c+7, c+8\rangle$; *or*
$\langle 9, 9u+3, 9u+6, c, c+2, c+3, c+5, c+6, c+8\rangle$, $1 \leq u \leq \frac{c-5}{9}$; *or*
$\langle 9, 9v+6, 9v+12, c, c+2, c+3, c+5, c+6, c+8\rangle$, $1 \leq v \leq \frac{c-14}{9}$.

Proof Let S be an Arf numerical semigroup with multiplicity 9 and conductor $c \equiv 5$ *(mod 9)*, where $c > 14$. Then $w(1) = c+5$, $w(8) = c+3$ by Lemma 2.3; and $\mathcal{M} = c+8 = w(4)$. Applying Proposition 2.3, $w(2) = c+6$, $w(7) = c+2$ and $w(5) = c$. It follows that

$$\mathcal{R} \in \{w(5) = c, w(3), w(6)\}.$$

The rest of the proof is similar to the proof of Proposition 3.1 and we omit it.

Proposition 3.6 *Let S be a numerical semigroup with multiplicity* 9 *and conductor c, where c* > 15 *and c* ≡ 6 (mod 9). *Then S is an Arf semigroup if and only if S is one of the following:*

$\langle 9, c-4, c-2, c, c+1, c+2, c+4, c+6, c+8\rangle$; *or*
$\langle 9, 9u+3, 9u+6, c+1, c+2, c+4, c+5, c+7, c+8\rangle$, $1 \leq u \leq \frac{c-15}{9}$; *or*
$\langle 9, c-2, c, c+1, c+2, c+4, c+5, c+6, c+8\rangle$; *or*
$\langle 9, 9v+6, 9v+12, c+1, c+2, c+4, c+5, c+7, c+8\rangle$, $1 \leq v \leq \frac{c-6}{9}$.

Proof Let S be an Arf numerical semigroup with multiplicity 9 and conductor $c \equiv 6$ *(mod 9)*, where $c > 15$. Then $w(1) = c+4$, $w(8) = c+2$ by Lemma 2.3; and $\mathcal{M} = c+8 = w(5)$. We have $w(2) \geq c-4$, $w(4) \geq c-2$ and $w(7) = c+1$

by Proposition 2.3. It follows that

$$\mathcal{R} \in \{w(2) = c - 4, w(3), w(4) = c - 2, w(6)\}.$$

The rest of the proof is similar to the proof of Proposition 3.1 and we omit it.

The proof of the next proposition is very similar to the proof of Proposition 3.4 and we omit it.

Proposition 3.7 *Let S be a numerical semigroup with multiplicity 9 and conductor c, where $c > 16$ and $c \equiv 7 \pmod 9$. Then S is an Arf semigroup if and only if S is one of the following:*

$\langle 9, c-3, c, c+1, c+3, c+4, c+5, c+7, c+8\rangle$; *or*
$\langle 9, c-2, c, c+1, c+3, c+4, c+5, c+6, c+8\rangle$; *or*
$\langle 9, c, c+1, c+3, c+4, c+5, c+6, c+7, c+8\rangle$.

Proposition 3.8 *Let S be a numerical semigroup with multiplicity 9 and conductor c, where $c > 17$ and $c \equiv 8 \pmod 9$. Then S is an Arf semigroup if and only if S is one of the following:*

$\langle 9, c-6, c-4, c-2, c, c+2, c+4, c+6, c+8\rangle$; *or*
$\langle 9, 9u+3, 9u+6, c, c+2, c+3, c+5, c+6, c+8\rangle$, $1 \le u \le \frac{c-8}{9}$; *or*
$\langle 9, c-4, c-2, c, c+2, c+3, c+4, c+6, c+8\rangle$, *or*
$\langle 9, c-4, c, c+2, c+3, c+4, c+6, c+7, c+8\rangle$; *or*
$\langle 9, c-3, c, c+2, c+3, c+4, c+5, c+7, c+8\rangle$; *or*
$\langle 9, 9v+6, 9v+12, c, c+2, c+3, c+5, c+6, c+8\rangle$, $1 \le v \le \frac{c-8}{9}$; *or*
$\langle 9, c, c+2, c+3, c+4, c+5, c+6, c+7, c+8\rangle$.

Proof Let S be an Arf numerical semigroup with multiplicity 9 and conductor $c \equiv 8 \pmod 9$, where $c > 17$. Then $w(1) = c+2$, $w(8) = c$ by Lemma 2.3; and $\mathcal{M} = c + 8 = w(7)$. Using Proposition 2.3, we get $w(2) \ge c-6$, $w(4) \ge c-4$ and $w(5) \ge c-3$. It follows that

$$\mathcal{R} \in \{w(2) = c-6, w(3), w(4) = c-4, w(5) = c-3, w(6), w(8) = c\}.$$

If $\mathcal{R} = w(2) = c-6$, then $w(3) = c+4$, $w(4) = c-4$, $w(5) = c+6$, and $w(6) = c-2$ by Proposition 2.4. Hence, in that case

$$S = \langle 9, c-6, c-4, c-3, c, c+4, c+6, c+8\rangle.$$

If $\mathcal{R} = w(3)$, then $w(3) = 9u+3$ for some $u \in \{1, \ldots, \frac{c-8}{9}\}$. In that case, $w(2) = c+3$, $w(4) = c+5$, $w(5) = c+6$ and $w(6) = 9u+6$ by Proposition 2.4. Hence

$$S = \langle 9, 9u+3, 9u+6, c, c+2, c+3, c+5, c+6, c+8\rangle,\ 1 \le u \le \frac{c-8}{9}.$$

If $\mathcal{R} = w(4) = c - 4$, then $w(3) = c + 4$ and $w(5) = c + 6$ by Proposition 2.4. We have $c - 4 = w(4) < w(2) \Longrightarrow w(2) = c + 3$; and using Proposition 2.3(i), $c + 4 = w(3) \leq w(6) + 6 \Longrightarrow w(6) \geq c - 2 \Longrightarrow w(6) = c - 2$ or $w(6) = c + 7$. Hence

$$S = \langle 9, c-4, c-2, c, c+2, c+3, c+4, c+6, c+8 \rangle, \text{ or}$$

$$S = \langle 9, c-4, c, c+2, c+3, c+4, c+6, c+7, c+8 \rangle.$$

If $\mathcal{R} = w(5) = c - 3$, then $w(4) = c + 5$ and $w(6) = c + 7$ by Proposition 2.4. We have $c - 3 = w(5) < w(3) \Longrightarrow w(3) = c + 4$; and using Proposition 2.3(ii), $c + 5 = w(4) \leq w(2) + 2 \Longrightarrow w(2) = c + 3$. Hence

$$S = \langle 9, c-3, c, c+2, c+3, c+4, c+5, c+7, c+8 \rangle.$$

If $\mathcal{R} = w(6)$, then $w(6) = 9v + 6$ for some $v \in \{1, \ldots, \frac{c-8}{9}\}$. We have $w(5) = c + 6$ by Proposition 2.4 and $w(3) = w(6) + 6 = 9v + 12$ by Proposition 2.5(i). On the other hand $w(6) < w(4) \Longrightarrow w(6) + 7 \leq w(4)$ by Lemma 2.4(ii). Therefore $w(3) < w(4)$ and this yields $w(4) = c + 5$ by Corollary 2.1. Now, applying Proposition 2.3(ii), we get $c + 5 = w(4) \leq w(2) + 2 \Longrightarrow w(2) = c + 3$. Hence

$$S = \langle 9, 9v+6, 9v+12, c, c+2, c+3, c+5, c+6, c+8 \rangle, \ 1 \leq v \leq \frac{c-8}{9}.$$

If $\mathcal{R} = w(8) = c$, then

$$S = \langle 9, c, c+2, c+3, c+4, c+5, c+6, c+7, c+8 \rangle.$$

We denote the number of Arf numerical semigroups with multiplicity m and conductor c by $N_{ARF}(m, c)$. For any rational number x, the greatest integer less than or equal to x will be denoted by $\lfloor x \rfloor$.

Corollary 3.9 *Let c be a positive integer such that $\lfloor \frac{c}{9} \rfloor > 1$. The number of Arf numerical semigroups with multiplicity 9 and conductor c is*

$$N_{ARF}(9, c) = \begin{cases} 2\frac{c}{9} + 2 & \text{if } c \equiv 0 \ (mod\ 9), \\ 2\frac{c-2}{9} & \text{if } c \equiv 2 \ (mod\ 9), \\ 2\frac{c-3}{9} - 1 & \text{if } c \equiv 3 \ (mod\ 9), \\ 3 & \text{if } c \equiv 4 \ (mod\ 9), \\ 2\frac{c-5}{9} & \text{if } c \equiv 5 \ (mod\ 9), \\ 2\frac{c-6}{9} + 2 & \text{if } c \equiv 6 \ (mod\ 9), \\ 3 & \text{if } c \equiv 7 \ (mod\ 9), \\ 2\frac{c-8}{9} + 5 & \text{if } c \equiv 8 \ (mod\ 9). \end{cases}$$

Example 3.10 There are 8 Arf numerical semigroups with multiplicity 9 and conductor 27, which are listed below.

$$\langle 9, 28, 29, 30, 31, 32, 33, 34, 35\rangle \,,\ \langle 9, 12, 15, 28, 29, 31, 32, 34, 35\rangle$$

$$\langle 9, 21, 24, 28, 29, 31, 32, 34, 35\rangle \,,\ \langle 9, 23, 25, 28, 29, 30, 31, 33, 35\rangle$$

$$\langle 9, 23, 28, 29, 30, 31, 33, 34, 35\rangle \,,\ \langle 9, 15, 21, 28, 29, 31, 32, 34, 35\rangle,$$

$$\langle 9, 24, 28, 29, 30, 31, 32, 34, 35\rangle \,,\ \langle 9, 25, 28, 29, 30, 31, 32, 33, 35\rangle.$$

Example 3.11 There are 3 Arf numerical semigroups with multiplicity 9 and conductor 40, which are listed below.

$$\langle 9, 40, 41, 42, 43, 44, 46, 47, 48\rangle \,,\ \langle 9, 38, 40, 41, 42, 43, 44, 46, 48\rangle \,,$$
$$\langle 9, 34, 38, 40, 41, 42, 44, 46, 48\rangle.$$

There are 3 Arf numerical semigroups with multiplicity 9 and conductor 940, too:

$$\langle 9, 940, 941, 942, 943, 944, 946, 947, 948\rangle \,,\ \langle 9, 938, 940, 941, 942, 943, 944, 946, 948\rangle,$$

$$\langle 9, 934, 938, 940, 941, 942, 944, 946, 948, \rangle.$$

Example 3.12 There are 11 Arf numerical semigroups with multiplicity 9 and conductor 35, which are listed below.

$$\langle 9, 29, 31, 33, 35, 37, 39, 41, 43\rangle \,,\ \langle 9, 12, 15, 35, 37, 38, 40, 41, 43\rangle \,,$$

$$\langle 9, 21, 24, 35, 37, 38, 40, 41, 43\rangle,\ \langle 9, 30, 33, 35, 37, 38, 40, 41, 43\rangle \,,$$

$$\langle 9, 31, 33, 35, 37, 38, 39, 41, 43\rangle \,,\ \langle 9, 31, 35, 37, 38, 39, 41, 42, 43\rangle,$$

$$\langle 9, 32, 35, 37, 38, 39, 40, 42, 43\rangle \,,\ \langle 9, 15, 21, 35, 37, 38, 40, 41, 43\rangle \,,$$

$$\langle 9, 24, 30, 35, 37, 38, 40, 41, 43\rangle,$$

$$\langle 9, 33, 30, 35, 37, 38, 39, 41, 43\rangle \,,\ \langle 9, 35, 37, 38, 39, 40, 41, 42, 43\rangle.$$

4 Arf Numerical Semigroups with Multiplicity 10

Let S be an Arf numerical semigroup with multiplicity 10 and conductor c. Then $c \equiv 0,\ 2,\ 3,\ 4,\ 5,\ 6,\ 7,\ 8$ or $9 \pmod{10}$.

The following proposition describes all Arf numerical semigroups S with multiplicity 10 and conductor c if $c \equiv 0 \pmod{10}$, $c > 10$.

Proposition 4.1 *Let S be a numerical semigroup with multiplicity* 10 *and conductor c, where $c > 10$ and $c \equiv 0 \pmod{10}$. Then S is an Arf semigroup if and only if S is one of the following:*

$\langle 10, c+1, c+2, c+3, c+4, c+5, c+6, c+7, c+8, c+9\rangle$; *or*

$\langle 10, 10u+2, 10u+4, 10u+6, 10u+8, c+1, c+3, c+5, c+7, c+9\rangle$, $1 \le u \le \frac{c-10}{10}$; *or*

$\langle 10, 10v+4, 10v+6, 10v+8, 10v+12, c+1, c+3, c+5, c+7, c+9\rangle$, $1 \le v \le \frac{c-10}{10}$, *or*
$\langle 10, 10v+4, 10v+8, 10v+12, 10v+16, c+1, c+3, c+5, c+7, c+9\rangle$, $1 \le v \le \frac{c-10}{10}$; *or*

$\langle 10, 10x+5, c-2, c+1, c+2, c+3, c+4, c+6, c+7, c+9\rangle$, $1 \le x \le \frac{c-10}{10}$, *or*
$\langle 10, 10x+5, c+1, c+2, c+3, c+4, c+6, c+7, c+8, c+9\rangle$, $1 \le x \le \frac{c-10}{10}$; *or*

$\langle 10, 10y+6, 10y+8, 10y+12, 10y+14, c+1, c+3, c+5, c+7, c+9\rangle$, $1 \le y \le \frac{c-10}{10}$, *or*
$\langle 10, 10y+6, 10y+12, 10y+14, 10y+18, c+1, c+3, c+5, c+7, c+9\rangle$, $1 \le y \le \frac{c-10}{10}$; *or*

$\langle 10, c-3, c+1, c+2, c+3, c+4, c+5, c+6, c+8, c+9\rangle$; *or*

$\langle 10, 10z+8, 10z+12, 10z+14, 10z+16, c+1, c+3, c+5, c+7, c+9\rangle$, $1 \le z \le \frac{c-10}{10}$.

Proof Let S be an Arf numerical semigroup with multiplicity 10 and conductor $c \equiv 0$ *(mod 10)*, where $c > 10$. Then $w(1) = c+1$ and $w(9) = c+9 = \mathcal{M}$ by Lemma 2.3. We have $w(3) = c+3$ and $w(7) \ge c-3$ by Proposition 2.3. It follows that

$$\mathcal{R} \in \{w(1) = c+1, w(2), w(4), w(5), w(6), w(7) = c-3, w(8)\}.$$

If $\mathcal{R} = w(1) = c+1$, then

$$S = \langle 10, c+1, c+2, c+3, c+4, c+5, c+6, c+7, c+8, c+9\rangle.$$

If $\mathcal{R} = w(2)$, then $w(2) = 10u + 2$ for some $u \in \{1, \ldots, \frac{c}{10} - 1\}$. Then we have $w(4) = 10u + 4$, $w(5) = c + 5$, $w(6) = 10u + 6$, $w(7) = c + 7$, $w(8) = 10u + 8$ by Proposition 2.4. In other words,

$$S = \langle 10, 10u{+}2, 10u{+}4, 10u{+}6, 10u{+}8, c{+}1, c{+}3, c{+}5, c{+}7, c{+}9, \rangle, \ 1 \leq u \leq \frac{c - 10}{10}.$$

If $\mathcal{R} = w(4)$, then $w(4) = 10v + 4$ for some $v \in \{1, \ldots, \frac{c}{10} - 1\}$. One can see by using Proposition 2.4 that $w(5) = c + 5$, $w(7) = c + 7$ and $w(8) = 10v + 8$. One can also see by Proposition 2.5 that $w(6) \in \{10v + 6, 10v + 16\}$. Moreover, we have $w(4) < w(2) \leq w(8) + 4 = w(4) + 8$ by Proposition 2.3, which implies $w(2) = 10v + 12$. Thus

$$S = \langle 10, 10v{+}4, 10v{+}6, 10v{+}8, 10v{+}12, c{+}1, c{+}3, c{+}5, c{+}7, c{+}9 \rangle, \ 1 \leq v \leq \frac{c - 10}{10}, \text{ or}$$

$$S = \langle 10, 10v{+}4, 10v{+}8, 10v{+}12, 10v{+}16, c{+}1, c{+}3, c{+}5, c{+}7, c{+}9 \rangle, \ 1 \leq v \leq \frac{c - 10}{10}.$$

If $\mathcal{R} = w(5)$, then $w(5) = 10x + 5$ for some $x \in \{1, \ldots, \frac{c}{10} - 1\}$. We have $w(4) = c + 4$ and $w(6) = c + 6$ by Proposition 2.4. By Proposition 2.3, we have $w(4) \leq w(2) + 2$ and $w(2) \leq w(8) + 4$. Hence $w(2) = c + 2$ and $w(8) \geq c - 2$. We have observed in the beginning of the proof that $w(7) \geq c - 3$. However $w(7) \neq c - 3$, because otherwise $^*w_7(5) = c - 5$ and $2(c - 3) - (c - 5) = c - 1 \notin S$. Therefore, $w(7) = c + 7$ and it follows that in this case

$S = \langle 10, 10x + 5, c - 2, c + 1, c + 2, c + 3, c + 4, c + 6, c + 7, c + 9 \rangle, 1 \leq x \leq \frac{c-10}{10}$,
or
$S = \langle 10, 10x + 5, c + 1, c + 2, c + 3, c + 4, c + 6, c + 7, c + 8, c + 9 \rangle, 1 \leq v \leq \frac{c-10}{10}$.

If $\mathcal{R} = w(6)$, then $w(6) = 10y + 6$ for some $y \in \{1, \ldots, \frac{c}{10} - 1\}$. In that case, $w(5) = c + 5$ and $w(7) = c + 7$ by Proposition 2.4. We see by Proposition 2.5 that $w(4) = 10y + 14$, $w(2) = 10y + 12$ and $w(8) \in \{10y + 8, 10y + 18\}$. Hence

$S = \langle 10, 10y + 6, 10y + 8, 10y + 12, 10y + 14, c + 1, c + 3, c + 5, c + 7, c + 9 \rangle$, $1 \leq y \leq \frac{c-10}{10}$, *or*
$S = \langle 10, 10y + 6, 10y + 12, 10y + 14, 10y + 18, c + 1, c + 3, c + 5, c + 7, c + 9 \rangle$, $1 \leq y \leq \frac{c-10}{10}$.

If $\mathcal{R} = w(7) = c - 3$, then $w(6) = c + 6$ and $w(8) = c + 8$ by Proposition 2.4. Note also that, $c - 3$ being the ratio, $w(2) = c + 2$, $w(4) = c + 4$ and $w(5) = c + 5$. Thus

$$S = \langle 10, c - 3, c + 1, c + 2, c + 3, c + 4, c + 5, c + 6, c + 8, c + 9 \rangle.$$

If $\mathcal{R} = w(8)$, then $w(8) = 10z + 8$ for some $z \in \{1, \ldots, \frac{c}{10} - 1\}$. We have $w(7) = c + 7$ by Proposition 2.4; $w(2) = w(8) + 4 = 10z + 12$, $w(4) = w(8) + 6 = 10z + 14$

and $w(6) = w(8) + 8 = 10z + 16$ by Proposition 2.5. On the other hand, $w(5)$ can not be less than c, because if we had $w(5) < c$, then we would have $w(5) \leq c - 5$ and $w(4) = c + 4$ by Corollary 2.1, and therefore $w(8) = c - 2$. This is impossible because $w(8)$ is the ratio. So $w(5) = c + 5$ and

$$S = \langle 10, 10z + 8, 10z + 12, 10z + 14, 10z + 16, c + 1, c + 3, c + 5, c + 7, c + 9\rangle,\ 1 \leq z \leq \tfrac{c-10}{10}.$$

The proof of each of the remaining propositions is very similar to the proof of Proposition 4.1. Therefore we state them without proof.

Proposition 4.2 *Let S be a numerical semigroup with multiplicity* 10 *and conductor c, where $c > 12$ and $c \equiv 2 \pmod{10}$. Then S is an Arf semigroup if and only if S is one of the following:*

$\langle 10, 10u + 2, 10u + 4, 10u + 6, 10u + 8, c + 1, c + 3, c + 5, c + 7, c + 9\rangle$, $1 \leq u \leq \frac{c-2}{10}$; *or*
$\langle 10, 10v + 4, 10v + 6, 10v + 8, 10v + 12, c + 1, c + 3, c + 5, c + 7, c + 9\rangle$, $1 \leq v \leq \frac{c-12}{10}$, *or*
$\langle 10, 10v + 4, 10v + 8, 10v + 12, 10v + 16, c + 1, c + 3, c + 5, c + 7, c + 9\rangle$, $1 \leq v \leq \frac{c-12}{10}$; *or*

$\langle 10, 10x + 5, c, c + 1, c + 2, c + 4, c + 5, c + 6, c + 7, c + 9\rangle$, $1 \leq x \leq \frac{c-12}{10}$; *or*

$\langle 10, 10y + 6, 10y + 8, 10y + 12, 10y + 14, c + 1, c + 3, c + 5, c + 7, c + 9\rangle$, $1 \leq y \leq \frac{c-12}{10}$, *or*
$\langle 10, 10y + 6, 10y + 12, 10y + 14, 10y + 18, c + 1, c + 3, c + 5, c + 7, c + 9\rangle$, $1 \leq y \leq \frac{c-12}{10}$; *or*
$\langle 10, c - 5, c, c + 1, c + 2, c + 3, c + 4, c + 6, c + 7, c + 9\rangle$; *or*

$\langle 10, 10z + 8, 10z + 12, 10z + 14, 10z + 16, c + 1, c + 3, c + 5, c + 7, c + 9\rangle$, $1 \leq z \leq \frac{c-12}{10}$.

Proposition 4.3 *Let S be a numerical semigroup with multiplicity* 10 *and conductor c, where $c > 13$ and $c \equiv 3 \pmod{10}$. Then S is an Arf semigroup if and only if S is one of the following:*

$\langle 10, c, c + 1, c + 2, c + 3, c + 4, c + 5, c + 6, c + 8, c + 9\rangle$; *or*

$\langle 10, 10u + 5, c, c + 1, c + 3, c + 4, c + 5, c + 6, c + 8, c + 9\rangle$, $1 \leq u \leq \frac{c-13}{10}$; *or*

$\langle 10, c - 6, c, c + 1, c + 2, c + 3, c + 5, c + 6, c + 8, c + 9\rangle$.

Proposition 4.4 *Let S be a numerical semigroup with multiplicity* 10 *and conductor c, where $c > 14$ and $c \equiv 4 \pmod{10}$. Then S is an Arf semigroup if and only if S is one of the following:*

$\langle 10, 10u+2, 10u+4, 10u+6, 10u+8, c+1, c+3, c+5, c+7, c+9\rangle$,
$1 \le u \le \frac{c-4}{10}$; *or*

$\langle 10, 10v+4, 10v+6, 10v+8, 10v+12, c+1, c+3, c+5, c+7, c+9\rangle$,
$1 \le v \le \frac{c-4}{10}$, *or*
$\langle 10, 10v+4, 10v+8, 10v+12, 10v+16, c+1, c+3, c+5, c+7, c+9\rangle$,
$1 \le v \le \frac{c-14}{10}$; *or*

$\langle 10, 10x+5, c, c+2, c+3, c+4, c+5, c+7, c+8, c+9\rangle$, $1 \le x \le \frac{c-14}{10}$, *or*
$\langle 10, 10x+5, c-2, c, c+2, c+3, c+4, c+5, c+7, c+9\rangle$, $1 \le x \le \frac{c-14}{10}$; *or*

$\langle 10, 10y+6, 10y+8, 10y+12, 10y+14, c+1, c+3, c+5, c+7, c+9\rangle$,
$1 \le y \le \frac{c-14}{10}$, *or*
$\langle 10, 10y+6, 10y+12, 10y+14, 10y+18, c+1, c+3, c+5, c+7, c+9\rangle$,
$1 \le y \le \frac{c-14}{10}$; *or*

$\langle 10, 10z+8, 10z+12, 10z+14, 10z+16, c+1, c+3, c+7, c+9\rangle$,
$1 \le z \le \frac{c-14}{10}$.

Proposition 4.5 *Let S be a numerical semigroup with multiplicity* 10 *and conductor c, where $c > 15$ and $c \equiv 5 \pmod{10}$. Then S is an Arf semigroup if and only if S is one of the following:*

$\langle 10, c-2, c, c+1, c+2, c+3, c+4, c+6, c+7, c+9\rangle$; *or*

$\langle 10, 10u+5, c-2, c+1, c+2, c+3, c+4, c+6, c+7, c+9\rangle$,
$1 \le u \le \frac{c-15}{10}$, *or*
$\langle 10, 10u+5, c+1, c+2, c+3, c+4, c+6, c+7, c+8, c+9\rangle$,
$1 \le u \le \frac{c-5}{10}$.

Proposition 4.6 *Let S be a numerical semigroup with multiplicity* 10 *and conductor c, where $c > 16$ and $c \equiv 6 \pmod{10}$. Then S is an Arf semigroup if and only if S is one of the following:*

$\langle 10, 10u+2, 10u+4, 10u+6, 10u+8, c+1, c+3, c+5, c+7, c+9\rangle$,
$1 \le u \le \frac{c-6}{10}$; *or*

$\langle 10, c-3, c, c+1, c+2, c+3, c+5, c+6, c+8, c+9\rangle$; *or*

$\langle 10, 10v+4, 10v+6, 10v+8, 10v+12, c+1, c+3, c+5, c+7, c+9\rangle$,
$1 \leq v \leq \frac{c-6}{10}$, *or*
$\langle 10, 10v+4, 10v+8, 10v+12, 10v+16, c+1, c+3, c+5, c+7, c+9\rangle$,
$1 \leq v \leq \frac{c-16}{10}$; *or*

$\langle 10, 10x+6, 10x+8, 10x+12, 10x+14, c+1, c+3, c+5, c+7, c+9\rangle$,
$1 \leq x \leq \frac{c-6}{10}$, *or*
$\langle 10, 10x+6, 10x+12, 10x+14, 10x+18, c+1, c+3, c+5, c+7, c+9\rangle$,
$1 \leq x \leq \frac{c-16}{10}$; *or*

$\langle 10, 10y+8, 10y+12, 10y+14, 10y+16, c+1, c+3, c+5, c+7, c+9\rangle$,
$1 \leq y \leq \frac{c-16}{10}$.

Proposition 4.7 *Let S be a numerical semigroup with multiplicity* 10 *and conductor c, where* $c > 17$ *and* $c \equiv 7 \pmod{10}$. *Then S is an Arf semigroup if and only if S is one of the following:*

$\langle 10, c-3, c, c+1, c+2, c+4, c+5, c+6, c+8, c+9\rangle$, *or*

$\langle 10, 10u+5, c, c+1, c+2, c+4, c+5, c+6, c+7, c+9\rangle$, $1 \leq u \leq \frac{c-7}{10}$, *or*

$\langle 10, c, c+1, c+2, c+4, c+5, c+6, c+7, c+8, c+9\rangle$.

Proposition 4.8 *Let S be a numerical semigroup with multiplicity* 10 *and conductor c, where* $c > 18$ *and* $c \equiv 8 \pmod{10}$. *Then S is an Arf semigroup if and only if S is one of the following:*

$\langle 10, 10u+2, 10u+4, 10u+6, 10u+8, c+1, c+3, c+5, c+7, c+9\rangle$,
$1 \leq u \leq \frac{c-8}{10}$; *or*

$\langle 10, c-5, c-2, c, c+1, c+3, c+4, c+6, c+7, c+9\rangle$; *or*

$\langle 10, 10v+4, 10v+6, 10v+8, 10v+12, c+1, c+3, c+5, c+7, c+9\rangle$,
$1 \leq v \leq \frac{c-8}{10}$, *or*
$\langle 10, 10v+4, 10v+8, 10v+12, 10v+16, c+1, c+3, c+5, c+7, c+9\rangle$,
$1 \leq v \leq \frac{c-8}{10}$; *or*

$\langle 10, 10x+5, c, c+1, c+3, c+4, c+5, c+6, c+8, c+9\rangle$, $1 \leq x \leq \frac{c-8}{10}$, *or*

$\langle 10, 10y+6, 10y+8, 10y+12, 10y+14, c+1, c+3, c+5, c+7, c+9\rangle$,
$1 \leq y \leq \frac{c-8}{10}$, *or*

$\langle 10, 10y+6, 10y+12, 10y+14, 10y+18, c+1, c+3, c+5, c+7, c+9\rangle$, $1 \le y \le \frac{c-18}{10}$; *or*

$\langle 10, 10z+8, 10z+12, 10z+14, 10z+16, c+1, c+3, c+5, c+7, c+9\rangle$, $1 \le z \le \frac{c-8}{10}$.

Proposition 4.9 *Let S be a numerical semigroup with multiplicity* 10 *and conductor c, where c* > 19 *and c* ≡ 9 (mod 10)*. Then S is an Arf semigroup if and only if S is one of the following:*

$\langle 10, c-6, c-3, c, c+2, c+3, c+5, c+6, c+8, c+9\rangle$; *or*

$\langle 10, 10u+5, c-2, c, c+2, c+3, c+4, c+5, c+8, c+9\rangle$, $1 \le u \le \frac{c-9}{10}$, *or*
$\langle 10, 10u+5, c, c+2, c+3, c+4, c+5, c+6, c+8, c+9\rangle$, $1 \le u \le \frac{c-9}{10}$, *or*
$\langle 10, c-3, c, c+2, c+3, c+4, c+6, c+7, c+8, c+9\rangle$; *or*

$\langle 10, c-2, c, c+2, c+3, c+4, c+5, c+6, c+7, c+9\rangle$; *or*

$\langle 10, c, c+2, c+3, c+4, c+5, c+6, c+7, c+8, c+9\rangle$.

Corollary 4.10 *Let c be a positive integer such that* $\lfloor \frac{c}{10} \rfloor > 1$. *The number of Arf numerical semigroups with multiplicity* 10 *and conductor c is*

$$N_{ARF}(10, c) = \begin{cases} \frac{4c}{5} - 6 & \textit{if } c \equiv 0 \pmod{10}, \\ \frac{7(c-2)}{10} - 5 & \textit{if } c \equiv 2 \pmod{10}, \\ \frac{c-3}{10} + 1 & \textit{if } c \equiv 3 \pmod{10}, \\ \frac{4(c-4)}{5} - 6 & \textit{if } c \equiv 4 \pmod{10}, \\ \frac{c-5}{5} & \textit{if } c \equiv 5 \pmod{10}, \\ \frac{3(c-6)}{5} - 2 & \textit{if } c \equiv 6 \pmod{10}, \\ \frac{c-7}{10} + 2 & \textit{if } c \equiv 7 \pmod{10}, \\ \frac{7(c-8)}{10} & \textit{if } c \equiv 8 \pmod{10}, \\ \frac{c-9}{5} + 4 & \textit{if } c \equiv 9 \pmod{10}. \end{cases}$$

Example 4.11 There are 10 Arf numerical semigroups with multiplicity 10 and conductor 24:

$\langle 10, 12, 14, 16, 18, 25, 27, 29, 31, 33\rangle$, $\langle 10, 22, 24, 26, 25, 27, 28, 29, 31, 33\rangle$,

$\langle 10, 14, 16, 18, 22, 25, 27, 29, 31, 33\rangle$, $\langle 10, 22, 24, 25, 26, 27, 28, 29, 31, 33\rangle$,

$\langle 10, 14, 18, 22, 25, 26, 27, 29, 31, 33\rangle$, $\langle 10, 15, 24, 25, 26, 27, 28, 31, 32, 33\rangle$,

$$\langle 10, 22, 24, 25, 26, 27, 28, 30, 31, 33\rangle, \langle 10, 16, 18, 22, 24, 25, 27, 29, 31, 33\rangle$$

$$\langle 10, 16, 22, 24, 25, 27, 28, 30, 31, 33\rangle, \langle 10, 18, 22, 24, 25, 26, 27, 29, 31, 33\rangle$$

Example 4.12 There are 6 Arf numerical semigroups with multiplicity 10 and conductor 35, which are listed below.

$$\langle 10, 33, 35, 36, 37, 38, 39, 41, 42, 44\rangle, \langle 10, 15, 33, 36, 37, 38, 39, 41, 42, 44\rangle,$$

$$\langle 10, 25, 33, 36, 37, 38, 39, 41, 42, 44\rangle, \langle 10, 15, 36, 37, 38, 39, 41, 42, 43, 44\rangle,$$

$$\langle 10, 25, 36, 37, 38, 39, 41, 42, 43, 44\rangle, \langle 10, 35, 36, 37, 38, 39, 41, 42, 43, 44\rangle$$

Example 4.13 There are 7 Arf numerical semigroups with multiplicity 10 and conductor 63:

$$\langle 10, 63, 64, 65, 66, 67, 68, 69, 71, 72\rangle, \langle 10, 15, 63, 64, 66, 67, 68, 69, 71, 72\rangle,$$

$$\langle 10, 25, 63, 64, 66, 67, 68, 69, 71, 72\rangle, \langle 10, 35, 63, 64, 66, 67, 68, 69, 71, 72\rangle,$$

$$\langle 10, 45, 63, 64, 66, 67, 68, 69, 71, 72\rangle, \langle 10, 55, 63, 64, 66, 67, 68, 69, 71, 72\rangle,$$

$$\langle 10, 57, 63, 66, 67, 68, 69, 71, 72\rangle.$$

Acknowledgements I would like to thank the referee for the careful reading and many useful and constructive suggestions about this work.

References

1. Arf, C.: Une interprétation algébrique de la suite ordres de multiplicité d'une branche algébrique. Proc. Lond. Math. Soc. **20**, 256–287 (1949)
2. Barucci, V., Dobbs, D.E., Fontana, M.: Maximality properties in numerical semigroups and applications to one-dimensional analytically irreducible local domains. Mem. Am. Math. Soc. **125/598**, 1–77 (1997)
3. Compillo, A., Farran, J.I., Munuera, C.: On the parameters of algebraic geometry codes related to arf semigroups. IEEE Trans. Inf. Theory **46**(7), 2634–2638 (2000)
4. Fröberg, R., Gottlieb, C., Häggkvist, R.: On numerical semigroups. Semigroup Forum **35**, 63–83 (1987)
5. Garcia-Sánchez, P.A., Heredia, B.A., Karakaş, H.I., Rosales, J.C.: Parametrizing Arf numerical semigroups. J. Algebra Appl. **16**(1), 31pp. (2017)
6. Lason, M.: On the relation between Betti number of an Arf semigroup and its blowup. Le Mathematiche **67**, 75–80 (2012)
7. Rosales, J.C., Garcia-Sánchez, P.A.: Numerical Semigroups. Springer, Berlin (2009)
8. Rosales, J.C., Garcia-Sánchez, P.A., Garcia-Garcia, J.I., Branco, M.B.: Arf numerical semigroups. J. Algebra **276**, 3–12 (2004)
9. Süer, M., Karakaş, H.I., İlhan, S.: The family of Arf numerical semigroups with multiplicity eight. Unpublished (2017), 18pp.

Numerical Semigroup Rings of Maximal Embedding Dimension with Determinantal Defining Ideals

Do Van Kien and Naoyuki Matsuoka

Abstract We give a criterion of a numerical semigroup ring for having the defining ideal generated by 2×2-minors of a $2 \times n$ matrix in terms of pseudo-Frobenius numbers when the numerical semigroup has maximal embedding dimension. The ring-theoretic properties of a symbolic Rees algebra of the defining ideal are also explored.

Keywords Cohen-Macaulay ring · Numerical semigroup · Numerical semigroup ring · Graded ring · Pseudo-Frobenius number · Minimal free resolution · Symbolic Rees algebra

1 Introduction

Let $S = k[x_1, x_2, \ldots, x_n]$ be the polynomial ring over a field k and I be an ideal of S. Exploring the graded minimal free resolution of $R = S/I$ is an important problem in commutative algebra because the resolution contains much information on R. In the present paper, we consider a relation between the generation of the defining ideal of a semigroup ring of a numerical semigroup H and the behavior of the pseudo-Frobenius numbers of H.

D. V. Kien
Department of Mathematics, Hanoi Pedagogical University 2, Vinh Phuc, Vietnam
e-mail: dovankien@hpu2.edu.vn

N. Matsuoka (✉)
Department of Mathematics, School of Science and Technology, Meiji University, Kawasaki, Tama-ku, Japan
e-mail: naomatsu@meiji.ac.jp

V. Barucci et al. (eds.), *Numerical Semigroups*, Springer INdAM Series 40,
https://doi.org/10.1007/978-3-030-40822-0_12

Let $a_1, a_2, \ldots, a_n$ be positive integers with $\gcd(a_1, a_2, \ldots, a_n) = 1$ and put

$$H = \langle a_1, a_2, \ldots, a_n \rangle = \left\{ \sum_{i=1}^{n} c_i a_i \;\middle|\; 0 \leq c_i \in \mathbb{Z} \text{ for all } 1 \leq i \leq n \right\}$$

be the numerical semigroup minimally generated by $a_1, a_2, \ldots, a_n$. Let k be a field. We set

$$R = k[H] = k[t^{a_1}, t^{a_2}, \ldots, t^{a_n}]$$

and we call it the numerical semigroup ring of H over k, where t is an indeterminate. Let $S = k[x_1, x_2, \ldots, x_n]$ be the weighted polynomial ring over k with $\deg x_i = a_i$ for each $1 \leq i \leq n$. Let $\varphi : S \to R$ denote the homomorphism of graded k-algebras defined by $\varphi(x_i) = t^{a_i}$ for all $1 \leq i \leq n$. Let $P = \operatorname{Ker} \varphi$ be the defining ideal of R.

When the number n is small, there are a few known results about the structure of P. The most important result is due to J. Herzog [14] when $n = 3$. When $n = 4$, there are some partial answers to describe the structure of the defining ideal: e.g. [2] for the symmetric case, [16] for the pseudo-symmetric case, [8] for the almost symmetric of type three case, and [19] for the case where H is generated by an almost arithmetic progression. On the other hand, results with no restriction on n also exist. P. Gimenez et al. [9] constructed an explicit form of the graded minimal free resolution of R when H is generated by an arithmetic progression. Besides, the authors, S. Goto, and H. L. Truong [13] found that the pseudo-Frobenius numbers of H affect the generation of the defining ideal P of $R = k[H]$, cf. Theorem 1.

We denote the Frobenius number of H by $\mathrm{f}(H) = \max(\mathbb{Z} \setminus H)$. We set

$$\mathrm{PF}(H) = \{\alpha \in \mathbb{Z} \setminus H \mid \alpha + a_i \in H \text{ for all } 1 \leq i \leq n\}$$

and call the elements in $\mathrm{PF}(H)$ pseudo-Frobenius numbers of H. Hence $\mathrm{f}(H) \in \mathrm{PF}(H)$ and

$$\mathrm{K}_R = \sum_{\alpha \in \mathrm{PF}(H)} Rt^{-\alpha}$$

[11], where K_R denotes the graded canonical module of R. Therefore, the a-invariant $\mathrm{a}(R)$ of R (resp. the Cohen-Macaulay type $\mathrm{r}(R)$ of R) is given by $\mathrm{a}(R) = \mathrm{f}(H)$ (resp. $\mathrm{r}(R) = \sharp\, \mathrm{PF}(H)$). For a given matrix A with entries in S, we denote by $\mathrm{I}_2(A)$ the ideal of S generated by 2×2 minors of A.

Theorem 1 ([13, Theorem 1.2]) *Let $H = \langle a_1, a_2, \ldots, a_n \rangle$ $(n \geq 3)$ be a numerical semigroup and assume that H is minimally generated by the n numbers $\{a_i\}_{1 \leq i \leq n}$. Then the following conditions are equivalent.*

(1) $P = \mathrm{I}_2\begin{pmatrix} f_1 & f_2 & \cdots & f_n \\ x_1 & x_2 & \cdots & x_n \end{pmatrix}$ *for some homogeneous elements* $f_1, f_2, \ldots, f_n \in S_+ = (x_i \mid 1 \leq i \leq n)$.

(2) *After suitable permutations of* $a_1, a_2, \ldots, a_n$ *if necessary, we have*

$$P = \mathrm{I}_2\begin{pmatrix} x_2^{\ell_2} & x_3^{\ell_3} & \cdots & x_n^{\ell_n} & x_1^{\ell_1} \\ x_1 & x_2 & \cdots & x_{n-1} & x_n \end{pmatrix}$$

for some positive integers $\ell_1, \ell_2, \ldots, \ell_n > 0$.

(3) *There exists an element* $\alpha \in \mathrm{PF}(H)$ *such that* $(n-1)\alpha \notin H$.

When this is the case, the following assertions hold true.

(a) *For each* $1 \leq i \leq n$, *we have* $\ell_i = \min\{\ell > 0 \mid \ell a_i \in H_i\} - 1$, *where*

$$H_i = \left\langle a_1, \ldots, \overset{\vee}{a_i}, \ldots, a_n \right\rangle.$$

(b) $\alpha = \deg f_i - a_i$ *for all* $1 \leq i \leq n$.

(c) $\mathrm{PF}(H) = \{\alpha, 2\alpha, \ldots, (n-1)\alpha\}$.

(d) *The numerical semigroup* H *is almost symmetric (For the definition, see [1]).*

In the light of Theorem 1 it is natural to ask when P is generated by 2×2 minors of a matrix

$$\begin{pmatrix} f_1 & f_2 & \cdots & f_n \\ g_1 & g_2 & \cdots & g_n \end{pmatrix}$$

where f_i and g_i are homogeneous elements in S. The main theorem of this paper gives an answer to this question when H has maximal embedding dimension as follows.

Theorem 2 *Let* $H = \langle a_1, a_2, \ldots, a_n \rangle$ *and we assume* H *has maximal embedding dimension* $n \geq 3$. *Then the following 4 conditions are equivalent to each other.*

(1) $H = \langle n, n+h+\alpha, n+h+2\alpha, \ldots, n+h+(n-1)\alpha \rangle$ *for some* $h \geq 0$ *and* $\alpha > 0$.

(2) $\mathrm{PF}(H) = \{h+\alpha, h+2\alpha, \ldots, h+(n-1)\alpha\}$ *for some* $h \geq 0$ *and* $\alpha > 0$.

(3) *There exist homogeneous elements* $f_1, f_2, \ldots, f_n, g_1, g_2, \ldots, g_n \in S_+$ *such that*

$$P = \mathrm{I}_2\left(\begin{smallmatrix} f_1 & f_2 & \cdots & f_n \\ g_1 & g_2 & \cdots & g_n \end{smallmatrix}\right).$$

(4) *After suitable permutations of* $a_1, a_2, \ldots, a_n$ *if necessary, we have*

$$P = \mathrm{I}_2\left(\begin{smallmatrix} x_2 & x_3 & \cdots & x_n & x_1^{s+\alpha} \\ x_1^s & x_2 & \cdots & x_{n-1} & x_n \end{smallmatrix}\right)$$

for some positive integers s *and* α.

When this is the case, we have the following.

(a) *The Eagon-Northcott complex associated to the matrix* $\left(\begin{smallmatrix} f_1 & f_2 & \cdots & f_n \\ g_1 & g_2 & \cdots & g_n \end{smallmatrix}\right)$ *gives rise to a graded minimal free resolution of* R.

(b) *The symbolic Rees algebra* $\mathcal{R}_S(P) = \bigoplus_{r \geq 0} P^{(r)}T^r \subseteq S[T]$ *is Noetherian and Cohen-Macaulay, where* $P^{(r)} = P^r S_P \cap S$ *for all integers* $r \geq 0$.

(c) $\mathcal{R}_S(P)$ *is Gorenstein if and only if* $n = 3$.

We clearly have the equivalence between (1) and (2), because $\mathrm{PF}(H) = \{a_2 - a_1, a_3 - a_1, \ldots, a_n - a_1\}$, if $H = \langle a_1, a_2, \ldots, a_n \rangle$ has maximal embedding dimension $n = a_1$. The implication (4)–(3) is obvious. In the next section, we give a proof of the implications (3)–(2) and (2)–(4) together with the assertions (a), (b), and (c). In the last section, we will explore examples to illustrate the meaning of the equivalence of (1)–(4).

Before we enter the next section, we recall history and known facts about the symbolic Rees algebras. R. Cowsik [4] asked whether the symbolic Rees algebra $\mathcal{R}_S(P)$ of a prime ideal P of a regular local ring (or the polynomial ring over a field) is Noetherian. This question has been studied by many researchers, even though counterexamples are already known. The first one is given by P. Roberts [18]. When P is the defining ideal $\mathfrak{p}_k(a, b, c)$ of a space monomial curve (t^a, t^b, t^c) in k^3, C. Huneke [15] and S. D. Cutkosky [5] found criterions for the Noetherian property of such rings, where k is a field and a, b, c are positive integers. In 1994, S. Goto et al. Watanabe [12] discovered integers a, b, c such that the symbolic Rees algebra of $\mathfrak{p}_k(a, b, c)$ is not finitely generated over k, if the characteristic of k is zero.

On the other hand, S. Goto [10, Theorem (7.4)] proved that the symbolic Rees algebra of the defining ideal of $k[t^a, t^{a+m}, \ldots, t^{a+(a-1)m}]$ is Noetherian and studied the Cohen-Macaulay property also. Recently, C. D'Cruz and S. K. Masuti [6] gave a complete description of the Cohen-Macaulay and Gorenstein properties of the same ring. The method to prove the assertions (b) and (c) in Theorem 2 is completely the same as their ones.

2 Proof of Theorem 2

2.1 *Proof of* (3) $\Rightarrow$ (2) *and the Assertion (a)*

In this subsection we give a proof of (3) $\Rightarrow$ (2) in Theorem 2 without the assumption that H has maximal embedding dimension.

Let $H = \langle a_1, a_2, \ldots, a_n \rangle$ $(n \geq 3)$ and we assume the embedding dimension of H is n. Let $f_1, f_2, \ldots, f_n, g_1, g_2, \ldots, g_n \in S_+$ be homogeneous elements. In this subsection, we always assume that the defining ideal P of $R = k[t^{a_1}, t^{a_2}, \ldots, t^{a_n}]$

is generated by 2×2 minors of the matrix

$$\begin{pmatrix} f_1 & f_2 & \cdots & f_n \\ g_1 & g_2 & \cdots & g_n \end{pmatrix}.$$

Notice that $P \subseteq (g_1, g_2, \ldots, g_n)(f_1, f_2, \ldots, f_n) \subseteq S_+(f_1, f_2, \ldots, f_n)$.

Lemma 1 *The difference* $\deg f_i - \deg g_i$ *is constant and independent of the choice of* $1 \le i \le n$.

Proof We begin with the following claim.

Claim $f_i, g_i \notin P$ for all $1 \le i \le n$.

Proof (Claim) It is enough to show this claim when $i = 1$. We put $M = S_+$. Suppose that $f_1 \in P$. Then we have

$$P = (f_1) + \mathrm{I}_2 \begin{pmatrix} 0 & f_2 & \cdots & f_n \\ g_1 & g_2 & \cdots & g_n \end{pmatrix},$$

whence $f_i g_1 \in P$ for all $2 \le i \le n$. In addition, we assume $g_1 \notin P$. Then $f_i \in P$ for all $2 \le i \le n$ because P is a prime ideal of S. Therefore we have $P \subseteq M(f_1, f_2, \ldots, f_n) \subseteq MP \subseteq P$. Thus $PS_M = (MS_M)(PS_M)$ which implies the contradiction that $P = (0)$ by Nakayama's lemma. Hence $g_1 \in P$. Then

$$P \subseteq \mathrm{I}_2 \begin{pmatrix} 0 & f_2 & \cdots & f_n \\ 0 & g_2 & \cdots & g_n \end{pmatrix} + M(f_1, g_1) \subseteq \mathrm{I}_2 \begin{pmatrix} f_2 & \cdots & f_n \\ g_2 & \cdots & g_n \end{pmatrix} + MP \subseteq P.$$

Again by Nakayama's lemma, we have

$$PS_M = \mathrm{I}_2 \begin{pmatrix} f_2 & \cdots & f_n \\ g_2 & \cdots & g_n \end{pmatrix} S_M.$$

Hence $\mathrm{ht}_S P = \mathrm{ht}_{S_M} PS_M \le n - 2$ (see [3, (2.1) Theorem]) which is impossible because $\dim S/P = 1$. Therefore $f_1 \notin P$. Similarly, we also have $g_1 \notin P$.

Now, suppose that $\deg f_i - \deg g_i \neq \deg f_j - \deg g_j$ for some $1 \le i < j \le n$. Then $\deg(f_i g_j) \neq \deg(f_j g_i)$. Since $f_i g_j - f_j g_i \in P$ and P is a homogeneous ideal of S, we have $f_i g_j, f_j g_i \in P$. However $f_i g_j$ can not be in P because $P \in \operatorname{Spec} S$ and $f_i, g_j \notin P$ by Claim 2.1. Therefore $\deg f_i - \deg g_i = \deg f_j - \deg g_j$ for all $1 \le i \le j \le n$.

Without loss of generality, we may assume $\deg f_i \ge \deg g_i$ for all $1 \le i \le n$. Let α denote the difference $\deg f_1 - \deg g_1$. We put $h = \sum_{i=1}^{n} (\deg g_i - a_i)$. The following theorem shows the implication (3)–(2) in Theorem 2.

Theorem 3 $\mathrm{PF}(H) = \{h + \alpha, h + 2\alpha, \ldots, h + (n-1)\alpha\}$.

Proof Although the method is almost the same as [13, Section 4], for the sake of completeness, let us include the construction of the Eagon-Northcott complex $\mathcal{C}$ associated to the matrix $\begin{pmatrix} f_1 & f_2 & \cdots & f_n \\ g_1 & g_2 & \cdots & g_n \end{pmatrix}$. Let F be a free S-module with a basis $\{T_i\}_{1 \le i \le n}$ and let $K = \bigwedge F$ be the exterior algebra of F over S. We denote ∂_1 and ∂_2 the differentials of the Koszul complexes $\mathrm{K}_\bullet(f_1, f_2, \ldots, f_n; S)$ and $\mathrm{K}_\bullet(g_1, g_2, \ldots, g_n; S)$, respectively. Let $U = S[y_1, y_2]$ be the polynomial ring over S with 2 variables and we regard U as a standard $\mathbb{Z}$-graded ring over S. Let

$$\begin{cases} C_0 = S \\ C_q = K_{q+1} \otimes_S U_{q-1} \quad 1 \le q \le n-1. \end{cases}$$

Then, for each $1 \le q \le n-1$, C_q is a finitely generated free S-module with a free basis

$$\{T_\Lambda \otimes y_1^{q-1-\ell} y_2^{\ell} \mid \Lambda \subseteq \{1, 2, \ldots, n\}, \sharp\Lambda = q+1, 0 \le \ell \le q-1\},$$

where $T_\Lambda = T_{i_1} T_{i_2} \cdots T_{i_{q+1}}$ with $\Lambda = \{i_1 < i_2 < \cdots < i_{q+1}\}$. Then the complex

$$\mathcal{C} : 0 \to C_{n-1} \xrightarrow{d_{n-1}} C_{n-2} \xrightarrow{d_{n-2}} \cdots \xrightarrow{d_2} C_1 \xrightarrow{d_1} C_0 \to 0$$

is called the Eagon-Northcott complex associated to $\begin{pmatrix} f_1 & f_2 & \cdots & f_n \\ g_1 & g_2 & \cdots & g_n \end{pmatrix}$ with the differentials

$$d_1(T_i T_j \otimes 1) = \det \begin{pmatrix} f_i & f_j \\ g_i & g_j \end{pmatrix}$$

for $1 \le i < j \le n$ and

$$d_q(T_\Lambda \otimes y_1^{q-1-\ell} y_2^{\ell}) = \begin{cases} \partial_1(T_\Lambda) \otimes y_1^{q-2-\ell} y_2^{\ell} & \text{if } 1 \le \ell \le q-2, \\ \qquad + \partial_2(T_\Lambda) \otimes y_1^{q-1-\ell} y_2^{\ell-1} & \\ \partial_1(T_\Lambda) \otimes y_1^{q-2} & \text{if } \ell = 0, \\ \partial_2(T_\Lambda) \otimes y_2^{q-2} & \text{if } \ell = q-1. \end{cases}$$

Because $\mathrm{ht}_S P = \dim S - \dim S/P = n-1$, $\mathcal{C}$ is a minimal S-free resolution of R ([7]). Next, we regard the complex $\mathcal{C}$ as a complex of graded S-modules by the grading

$$\deg(T_\Lambda \otimes y_1^{q-1-\ell} y_2^{\ell}) = \sum_{i \in \Lambda} \deg f_i - (\ell + 1)\alpha.$$

Therefore

$$C_{n-1} = \bigoplus_{i=1}^{n-1} S(i\alpha - b)$$

as a graded S-module, where $b = \sum_{i=1}^{n} \deg f_i$. Let $\mathrm{K}_S = S(-\sum_{i=1}^{n} a_i)$ denote the graded canonical module of S. By taking the K_S-dual of $\mathcal{C}$, we get a minimal presentation

$$\bigoplus_{i=1}^{n-1} S(-i\alpha + b - \sum_{j=1}^{n} a_j) \to \mathrm{K}_R \to 0$$

of K_R. Since $\alpha = \deg f_j - \deg g_j$ for all $1 \leq j \leq n$ and $b = \sum_{i=1}^{n} \deg f_i$,

$$-i\alpha + b - \sum_{j=1}^{n} a_j = \sum_{j=1}^{n} (\deg g_j - a_j) + (n-i)\alpha.$$

Hence $\mathrm{PF}(H) = \{h+\alpha, h+2\alpha, \ldots, h+(n-1)\alpha\}$ where $h = \sum_{i=1}^{n} (\deg g_i - a_i)$, because $\mathrm{K}_R = \sum_{p \in \mathrm{PF}(H)} Rt^{-p}$ ([11]).

2.2 *Proof of* (2) ⇒ (4)

Let $H = \langle a_1, a_2, \ldots, a_n \rangle$ and assume the embedding dimension of H is $n \geq 3$. We begin with the following lemma.

Lemma 2 *Suppose* $\mathrm{PF}(H) = \{h+\alpha, h+2\alpha, \ldots, h+(n-1)\alpha\}$ *for some* $h \geq 0$ *and* $\alpha > 0$. *Then* $\alpha \notin H$ *and* $h \in H$.

Proof Since $n \geq 3$, we have $h+\alpha, h+2\alpha \in \mathrm{PF}(H)$. Hence $\alpha \notin H$. This implies that $p - \alpha \in H$ for some $p \in \mathrm{PF}(H)$. Therefore we can find $1 \leq i \leq n-1$ such that $(h+i\alpha) - \alpha = h + (i-1)\alpha \in H$. Then we must have $i = 1$, because $h + (i-1)\alpha \in \mathrm{PF}(H)$ if $2 \leq i \leq n-1$. Thus $h \in H$ as desired.

We are now in a position to prove the implication (2)–(4) in Theorem 2.

Proof ((2) ⇒ (4) ***in Theorem 2***)
We assume the condition (2). After suitable permutations of $a_1, a_2, \ldots, a_n$, we may assume $n = a_1 < a_2 < \cdots < a_n$. Since H has maximal embedding dimension, we have

$$\begin{aligned} \mathrm{PF}(H) &= \{a_2 - a_1, a_3 - a_1, \ldots, a_n - a_1\} \\ &= \{h+\alpha, h+2\alpha, \ldots, h+(n-1)\alpha\}. \end{aligned}$$

Therefore $a_i = a_1 + h + (i-1)\alpha$ for all $2 \le i \le n$. Because $h < a_2 = a_1 + h + \alpha$ and $h \in H$ by Lemma 2, h must be divided by $a_1 = n$. We put $s = \frac{h}{n} + 1$.

Claim $P = \mathrm{I}_2 \begin{pmatrix} x_2 & x_3 & \cdots & x_n & x_1^{s+\alpha} \\ x_1^s & x_2 & \cdots & x_{n-1} & x_n \end{pmatrix}$.

Proof (Claim) We put $B = (b_{ij}) = \begin{pmatrix} x_2 & x_3 & \cdots & x_n & x_1^{s+\alpha} \\ x_1^s & x_2 & \cdots & x_{n-1} & x_n \end{pmatrix}$ and $Q = \mathrm{I}_2(B)$. Then, because $a_i = a_1 + h + (i-1)\alpha$ for all $2 \le i \le n$ and $a_1 = n$, we get $\deg b_{1j} - \deg b_{2j} = \alpha$ for all $1 \le j \le n$. Hence $Q \subseteq P$. It implies that $(x_1) + Q \subseteq (x_1) + P$. Since

$$\begin{aligned}(x_1) + Q &= (x_1) + \mathrm{I}_2 \begin{pmatrix} x_2 & x_3 & \cdots & x_n & 0 \\ 0 & x_2 & \cdots & x_{n-1} & x_n \end{pmatrix} \\ &= (x_1) + (x_2, x_3, \ldots, x_n)^2,\end{aligned}$$

we have $\ell_S(S/(x_1) + Q) = n$. On the other hand,

$$\ell_S(S/(x_1) + P) = \ell_R(R/(t^{a_1})) = a_1 = n.$$

Hence $(x_1) + P = (x_1) + Q$. Consider the following commutative diagram

$$\begin{array}{ccccccccc} 0 & \longrightarrow & P/Q & \longrightarrow & S/Q & \longrightarrow & S/P & \longrightarrow & 0 \\ & & \downarrow{\scriptstyle \widehat{x_1}} & & \downarrow{\scriptstyle \widehat{x_1}} & & \downarrow{\scriptstyle \widehat{x_1}} & & \\ 0 & \longrightarrow & P/Q & \longrightarrow & S/Q & \longrightarrow & S/P & \longrightarrow & 0 \end{array}$$

where $\widehat{x_1}$ denotes the map of multiplication by x_1. By Snake Lemma and the fact that $R \cong S/P$ is an integral domain, we have the exact sequence

$$0 \to (P/Q)/x_1(P/Q) \to S/[(x_1) + Q] \to S/[(x_1) + P] \to 0.$$

Hence $P/Q = x_1(P/Q)$. Therefore, by Nakayama's lemma, we get $P = Q$ as desired.

The proof of the implication (2)–(4) in Theorem 2 is now completed.

Remark 1 As we proved, the implication (3)–(2) holds without our assumption that H has maximal embedding dimension. The authors conjectured the implication (2)–(4) also holds without the assumption through the discussion with D. T. Cuong and H. L. Truong. At this moment, we have no proof for the general case and also no counter-example for the conjecture.

2.3 *Proof of (b) and (c)*

Now recall our notation. Let k be a field. Let $S = k[x_1, x_2, \dots, x_n]$ be the polynomial ring. We assume the condition (4) in Theorem 2, namely,

$$P = \mathrm{I}_2\begin{pmatrix} x_2 & x_3 & \cdots & x_n & x_1^{s+\alpha} \\ x_1^s & x_2 & \cdots & x_{n-1} & x_n \end{pmatrix}$$

is the defining ideal of the numerical semigroup ring R. Let $P^{(r)} = P^r S_P \cap S$ be the r-th symbolic power of P and $\mathcal{R}_S(P) = \bigoplus_{r \geq 0} P^{(r)} T^r \subseteq S[T]$ the symbolic Rees algebra of P, where T is an indeterminate. Notice that all the proofs in this subsection are deeply inspired by works by S. Goto [10] and by C. D'Cruz and S. K. Masuti [6] and the method is the same as theirs. Although the results in [10] hold for local rings, we can apply the results to our situation by passing to the formal power series ring $k[[x_1, x_2, \dots, x_n]] = \widehat{S_M}$ where $M = (x_1, x_2, \dots, x_n) \subseteq S$ and $\widehat{S_M}$ denotes the MS_M-adic completion of S_M.

First, we prove that $\mathcal{R}_S(P)$ is a Noetherian ring. Let

$$Y = (y_{ij}) := \begin{pmatrix} x_1^s & x_2 & x_3 & \cdots & x_{n-2} & x_{n-1} & x_n \\ x_2 & x_3 & x_4 & \cdots & x_{n-1} & x_n & x_1^{s+\alpha} \\ x_3 & x_4 & x_5 & \cdots & x_n & x_1^{s+\alpha} & x_1^{\alpha} x_2 \\ \cdots & \cdots & \cdots & \cdots & \cdots & \cdots & \cdots \\ x_n & x_1^{s+\alpha} & x_1^{\alpha} x_2 & \cdots & x_1^{\alpha} x_{n-3} & x_1^{\alpha} x_{n-2} & x_1^{\alpha} x_{n-1} \end{pmatrix}.$$

Then we have the following.

Lemma 3 $\mathrm{I}_{m+1}(Y) \subseteq P^{(m)}$ *for all* $1 \leq m \leq n-1$.

Proof We prove by induction on m. Indeed, because $\deg y_{ij} - \deg y_{kj} = (i-k)\alpha$ for all $1 \leq i < k \leq n$, $1 \leq j \leq n$, we get $\mathrm{I}_2(Y) \subseteq P$. Therefore, the assertion is true with $m = 1$. Now suppose that $m \geq 2$ and that our assertion holds true for $m-1$. Let $Z = (z_{ij})$ be the $(m+1) \times (m+1)$ submatrix of Y and $\Delta = \det Z$. We want to show that $\Delta \in P^{(m)}$. For each $1 \leq i \leq m+1$, Z_i denotes the $m \times m$ matrix obtained from Z by deleting i-th row and the last column. Then $\Delta_i = (-1)^{i+m+1} \det Z_i \in P^{(m-1)}$ for each $1 \leq i \leq m+1$ by the hypothesis of induction. Moreover, we have $\Delta = \sum_{i=1}^{m+1} z_{i,m+1} \Delta_i$ and $\sum_{i=1}^{m+1} z_{im} \Delta_i = 0$. Hence

$$z_{1m}\Delta = \sum_{i=1}^{m+1} (z_{1m} z_{i,m+1} - z_{im} z_{1,m+1}) \Delta_i \in \mathrm{I}_2(Y) P^{(m-1)}.$$

It follows that $\Delta \in P^{(m)}$ since $z_{1m} \notin P$ and $P^{(m)}$ is a P-primary ideal. Thus $\mathrm{I}_{m+1}(Y) \subseteq P^{(m)}$ for all $1 \leq m \leq n-1$.

For each $0 \leq i \leq n-1$, $Y(i)$ denotes the submatrix of Y consiting of the first $(i+1)$ rows and columns of Y and we put $\xi_i = \det Y(i)$.

Theorem 4 *$\mathcal{R}_S(P)$ is a Noetherian ring.*

Proof By Lemma 3 we have $\xi_i \in \mathrm{I}_{i+1}(Y) \subseteq P^{(i)}$ for all $1 \leq i \leq n-1$. Notice that $\xi_0 = x_1^s$ and $\xi_i \equiv (-1)^{\frac{i(i+1)}{2}} x_{i+1}^{i+1} \mod (x_1, x_2, \ldots, x_i)S$ for all $1 \leq i \leq n-1$. We now consider $\ell_S(S/(\xi_0, \xi_1, \ldots, \xi_{n-1}))$. Let $\mathfrak{a} = (\xi_1, \xi_2, \ldots, \xi_{n-1})$. Since

$$\begin{aligned}[(x_1^m) + \mathfrak{a}]/[(x_1^{m+1}) + \mathfrak{a}] &\cong (x_1^m)/[(x_1^m) \cap ((x_1^{m+1}) + \mathfrak{a})] \\ &= (x_1^m)/[(x_1^{m+1}) + ((x_1^m) \cap \mathfrak{a})] \\ &= (x_1^m)/x_1^m[(x_1) + \mathfrak{a}] \\ &\cong S/[(x_1) + \mathfrak{a}]\end{aligned}$$

for all $m \geq 1$, we have $\ell_S(S/(\xi_0, \xi_1, \ldots, \xi_{n-1})) = \ell_S(S/[(x_1^s) + \mathfrak{a}]) = s \cdot \ell_S(S/[(x_1) + \mathfrak{a}])$. Thanks to the form of ξ_i's, the same argument shows that

$$\ell_S(S/(\xi_0, \xi_1, \ldots, \xi_{n-1})) = s \cdot 2 \cdot 3 \cdots n = s \cdot n!.$$

Since the coefficient fields of R and S coincide and $R/(t^{a_1 s}) \cong S/[(x_1^s) + P]$ as S-modules, we know that $\ell_S(S/[(x_1^s) + P]) = \ell_R(R/(t^{a_1 s})) = a_1 s = ns$. Therefore $\ell_S(S/(\xi_0, \xi_1, \ldots, \xi_{n-1})) = \ell_S(S/[(x_1^s) + P]) \cdot (n-1)!$. Thus $\mathcal{R}_S(P)$ is a Noetherian ring by [10, Theorem (1.1)].

Theorem 5 *$\mathcal{R}_S(P)$ is a Cohen-Macaulay ring.*

Proof Thanks to [10, Corollary (6.9)], to show that $\mathcal{R}_S(P)$ is a Cohen-Macaulay ring it is sufficient to show that $S/\left[(\xi_1, \ldots, \xi_{n-1}) + P^{(i)}\right]$ is Cohen-Macaulay for all $1 \leq i \leq \binom{n-1}{2} = \sum_{i=1}^{n-1} i - n + 1$. By applying the proof of Theorem 6.5 in [6] for the matrix Y, we can see that

$$e\left(x_1; S/[(\xi_1, \ldots, \xi_{n-1}) + P^{(i)}]\right) = \ell_S\left(S/[(x_1, \xi_1, \ldots, \xi_{n-1}) + P^{(i)}]\right)$$

for all $i \geq 1$. Here $e\left(x_1; S/[(\xi_1, \ldots, \xi_{n-1}) + P^{(i)}]\right)$ denotes the multiplicity of $S/[(\xi_1, \ldots, \xi_{n-1}) + P^{(i)}]$ with respect to (x_1). Therefore we obtain the assertion by [17, Theorem 17.11].

We put $\mathrm{G}_S(P) = \bigoplus_{i \geq 0} P^{(i)}/P^{(i+1)}$. Then, by [10, Lemma (6.1)] and $\mathrm{G}(PS_P)$ is a polynomial ring in $n-1$ variables, we get $\mathrm{a}(\mathrm{G}_S(P)) = \mathrm{a}(\mathrm{G}(PS_P)) = -(n-1)$, where $\mathrm{a}(\mathrm{G}_S(P))$ denotes the a-invariant of $\mathrm{G}_S(P)$. Moreover, thanks to [10, Corollary (5.9)], $\mathrm{G}_S(P)$ is Gorenstein because $S/[(f_1, f_2, \ldots, f_n) + P^{(i)}]$ is Cohen-Macaulay for all $1 \leq i \leq \binom{n-1}{2}$ by the proof of the previous theorem. Therefore

$\mathcal{R}_S(P)$ is Gorenstein if and only if $\mathrm{a}(\mathrm{G}_S(P)) = -2$ by [10, Theorem (6.6)] and the latter condition is equivalent to $n = 3$. Here, notice that we assume $n \geq 3$. Now we get the following and finish to prove Theorem 2.

Theorem 6 *$\mathcal{R}_S(P)$ is a Gorenstein ring if and only if $n = 3$.*

3 Examples

Here let us show 2 examples. One of them satisfies the condition in Theorem 2 and the other one does not.

Example 1 Let $H = \langle 4, 11, 14, 17\rangle$. Then H satisfies the condition (1) by taking $n = 4$, $h = 4$, and $\alpha = 3$. Thanks to the proof of the implication (2)–(4) of Theorem 2, if we take $s = \frac{h}{n} + 1 = 2$, we have

$$P = \mathrm{I}_2\begin{pmatrix} y & z & w & x^{s+\alpha} \\ x^s & y & z & w \end{pmatrix}$$

$$= \mathrm{I}_2\begin{pmatrix} y & z & w & x^5 \\ x^2 & y & z & w \end{pmatrix},$$

where $P \subseteq k[x, y, z, w]$ is the defining ideal of $k[H]$. Notice that $\deg y - \deg x^2 = \deg z - \deg y = \deg w - \deg z = \deg x^5 - \deg w = 3 = \alpha$.

Example 2 Let $H = \langle 4, 10, 11, 13\rangle$. Then H does not satisfy the condition (1) in Theorem 2. Hence the defining ideal $P \subseteq k[x, y, z, w]$ of $k[H]$ can not have the form as in the condition (3). In fact, we can check that

$$P = \mathrm{I}_2\begin{pmatrix} y & x^3 & w & xz \\ x^2 & y & z & w \end{pmatrix} + \mathrm{I}_2\begin{pmatrix} z & w & xy & x^4 \\ x^2 & y & z & w \end{pmatrix}.$$

Notice that $\deg y - \deg x^2 = \deg x^3 - \deg y = \deg w - \deg z = \deg xz - \deg w = 2$ and $\deg z - \deg x^2 = \deg w - \deg y = \deg xy - \deg z = \deg x^4 - \deg w = 3$. These numbers coincide with the numbers in $\{\mathrm{f}(H) - \alpha \mid \alpha \in \mathrm{PF}(H) \setminus \{\mathrm{f}(H)\}\}$. Notice that $\mathrm{PF}(H) = \{6, 7, \mathrm{f}(H) = 9\}$.

Acknowledgements The authors are grateful to Professor Shiro Goto for giving advices to improve the results, especially for noticing a question about the symbolic Rees algebras and suggesting the proof. The authors also would like to warmly thank the referee for carefully reading our manuscript and for giving valuable remarks as well as detailed suggestions which helped us to improve this article.

The first author was supported by the Hanoi Pedagogical University 2 under grant number C.2019-18-02. The second author was partially supported by JSPS Grant-in-Aid for Scientific Research (C) 18K03227.

References

1. Barucci, V., Fröberg, R.: One-dimensional almost Gorenstein rings. J. Algebra **188**, 418–442 (1997)
2. Bresinsky, H.: Symmetric semigroups of integers generated by 4 elements. Manuscr. Math. **17**, 205–219 (1975)
3. Bruns, W., Vetter, U.: Determinantal Rings. Lecture Notes in Mathematics, vol. 1327. Springer, Berlin (1988)
4. Cowsik, R.: Symbolic powers and the number of defining equations, algebras and its applications. Lect. Notes Pure Appl. Math. **91**, 13–14 (1985)
5. Cutkosky, S.D.: Symbolic algebras of monomial primes. J. Reine Angew. Math. **416**, 71–89 (1991)
6. D'Cruz, C., Masuti, S.K.: Cohen-Macaulayness and Gorensteinness of symbolic blowup algebras of certain monomial curves (2016). arXiv:1610.03658
7. Eagon, J.A., Northcott, D.G.: Ideals defined by matrices and a certain complex associated with them. Proc. R. Soc. Ser. A **269**, 188–204 (1962)
8. Eto, K.: Almost Gorenstein monomial curves in affine four space. J. Algebra **488**, 362–387 (2017)
9. Gimenez, P., Sengupta, I., Srinivasan, H.: Minimal graded free resolutions for monomial curves defined by arithmetic sequences. J. Algebra **388**, 294–310 (2013)
10. Goto, S.: The Cohen-Macaulay symbolic Rees algebras for curve singularities. The Cohen-Macaulay and Gorenstein Rees algebras associated to filtrations. Mem. Am. Math. Soc. **110**, 1–68 (1994)
11. Goto, S., Watanabe, K.-I.: On graded rings I. J. Math. Soc. Jpn **30**, 179–213 (1978)
12. Goto, S., Nishida, K., Watanabe, K.-I.: Non-Cohen-Macaulay symbolic blow-ups for space monomial curves and counterexamples to Cowsik's question. Proc. Am. Math. Soc. **120**, 383–392 (1994)
13. Goto, S., Kien, D.V., Matsuoka, N., Truong, H.L.: Pseudo-Frobenius numbers versus defining ideals in numerical semigroup rings. J. Algebra **508**, 1–15 (2018)
14. Herzog, J.: Generators and relations of Abelian semigroups and semigroup rings. Manuscr. Math. **3**, 175–193 (1970)
15. Huneke, C.: Hilbert functions and symbolic powers. Mich. Math. J. **34**, 293–318 (1987)
16. Komeda, J.: On the existence of Weierstrass points with a certain semigroup generated by 4 elements. Tsukuba J. Math. **6**, 237–270 (1982)
17. Matsumura, H.: Commutative Ring Theory. Cambridge Studies in Advanced Mathematics. Translated from the Japanese by M. Reid, 2nd edn. Cambridge University Press, Cambridge (1989)
18. Roberts, P.: A prime ideal in a polynomial ring whose symbolic blow-up is not Noetherian. Proc. Am. Math. Soc. **94**, 589–592 (1985)
19. Roy, A.K., Sengupta, I., Tripathi, G.: Minimal graded free resolutions for monomial curves in $\mathbb{A}^4$ defined by almost arithmetic sequences. Commun. Algebra **45**, 521–551 (2017)

Embedding Dimension of a Good Semigroup

Nicola Maugeri and Giuseppe Zito

Abstract In this paper, we study good semigroups of $\mathbb{N}^n$, a class of semigroups that contains the value semigroups of algebroid curves with n branches. We give the definition of embedding dimension of a good semigroup showing that, in the case of good semigroups of $\mathbb{N}^2$, some of its properties agree with the analogue concepts defined for numerical semigroups.

Keywords Good semigroups · Embedding dimension · Semigroup of a ring

1 Introduction

The concept of good subsemigroup of $\mathbb{N}^n$ was formally introduced in [1]. Its definition arises from the properties of the value semigroups of one dimensional analytically unramified rings (for example the local rings of an algebraic curve) that were initially studied in [2, 4, 5, 7, 10, 11, 13]. In [1], the authors proved that the class of good semigroups is actually larger than the one of value semigroups. Thus, such semigroups can be seen as a natural generalization of numerical semigroups and studied without necessarily referring to the ring theory context, using a more combinatorial approach.

Although, as we have already pointed out, good semigroups share traits with the numerical semigroups, there are some important properties of the latter that cannot be generalized to them. For instance, they are not finitely generated as monoids, and they are not closed under finite intersections. This makes the study of good semigroups much more difficult than the numerical ones.

The authors "N. Maugeri and G. Zito" contributed equally to this work.

N. Maugeri (✉) · G. Zito
Università di Catania, Catania, Italy

V. Barucci et al. (eds.), *Numerical Semigroups*, Springer INdAM Series 40,
https://doi.org/10.1007/978-3-030-40822-0_13

Thus, a relevant part of the literature dedicated to these objects is concerned to find a suitable way to represent them by means of a finite set of data.

For instance, for what concerns good semigroups which are also value semigroups, in [13, 18] singularities with only two branches are studied. In these papers, the finite set considered is the set of maximal elements (in [10], it is possible to find a generalization of this approach to the case of more than two branches). In [8], the authors considered a new approach that is still valid for good semigroups not realizable as value semigroups of curves. They firstly notice that the set of *small elements* of the semigroup, that is, the finite set of elements between 0 and the conductor of the semigroup with the usual partial order, completely describes it. Then they proved the uniqueness of the minimal subset $G \subsetneq \mathrm{Small}(S)$, called *minimal good generating system*, from which is possible to recover completely the semigroup S, if also the conductor of S is known. Another interesting approach is the one presented in [6], where the authors introduced the semiring of values Γ of an algebroid curve R where also the values of the zero-divisors elements are considered ($v(0) = (\infty, \dots, \infty)$). Thus Γ contains the value semigroup of R and $(\Gamma, +)$ is a semigroup setting $\gamma + \underline{\infty} = \underline{\infty}$ for all $\gamma \in \Gamma$. The key point is that Γ, equipped with the tropical operations

$$\boldsymbol{\alpha} \oplus \boldsymbol{\beta} = \min\{\boldsymbol{\alpha}, \boldsymbol{\beta}\} := (\min\{\alpha_1, \beta_1\}, \dots, \min\{\alpha_n, \beta_n\}) \quad \text{and} \quad \boldsymbol{\alpha} \odot \boldsymbol{\beta} = \boldsymbol{\alpha} + \boldsymbol{\beta},$$

is a *finitely generated semiring*. This leads the authors to introduce the concept of minimal standard basis.

The aim of this paper is to continue this kind of investigation, in order to find the smallest possible finite set that is able to encode some of the information of a good semigroup with two branches. Specifically, we introduce the concept of minimal set of representatives of a good subsemigroup S of $\mathbb{N}^2$. Although a minimal set of representatives η of S does not univocally describe the semigroup (however S is still among the minimal good semigroups containing η), it is possible to show that it stores relevant data. For instance, in the case of value semigroup, a system of representatives contains all the information regarding the value of a minimal system of generators of the corresponding ring. This leads us to generalize in a reasonable way, to the good semigroups of $\mathbb{N}^2$, the concept of *embedding dimension* that plays an important role in the numerical case.

The structure of the paper is the following.

In Sect. 2 we give all the basic definitions and we introduce all the main tools of the paper. In particular, in Sect. 2.1 we recall the definition of good semigroup and we explain how to associate to a good semigroup S of $\mathbb{N}^2$ a semiring Γ_S. Then, in Proposition 3, we prove that, in the case of value semigroups, our semiring coincides with the one given in [6]. In Sect. 2.2 we define the concept of *irreducible* and *absolute element* of Γ_S, and in Theorem 10, we prove that Γ_S is generated as a semiring by its set I_A of irreducible absolute elements generalizing to all good semigroups a result proved by Carvalho E. and Hernandes M.E. [6, Thm 11, Cor 20] for the value semigroups of a ring.

In Sect. 3, we introduce the notation S_η for the set of the minimal good semigroups containing η. In Proposition 13 we give some conditions on η in order to have finitely many elements in S_η. Then, given a good semigroup S, a set η is called a *system of representatives* of S if $S \in S_\eta$. This lets us to define the embedding dimension of a good semigroup S as the smallest cardinality of a system of representatives of S. Starting from this point we work on good semigroups of $\mathbb{N}^2$ in order to study the property of the embedding dimension. In Sect. 3.1 we introduce the definition of *track* of a good semigroup S and with Lemma 21 we show how to obtain a good semigroup S' contained in S by removing one of its tracks. Using this lemma we can compute an inferior bound for the embedding dimension. In Sect. 3.2 it is given the definition of reducibility of an element of $I_A(S)$ with respect to a subset $\eta \subseteq I_A(S)$. Then, Theorem 31 gives a way to use this concept in order to develop a strategy to find a superior bound for the embedding dimension. In Sect. 3.3 we present a series of functions implemented in GAP [17] that, using the computational vantages of calculating the previous bounds, allow us to describe a fast algorithm to find the embedding dimension. In the examples proposed in this section, for reasons of legibility and space, some verifications are not reported; these were made using functions written in GAP [17].

Finally, Sect. 4 is dedicated to studying whether the embedding dimension defined in $\mathbb{N}^2$ retains some of the features of the numerical case. In particular in Theorem 39 we prove that a good semigroup S, realizable as a value semigroup, has embedding dimension greater or equal than the corresponding ring (as in the numerical case). Then we give some examples, when the previous inequality is strict, where it is possible to observe the limits of the combinatorial structure of a good semigroup that is not always able to store all the information contained in the ring in the same amount of data given by a system of generators. In Sect. 4.2 we give the definition of levels of the Apéry set of a good semigroup as in [9], and we use it to prove that $\mathrm{edim}(S) \leq e_1 + e_2$, where $\boldsymbol{e} = (e_1, e_2)$ is the multiplicity vector of S (extending the relation $\mathrm{edim}(S) \leq e$ of the numerical case and the corresponding relation for one-dimensional rings). This result also lets us to prove Corollary 50, where we show that the Arf good semigroups of $\mathbb{N}^2$ have maximal embedding dimension, generalizing another important property valid in the numerical case.

2 Semiring Associated to a Good Semigroup and Irreducible Absolutes

2.1 Semiring Γ_S and Basic Properties

We start this section recalling the definition of good semigroup introduced in [1].

Definition 1 A submonoid S of $(\mathbb{N}^n, +)$ is a *good semigroup* if it satisfies the following properties:

(G1) If $\boldsymbol{\alpha}, \boldsymbol{\beta} \in S$, then $\min(\boldsymbol{\alpha}; \boldsymbol{\beta}) = (\min\{\alpha_1, \beta_1\}, \ldots, \min\{\alpha_n, \beta_n\}) \in S$;

(G2) There exists $\boldsymbol{\delta} \in \mathbb{N}^n$ such that $S \supseteq \boldsymbol{\delta} + \mathbb{N}^n$;
(G3) If $(\boldsymbol{\alpha}, \boldsymbol{\beta}) \in S$; $\boldsymbol{\alpha} \neq \boldsymbol{\beta}$ and $\alpha_i = \beta_i$ for some $i \in \{1, \ldots, n\}$; then there exists $\boldsymbol{\epsilon} \in S$ such that $\epsilon_i > \alpha_i = \beta_i$ and $\epsilon_j \geq \min\{\alpha_j, \beta_j\}$ for each $j \neq i$ (and if $\alpha_j \neq \beta_j$, the equality holds).

Furthermore, we always suppose to work with a *local* good semigroup S, i.e. if $\boldsymbol{\alpha} = (\alpha_1, \ldots, \alpha_n) \in S$ and $\alpha_i = 0$ for some $i \in \{1, \ldots, n\}$, then $\boldsymbol{\alpha} = \mathbf{0}$. As a consequence of property (G2), the element $\boldsymbol{c} = \min\{\delta | S \supseteq \boldsymbol{\delta} + \mathbb{N}^n\} = (c_1, \ldots, c_n)$ is well defined and it is called *conductor* of the good semigroup. We denote by $\leq$, the partial order on the elements of S induced by the standard order on $\mathbb{N}^n$. Furthermore, we denote by $\boldsymbol{e} = \min(S \setminus \{\mathbf{0}\})$ the *multiplicity vector* of the good semigroup. In order to simplify the notation and some proofs, in this paper, we often work with good semigroups $S \subseteq \mathbb{N}^2$ but most of the definitions and proofs remain true also in the general case.

According to the work of Carvalho and Hernandes [6], we wish to introduce a semiring Γ_S associated with the good semigroup $S \subseteq \mathbb{N}^2$.

We set $\overline{\mathbb{N}} = \mathbb{N} \cup \{\infty\}$, where ∞ is just a symbol that will correspond to the value of the element 0 if the semigroup is the value semigroup of a ring. We extend the natural order and the sum over $\mathbb{N}$ to $\overline{\mathbb{N}}$, setting respectively, $a < \infty$ for all $a \in \mathbb{N}$ and $x + \infty = \infty + x = \infty$.

We set:

$$S_1^\infty = \{(a, \infty) \mid \exists \tilde{y} \in \mathbb{N} : (a, y) \in S \ \forall y \geq \tilde{y}\};$$

$$S_2^\infty = \{(\infty, b) \mid \exists \tilde{x} \in \mathbb{N} : (x, b) \in S \ \forall x \geq \tilde{x}\};$$

$$S^\infty = S_1^\infty \cup S_2^\infty \cup \{(\infty, \infty)\};$$

$$\Gamma_S = S \cup S^\infty.$$

If $\boldsymbol{\alpha} = (\alpha_1, \alpha_2)$, $\boldsymbol{\beta} = (\beta_1, \beta_2) \in \Gamma_S$, we set $\min\{\boldsymbol{\alpha}, \boldsymbol{\beta}\} := (\min\{\alpha_1, \beta_1\}, \min\{\alpha_2, \beta_2\})$.

Now we define over Γ_S the following tropical operations:

$$\oplus : \boldsymbol{\alpha} \oplus \boldsymbol{\beta} = \min\{\boldsymbol{\alpha}, \boldsymbol{\beta}\}$$

$$\odot : \boldsymbol{\alpha} \odot \boldsymbol{\beta} = \boldsymbol{\alpha} + \boldsymbol{\beta}$$

It is easy to prove that, with these operations, $(\Gamma_S, \oplus, \odot)$ is a semiring.

We observe that, with the symbols $+$ and $\odot$, we denoted exactly the same operation on Γ_S. For this reason these two symbols will be used with the same meaning in the following.

Now we recall some facts and fix some notations that will be useful for the following.

Let be $R = \mathbb{K}[\![x_1, \ldots, x_n]\!]/Q$ a two-branches algebroid curve, where $Q = P_1 \cap P_2$ is an ideal of $\mathbb{K}[\![x_1, \ldots, x_n]\!]$ such that P_1,P_2 are prime ideals.

We can embed $R \hookrightarrow R_1 \times R_2$ where $R_i = \mathbb{K}[[x_1, \ldots, x_n]]/P_i$, $i = 1, 2$. Furthermore $R \hookrightarrow \overline{R} \cong \overline{R_1} \times \overline{R_2} \cong \mathbb{K}[[t_1]] \times \mathbb{K}[[t_2]]$. Given $r \in R$, $r = (r_1, r_2) \in \mathbb{K}[[t_1]] \times \mathbb{K}[[t_2]]$ that is a product of DVRs, so we can associate to each element of R a valuation. If v_i is the valuation function on $\mathbb{K}[[t_i]]$, we set:

$$v_i(r) = \begin{cases} v_i(r_i) & \text{if } r_i \neq 0 \\ \infty & \text{if } r_i = 0 \end{cases}$$

and $v(r) = (v_1(r), v_2(r))$.

According to the notation of Carvalho and Hernandes [6], we introduce the following sets:

$$\Gamma_{S_i} = \{v_i(r) \mid r \in R\} \subseteq \overline{\mathbb{N}};$$

$$S_i = \{v_i(r) | r \text{ is not a zerodivisor in } R\} \subseteq \mathbb{N};$$

$$\Gamma_R = \{\boldsymbol{v(r)} := (v_1(r), v_2(r)), r \in R\} \subseteq \overline{\mathbb{N}}^2;$$

$$S = \{\boldsymbol{v(r)} := (v_1(r), v_2(r)) \mid r \text{ is not a zerodivisor in } R\} \subseteq \mathbb{N}^2.$$

Γ_R and S will be called respectively *semiring of values* and *semigroup of values* associated to R. It is easy to observe that $S = \Gamma_R \cap \mathbb{N}^2$.

At this point, we wish to prove that, if R is a two-branches algebroid curve, and S is its semigroup of values, then $\Gamma_S = \Gamma_R$.

Lemma 2 *The following statements hold:*

(i) $(a, \infty) \in \Gamma_S$ *if and only if* $(a, y) \in S$ *for any* $y \geq c_2$.
(ii) $(\infty, b) \in \Gamma_S$ *if and only if* $(x, b) \in S$ *for any* $x \geq c_1$.

Proof We prove (i), the other statement is analogue. If $(a, \infty) \in \Gamma_S$, then there exists $\tilde{y} \in \mathbb{N}$ such that $(a, \tilde{y}), \ldots, (a, \tilde{y} + n) \in S$ for any $n \in \mathbb{N}$. If $\tilde{y} \leq c_2$ the statement is proved, otherwise $\tilde{y} = c_2 + n$, with $n \in \mathbb{N}$. Since S is a good semigroup, for all $i < n$, $a < c_1$, we have that $(a, c_2 + i) = \min\{(a, \tilde{y}), (c_1, c_2 + i)\} \in S$.

Proposition 3 *If R is a two-branches algebroid curve and S is its semigroup of values, then* $\Gamma_S = \Gamma_R$.

Proof We have observed that $S = \Gamma_R \cap \mathbb{N}^2$, thus we need to prove that $\Gamma_R \setminus S = S^\infty$. If $\boldsymbol{\alpha} \in \Gamma_R \setminus S$, we can write $\boldsymbol{\alpha} = v(r)$, where r is a zerodivisor in R or $r = 0$; in both cases we have $r \in P_1 \cup P_2$. If $r = 0$, $v(r) = (\infty, \infty)$; if $r \in P_1$, then $r = 0$ in R_1, $v_1(r) = \infty$, hence $\boldsymbol{\alpha} \in S_2^\infty$; if $r \in P_2$, then $r = 0$ in R_2, $v_2(r) = \infty$, hence $\boldsymbol{\alpha} \in S_1^\infty$. If $\boldsymbol{\alpha} \in S^\infty$, without loss of generality, we can suppose $\boldsymbol{\alpha} \in S_2^\infty$, we can write $\boldsymbol{\alpha} = (\infty, b)$, and, as a consequence of Lemma 1, $(c_1, b) \in S$. Since $S = v(R)$ and the conductor ideal is $\mathcal{C} = (t^{c_1}, u^{c_2})(\mathbb{K}[[t]] \times \mathbb{K}[[u]])$, there exists an element in R of the form $(t^{c_1}, b_y(u))$ with $v(b_y(u)) = b$. Since the element $(t^{c_1}, 0) \in R$, we have that the element $(0, b_y(u)) \in R$, thus $(\infty, b) \in \Gamma_R$.

2.2 A System of Generators of Γ_S as a Semiring

Definition 4 We will say that an element $\boldsymbol{\alpha} \in \Gamma_S \setminus \{\mathbf{0}\}$ is irreducible if, from $\boldsymbol{\alpha} = \boldsymbol{\beta}+\boldsymbol{\gamma}$, with $\boldsymbol{\beta}, \boldsymbol{\gamma} \in \Gamma_S$, it follows $\boldsymbol{\alpha} = \boldsymbol{\beta}$ or $\boldsymbol{\alpha} = \boldsymbol{\gamma}$. An element that is not irreducible will be said reducible.

We denote by $I(S)$ the set of irreducible elements of Γ_S.

Remark 5 We observe that:

1. If $\boldsymbol{\alpha} = (a, b) \in \Gamma_S$ with $a \geq c_1 + e_1$ and $b \geq c_2 + e_2$, then $\boldsymbol{\alpha}$ is reducible.
2. If $\boldsymbol{\alpha} = (a, \infty) \in \Gamma_S$ with $a \geq c_1 + e_1$, then $\boldsymbol{\alpha}$ is reducible.
3. If $\boldsymbol{\alpha} = (\infty, b) \in \Gamma_S$ with $b \geq c_2 + e_2$, then $\boldsymbol{\alpha}$ is reducible.

Given a good semigroup $S \subseteq \mathbb{N}^2$, and an element $\boldsymbol{\alpha} \in \mathbb{N}^2$, following the notation in [1], we set:

$$\begin{aligned}
\Delta_i(\boldsymbol{\alpha}) &:= \{\boldsymbol{\beta} \in \mathbb{Z}^2 | \alpha_i = \beta_i \text{ and } \alpha_j < \beta_j \text{ for } j \neq i\} \\
\Delta(\boldsymbol{\alpha}) &:= \Delta_1(\boldsymbol{\alpha}) \cup \Delta_2(\boldsymbol{\alpha}) \\
\Delta_i^S(\boldsymbol{\alpha}) &:= S \cap \Delta_i(\boldsymbol{\alpha}) \\
\Delta^S(\boldsymbol{\alpha}) &:= S \cap \Delta(\boldsymbol{\alpha}).
\end{aligned}$$

Furthermore we define:

$$\begin{aligned}
{}_i\Delta(\boldsymbol{\alpha}) &:= \{\boldsymbol{\beta} \in \mathbb{Z}^2 | \alpha_i = \beta_i \text{ and } \beta_j < \alpha_j \text{ for } j \neq i\} \\
{}_i\Delta^S(\boldsymbol{\alpha}) &:= S \cap_i \Delta(\boldsymbol{\alpha}).
\end{aligned}$$

Extending the previous definitions to infinite elements of $\overline{\mathbb{N}}^2$, we set

$$\begin{aligned}
{}_1\Delta((\alpha_1, \infty)) &:= \{\boldsymbol{\beta} \in \mathbb{Z}^2 | \beta_1 = \alpha_1\} \\
{}_2\Delta((\alpha_1, \infty)) &:= \emptyset \\
{}_1\Delta((\infty, \alpha_2)) &:= \emptyset \\
{}_2\Delta((\infty, \alpha_2)) &:= \{\boldsymbol{\beta} \in \mathbb{Z}^2 | \beta_2 = \alpha_2\} \\
{}_i\Delta^S(\boldsymbol{\alpha}) &:= S \cap {}_i\Delta(\boldsymbol{\alpha}).
\end{aligned}$$

Definition 6 An element $\boldsymbol{\alpha} \in \Gamma_S$ will be said absolute in Γ_S if $\boldsymbol{\alpha} \in S$ and $\Delta^S(\boldsymbol{\alpha}) = \emptyset$ (finite absolute), or if $\boldsymbol{\alpha} \in S^\infty$ (infinite absolute).

Remark 7 We observe that an element $\boldsymbol{\alpha} \in \Gamma_S$ is an absolute in Γ_S if and only if it is irreducible with respect to the operation $\oplus$. If we suppose that $\boldsymbol{\alpha} \in \Gamma_S$ is not an absolute, then $\Delta^S(\boldsymbol{\alpha}) \neq \emptyset$, hence there exists $\boldsymbol{\beta} \in \Delta_i^S(\boldsymbol{\alpha})$, with $i \in \{1, 2\}$. Therefore, by property (G3) of the good semigroups, there exists $\boldsymbol{\gamma} \in \Delta_{3-i}^S(\boldsymbol{\alpha})$, hence we would have $\boldsymbol{\alpha} = \boldsymbol{\beta} \oplus \boldsymbol{\gamma}$, which is a contradiction. If we suppose that an element $\boldsymbol{\alpha} \in \Gamma_S$ is such that $\boldsymbol{\alpha} = \boldsymbol{\beta} \oplus \boldsymbol{\gamma}$ with $\boldsymbol{\beta}, \boldsymbol{\gamma} \neq \boldsymbol{\alpha}$, then $\boldsymbol{\alpha} \in S$ and $\Delta^S(\boldsymbol{\alpha}) \neq \emptyset$.

We denote by $A_f(\Gamma_S)$ the set of *finite absolutes* in Γ_S, by $A^\infty(\Gamma_S)$ the set of *infinite absolutes* in Γ_S and by $A(\Gamma_S)$ the set of all *absolutes* in Γ_S. We call $I_{A_f}(\Gamma_S)$ the set of *finite irreducible absolutes* in Γ_S, $I_A^\infty(\Gamma_S)$ the set of *infinite irreducible absolutes* in Γ_S and $I_A(\Gamma_S)$ the set of all *irreducible absolutes* in Γ_S.

Remark 8 By Remark 7, $I_A(S)$ can be seen as the set of the elements of Γ_S that are irreducible with respect to both the operations defined in it. Notice that this interpretation lets us to naturally generalize the concept of irreducible absolute elements to good subsemigroups of $\mathbb{N}^n$, with $n > 2$.

As a consequence of the Remark 5, the set of irreducible absolutes is finite. Now we introduce other sets that will be considered in the following:

$$\begin{aligned}
\text{small}(S) &= \{(a,b) \in S | a \leq c_1, b \leq c_2\};\\
\text{small}(\Gamma_S) &= \text{small}(S) \cup \{(\infty,b) \in S_2^\infty, b \leq c_2\} \cup \{(a,\infty) \in S_1^\infty, a \leq c_1\};\\
B_1^\infty(\Gamma_S) &= \{(a,\infty) \in \Gamma_S | c_1 < a \leq c_1 + e_1\} \subseteq S_1^\infty;\\
B_2^\infty(\Gamma_S) &= \{(\infty,b) \in \Gamma_S | c_2 < b \leq c_2 + e_2\} \subseteq S_2^\infty;\\
B^\infty(\Gamma_S) &= B_1^\infty(\Gamma_S) \cup B_2^\infty(\Gamma_S) \subseteq S^\infty(C).
\end{aligned}$$

The sets small(S), small(Γ_S), $B^\infty(\Gamma_S)$ will be said respectively: *small elements* of S, *small elements* of Γ_S and *beyond elements* of Γ_S (Fig. 1).

Remark 9 It is easy to observe the following facts:

(i) Each element in the semiring can be written as a tropical product of irreducible elements, i.e. if $\boldsymbol{\alpha} \in \Gamma_S$, $\boldsymbol{\alpha} = \boldsymbol{\alpha_1} \odot \ldots \odot \boldsymbol{\alpha_n}$ where $\boldsymbol{\alpha_i} \in I(\Gamma_S)$.

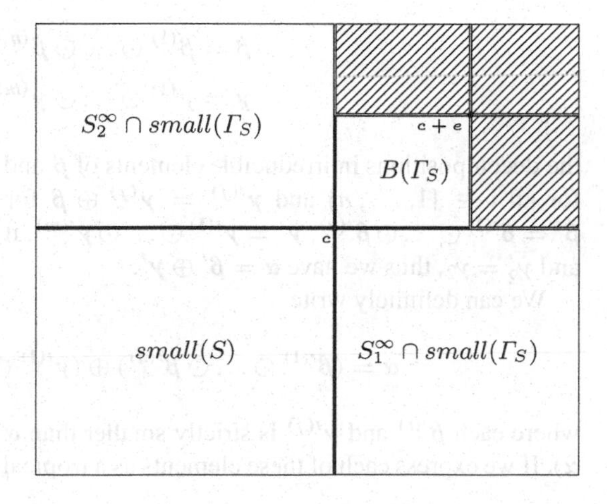

Fig. 1 A graphic representation of Γ_S's elements

(ii) Each element in the semiring can be written as a tropical sum of two absolute elements, i.e. if $\boldsymbol{\beta} \in \Gamma_S$, $\boldsymbol{\beta} = \boldsymbol{\beta}_1 \oplus \boldsymbol{\beta}_2$ where $\boldsymbol{\beta}_1, \boldsymbol{\beta}_2 \in A(\Gamma_S)$.

Now we prove that the set of irreducible absolutes generates Γ_S as a semiring.

Theorem 10 *$(\Gamma_S, \oplus, \odot)$ is generated as a semiring by the irreducible absolutes, i.e. if $\boldsymbol{\alpha} \in \Gamma_S \setminus \{\mathbf{0}\}$,*

$$\boldsymbol{\alpha} = \bigoplus_{i=1}^{m} (\bigodot_{j=1}^{n} \boldsymbol{\gamma}_{j_i}),$$

with $\boldsymbol{\gamma}_{j_i} \in I_A(S)$.

Proof First of all, we observe that we can reduce to prove the thesis only for the elements $\boldsymbol{\alpha} \in \text{small}(\Gamma_S) \cup B(\Gamma_S)$. Indeed, if $\boldsymbol{\alpha} \notin \text{small}(\Gamma_S) \cup B(\Gamma_S)$, then there exists $k \in \mathbb{N}$ such that $\boldsymbol{\beta} = \boldsymbol{\alpha} - k\boldsymbol{e} \in \text{small}(\Gamma_S) \cup B(\Gamma_S)$. In this case we would have $\boldsymbol{\alpha} = \boldsymbol{\beta} \odot k\boldsymbol{e}$, where $\boldsymbol{\beta} \in \text{small}(\Gamma_S) \cup B_S$ and $\boldsymbol{e}$ is trivially irreducible.

We can reduce again the proof only for the elements $\boldsymbol{\alpha} \in I(\Gamma_S) \cap S$ (finite irreducibles). In fact, if $\boldsymbol{\alpha}$ is reducible, by Remark 9, we can write $\boldsymbol{\alpha} = \boldsymbol{\alpha}^{(1)} \odot \ldots \odot \boldsymbol{\alpha}^{(n)}$, with $\boldsymbol{\alpha}^{(i)}$ irreducibles. Furthermore, we observe that if $\boldsymbol{\alpha}^{(i)} \in S^{\infty}$, then $\boldsymbol{\alpha}^{(i)} \in I_A(\Gamma_S)$; thus we can write:

$$\boldsymbol{\alpha} = \boldsymbol{\alpha}^{(1)} \odot \ldots \odot \boldsymbol{\alpha}^{(f)} \odot \left(\odot_{\boldsymbol{\gamma} \in I_A(S)} \boldsymbol{\gamma}\right)$$

where $\boldsymbol{\alpha}^{(i)} \in I(\Gamma_S) \cap S$. If we prove the thesis for the elements $\boldsymbol{\alpha}^{(i)}$ with $i \in \{1, \ldots, f\}$, using the distributive property of $\odot$ with respect to $\oplus$, the result is true also for $\boldsymbol{\alpha}$. Therefore we can suppose $\boldsymbol{\alpha} \in I(\Gamma_S) \cap S$ and prove the thesis. By Remark 9, we can write $\boldsymbol{\alpha} = \boldsymbol{\beta} \oplus \boldsymbol{\gamma}$ with $\boldsymbol{\beta} = (\beta_1, \beta_2) \in A$, $\boldsymbol{\gamma} = (\gamma_1, \gamma_2) \in A$ and we can assume $\beta_1 = \alpha_1 \leq \gamma_1$ and $\gamma_2 = \alpha_2 \leq \beta_2$.

We consider

$$\boldsymbol{\beta} = \boldsymbol{\beta}^{(1)} \odot \ldots \odot \boldsymbol{\beta}^{(n)},$$
$$\boldsymbol{\gamma} = \boldsymbol{\gamma}^{(1)} \odot \ldots \odot \boldsymbol{\gamma}^{(m)},$$

the decompositions in irreducible elements of $\boldsymbol{\beta}$ and $\boldsymbol{\gamma}$. We define $\boldsymbol{\beta}'^{(i)} = \boldsymbol{\beta}^{(i)} \oplus \boldsymbol{\gamma}$, for all $i \in \{1, \ldots, n\}$ and $\boldsymbol{\gamma}'^{(j)} = \boldsymbol{\gamma}^{(j)} \oplus \boldsymbol{\beta}$ for all $j \in \{1, \ldots, m\}$. Defining $\boldsymbol{\beta}' = \boldsymbol{\beta}'^{(1)} \odot \ldots \odot \boldsymbol{\beta}'^{(n)}$, $\boldsymbol{\gamma}' = \boldsymbol{\gamma}'^{(1)} \odot \ldots \odot \boldsymbol{\gamma}'^{(m)}$, it is easy to observe that $\beta_1' = \beta_1$ and $\gamma_2' = \gamma_2$, thus we have $\boldsymbol{\alpha} = \boldsymbol{\beta}' \oplus \boldsymbol{\gamma}'$.

We can definitely write:

$$\boldsymbol{\alpha} = (\boldsymbol{\beta}'^{(1)} \odot \ldots \odot \boldsymbol{\beta}'^{(n)}) \oplus (\boldsymbol{\gamma}'^{(1)} \odot \ldots \odot \boldsymbol{\gamma}'^{(m)}),$$

where each $\boldsymbol{\beta}'^{(i)}$ and $\boldsymbol{\gamma}'^{(j)}$ is strictly smaller than $\boldsymbol{\alpha}$ (that is $\boldsymbol{\gamma}'^{(j)} \leq \boldsymbol{\alpha}$ and $\boldsymbol{\gamma}'^{(j)} \neq \boldsymbol{\alpha}$). If we express each of these elements as a tropical product of irreducibles, we can write $\boldsymbol{\alpha}$ as a tropical sum of tropical products, where all the terms are irreducible and strictly smaller than $\boldsymbol{\alpha}$. This means that if we repeat the same argument on each element in this expression, in a finite number of iteration we will surely obtain the required expression.

Remark 11 In the case of good semigroups that are value semigroup of a ring, the theorem above follows by [6, Thm 11] and [6, Thm 19].

But we recall that not all good semigroups are value semigroup of a ring (for an example cf.[1, Example 2.16]).

Thus, the previous theorem generalizes this property to all semirings obtained by semigroups of $\mathbb{N}^2$, also if they are not value semigroup of a ring.

3 Embedding Dimension of a Good Semigroup

It is a well known fact that every numerical semigroup $S \subseteq \mathbb{N}$ admits a unique minimal system of generators as a monoid and the embedding dimension of the numerical semigroup is defined as the number of these generators. This name follows from the fact that it is equal to the embedding dimension of the monomial curve associated with the numerical semigroup.

Now we will define a set of vectors that, although it does not uniquely determine a good semigroup, will allow us to give a definition of embedding dimension of a good semigroup. This embedding dimension, in the case of good semigroup of $\mathbb{N}^2$, will satisfy some of the properties that are valid in the case of numerical semigroups.

Starting from this point, in order to lighten the notations, when we consider a good semigroup S, we suppose that it coincides with the semiring Γ_S, i.e. we treat the infinite elements as elements of S. Given a set of vectors $\eta \subseteq \overline{\mathbb{N}}^n$, we denote by $\langle \eta \rangle_{\oplus}$ the semiring generated by η. Furthermore, given a set of vectors $\eta \subseteq \overline{\mathbb{N}}^n$, we denote by S_η the family of all the good semigroups containing η and that are minimal with respect to the set inclusion. S_η can be finite, infinite or empty as in the following example.

Example 12 Let us consider $\eta = \{[2, 2], [3, 3]\} \subseteq \mathbb{N}^2$, and suppose that there exists a good semigroup $S \in S_\eta$.

First of all we prove that, for any $n \in \mathbb{N}\backslash\{1\}$, we have $(n, n) \in S$. In fact, it is easy to observe that each natural number $n \neq 1$ can be written as $n = 2\alpha+3\beta$, with $\alpha, \beta \in \mathbb{N}$. Hence we can write $(n, n) = (2\alpha+3\beta, 2\alpha+3\beta) = \alpha(2, 2)+\beta(3, 3) \in S$.

We denote by $\boldsymbol{c}(S) = (c_1, c_2)$ the conductor of S. If $c_1 = 1$, we have that $(1, 2) = \min\{(1, c_2), (2, 2)\} \in S$; hence, as a consequence of properties (G1) and (G3) of the good semigroups, either $\boldsymbol{c}(S) = (1, 2)$ or $S = \mathbb{N}^2$. In both cases, if we consider S' such that $\text{small}(S') = \{(0, 0), (2, 2)\}$ we have that S' is a good semigroup containing η and such that $S' \subset S$; but this contradicts the minimality of S. Therefore we have obtained $c_1 \neq 1$ and, using the same argument, we can suppose $c_2 \neq 1$.

If $c_1 > 1$ and $c_2 > 1$ we prove that $\boldsymbol{c}(S) = (c, c)$, with $c \in \mathbb{N}$. Let us assume by contradiction that $\boldsymbol{c}(S) = (c_1, c_2)$ with $c_1 < c_2$; in this case, there exists $\boldsymbol{\alpha} = (\alpha_1, \alpha_2)$ with $\alpha_1 \geq c_1, c_1 \leq \alpha_2 < c_2$ such that $\boldsymbol{\alpha} \notin S$. If $\alpha_1 \leq \alpha_2$, we would have $\boldsymbol{\alpha} = \min\{(\alpha_1, c_2), (\alpha_2, \alpha_2)\} \in S$, hence we necessarily have $\alpha_1 > \alpha_2$. Now we observe that $(c_1, \alpha_2) = \min\{\boldsymbol{c}(S), (\alpha_2, \alpha_2)\} \in S$ and by property (G3) of good semigroups applied to $\boldsymbol{c}(S)$ and (c_1, α_2), there exists $(x_1, \alpha_2) \in S$ with $x_1 > c_1$. If $x_1 \geq \alpha_1$, we would have $\boldsymbol{\alpha} = \min\{(x_1, \alpha_2), (\alpha_1, c_2)\} \in S$ that is a contradiction. Thus we necessarily have $x_1 < \alpha_1$. Now, if we consider $(x_1, \alpha_2), (x_1, c_2) \in S$, using again property (G3), we observe that there exists $(x_2, \alpha_2) \in S$ with $x_2 > x_1$. We can repeat this argument until we find an element $(x_i, \alpha_2) \in S$ with $x_i \geq \alpha_1$. In this case we obtain $\boldsymbol{\alpha} = \min\{(x_i, \alpha_2), (\alpha_1, c_2)\} \in S$, that is a contradiction.

Now, by repeatedly using the properties (G2) and (G3), it is easy to observe that, $\text{small}(S) = \{(0, 0), (2, 2), (3, 3), \ldots, (c-1, c-1), (c, c)\}$. If we define S' such that $\text{small}(S') = \{(0, 0), (2, 2), (3, 3), \ldots, (c, c), (c+1, c+1)\}$, we have found a minimal good semigroup containing $(2, 2)$, $(3, 3)$ and strictly contained in S, in contradiction with the minimality of S.

The following proposition gives a condition that guarantees that S_η is finite.

Proposition 13 *Suppose we have* $\eta = \{\boldsymbol{\eta}^{(1)} = (\eta_1^1, \ldots, \eta_n^1), \ldots, \boldsymbol{\eta}^{(k)} = (\eta_1^k, \ldots, \eta_n^k)\} \subseteq \mathbb{N}^n$.

Then the set S_η *is finite if the following conditions hold:*

- $\gcd\{\eta_i^h, h = 1, \ldots, k\} = 1$ *for* $i = 1, \ldots, n$;
- *For all* $i, j \in \{1, \ldots, n\}$ *with* $i \neq j$ *there exists a* $l \in \{1, \ldots, k\}$ *such that* $\eta_i^l \neq \eta_j^l$.

Proof We denote by $\langle \eta \rangle_\oplus$ the semiring generated by η. We claim that for each $i = 1, \ldots, n$, we can obtain two vectors $\boldsymbol{\alpha}^{(i)} = (\alpha_1^i, \ldots, \alpha_n^i)$ and $\boldsymbol{\beta}^{(i)} = (\beta_1^i, \ldots, \beta_n^i)$ in $\langle \eta \rangle_\oplus$ such that

$$\alpha_i^i = \beta_i^i \text{ and } \alpha_i^j < \beta_i^j \text{ for all } j \neq i.$$

We will prove this fact by induction on n.

- **Base case** $n = 2$. Suppose that $i = 1$. By the second property assumed on the set η, there exists a $\boldsymbol{\eta}^{(l)} \in \eta$ such that $\eta_1^l \neq \eta_2^l$. Then η must contain a vector $\boldsymbol{\eta}^{(m)}$ such that $\frac{\eta_2^m}{\eta_1^m} \neq \frac{\eta_2^l}{\eta_1^l}$. We assume by contradiction that $\frac{\eta_2^h}{\eta_1^h} = \frac{\eta_2^l}{\eta_1^l} \neq 1$ for all $h = 1, \ldots, k$. If η_1^l did not divide η_2^l, it would follow from $\eta_2^h = \frac{\eta_2^l}{\eta_1^l}\eta_1^h$ that η_1^l divides η_1^h for all $h = 1, \ldots, k$. Hence η_1^l would divide $\gcd\{\eta_1^h, h = 1, \ldots, k\} = 1$; but this contradicts the first assumption on the set η. Therefore we have $\frac{\eta_2^l}{\eta_1^l} \in \mathbb{N}$. Since the integer $\frac{\eta_2^l}{\eta_1^l}$, divides η_2^h for all $h = 1, \ldots, k$; it divides $\gcd\{\eta_2^h, h = 1, \ldots, k\} = 1$ but this is a contradiction.

Then, we consider $\boldsymbol{\eta}^{(m)}$ such that $\frac{\eta_2^m}{\eta_1^m} \neq \frac{\eta_2^l}{\eta_1^l}$ and the vectors

$$\boldsymbol{\alpha}^{(1)} = (\eta_1^l \eta_1^m, \eta_2^l \eta_1^m), \qquad \boldsymbol{\beta}^{(1)} = (\eta_1^l \eta_1^m, \eta_1^l \eta_2^m)$$

satisfy our condition because $\eta_2^l \eta_1^m \neq \eta_1^l \eta_2^m$ and they belong to $\langle \eta \rangle_{\oplus}$. For $i = 2$ we can use the same strategy.

– **Inductive step**: Let us suppose that the claim is true for $n-1$ and we prove it for n. We suppose that $i = 1$ (the other cases can be treated in the same way). We consider the set $\tilde{\eta} = \{\boldsymbol{\eta}^{(h)} = (\eta_1^h, \ldots, \eta_{n-1}^h), h = 1, \ldots, k\}$ that satisfies the conditions of the theorem. Then, by the inductive step, it easily follows that in $\langle \eta \rangle_{\oplus}$ there exist two vectors $\boldsymbol{\gamma}^{(1)} = (\gamma_1^1, \ldots, \gamma_n^1)$ and $\boldsymbol{\delta}^{(1)} = (\delta_1^1, \ldots, \delta_n^1)$ such that

$$\gamma_1^1 = \delta_1^1 \text{ and } \gamma_j^1 < \delta_j^1 \text{ for all } j = 2, \ldots, n-1.$$

If $\gamma_n^1 < \delta_n^1$, then the claim is true for $\boldsymbol{\alpha}^{(1)} = \boldsymbol{\gamma}^{(1)}$ and $\boldsymbol{\beta}^{(1)} = \boldsymbol{\delta}^{(1)}$. If $\gamma_n^1 > \delta_n^1$, we consider $\boldsymbol{\alpha}^{(1)} = \min(2\boldsymbol{\gamma}^{(1)}, 2\boldsymbol{\delta}^{(1)})$ and $\boldsymbol{\beta}^{(1)} = \boldsymbol{\gamma}^{(1)} \odot \boldsymbol{\delta}^{(1)}$. In fact we have $\alpha_1^1 = 2\gamma_1^1 = \beta_1^1$. If $j \in \{2, \ldots, n-1\}$, then $\alpha_j^1 = 2\gamma_j^1 < \gamma_j^1 + \delta_j^1 = \beta_j^1$. Finally, we have $\alpha_n^1 = 2\delta_n^1 < \gamma_n^1 + \delta_n^1 = \beta_n^1$. Thus suppose that $\gamma_n^1 = \delta_n^1$. In this case we can consider $\overline{\eta} = \{\boldsymbol{\eta}^{(h)} = (\eta_1^h, \eta_3^h, \ldots, \eta_n^h), h = 1, \ldots, k\}$. By the inductive step there exist two vectors $\boldsymbol{\gamma}^{(2)} = (\gamma_1^2, \ldots, \gamma_n^2)$ and $\boldsymbol{\delta}^{(2)} = (\delta_1^2, \ldots, \delta_n^2) \in \langle \eta \rangle_{\oplus}$ such that

$$\gamma_1^2 = \delta_1^2 \text{ and } \gamma_j^2 < \delta_j^2 \text{ for all } j = 3, \ldots, n.$$

Then, as we have just seen, if $\gamma_2^2 \neq \delta_2^2$ the claim is true. Therefore we suppose that $\gamma_2^2 = \delta_2^2$. Then, it is very easy to check that the claim is true with $\boldsymbol{\alpha}^{(1)} = \boldsymbol{\gamma}^{(1)} \odot \boldsymbol{\gamma}^{(2)}$ and $\boldsymbol{\beta}^{(1)} = \boldsymbol{\delta}^{(1)} \odot \boldsymbol{\delta}^{(2)}$.

Now, we denote by $c^{(i)}$ the conductor of the numerical semigroup generated by $\{\eta_i^h : h = 1, \ldots, k\}$ and we choose $\boldsymbol{\alpha}^{(i)} = (\alpha_1^i, \ldots, \alpha_n^i)$ and $\boldsymbol{\beta}^{(i)} = (\beta_1^i, \ldots, \beta_n^i)$ in $\langle \eta \rangle_{\oplus}$ as in the previous claim. We will prove that for each $i \in \{1, \ldots, n\}$ there exist $c_{i,j}$ for $j = 1, \ldots, i-1, i+1, \ldots, n$ such that the vectors

$$\boldsymbol{\sigma}^i(y) = (c_{i,1}, \ldots, c_{i,i-1}, c^{(i)} + \alpha_i^i + y, c_{i,i+1}, \ldots, c_{i,n}) \in S,$$

for each $S \in S_\eta$, and $y \in \mathbb{N}$. If this is true then it is clear that

$$c_\eta = \bigodot_{i=1}^{n} \boldsymbol{\sigma}^i(0) + \mathbb{N}^n \subseteq S,$$

for all $S \in S_\eta$.

Suppose that $i = 1$ (the proof is identical in the other cases). Let us consider an arbitrary $S \in S_\eta$. We obviously have $\langle \eta \rangle_\oplus \subseteq S$. We will denote by $m = \alpha_1^1 = \beta_1^1$. Since $c^{(1)}$ is the conductor of $\langle \{\eta_1^h : h = 1, \ldots, k\} \rangle$, we can find the vectors:

$$\boldsymbol{\sigma}^{(h)} = (\sigma_1^h, \ldots, \sigma_n^h) \in \langle \eta \rangle_\oplus, \qquad \text{for } h = 0, \ldots, m-1,$$

such that $\sigma_1^h = c^{(1)} + h$ for all $h = 0, \ldots, m-1$.

For each $i = 0, \ldots, m-1$ we consider $\boldsymbol{\lambda}^{(i)} = \bigoplus_{k=i}^{m-1} \boldsymbol{\sigma}^{(k)}$. Then we have $\boldsymbol{\lambda}^{(0)} \leq \ldots \leq \boldsymbol{\lambda}^{(m-1)}$ and, if $\boldsymbol{\lambda}^{(h)} = (\lambda_1^h, \ldots, \lambda_n^h)$, then $\lambda_1^h = c^{(1)} + h$.

Now we want to show that $(c^{(1)} + m + y, \lambda_2^0 + \alpha_2^1, \ldots, \lambda_n^0 + \alpha_n^1) \in S$ for each $y \in \mathbb{N}$.

We notice that

$$\boldsymbol{\lambda}^{(0)} \odot \boldsymbol{\alpha}^{(1)} = (c^{(1)} + m, \lambda_2^0 + \alpha_2^1, \ldots, \lambda_n^0 + \alpha_n^1) \in S,$$
$$\boldsymbol{\lambda}^{(0)} \odot \boldsymbol{\beta}^{(1)} = (c^{(1)} + m, \lambda_2^0 + \beta_2^1, \ldots, \lambda_n^0 + \beta_n^1) \in S,$$

thus, recalling that $\alpha_j^1 < \beta_j^1$ for all $j = 2, \ldots, n$, it follows by (G3) that there exists $x > c^{(1)} + m$ such that $(x, \lambda_2^0 + \alpha_2^1, \ldots, \lambda_n^0 + \alpha_n^1) \in S$.

Now we consider

$$\boldsymbol{\lambda}^{(1)} \odot \boldsymbol{\beta}^{(1)} = (c^{(1)} + m + 1, \lambda_2^1 + \beta_2^1, \ldots, \lambda_n^1 + \beta_n^1) \in S.$$

Since $\lambda_h^0 \leq \lambda_h^1$ for all $h = 2, \ldots, n$ and $\alpha_j^1 < \beta_j^1$ for all $j = 2, \ldots, n$, we have

$$(x, \lambda_2^0 + \alpha_2^1, \ldots, \lambda_n^0 + \alpha_n^1) \oplus (\boldsymbol{\lambda}^{(1)} \odot \boldsymbol{\beta}^{(1)}) = (c^{(1)} + m + 1, \lambda_2^0 + \alpha_2^1, \ldots, \lambda_n^0 + \alpha_n^1) \in S.$$

Now, as before, from

$$(c^{(1)} + m + 1, \lambda_2^0 + \alpha_2^1, \ldots, \lambda_n^0 + \alpha_n^1) \in S,$$
$$(c^{(1)} + m + 1, \lambda_2^0 + \beta_2^1, \ldots, \lambda_n^0 + \beta_n^1) \in S,$$

we can deduce that there exists $x > c^{(1)} + m + 1$ such that $(x, \lambda_2^0 + \alpha_2^1, \ldots, \lambda_n^0 + \alpha_n^1) \in S$.

Repeating the previous considerations and using the fact that $\boldsymbol{\lambda}^{(0)} \leq \boldsymbol{\lambda}^{(i)}$ for each $i \leq m-1$, we can show that

$$(c^{(1)} + m + y, \lambda_2^0 + \alpha_2^1, \ldots, \lambda_n^0 + \alpha_n^1) \in S,$$

for all $y = 0, \dots, m-1$. Now, we can consider

$$\boldsymbol{\lambda}^{(0)} \odot \boldsymbol{\beta}^{(1)} = (c^{(1)} + m, \lambda_2^0 + \beta_2^1, \dots, \lambda_n^0 + \beta_n^1) \in S$$

$$\dots$$

$$\boldsymbol{\lambda}^{(m-1)} \odot \boldsymbol{\beta}^{(1)} = (c^{(1)} + 2m - 1, \lambda_2^{m-1} + \beta_2^1, \dots, \lambda_n^{m-1} + \beta_n^1) \in S$$

and since $\lambda_h^j + \beta_h^1 > \lambda_h^0 + \alpha_h^1$ for all $j = 0, \dots, m-1$ and $h = 2, \dots, n$, we can use the same strategy to show that

$$(c^{(1)} + m + y, \lambda_2^0 + \alpha_2^1, \dots, \lambda_n^0 + \alpha_n^1) \in S,$$

for all $y = 0, \dots, 2m - 1$. Now it is clear that we can endlessly repeat the strategy and we finally proved that

$$(c^{(1)} + m + y, \lambda_2^0 + \alpha_2^1, \dots, \lambda_n^0 + \alpha_n^1) \in S,$$

for all $y \in \mathbb{N}$ and for all the $S \in S_\eta$ (S was arbitrarily chosen). Therefore we proved that if $S \in S_\eta$, then the conductor of S is smaller than c_η. Now we know that a good semigroup is completely characterized by its small elements. This implies that the set of good semigroups with a conductor smaller than c_η is finite and therefore also S_η must be finite.

Remark 14 Let us consider a set of vector $\eta \subseteq \mathbb{N}^n$ which satisfies the hypothesis of the previous theorem. The proof of the theorem gives us also a way to determine a bound for the conductor of all good semigroups containing η.

Definition 15 Given a good semigroup $S \subseteq \mathbb{N}^2$ and a set of vector $\eta \subseteq I_A(S)$, we say that η is a *system of representatives* of S, or more simply *sor*, if $S \in S_\eta$.

Remark 16 As a consequence of the Theorem 10, $I_A(S)$ is a *sor* of S, because every semigroup containing the elements of $I_A(S)$ must contain S.

Definition 17 A system of representatives η of S is minimal, if given another set of representatives $\eta' \subseteq \eta$, it follows $\eta' = \eta$. We call such a set a *msor* of S.

It is possible to show that two *msor* can have different cardinalities (see Example 35).

Definition 18 Given a good semigroup S, we define embedding dimension of S:

$$\mathrm{edim}(S) = \min\{|\eta| : S \in S_\eta \text{ and } \eta \subseteq I_A(S)\}.$$

From this point onwards we will start to analyze the properties of the embedding dimension. We will consider only good semigroups $S \subseteq \mathbb{N}^2$. Computing all the minimal good semigroups containing a set of vectors is computationally very dispensing, also in the two-branches case. At this point, our first aim is to produce a "fast" algorithm that, in the case of good semigroup $S \subseteq \mathbb{N}^2$, returns a *msor* of S. In order to do this we will calculate two bounds for the embedding dimension.

3.1 An Inferior Bound for the Embedding Dimension

First of all we want to produce an inferior bound for the embedding dimension. We give the following definitions.

Definition 19 Given $\boldsymbol{\alpha}, \boldsymbol{\beta} \in I_A(S)$ we say that $\boldsymbol{\alpha}$ and $\boldsymbol{\beta}$ are connected by a *piece of track* if they are not comparable, i.e. $\boldsymbol{\alpha} \not\leq \boldsymbol{\beta}$ and $\boldsymbol{\beta} \not\leq \boldsymbol{\alpha}$, and denoted by $\boldsymbol{\gamma} = \boldsymbol{\alpha} \oplus \boldsymbol{\beta}$, we have $\Delta^S(\boldsymbol{\gamma}) \cap (S \setminus I(S)) = \emptyset$.

Definition 20 Given $\boldsymbol{\alpha_1}, \ldots, \boldsymbol{\alpha_n} \in I_A(S)$, with $\alpha_{11} < \ldots < \alpha_{n1}$ we say that $\boldsymbol{\alpha_1}, \ldots, \boldsymbol{\alpha_n}$ are connected by a *track* if we have:

- ${}_2\Delta^S(\boldsymbol{\alpha_1}) \cap (S \setminus I(S)) = \emptyset$;
- ${}_1\Delta^S(\boldsymbol{\alpha_n}) \cap (S \setminus I(S)) = \emptyset$;
- $\boldsymbol{\alpha_i}$ and $\boldsymbol{\alpha_{i+1}}$ are connected by a *piece of track* for all $i \in \{1, \ldots, n-1\}$.

In this case, denoted with $\boldsymbol{\gamma_i} = \boldsymbol{\alpha_i} \oplus \boldsymbol{\alpha_{i+1}}$ for $i \in \{1, \ldots, n-1\}$, we set:

$$T((\boldsymbol{\alpha_1}, \ldots, \boldsymbol{\alpha_n})) = \{\boldsymbol{\alpha_1}\} \cup {}_2\Delta^S(\boldsymbol{\alpha_1}) \cup \left(\cup_{i=1}^{n-1} \Delta^S(\boldsymbol{\gamma_i})\right) \cup {}_1\Delta^S(\boldsymbol{\alpha_n}) \cup \{\boldsymbol{\alpha_n}\},$$

and we call this set the *track* connecting $\boldsymbol{\alpha_1}, \ldots, \boldsymbol{\alpha_n}$.

We will simply say that $T \subseteq S$ is a *track* in S if there exist $\boldsymbol{\alpha_1}, \ldots, \boldsymbol{\alpha_n} \in I_A(S)$ such that T is the track connecting $\boldsymbol{\alpha_1}, \ldots, \boldsymbol{\alpha_n}$. Notice that the previous definition implies that a track T of S never contains elements $\boldsymbol{\alpha}$ such that $\boldsymbol{\alpha} \geq \boldsymbol{c(S)} + \boldsymbol{e(S)}$.

In the following lemma we will show how these definitions are related to the embedding dimension.

Lemma 21 *Given a good semigroup S, and a track $T = T((\boldsymbol{\alpha_1}, \ldots, \boldsymbol{\alpha_n}))$ in S, then, $S' = S \setminus T$ is a good semigroup strictly contained in S.*

Proof If $\boldsymbol{\alpha}, \boldsymbol{\beta} \in S'$, since $\boldsymbol{\alpha}, \boldsymbol{\beta} \in S$ and $T \cap (S \setminus I(S)) = \emptyset$, we have $\boldsymbol{\alpha} + \boldsymbol{\beta} \in S'$, thus S' is a semigroup. Now, we have to check that S' satisfies the property (G1); therefore, considering $\boldsymbol{\alpha}, \boldsymbol{\beta} \in S'$, we have to prove that $\boldsymbol{\alpha} \oplus \boldsymbol{\beta} \in S'$. If we suppose $\boldsymbol{\alpha} \oplus \boldsymbol{\beta} \in T$, then: there exists a $\boldsymbol{\gamma_i} = \boldsymbol{\alpha_i} \oplus \boldsymbol{\alpha_{i+1}}$ such that $\boldsymbol{\alpha} \oplus \boldsymbol{\beta} \in \Delta^S(\boldsymbol{\gamma_i})$; or $\boldsymbol{\alpha} \oplus \boldsymbol{\beta} \in {}_1\Delta^S(\boldsymbol{\alpha_1})$; or $\boldsymbol{\alpha} \oplus \boldsymbol{\beta} \in {}_2\Delta^S(\boldsymbol{\alpha_n})$. But, in all the previous cases, by the definition of track, this would imply that $\boldsymbol{\alpha}, \boldsymbol{\beta} \in T$. Furthermore, for all $\boldsymbol{\alpha} \in S$ with $\boldsymbol{\alpha} \geq \boldsymbol{c(S)} + \boldsymbol{e(S)}$ we have $\boldsymbol{\alpha} \in S'$, thus S' satisfies property (G2). We complete the proof verifying the property (G3). Therefore, we take $\boldsymbol{\alpha}, \boldsymbol{\beta} \in S'$ and suppose that $\boldsymbol{\beta} \in \Delta_i^{S'}(\boldsymbol{\alpha})$, we need to show that $\Delta_j^{S'}(\boldsymbol{\alpha}) \neq \emptyset$, where $j \in \{1, 2\} \setminus \{i\}$. Since $\boldsymbol{\alpha}, \boldsymbol{\beta} \in S$, for property (G3), there exists $\boldsymbol{\delta} \in \Delta_j^S(\boldsymbol{\alpha})$. If $\boldsymbol{\delta} \in \Delta_j^{S'}(\boldsymbol{\alpha})$, the thesis is proved; hence we suppose the converse, in this case $\boldsymbol{\delta}$ necessarily belongs to T'. We have two cases. Case 1: there exists $\boldsymbol{\gamma_k} = \boldsymbol{\alpha_k} \oplus \boldsymbol{\alpha_{k+1}}$ such that $\boldsymbol{\delta} \in \Delta_j^S(\boldsymbol{\gamma_k})$, but this implies $\boldsymbol{\gamma_k} \in \Delta_j^{S'}(\boldsymbol{\alpha})$. Case 2: there exists $\boldsymbol{\gamma_k} = \boldsymbol{\alpha_k} \oplus \boldsymbol{\alpha_{k+1}}$ such that $\boldsymbol{\delta} \in \Delta_i^S(\boldsymbol{\gamma_k})$. We notice that, if $\boldsymbol{\delta} \in I_A(S)$ we can reduce to the previous case, hence we can suppose that there exists $\boldsymbol{\rho} \neq \boldsymbol{\delta}$ with $\boldsymbol{\rho} \in \Delta_i^S(\boldsymbol{\gamma_k}) \cap I_A(S)$. But, since $\boldsymbol{\rho}, \boldsymbol{\delta} \in S$, by property

(G3) in S and by definition of track, $\Delta_j^{S'}(\boldsymbol{\delta}) \neq \emptyset$, then we also have $\Delta_j^{S'}(\boldsymbol{\alpha}) \neq \emptyset$. Case 3: $\boldsymbol{\delta} \in \Delta_i^S(\boldsymbol{\alpha_1})$ if $i = 2$ or $\boldsymbol{\delta} \in \Delta_i^S(\boldsymbol{\alpha_n})$ if $i = 1$; in this case we can conclude the proof with the same argument of Case 2.

Definition 22 Given $M \subset I_A(S)$, we say that M is an *hitting set* (HS) of S, if for any track T in S there exists an element $\boldsymbol{\alpha} \in M$ such that $\boldsymbol{\alpha} \in T$. We say that M is a minimal hitting set (MHS), if for any hitting set M such that $M' \subseteq M$, we have $M' = M$.

Remark 23 Given a hypergraph (V, E), with $E = \{E_1, \dots E_n\}$, $E_i \subseteq V$, a set of vertices $H \subset V$ such that $H \cap E_i \neq \emptyset$ for all $i = 1, \dots, n$ is called *transversal* or *hitting set* [3].

If we consider the hypergraph with vertices $V = I_A(S) \subset \Gamma_S$ and edges $E = \{T \subset S : T \text{ is a track}\}$, then the hitting sets of the good semigroup S correspond exactly to the hitting sets of this hypergraph. The problem of finding the minimal hitting set of an hypergraph is an NP-hard problem and there are several algorithms related to its computation (see for example [12, 15]).

We set $\mathfrak{H} = \{M \mid M \text{ is a HS}\}$.

Proposition 24 *If M is a sor, then $M \in \mathfrak{H}$.*

Proof If we suppose that M is not a HS, then it would exist a track T in S that does not contain elements of M. Using the same construction of Lemma 21 we could build a good semigroup S' such that $M \subseteq S' \subsetneq S$, but it is a contradiction.

The converse of the previous theorem is not true in general as it is shown by the following example.

Example 25 Let us consider the good semigroup S with the following set of irreducible absolute elements:

$$I_A(S) = \{(6, 3), (12, 17), (18, 25), (19, 6), (24, \infty), (25, 28), (27, 9), (31, \infty),$$
$$(33, 20), (39, \infty), (41, \infty), (44, \infty), (46, \infty), (\infty, 15), (\infty, 23), (\infty, 31)\}.$$

From Fig. 2 we can easily deduce that S contains only the following tracks:

- $T_1 = T((6, 3))$;
- $T_2 = T((12, 17), (19, 6))$;
- $T_3 = T((39, \infty), (\infty, 31))$;
- $T_4 = T((41, \infty), (\infty, 23))$;
- $T_5 = T((41, \infty), (\infty, 31))$;
- $T_6 = T((41, \infty))$;
- $T_7 = T((46, \infty), (\infty, 15))$;
- $T_8 = T((46, \infty), (\infty, 23))$;
- $T_9 = T((46, \infty), (\infty, 31))$.

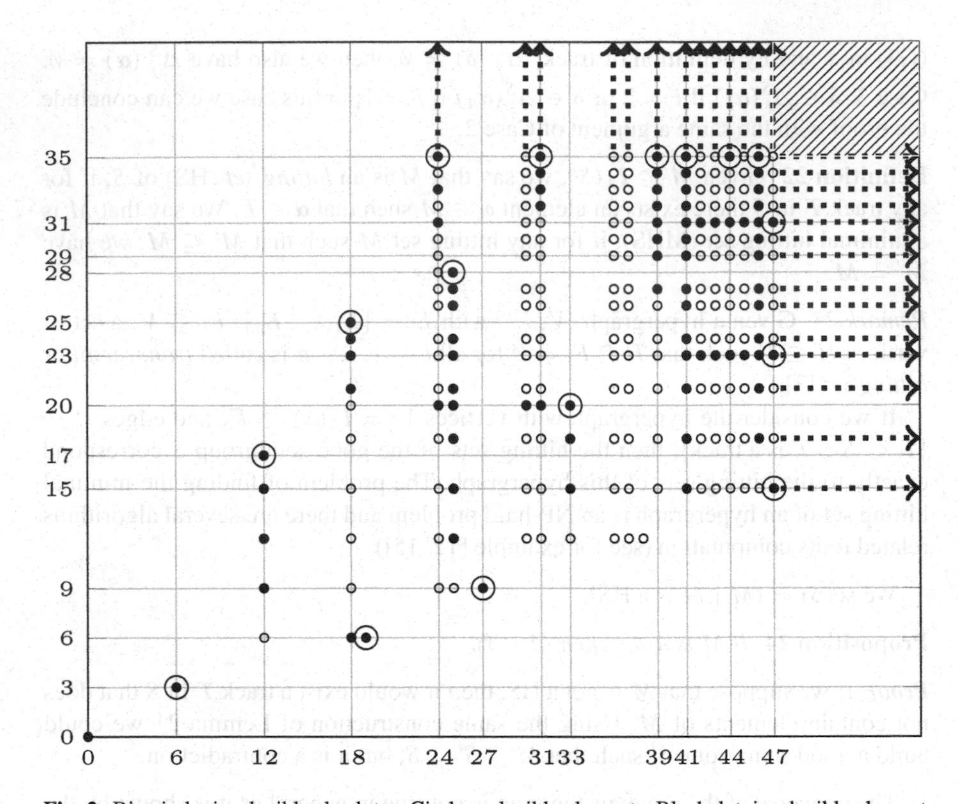

Fig. 2 Big circle: irreducible absolutes; Circle: reducible elements; Black dot: irreducible element

Then, it is easy to verify that $M = \{(6, 3), (12, 17), (39, \infty), (41, \infty), (46, \infty)\}$ is a MHS for S.

However, M is not a sor for S, in fact it is possible to check that there exists a good semigroup S' with

$$I_A(S') = \{(6, 3), (12, 17), (19, 6), (24, \infty), (39, \infty), (41, \infty), (46, \infty), (50, \infty), (\infty, 18), (\infty, 29), (\infty, 34)\},$$

such that S' is strictly contained in S and we have $M \subseteq S'$.

Now we define: $\text{bedim}(S) = \min\{|M|,\ M \in \mathfrak{H}\}$.

Corollary 26 *Given a good semigroup* $S \subseteq \mathbb{N}^2$, $\text{bedim}(S) \leq \text{edim}(S)$.

Example 27 The inequality of Corollary 26 can be strict. In fact, for instance, it is possible to check that each minimal hitting set of the semigroup S described in Example 25 is not a *sor* for S, implying that $\text{bedim}(S) < \text{edim}(S)$.

3.2 A Superior Bound for the Embedding Dimension

Let $S \subseteq \mathbb{N}^2$ be a good semigroup; given $\eta \subseteq I_A(S)$, and $\boldsymbol{\alpha} \in I_A(S)$, we want to define the reducibility of $\boldsymbol{\alpha}$ with respect to η. By convention we will say that all the elements $\boldsymbol{\alpha} \in \eta$ are reducible by η. We take $\boldsymbol{\alpha} \in I_A(S)\backslash\eta$ and we will treat the finite and infinite elements separately.

Finite Case We suppose $\boldsymbol{\alpha} = (\alpha_1, \alpha_2) \in I_{A_f}(S)\backslash\eta$.

Given a semiring $\Gamma \subseteq \overline{\mathbb{N}}^2$, we introduce the following notations:

$$
\begin{aligned}
{}_i\Delta^{\Gamma}(\boldsymbol{\alpha}) &:= \Gamma \cap {}_i\Delta(\boldsymbol{\alpha}) \\
{}_1\delta^{\Gamma}(\boldsymbol{\alpha}) &:= \max\{y \mid (a, y) \in {}_1\Delta^{\Gamma}(\boldsymbol{\alpha})\} \text{ if } {}_1\Delta^{\Gamma}(\boldsymbol{\alpha}) \neq \emptyset \\
{}_2\delta^{\Gamma}(\boldsymbol{\alpha}) &:= \max\{x \mid (x, b) \in {}_2\Delta^{\Gamma}(\boldsymbol{\alpha})\} \text{ if } {}_2\Delta^{\Gamma}(\boldsymbol{\alpha}) \neq \emptyset.
\end{aligned}
$$

Notice that the fact that $\boldsymbol{\alpha}$ is an absolute finite element implies that ${}_i\delta^{\Gamma}(\boldsymbol{\alpha})$ is finite. In the following, given $\eta \subseteq I_A(S)$, we will work with the semiring $\langle\eta\rangle_{\oplus}$. In order to simplify the notation we will write ${}_i\Delta^{\eta}(\boldsymbol{\alpha})$ instead of ${}_i\Delta^{\langle\eta\rangle_{\oplus}}(\boldsymbol{\alpha})$

Remark 28 If ${}_i\Delta^{\eta}(\boldsymbol{\alpha}) \neq \emptyset$, we have ${}_i\delta^{\eta}(\boldsymbol{\alpha}) \leq {}_i\delta^{S}(\boldsymbol{\alpha})$.

If ${}_1\Delta^{\eta}(\boldsymbol{\alpha}) \neq \emptyset$, we define $Y^{\eta}(\boldsymbol{\alpha}) = \{y \in \{{}_1\delta^{\eta}(\boldsymbol{\alpha}), \ldots, {}_1\delta^{S}(\boldsymbol{\alpha})\} \mid (\alpha_1, y) \in S\}$ and similarly if ${}_2\Delta^{\eta}(\boldsymbol{\alpha}) \neq \emptyset$, we define $X^{\eta}(\boldsymbol{\alpha}) = \{x \in \{{}_2\delta^{\eta}(\boldsymbol{\alpha}), \ldots, {}_2\delta^{S}(\boldsymbol{\alpha})\} \mid (x, \alpha_2) \in S\}$.

Definition 29 We say that $\boldsymbol{\alpha} = (\alpha_1, \alpha_2) \in I_{A_f}(S)\backslash\eta$ is reducible by η, if ${}_1\Delta^{\eta}(\boldsymbol{\alpha}) \cup {}_2\Delta^{\eta}(\boldsymbol{\alpha}) \neq \emptyset$ and one of the following conditions is satisfied:

1. ${}_1\Delta^{\eta}(\boldsymbol{\alpha}) \neq \emptyset$, and for all $y \in Y^{\eta}(\boldsymbol{\alpha})$, there exists $(x, y) \in \langle\eta\rangle_{\oplus}$ such that $x > \alpha_1$.
2. ${}_2\Delta^{\eta}(\boldsymbol{\alpha}) \neq \emptyset$, and for all $x \in X^{\eta}(\boldsymbol{\alpha})$, there exists $(x, y) \in \langle\eta\rangle_{\oplus}$ such that $y > \alpha_2$.

Infinite Case If $\boldsymbol{\alpha} = (\alpha_1, \infty) \in I_A(S)^{\infty}\backslash\eta$, then we consider $\tilde{y}$ such that $(\alpha_1, y) \in S$ for all $y \geq \tilde{y}$ (it exists by Lemma 2). Let us consider the set:

$$
Y^{\eta}(\boldsymbol{\alpha}) = \{y \in \{{}_1\delta^{\eta}(\boldsymbol{\alpha}), \ldots, \max\{\tilde{y}, {}_1\delta^{\eta}(\boldsymbol{\alpha})\} + e_2 - 1\} \mid (\alpha_1, y) \in S\}.
$$

If $\boldsymbol{\alpha} = (\infty, \alpha_2) \in I_A(S)^{\infty}\backslash\eta$, then we consider $\tilde{x}$ such that $(x, \alpha_2) \in S$ for all $x \geq \tilde{x}$ (it exists by Lemma 2). Let us consider the set:

$$
X^{\eta}(\boldsymbol{\alpha}) = \{x \in \{{}_2\delta^{\eta}(\boldsymbol{\alpha}), \ldots, \max\{\tilde{x}, {}_2\delta^{\eta}(\boldsymbol{\alpha})\} + e_1 - 1\} \mid (x, \alpha_2) \in S\}.
$$

Definition 30 We say that $\boldsymbol{\alpha} = (\alpha_1, \infty)$ is reducible by η, if ${}_1\Delta^{\eta}(\boldsymbol{\alpha}) \neq \emptyset$ and for all $y \in Y^{\eta}(\boldsymbol{\alpha})$, there exists an element $(x, y) \in \langle\eta\rangle_{\oplus}$ with $x > \alpha_1$. We say that $\boldsymbol{\alpha} = (\infty, \alpha_2)$ is reducible by η, if ${}_2\Delta^{\eta}(\boldsymbol{\alpha}) \neq \emptyset$ and for all $x \in X^{\eta}(\boldsymbol{\alpha})$, there exists an element $(x, y) \in \langle\eta\rangle_{\oplus}$ with $y > \alpha_2$.

As we will see in detail in the proof of Theorem 31, the previous definitions are motivated by the fact that the reducibility of an element $\boldsymbol{\alpha} \in I_A(S)$ by a set $\eta \subseteq I_A(S)$ essentially ensures that the presence of $\boldsymbol{\alpha}$ in $I_A(S)$ is forced by η as a consequence of property (G3) of good semigroups.

Given $\eta \subseteq I_A(S)$, we set:

$$\langle\langle \eta \rangle\rangle := \{\boldsymbol{\alpha} \in I_A(S) \mid \boldsymbol{\alpha} \text{ is reducible by } \eta\}.$$

Let us consider the following algorithm:

```
input : η ⊆ I_A(S)
output: A subset η', with η ⊆ η' ⊆ I_A
η' ⟵ ⟨⟨η⟩⟩
while η' ≠ η do
  | η ⟵ η'
  | η' ⟵ ⟨⟨η⟩⟩
end
return η'
```

Algorithm 1: Algorithm to find red(η)

The input of Algorithm 1 is a subset η of $I_A(S)$. As long as we can, we expand η by including elements of $I_A(S) \setminus \eta$ that are reducible by it. Notice that the algorithm produces an output in finite time, since $I_A(S)$ is finite. We denote by red(η) the output of the previous algorithm and we introduce the set $\mathfrak{R}(S) = \{\eta \subseteq I_A(S) \mid \text{red}(\eta) = I_A(S)\}$. We will say that $\eta \subseteq I_A(S)$ satisfy the *reducibility condition* if $\eta \in \mathfrak{R}(S)$.

We have the following statement:

Theorem 31 *If $\eta \in \mathfrak{R}(S)$, then η is a sor.*

Proof From $\eta \in \mathfrak{R}(S)$ it follows that there exists a chain of subset of $I_A(S)$:

$$\eta \subset \eta_1 \subset \ldots \subset \eta_{n-1} \subset \eta_n = \text{red}(\eta) = I_A(S)$$

such that $\eta_i = \langle\langle \eta_{i-1} \rangle\rangle$ We prove that η is a *sor* using a decreasing induction on this chain. We have that $\eta_n = I_A(S)$ is a *sor* for Remark 16, now we prove that if η_{i+1} is a *sor*, then η_i is a *sor*.

We assume by contradiction that $S \notin \mathcal{S}_{\eta_i}$; in this case there exists a good semigroup S_i such that $\eta_i \subseteq S_i \subsetneq S$.

If we suppose $\eta_{i+1} \subseteq I_A(S_i)$, we would have $\eta_{i+1} \subseteq \langle \eta_{i+1} \rangle_\oplus \subseteq \langle I_A(S_i) \rangle_\oplus = S_i \subsetneq S$, against the fact that η_{i+1} is a *sor* for S. For this reason, we can always suppose that there exists $\boldsymbol{\alpha} = (\alpha_1, \alpha_2) \in \eta_{i+1} \backslash I_A(S_i)$. Furthermore we observe that $\boldsymbol{\alpha} \notin \eta_i$, indeed, assuming the opposite, we should have $\boldsymbol{\alpha} \in S_i$ and since $S_i \subseteq S$ and $\boldsymbol{\alpha} \in \eta_{i+1} \subseteq I_A(S)$, it would imply that $\boldsymbol{\alpha} \in I_A(S_i)$. We distinguish two case: $\boldsymbol{\alpha} \in \eta_{i+1} \cap I_{A_f}(S)$ and $\boldsymbol{\alpha} \in \eta_{i+1} \cap I_A^\infty(S)$.

Case 1 $\boldsymbol{\alpha} \in \eta_{i+1} \cap I_{A_f}(S)$.

Since $\langle\langle \eta_i \rangle\rangle = \eta_{i+1}$, $\boldsymbol{\alpha}$ is reducible by η_i. Without loss of generality we can assume ${}_1\Delta^{\eta_i}(\boldsymbol{\alpha}) \neq \emptyset$; in this case there exists $(\alpha_1, {}_1\delta^{\eta_i}(\boldsymbol{\alpha})) \in \langle \eta_i \rangle_{\oplus} \subseteq S_i$. We have ${}_1\delta^{\eta_i}(\boldsymbol{\alpha}) \in Y^{\eta_1}(\boldsymbol{\alpha})$ and, from the reducibility of $\boldsymbol{\alpha}$ by η_i, there exists $(x^{\eta}(\boldsymbol{\alpha}), {}_1\delta^{\eta_i}(\boldsymbol{\alpha})) \in \langle \eta_i \rangle_{\oplus} \subseteq S_i$. We have obtained two elements $(\alpha_1, {}_1\delta^{\eta_i}(\boldsymbol{\alpha}))$, $(x^{\eta}(\boldsymbol{\alpha}), {}_1\delta^{\eta_i}(\boldsymbol{\alpha})) \in S_i$, by property (G3), there exists $(\alpha_1, y) \in S_i$, with $y > {}_1\delta^{\eta_i}(\boldsymbol{\alpha})$. We observe that, from the definition of ${}_1\delta^{S}(\boldsymbol{\alpha})$, $y \leq {}_1\delta^{S}(\boldsymbol{\alpha})$. Hence $y \in Y^{\eta_i}(\boldsymbol{\alpha})$. We can repeat the same argument until we obtain that $(\alpha_1, {}_1\delta^{S}(\boldsymbol{\alpha})) \in S_i$. Using again the property (G3) we should obtain $\boldsymbol{\alpha} \in S_i$ (notice that $\Delta_1^{S_i}(\boldsymbol{\alpha}) = \emptyset$), but this is a contradiction.

Case 2 $\boldsymbol{\alpha} \in \eta_{i+1} \cap I_{A^{\infty}}(S)$.

Without loss of generality we can suppose $\boldsymbol{\alpha} = (\alpha_1, \infty)$. Since $\boldsymbol{\alpha}$ is reducible by η_i, we have ${}_1\Delta^{\eta_i}(\boldsymbol{\alpha}) \neq \emptyset$. We set $M(\boldsymbol{\alpha}) := \max\{\tilde{y}, u\} + e_2 - 1$, where $\tilde{y}$ is such that $(\alpha_1, y) \in S$ for any $y > \tilde{y}$. Now, using the same argument of the finite case, we obtain that $(\alpha_1, M(\boldsymbol{\alpha})) \in S_i$, but, by Lemma 2, this implies $(\alpha_1, \infty) \in S_i$ which is a contradiction.

The following example shows that the converse of the previous theorem is not true in general.

Example 32 Let us consider the good semigroup S with the following set of irreducible absolute elements (Fig. 3):

$$I_A(S) = \{(3,4), (6,\infty), (7,8), (10,15), (14,18), (17,25), (\infty,12), (\infty,19), (\infty,22), (\infty,29)\}.$$

Notice that, since S contains only the tracks $T_1 = T((3,4))$, $T_2 = T((6,\infty),(7,8))$, $T_3 = T((6,\infty),(10,15),(\infty,12))$ and $T_4 = T((10,15),(\infty,12))$, we have that $\eta = \{(3,4),(7,8),(10,15),(14,18),(\infty,12),(\infty,22)\}$ is

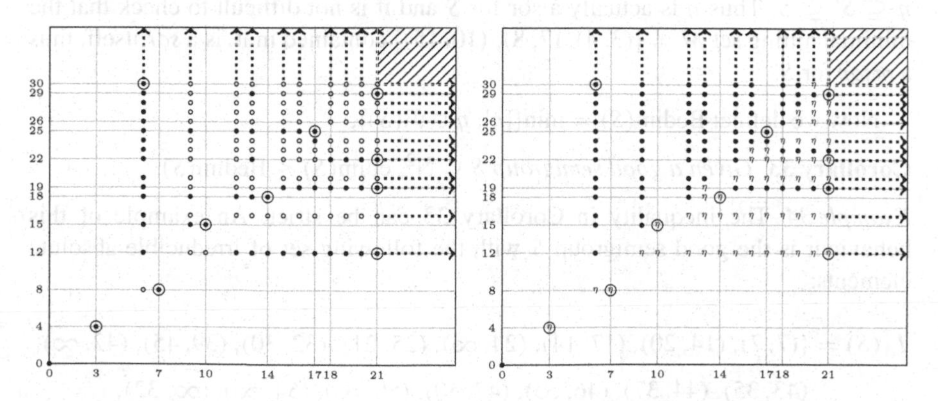

Fig. 3 Big circle: irreducible absolutes; Circle: reducible elements; Black dot: irreducible elements, η: Elements of $\langle \text{red}(\eta) \rangle_{\oplus}$

a HS for S. Let us show that $\mathrm{red}(\eta) \neq I_A(S)$. It suffices to show that $\langle\langle\eta\rangle\rangle = \eta$, i.e. all the elements in $I_A(S) \setminus \eta$ are not reducible by η. We have

- $\boldsymbol{\alpha_1} = (6, \infty)$ is not reducible by η. Notice that there exists $(6, 8) = 2(3, 4) \in {}_1\Delta^\eta(\boldsymbol{\alpha_1})$, thus we have ${}_1\delta^\eta(\boldsymbol{\alpha_1}) = 8$. Furthermore, $\tilde{y} = 22$ and we have:

$$Y^\eta(\boldsymbol{\alpha_1}) = \{y \in \{{}_1\delta^\eta(\boldsymbol{\alpha_1}), \ldots, \max\{\tilde{y}, {}_1\delta^\eta(\boldsymbol{\alpha_1})\} + e_2 - 1 | (6, y) \in S\} =$$
$$= \{8, 12, 15, 16, 18, 19, 20, 22, 23, 24, 25\}.$$

  For each element y in $Y^\eta(\boldsymbol{\alpha_1})$ we need to find $(x, y) \in \langle\eta\rangle_\oplus$ with $x > 6$. It is not difficult to notice that for $y = 25 \in Y^\eta(\boldsymbol{\alpha_1})$, it is not possible to produce such an element in $\langle\eta\rangle_\oplus$.
- $\boldsymbol{\alpha_2} = (17, 25)$ is not reducible by η. Notice that there exists $(17, 23) = (7, 8) \odot (10, 15) \in {}_1\Delta^\eta(\boldsymbol{\alpha_2})$, thus we have ${}_1\delta^\eta(\boldsymbol{\alpha_2}) = 23$ (while ${}_2\Delta^\eta(\boldsymbol{\alpha_2}) = \emptyset$). Furthermore, ${}_1\delta^S(\boldsymbol{\alpha_2}) = 24$ and we have:

$$Y^\eta(\boldsymbol{\alpha_2}) = \{y \in \{{}_1\delta^\eta(\boldsymbol{\alpha_2}), \ldots, {}_1\delta^S(\boldsymbol{\alpha_2}) = 24 | (17, y) \in S\} = \{23, 24\}.$$

  For each element y in $Y^\eta(\boldsymbol{\alpha_2})$ we need to find $(x, y) \in \langle\eta\rangle_\oplus$ with $x > 17$. However for $y = 23 \in Y^\eta(\boldsymbol{\alpha_2})$, it is not possible to produce such an element in $\langle\eta\rangle_\oplus$
- $\boldsymbol{\alpha_3} = (\infty, 19)$ is not reducible by η. Notice that there exists $(13, 19) = (3, 4) \odot (10, 15) \in {}_2\Delta^\eta(\boldsymbol{\alpha_3})$, thus we have ${}_2\delta^\eta(\boldsymbol{\alpha_3}) = 13$. Furthermore, $\tilde{x} = 15$ and we have:

$$X^\eta(\boldsymbol{\alpha_3}) = \{x \in \{{}_2\delta^\eta(\boldsymbol{\alpha_3}), \ldots, \max\{\tilde{x}, {}_2\delta^\eta(\boldsymbol{\alpha_3})\} + e_1 - 1 | (x, 19) \in S\} = \{13, 15, 16, 17\}.$$

  For each element x in $X^\eta(\boldsymbol{\alpha_3})$ we need to find $(x, y) \in \langle\eta\rangle_\oplus$ with $y > 19$. It is not difficult to notice that for $x = 13 \in X^\eta(\boldsymbol{\alpha_3})$, it is not possible to do that.
- $\boldsymbol{\alpha_4} = (\infty, 29)$ is not reducible by η, since ${}_2\Delta^\eta(\boldsymbol{\alpha_4}) = \emptyset$.

However it is possible to check that there are no good semigroups S' such that $\eta \subseteq S' \subsetneq S$. Thus η is actually a sor for S and it is not difficult to check that the minimal hitting set $M = \{(3, 4), (7, 8), (10, 15)\}$ contained in it, is a *sor* itself, thus a *msor* for S.

Now we define: $\mathrm{Bedim}(S) = \min\{|\eta|, \eta \in \mathfrak{R}(S)\}$,

Corollary 33 *Given a good semigroup $S \subseteq \mathbb{N}^2$,* $\mathrm{edim}(S) \leq \mathrm{Bedim}(S)$.

Example 34 The inequality in Corollary 33 can be strict. An example of this behaviour is the good semigroup S with the following set of irreducible absolute elements:

$$I_A(S) = \{(7, 7), (14, 20), (17, 14), (24, \infty), (25, 21), (32, 30), (39, 45), (42, \infty),$$
$$(43, 35), (44, 37), (46, \infty), (47, 50), (50, \infty), (54, \infty), (\infty, 32),$$
$$(\infty, 34), (\infty, 42), (\infty, 51), (\infty, 57)\}.$$

It is possible to prove that for each MHS η of S we have that η is a sor for S but $\text{red}(\eta) \neq I_A(S)$. This easily implies that $\text{edim}(S) < \text{Bedim}(S)$.

Example 35 Let us consider the good semigroup S, with

$$I_A(S) = \{(4,3), (6,7), (8,8), (9,6), (11,\infty), (12,\infty), (13,\infty), (14,\infty), (\infty,9), (\infty,11), (\infty,13)\}.$$

This is an example of good semigroup having *msor* with distinct cardinalities. In fact, it is possible to prove that the sets $\eta_1 = \{(4,3), (6,7), (8,8), (11,\infty), (13,\infty)\}$ and $\eta_2 = \{(4,3), (6,7), (8,8), (11,\infty), (\infty,9), (\infty,11)\}$ are both MHS of S satisfying the reducibility condition. In particular $\text{edim}(S) = 5$.

3.3 An Algorithm for the Computation of the Embedding Dimension of a Semigroup $S \subseteq \mathbb{N}^2$

We will conclude this section presenting an algorithm for the computation of the embedding dimension and with some remarks concerning the definition that we have given.

We proved that:

$$\text{bedim}(S) \leq \text{edim}(S) \leq \text{Bedim}(S)$$

and both inequalities are sharp as we will see in Example 38.

We implemented in GAP [17] the following functions:

- ComputeMHS(S): it takes in input a good semigroup and returns the set of its MHS.
- VerifyReducibility(list): it takes in input a list of subsets of $I_A(S)$ and returns the first set that satisfy the condition of reducibility if there exists, otherwise it returns "fail".
- IsThereAMGSContainedInAndContaining(S,V): it takes in input a good semigroup S and a subset V of $I_A(S)$ and returns "true" if there exists a good semigroup S' such that $V \subseteq S' \subsetneq S$

Remark 36 Testing in GAP these functions on a sample of about 200,000 semigroups, we observed empirically that VerifyReducibility is about seventy times faster than IsThereAMGSContainedInAndContaining.

We introduce the following algorithm to compute the embedding dimension and a set of representatives with minimal cardinality.

```
input : A good Semigroup S
output: A minimal system of representatives of minimal cardinality
𝔐 ⟵ ComputeMHS(S)
𝔋 ⟵ 𝔐
n ⟵ bedim(S)
Stop ⟵ false
while Stop=false do
    𝔋 ⟵ {η ⊆ I_A(S) | |η| = n and H ⊆ η for some H ∈ 𝔋} ∪ {η ∈ 𝔐||η| = n}
    if VerifyReducibility(𝔋)=η then
        Stop=true, return η
    end
    if VerifyReducibility(𝔋)=fail then
        if ForAny η ∈ 𝔋, IsThereAMGSContainedInAndContaining(S,η)=false then
            Stop=true, return η
        else
            n ⟵ n + 1
        end
    end
end
```

Algorithm 2: Algorithm to find an *msor* of minimal cardinality

Remark 37 We tested the algorithm on a sample of 200,000 good semigroups and we noticed that, for $n = \text{bedim}(S)$:

- The condition "*VerifyReducibility*($\mathfrak{H}$) $= fail$" occurred only in 82 cases.
- Both the conditions "*VerifyReducibility*($\mathfrak{H}$) $= fail$" and "*IsThereAMGSContainedInAndContaining*$(S, \eta) = true$ for all $\eta \in \mathfrak{H}$" occurred only in 2 cases. In these cases $\text{bedim}(S) \neq \text{edim}(S)$.
- The situation which all MSH of minimal cardinality are not reducible and at least one of them is a *sor* occurred only in one case. In this case $\text{Bedim}(S) \neq \text{edim}(S)$).

For this reasons and by Remark 36 this algorithm is considerably faster than to computing the embedding dimension using the definition.

Example 38 Let us consider the good semigroup S, represented in Fig. 4, we want to find a *msor* for S and the embedding dimension of S.

We have that

$$I_A(S) = \{(4, 3), (7, 13), (11, 17), (14, \infty), (15, \infty), (16, 20), (24, \infty),$$
$$(\infty, 12), (\infty, 16), (\infty, 26)\}.$$

First of all we need to compute the minimal hitting sets of S. It contains the following tracks:

- $T_1 = T((4, 3))$;
- $T_2 = T((7, 13))$;
- $T_3 = T((11, 17), (\infty, 16))$;
- $T_4 = T((15, \infty), (16, 20), (\infty, 12))$;

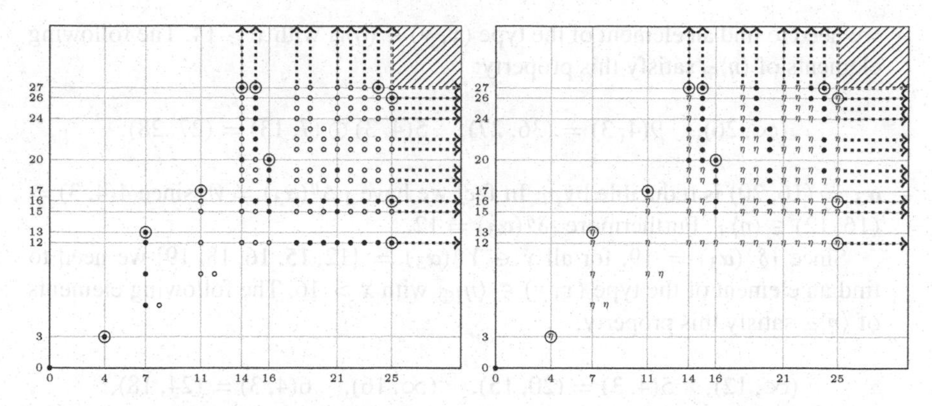

Fig. 4 Big circle: irreducible absolutes; Circle: reducible elements; Black dot: irreducible elements, η: Elements of $\langle \mathrm{red}(\eta)\rangle_\oplus$

- $T_5 = T((15, \infty), (16, 20), (\infty, 16))$;
- $T_6 = T((24, \infty), (\infty, 26))$.

Thus the following is the complete list of the MHS of S.

- $\eta_1 = \{(4, 3), (7, 13), (\infty, 12), (\infty, 16), (\infty, 26)\}$;
- $\eta_2 = \{(4, 3), (7, 13), (11, 17), (15, \infty), (\infty, 26)\}$;
- $\eta_3 = \{(4, 3), (7, 13), (11, 17), (16, 20), (24, \infty)\}$;
- $\eta_4 = \{(4, 3), (7, 13), (11, 17), (16, 20), (\infty, 26)\}$;
- $\eta_5 = \{(4, 3), (7, 13), (15, \infty), (24, \infty), (\infty, 16)\}$;
- $\eta_6 = \{(4, 3), (7, 13), (15, \infty), (\infty, 16), (\infty, 26)\}$;
- $\eta_7 = \{(4, 3), (7, 13), (16, 20), (24, \infty), (\infty, 16)\}$;
- $\eta_8 = \{(4, 3), (7, 13), (16, 20), (\infty, 16), (\infty, 26)\}$;
- $\eta_9 = \{(4, 3), (7, 13), (24, \infty), (\infty, 12), (\infty, 16)\}$;
- $\eta_{10} = \{(4, 3), (7, 13), (11, 17), (15, \infty), (24, \infty)\}$.

Thus for this semigroup $\mathrm{bedim}(S) = 5$. We consider $\eta = \eta_1 = \{(4, 3), (7, 13), (\infty, 12), (\infty, 16), (\infty, 26)\}$ and we want to show that $\eta \in \mathfrak{R}(S)$.

We have $\eta^1 = \langle\langle \eta \rangle\rangle = \{(4, 3), (7, 13), (11, 17), (14, \infty), (16, 20), (24, \infty), (\infty, 12), (\infty, 16), (\infty, 26)\}$.

In fact

- $\boldsymbol{\alpha_1} = (11, 17)$ is reducible by η because we have ${}_1\Delta^\eta(\boldsymbol{\alpha_1}) \neq \emptyset$ since $(4, 3) \odot (7, 13) = (11, 16) \in \langle \eta \rangle_\oplus$. Furthermore ${}_1\delta^\eta(\boldsymbol{\alpha_1}) = 16$.

  Since ${}_1\delta^S(\boldsymbol{\alpha_1}) = 16$ we need only to find an element of the type $(x, 16) \in \langle \eta \rangle_\oplus$ with $x > 11$. The element $(\infty, 16) \in \eta$ satisfies this property.
- $\boldsymbol{\alpha_2} = (14, \infty)$ is reducible by η because we have ${}_1\Delta^\eta(\boldsymbol{\alpha_2}) \neq \emptyset$; in fact we have $2(7, 13) = (14, 26) \in \langle \eta \rangle_\oplus$. Furthermore ${}_1\delta^\eta(\boldsymbol{\alpha_2}) = 26$.

  Since $\tilde{y} = 18$, for all

$$y \in Y^\eta(\boldsymbol{\alpha_2}) = \{y \in \{{}_1\delta^\eta(\boldsymbol{\alpha_2}) = 26, \ldots, \max\{\tilde{y}, {}_1\delta^\eta(\boldsymbol{\alpha_2})\} + e_2 - 1 = 28 | (14, y) \in S\} =$$

$$= \{26, 27, 28\},$$

we need to find an element of the type $(x, y) \in \langle\eta\rangle_\oplus$ with $x > 14$. The following elements of $\langle\eta\rangle_\oplus$ satisfy this property:

$$(\infty, 26), \quad 9(4,3) = (36, 27), \quad 5(4,3) \odot (7, 13) = (27, 28).$$

- $\boldsymbol{\alpha_3} = (16, 20)$ is reducible by η. In fact we have ${}_1\Delta^\eta(\boldsymbol{\alpha_3}) \neq \emptyset$; since $4(4, 3) = (16, 12) \in \langle\eta\rangle_\oplus$. Furthermore ${}_1\delta^\eta(\boldsymbol{\alpha_3}) = 12$.

  Since ${}_1\delta^S(\boldsymbol{\alpha_3}) = 19$, for all $y \in Y^\eta(\boldsymbol{\alpha_3}) = \{12, 15, 16, 18, 19\}$ we need to find an element of the type $(x, y) \in \langle\eta\rangle_\oplus$ with $x > 16$. The following elements of $\langle\eta\rangle_\oplus$ satisfy this property:

$$(\infty, 12), \quad 5(4,3) = (20, 15), \quad (\infty, 16), \quad 6(4,3) = (24, 18),$$
$$(4,3) \odot (\infty, 16) = (\infty, 19).$$

- $\boldsymbol{\alpha_4} = (24, \infty)$ is reducible by η. In fact ${}_1\Delta^\eta(\boldsymbol{\alpha_4}) \neq \emptyset$ since $6(4, 3) = (24, 18) \in \langle\eta\rangle_\oplus$. Thus ${}_1\delta^\eta(\boldsymbol{\alpha_4}) = 18$. Since $\tilde{y} = 24$, for all

$$y \in Y^\eta(\boldsymbol{\alpha_4}) = \{y \in \{{}_1\delta^\eta(\boldsymbol{\alpha_4}) = 18, \ldots, \max\{\tilde{y}, {}_1\delta^\eta(\boldsymbol{\alpha_4})\} + e_2 - 1 = 26 | (24, y) \in S\} =$$
$$= \{18, 19, 21, 22, 24, 25, 26\},$$

we need to find an element of the type $(x, y) \in \langle\eta\rangle_\oplus$ with $x > 24$. The following elements of $\langle\eta\rangle_\oplus$ satisfy this property:

$$2(4,3) \odot (\infty, 12) = (\infty, 18) \quad (4,3) \odot (\infty, 16) = (\infty, 19),$$
$$3(4,3) \odot (\infty, 12) = (\infty, 21), 2(4,3) \odot (\infty, 16) = (\infty, 22),$$
$$2(\infty, 12) = (\infty, 24), \quad (7, 13) \odot (\infty, 12) = (\infty, 25), \quad (\infty, 26).$$

Notice that $\boldsymbol{\alpha_5} = (15, \infty)$ is not reducible by η, but it is reducible by η^1. In fact ${}_1\Delta^{\eta^1}(\boldsymbol{\alpha_5}) \neq \emptyset$ since $(4, 3) \odot (11, 17) = (15, 20) \in \langle\eta^1\rangle_\oplus$. Thus ${}_1\delta^{\eta^1}(\boldsymbol{\alpha_5}) = 20$. Since $\tilde{y} = 18$, for all

$$y \in Y^{\eta^1}(\boldsymbol{\alpha_5}) = \{y \in \{{}_1\delta^{\eta^1}(\boldsymbol{\alpha_5}) = 20, \ldots, \max\{\tilde{y}, {}_1\delta^{\eta^1}(\boldsymbol{\alpha_5})\} + e_2 - 1 = 22 | (14, y) \in S\} =$$
$$= \{20, 21, 22\},$$

we need to find an element of the type $(x, y) \in \langle\eta^1\rangle_\oplus$ with $x > 15$. The following elements of $\langle\eta^1\rangle_\oplus$ satisfy this property:

$$(16, 20), \quad 3(4,3) \odot (\infty, 12) = (\infty, 21), \quad 3(4,3) \odot (7, 13) = (19, 22).$$

Thus $\langle\langle \eta^1 \rangle\rangle = I_A(S)$, and this means $\eta \in \mathfrak{R}(S)$ since $\mathrm{red}(\eta) = I_A(S)$. Hence $\mathrm{Bedim}(S) \leq 5 = |\eta|$. Since we have

$$5 = \mathrm{bedim}(S) \leq \mathrm{edim}(S) \leq \mathrm{Bedim}(S) \leq 5,$$

we can finally deduce that $\mathrm{edim}(S) = 5$ and η is an *msor*.

It is possible to check that all the minimal hitting sets previously found satisfy the reducibility condition, thus they are all *msor* for S.

All the previous computations were realized implementing all the previous algorithms in GAP [17].

4 Properties of Embedding Dimension

4.1 Relationship Between Embedding Dimension of a Ring and Embedding Dimension of Its Value Semigroup

Theorem 39 *Let S be a good semigroup of $\mathbb{N}^2$ such that there exists an algebroid curve R with $v(R) = S$. Then* $\mathrm{edim}(S) \geq \mathrm{edim}(R)$.

Proof Let us consider an algebroid curve R such that $v(R) = S$ and denote by ε the embedding dimension of S. Thus there exists $\eta \subset I_A(S)$, *msor* of S, with $|\eta| = \varepsilon$. We want to prove $\mathrm{edim}(R) \leq \varepsilon$.

We denote by

$$\eta = \{\boldsymbol{\alpha}_1, \ldots, \boldsymbol{\alpha}_\varepsilon\},$$

and we want to show that it is possible to choose elements $\phi_1, \ldots, \phi_\varepsilon$ in R, such that:

- $v(\phi_j) = \boldsymbol{\alpha}_j$ for each $j = 1, \ldots, \varepsilon$;
- $v(\mathbb{K}[\![\phi_1, \ldots, \phi_\varepsilon]\!])$ is a good semigroup.

Denote by $R_1 = \mathbb{K}[\![\phi_1, \ldots, \phi_\varepsilon]\!]$. By construction, for each choice of the elements ϕ_j, the subsemigroup $v(R_1) \subseteq \mathbb{N}^2$ always satisfies the properties (G1) and (G3) of good semigroups, thus we need to guarantee the existence of a conductor. This can be done by forcing in v the presence of vectors that fulfil the conditions of Proposition 13 (it is not difficult to do that by accordingly adding to the ϕ_i elements of R with value greater than its conductor).

Now,

$$\eta \subseteq v(R_1) \subseteq v(R) = S,$$

and, since η is a *msor* of S and $v(R_1)$ is a good semigroup, we have $v(R_1) = S$. Notice that $R_1 \subseteq R$ with $v(R) = v(R_1)$ implies that $R_1 = R$. In fact, considered an element $r \in R$, there exists an element $r_1 \in R_1$ such that $v(r) = v(r_1)$. Thus we can fin a $k_1 \in \mathbb{K}$ such that $v(r - k_1 r_1)$ is strictly greater than $v(r)$. We eventually find $k_j \in \mathbb{K}$ and $r_j \in R_1$ such that $v(r - \sum k_j r_j) \geq c(S) = c(v(R_1))$, implying that $r - \sum k_j r_j \in R_1$, and $r \in R_1$.

Thus $\text{edim}(R) = \text{edim}(R_1) \leq \varepsilon = \text{edim}(S)$.

We want to show that the inequality can be strict and we want to analyze the cases when this happens.

Example 40 Let us consider the ring $R \cong \mathbb{K}[\![(t^4, u^4), (t^6 + t^9, u^6 + u^7), (2t^{15} + t^{18}, 2u^{13} + u^{14})]\!]$, and the corresponding semigroup $v(R)$ (Fig. 5).

We observe that $R = \mathbb{K}[\![(t^4, u^4), (t^6 + t^9, u^6 + u^7), (2t^{15} + t^{18}, 2u^{13} + u^{14})]\!] = \mathbb{K}[\![(t^4, u^4), (t^6 + t^9, u^6 + u^7)]\!]$, in fact:

$$(2t^{15} + t^{18}, 2u^{13} + u^{14}) = (t^6 + t^9, u^6 + u^7)^2 - (t^4, u^4)^3.$$

We have that $\text{edim}(R) = 2$, but $\text{edim}(v(R)) = 3$, since $M = \{(4, 4), (6, 6), (15, 13)\}$ is the only hitting set of the semigroup $v(R)$

This fact happens because in the ring R the element of value $(15, 13)$ is obtained by the sum of the elements of value $(4, 4)$ and $(6, 6)$ because of a cancellation.

This situation cannot be controlled by the property (G3) of the good semigroups. This gap in embedding dimension can be justified by the fact that this piece of information is lost in the passage from the ring to the semigroup. For this value semigroup it is possible to find a ring, namely $T = \mathbb{K}[\![(t^4, u^4), (t^6, u^6), (t^{15}, u^{13})]\!]$ with $v(T) = v(R)$, and such that $\text{edim}(T) = \text{edim}(v(T))$. This situation is not guaranteed to happen in general, as it is shown in the following example.

Example 41 Let us consider the ring $R = [\![(t^4, u^3), (t^7, u^{13}), (t^{11}, u^{17}), (t^{16}, u^{20})]\!]$ that has embedding dimension 4. Its value semigroup is the good semigroup that appeared in Example 38, where we proved that its embedding dimension is five.

We focus on one of its *msor*, namely $\eta = \{(4, 3), (7, 13), (11, 17), (16, 20), (\infty, 26)\}$. If we analyze in detail what happens, we observe that $(t^{23}, u^{33}) = (t^7, u^{13}) \cdot (t^{16}, u^{20}) \in R$ and $(t^{23}, u^{26}) = (t^{11}, u^{17}) \cdot (t^{12}, u^9) \in R$, thus $(0, u^{26} - u^{33}) \in R$. But $\eta = \{(4, 3), (7, 13), (11, 17), (16, 20)\}$ is not a *sor*, since we have seen in the Example 38 that all MHS have to contain either $(\infty, 26)$ or $(24, \infty)$. This fact happens because in the ring R all the elements of value $(x, 26)$ with $x \geq 25$ appear because we have a complete cancellation on the first component (i.e. we obtain 0 on the first component).

In the semiring $\langle \eta \rangle_{\oplus}$ the existence of the elements $(23, 33)$ and $(23, 26)$ guarantees, by property (G3) of the good semigroups, only the existence of one element of value $(>23, 26)$, but not the presence of all elements $(x, 26)$, with $x \geq 24$.

Also in this case in the semigroup we lose a piece of information present in the ring.

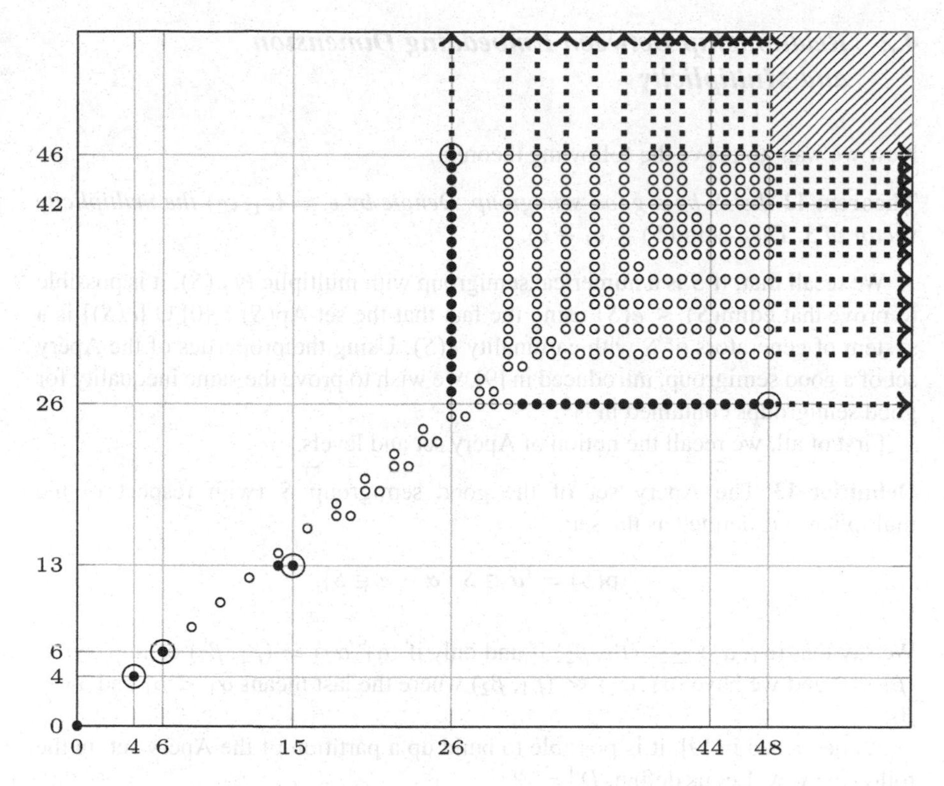

Fig. 5 Semigroup $v(R)$ of the Example 40

Differently from the previous example, it is not possible to find a ring T such that $v(T) = v(R)$ and $\mathrm{edim}(T) = \mathrm{edim}(v(T)) = 5$. To see this, let us suppose by contradiction that such a ring T exists. Let us consider $\psi_1, \ldots, \psi_5 \in T$, such that

- $v(\psi_1) = (4, 3)$;
- $v(\psi_2) = (7, 13)$;
- $v(\psi_3) = (11, 17)$;
- $v(\psi_4) = (16, 20)$;
- $v(\psi_5) = (\infty, 26)$.

From the proof of Theorem 39 we have that $T \cong \mathbb{K}[[\psi_1, \psi_2, \psi_3, \psi_4, \psi_5]]$.

Let us consider the ring $T' \cong \mathbb{K}[[\psi_1, \psi_2, \psi_3, \psi_4]]$. We must have $v(T') \subsetneq v(T)$ because otherwise $T = T'$, against the fact that $\mathrm{edim}(T) = 5$. Now we have that $\{(4, 3), (7, 13), (11, 17), (16, 20)\} \subseteq v(T')$ and it is not difficult to show that there exists only one good semigroup D containing these vectors and contained in $v(T)$. The good semigroup D is the one appeared in [1, Example 2.16] as the first example of a good semigroup that cannot be a value semigroup of a ring. Thus $v(T') = v(T)$ and we have a contradiction.

4.2 *Relationship Between Embedding Dimension and Multiplicity*

Now we want to prove the following theorem.

Theorem 42 *Let S be a good semigroup. Denote by $\boldsymbol{e} = (e_1, e_2)$ the multiplicity vector of S. Then* $\text{edim}(S) \leq e_1 + e_2$.

We recall that, if S is a numerical semigroup with multiplicity $e(S)$, it is possible to prove that $\text{edim}(S) \leq e(S)$ using the fact that the set $\text{Ap}(S) \setminus \{0\} \cup \{e(S)\}$ is a system of generators of S with cardinality $e(S)$. Using the properties of the Apéry set of a good semigroup, introduced in [9], we wish to prove the same inequality for good semigroups contained in $\mathbb{N}^2$.

First of all, we recall the notion of Apéry set and levels.

Definition 43 The Apéry set of the good semigroup S (with respect to the multiplicity) is defined as the set:

$$\text{Ap}(S) = \{\boldsymbol{\alpha} \in S : \boldsymbol{\alpha} - \boldsymbol{e} \notin S\}.$$

We say that $(\alpha_1, \alpha_2) \leq\leq (\beta_1, \beta_2)$ if and only if $(\alpha_1, \alpha_2) = (\beta_1, \beta_2)$ or $(\alpha_1, \alpha_2) \neq (\beta_1, \beta_2)$ and we have $(\alpha_1, \alpha_2) \ll (\beta_1, \beta_2)$ where the last means $\alpha_1 < \beta_1$ and $\alpha_2 < \beta_2$.

As described in [9], it is possible to build up a partition of the Apéry set, in the following way. Let us define, $D^0 = \emptyset$:

$$B^{(i)} = \{\boldsymbol{\alpha} \in \text{Ap}(S) \backslash (\cup_{j<i} D^{(j)}) : \boldsymbol{\alpha} \text{ is maximal with respect to } \leq\leq\}$$

$$C^{(i)} = \{\boldsymbol{\alpha} \in B^{(i)} : \boldsymbol{\alpha} = \boldsymbol{\beta}_1 \oplus \boldsymbol{\beta}_2 \text{ for some } \boldsymbol{\beta}_1, \boldsymbol{\beta}_2 \in B^{(i)} \setminus \{\boldsymbol{\alpha}\}\}$$

$$D^{(i)} = B^{(i)} \backslash C^{(i)}.$$

For a certain $N \in \mathbb{N}$, we have $\text{Ap}(S) = \cup_{i=1}^{N} D^{(i)}$ and $D^{(i)} \cap D^{(j)} = \emptyset$. In according to notation of [9], we rename these sets in an increasing order setting $A_i = D^{(N+1-i)}$. Thus we have

$$\text{Ap}(S) = \cup_{i=1}^{N} A_i.$$

Notice that the first level A_1 of $\text{Ap}(S)$ consists only of the zero vector. It was proved [9, Thm. 3.4] that $N = e_1 + e_2$, a key result in the proof of our inequality.

In order to simplify the notation in the following results we define the set $\overline{\text{Ap}(S)} = (\text{Ap}(S) \setminus \{\boldsymbol{0}\}) \cup \{\boldsymbol{e}\}$. Since we are only interchanging the role of the multiplicity vector and the zero vector, we have

$$\overline{\text{Ap}(S)} = \cup_{i=1}^{N} A'_i,$$

where $A_i = A'_i$ for $i = 2, \ldots, N$, and $A'_1 = \{\boldsymbol{e}\}$.

In order to prove Theorem 42, it is useful to introduce the following new definition of reducibility of an element of $I_A(S)$ by a subset $\eta \subseteq I_A(S)$.

Definition 44 Let $\boldsymbol{\alpha} = (\alpha_1, \alpha_2) \in I_A(S)$ and $\eta \subseteq I_A(S)$.

- **Case** $(\alpha_1, \alpha_2) \in I_{A_f}(S)$. Then $\boldsymbol{\alpha}$ is ρ-reducible by η if
  1. $\exists \boldsymbol{h}_1, \ldots, \boldsymbol{h}_k \in \eta$ such that $\boldsymbol{h}_1 \odot \cdots \odot \boldsymbol{h}_k = (\beta_1, \alpha_2)$ with $\beta_1 < \alpha_1$.
  2. $\forall x \in \{\beta_1, \ldots, 2\delta^S(\boldsymbol{\alpha})\}$ such that $(x, \alpha_2) \in S$ we can find $\boldsymbol{j}_1, \ldots, \boldsymbol{j}_l \in \eta$ such that $\boldsymbol{j}_1 \odot \cdots \odot \boldsymbol{j}_l = (x, \beta_2)$ with $\beta_2 > \alpha_2$.
- **Case** $\boldsymbol{\alpha} = (\infty, \alpha_2) \in I_A(S)^\infty$. Denote, as we did before, by $\tilde{x}$ the minimal element such that $(x, \alpha_2) \in S$ for all $x \geq \tilde{x}$. Then (∞, α_2) is ρ-reducible by η if
  1. $\exists \boldsymbol{h}_1, \ldots, \boldsymbol{h}_k \in \eta$ such that $\boldsymbol{h}_1 \odot \cdots \odot \boldsymbol{h}_k = (\beta_1, \alpha_2)$ with $\beta_1 < \infty$.
  2. $\forall \tilde{x} \in \{x \in \{\beta_1, \ldots, \max(\beta_1, \tilde{x}) + e_1 - 1\} : (x, \alpha_2) \in S\}$ we can find $\boldsymbol{j}_1, \ldots, \boldsymbol{j}_l \in \eta$ such that $\boldsymbol{j}_1 \odot \cdots \odot \boldsymbol{j}_l = (\tilde{x}, \beta_2)$ with $\beta_2 > \alpha_2$.
- **Case** $\boldsymbol{\alpha} = (\alpha_1, \infty) \in I_A(S)^\infty$. Such an element is never ρ-reducible by η.

Remark 45 If an element of $I_A(S)$ is ρ-reducible by η, it is also reducible by η.

Remark 46 If an element (α_1, α_2) of $I_A(S)$ is ρ-reducible by η, then it is also ρ-reducible by $\eta_{\alpha_1} = \{(x, y) \in \eta : x < \alpha_1\}$. In fact, the elements required to satisfy the condition 1. and 2. of Definition 44 cannot be obtained by using irreducible absolute elements of S with first component bigger than α_1 (because we only allow the operation $\odot$ to produce them).

Now we write

$$I_A(S) = \{\boldsymbol{\alpha}^{(1)} = (\alpha_1^{(1)}, \alpha_2^{(1)}), \ldots, \boldsymbol{\alpha}^{(n)} = (\alpha_1^{(n)}, \alpha_2^{(n)})\},$$

where the elements are ordered in decreasing order with respect to the first coordinate, i.e. if $j < l$, then $\alpha_1^{(j)} > \alpha_1^{(l)}$ or $\alpha_1^{(j)} = \alpha_1^{(l)} = \infty$ and $\alpha_2^{(j)} > \alpha_2^{(l)}$. Let us consider the following algorithm to produce, starting from $I_A(S)$, a set η that is still a *sor* for S.

```
input : The set of irreducible absolute elements I_A(S)
output: A subset η ⊆ I_A(S)
η ← I_A(S)
for k ← 1 to n do
    if α^(k) is ρ-reducible by I_A(S) \ {α^(k)} then
        η ← η \ {α^(k)}
    end
end
return η
```

Algorithm 3: A way to produce a *sor* using ρ-reducibility

Proposition 47 *The output η of Algorithm 3 is a* sor *for S*

Proof Let us prove by induction on k that the subset η produced by the algorithm is a *sor* for S. By Theorem 31, we can do it by showing that it satisfies the reducibility condition. At the first step $\eta = I_A(S)$, hence we have a *sor* for S. Suppose that at the kth step of the algorithm $\eta \in \mathfrak{R}(S)$ and let us show that it still satisfies the reducibility condition after the $k+1$th step. If $\boldsymbol{\alpha}^{(k+1)}$ is not ρ-reducible by $I_A(S) \setminus \{\boldsymbol{\alpha}^{(k+1)}\}$, then we have nothing to prove since η remains unchanged. Now let us suppose that $\boldsymbol{\alpha}^{(k+1)}$ is ρ-reducible by $I_A(S) \setminus \{\boldsymbol{\alpha}^{(k+1)}\}$. We need to prove that $\eta \setminus \{\boldsymbol{\alpha}^{(k+1)}\} = \eta' \in \mathfrak{R}(S)$. By Remark 46, $\boldsymbol{\alpha}^{(k+1)}$ is ρ-reducible by the set $W = \{(\alpha_1, \alpha_2) \in I_A(S) \setminus \{\boldsymbol{\alpha}^{(k+1)}\} : \alpha_1 < \alpha_1^{(k+1)}\} = \{\boldsymbol{\alpha}^{(k+2)}, \ldots, \boldsymbol{\alpha}^{(n)}\}$. But at this step of the algorithm $W \subseteq \eta'$, thus $\boldsymbol{\alpha}^{(k+1)}$ is ρ-reducible by η', thus also reducible by η and this means that $\eta \subseteq \mathrm{red}(\eta')$. By the inductive step $I_A(S) = \mathrm{red}(\eta) \subseteq \mathrm{red}(\eta')$, hence $\eta' \in \mathfrak{R}(S)$ and it is still a *sor*.

Proposition 48 *If $\boldsymbol{\alpha} = (\alpha_1, \alpha_2) \in I_A(S)$ is such that ${}_2\Delta^S(\boldsymbol{\alpha}) \nsubseteq \overline{\mathrm{Ap}(S)}$, then $\boldsymbol{\alpha}$ is ρ-reducible by $I_A(S) \setminus \{\boldsymbol{\alpha}\}$.*

Proof Let us choose $(\beta_1, \alpha_2) \notin \overline{\mathrm{Ap}(S)}$ with the largest possible β_1. Thus, there exists an integer $k \geq 1$ such that $(\tilde{\alpha}_1, \tilde{\alpha}_2) \odot k(e_1, e_2) = (\beta_1, \alpha_2)$, where $(\tilde{\alpha}_1, \tilde{\alpha}_2) \in \overline{\mathrm{Ap}(S)} \cup \{\mathbf{0}\}$. Notice that, if $(\tilde{\alpha}_1, \tilde{\alpha}_2) = \mathbf{0}$, then $k \geq 2$, otherwise we would have $(\beta_1, \alpha_2) = (e_1, e_2) \in \overline{\mathrm{Ap}(S)}$.

If $(\tilde{\alpha}_1, \tilde{\alpha}_2) \neq \mathbf{0}$, we write it as

$$(\tilde{\alpha}_1, \tilde{\alpha}_2) = \boldsymbol{h}_1 \odot \cdots \odot \boldsymbol{h}_l,$$

where the $\boldsymbol{h}_j$ are irreducible elements of S.

Each $\boldsymbol{h}_j = (\alpha_1^j, \alpha_2^j)$ is an absolute element. In fact, if it were possible to write it as

$$(x, \alpha_2^j) \oplus (\alpha_1^j, y), \text{ with } x > \alpha_1^j \text{ and } y > \alpha_2^j,$$

and $(x, \alpha_2^j), (\alpha_1^j, y) \in S$, then it would follow that

$$\boldsymbol{h}_1 \odot \cdots \odot (x, \alpha_2^j) \odot \cdots \odot \boldsymbol{h}_l \odot k(e_1, e_2) = (\gamma_1, \alpha_2) \notin \overline{\mathrm{Ap}(S)},$$

and $\gamma_1 > \beta_1$, this is against the maximality of β_1.

Thus $\boldsymbol{h}_i \in I_A(S)$ for all i (and they are clearly distinct from (α_1, α_2)).

Now, if $(e_1, e_2) \in I_A(S)$, then

$$(\beta_1, \alpha_2) = k(e_1, e_2) \odot \boldsymbol{h}_1 \odot \cdots \odot \boldsymbol{h}_l$$

is already the element required to fulfill condition 1. in Definition 44.

Thus, let us suppose that $(e_1, e_2) = (\tilde{e_1}, e_2) \oplus (e_1, \tilde{e_2})$, where $\tilde{e_1} > e_1$, $\tilde{e_2} > e_2$ and $(\tilde{e_1}, e_2), (e_1, \tilde{e_2}) \in I_A(S) \setminus \{(\alpha_1, \alpha_2)\}$. Notice that $\boldsymbol{\alpha}$ cannot be of the type $(\tilde{e_1}, e_2)$ or $(e_1, \tilde{e_2})$ because in both cases we would have ${}_2\Delta^S(\boldsymbol{\alpha}) \subseteq \overline{\mathrm{Ap}(S)}$ against our hypothesis.

First of all notice that $\tilde{e_1} \neq \infty$. In fact, if it were equal to ∞, then there would exist $\overline{x}$ such that $(x, e_2) \in S$ for all $x \geq \overline{x}$. This implies that

$$k(x, e_2) \odot \boldsymbol{h}_1 \odot \cdots \odot \boldsymbol{h}_l = (kx + \tilde{\alpha_1}, \alpha_2) \in S$$

for all $x \geq \overline{x}$. Thus $(\alpha_1, \alpha_2) = (\infty, \alpha_2)$ and this is a contradiction since

$$(\alpha_1, \alpha_2) = (\infty, \alpha_2) = k(\infty, e_2) \odot \boldsymbol{h}_1 \odot \cdots \odot \boldsymbol{h}_l,$$

is not an element of $I_A(S)$ being reducible (recall that if $\boldsymbol{h}_1 \odot \cdots \odot \boldsymbol{h}_l = \boldsymbol{0}$, then $k \geq 2$). Thus $\tilde{e_1} \neq \infty$, and the element

$$(\overline{\alpha}_1, \alpha_2) = k(\tilde{e_1}, e_2) \odot \boldsymbol{h}_1 \odot \cdots \odot \boldsymbol{h}_l,$$

is the required element that satisfies the condition 1. of Definition 44.

Now we want to show that we can satisfy the condition 2. of ρ-reducibility. Let us suppose that $\boldsymbol{\alpha} = (\alpha_1, \alpha_2) \in I_{A_f}(S)$ (all the following considerations can be adapted to the case $(\alpha_1, \alpha_2) = (\infty, \alpha_2)$).

We have to show that for each $\tilde{x} \in X = \{x \in \{\beta_1, \ldots, {}_2\delta^S(\boldsymbol{\alpha})\} : (x, \alpha_2) \in S\}$ we can find $\boldsymbol{j}_1, \ldots, \boldsymbol{j}_l \in \eta$ such that $\boldsymbol{j}_1 \odot \cdots \odot \boldsymbol{j}_l = (\tilde{x}, \beta_2)$ with $\beta_2 > \alpha_2$.

Thus, let us consider an arbitrary $\tilde{x} \in X$. Since $(\tilde{x}, \alpha_2), (\alpha_1, \alpha_2) \in S$, by the (G3) property of Definition 1, there exists $\beta_2 > \alpha_2$ such that $(\tilde{x}, \beta_2) \in S$.

Theorem 10 ensures that we can write

$$(\tilde{x}, \beta_2) = \bigoplus_{i=1}^{m} (\bigodot_{j=1}^{n} \boldsymbol{\gamma}_{j_i}), \boldsymbol{\gamma}_{j_i} \in I_A(S).$$

It must exist an index $\overline{i}$ such that

$$\bigodot_{j=1}^{n} \boldsymbol{\gamma}_{j_{\overline{i}}} = (\tilde{x}, \tilde{\beta_2}).$$

Notice that $\boldsymbol{\gamma}_{j_{\overline{i}}} \in I_A(S) \setminus \{(\alpha_1, \alpha_2)\}$ for all $j = 1, \ldots, n$ (they all have first coordinate less than $\tilde{x} \leq \alpha_1$). Furthermore $\tilde{\beta_2} \geq \beta_2 > \alpha_2$, thus it is the element which we were looking for in order to satisfy the condition 2. of ρ-reducibility.

As a consequence of Proposition 48 and Algorithm 3, we can immediately deduce the following Corollary.

Corollary 49 *Let S be a good semigroup. Then the set*

$$\eta_S = \{\boldsymbol{\alpha} \in I_A(S) : \quad {}_2\Delta^S(\boldsymbol{\alpha}) \subseteq \overline{\mathrm{Ap}(S)}\}$$

is a sor *for S.*

Now we are ready to give a proof of Theorem 42.

Proof *(Proof of Theorem 42)* Using Corollary 49 and the definition of embedding dimension, it suffices to show that $|\eta_S| \leq e_1 + e_2$.

Let us write $\eta_S = \{\boldsymbol{h}^{(1)} = (\alpha_1^{(1)}, \alpha_2^{(1)}), \dots, \boldsymbol{h}^{(k)} = (\alpha_1^{(k)}, \alpha_2^{(k)})\}$ where if $i < j$ then $\alpha_2^{(i)} < \alpha_2^{(j)}$ or $\alpha_2^{(i)} = \alpha_2^{(j)} = \infty$ with $\alpha_1^{(i)} < \alpha_2^{(j)}$. Furthermore we denote by $c = (c_1, c_2)$ the conductor of S. Now to each element $\boldsymbol{h}^{(i)}$ of η_S we associate an element $\overline{\boldsymbol{h}}^{(i)}$ in the following way:

- **Case** $\boldsymbol{h}^{(i)} = (\alpha_1, \infty)$. Then we set $\overline{\boldsymbol{h}}^{(i)} = (\alpha_1, c_2 + i)$.
- **Case** $\boldsymbol{h}^{(i)} = (\alpha_1, \alpha_2)$, with $\alpha_2 \neq \infty$. Then we set $\overline{\boldsymbol{h}}^{(i)} = \min({}_2\Delta^S(\boldsymbol{h}^{(i)}) \cup \{\boldsymbol{h}^{(i)}\})$.

We consider the set $\eta' = \{\overline{\boldsymbol{h}}^{(1)}, \dots, \overline{\boldsymbol{h}}^{(k)}\}$, and we want to show that distinct elements of η' belong to distinct levels of the Apéry set of S. In order to do that we consider two arbitrary elements $\overline{\boldsymbol{h}}^{(i)}$ and $\overline{\boldsymbol{h}}^{(j)}$ of η' and we prove that they cannot belong to the same level of the Apéry set. We have four possible configurations:

- **Case** $\overline{\boldsymbol{h}}^{(i)} = (\alpha_1^{(i)}, \alpha_2^{(i)})$ and $\overline{\boldsymbol{h}}^{(j)} = (\alpha_1^{(j)}, \alpha_2^{(j)})$, with $\alpha_1^{(i)} < \alpha_1^{(j)}$ and $\alpha_2^{(i)} < \alpha_2^{(j)}$.
  In this case $\overline{\boldsymbol{h}}^{(i)} \ll \overline{\boldsymbol{h}}^{(j)}$ and from definition of Apéry levels it follows that $\overline{\boldsymbol{h}}^{(j)} \in A_n$ and $\overline{\boldsymbol{h}}^{(i)} \in A_m$ with $m < n$.
- **Case** $\overline{\boldsymbol{h}}^{(i)} = (\alpha_1^{(i)}, \alpha_2^{(i)})$ and $\overline{\boldsymbol{h}}^{(j)} = (\alpha_1^{(j)}, \alpha_2^{(j)})$, with $\alpha_1^{(i)} < \alpha_1^{(j)}$ and $\alpha_2^{(i)} = \alpha_2^{(j)}$.
  This configuration is not possible, because it is against the minimality of the element $\overline{\boldsymbol{h}}^{(j)}$ (it is easy to check that this situation cannot involve elements that come from $\boldsymbol{h}^{(i)}$ of the type (α_1, ∞)).
- **Case** $\overline{\boldsymbol{h}}^{(i)} = (\alpha_1^{(i)}, \alpha_2^{(i)})$ and $\overline{\boldsymbol{h}}^{(j)} = (\alpha_1^{(j)}, \alpha_2^{(j)})$, with $\alpha_1^{(i)} < \alpha_1^{(j)}$ and $\alpha_2^{(i)} > \alpha_2^{(j)}$.
  This configuration is not possible, since the element $\overline{\boldsymbol{h}}^{(i)} \oplus \overline{\boldsymbol{h}}^{(j)} \in S$ is against the minimality of the element $\overline{\boldsymbol{h}}^{(j)}$ (it is also easy to check that this situation cannot involve elements that come from $\boldsymbol{h}^{(i)}$ of the type (α_1, ∞)).
- **Case** $\overline{\boldsymbol{h}}^{(i)} = (\alpha_1^{(i)}, \alpha_2^{(i)})$ and $\overline{\boldsymbol{h}}^{(j)} = (\alpha_1^{(j)}, \alpha_2^{(j)})$, with $\alpha_1^{(i)} = \alpha_1^{(j)}$ and $\alpha_2^{(i)} > \alpha_2^{(j)}$.
  Suppose by contradiction that there exists $n \in \mathbb{N}$ such that $\overline{\boldsymbol{h}}^{(i)}, \overline{\boldsymbol{h}}^{(j)} \in A_n$. From the definition of η_S it follows that $\Delta_2^S(\overline{\boldsymbol{h}}^{(j)}) \subseteq \overline{\mathrm{Ap}(S)}$. Thus from Lemma 3.3 (3) of [9], the minimal element $\boldsymbol{\beta}$ of $\Delta_2^S(\overline{\boldsymbol{h}}^{(j)}) \in A_m$ with $m \leq n$. On the other

hand $\overline{h}^{(j)} \leq \beta$, thus $\beta \in A_l$ with $l \geq n$. Thus $\beta \in A_n$ and this is a contradiction because we have

$$\overbrace{\overline{h}^{(i)}}^{\in A_n} \oplus \overbrace{\beta}^{\in A_n} = \overline{h}^{(j)} \in A_n,$$

that is against the definition of Apéry set level. Since Theorem 3.4 of [9], states that the levels of the Apéry Set are exactly $e_1 + e_2$, it follows that

$$\mathrm{edim}(S) \leq |\eta_S| = |\eta'| \leq e_1 + e_2,$$

and the proof of Theorem 42 is complete.

We recall that a good semigroup is said to be Arf if and only if $S(\alpha) = \{\beta \in S | \beta \geq \alpha\}$ is a semigroup for any $\alpha \in S$. In [1, Proposition 3.19 and Corollary 5.8] the authors proved that an Arf semigroup can be always seen as the value semigroup of an Arf ring. From this result and Theorem 42 we can deduce the following corollary.

Corollary 50 *Let S be an Arf good subsemigroup of $\mathbb{N}^2$. Then, denoted as usual by $e = (e_1, e_2)$ the multiplicity vector of S, we have* $\mathrm{edim}(S) = e_1 + e_2$.

Proof By Theorem 42 we have $\mathrm{edim}(S) \leq e_1 + e_2$. Denote by R an Arf ring such that $v(R) = S$. By Theorem 39 we have $\mathrm{edim}(S) \geq \mathrm{edim}(R)$. But R is an Arf ring, thus its embedding dimension is equal to its multiplicity (cf.[14, Theorem 2.2]). Since the multiplicity of R is also equal to $e_1 + e_2$, we have

$$e_1 + e_2 = \mathrm{edim}(R) \leq \mathrm{edim}(S) \leq e_1 + e_2,$$

and the proof of the corollary is complete.

We say that a good semigroup $S \subseteq \mathbb{N}^2$ is *maximal embedding dimension* if $\mathrm{edim}(S) = e_1 + e_2$. Thus, Arf good semigroups constitute a particular class of maximal embedding dimension semigroups. It is known that a numerical semigroup is maximal embedding dimension if and only if $M + M = e + M$ where $M = S \setminus \{0\}$ is its maximal ideal and e is its multiplicity (cf.[16]).

Thus, we propose the following conjecture.

Conjecture 51 Let S be a good subsemigroup of $\mathbb{N}^2$. Then S is maximal embedding dimension if and only if $M + M = e + M$, where e is its multiplicity vector and $M = S \setminus \{\mathbf{0}\}$.

At the moment we have tested Conjecture 51 for a large number of good semigroup, and we have a proof of the fact that $M + M = e + M$ implies $\mathrm{edim}(S) = e_1 + e_2$.

Acknowledgements The authors would like to thank Marco D'Anna and Pedro A. García Sánchez for their helpful comments and suggestions during the development of this paper. They also thank the "INdAM" and the organizers of the "International meeting on numerical semigroups (Cortona 2018)" for inviting them to attend the conference, which was of great help in developing some of the ideas in this paper. Finally, a special thank goes to the anonymous referee for the interesting and extensive comments on an earlier version of this paper.

References

1. Barucci, V., D'Anna, M., Fröberg, R.: Analytically unramified one-dimensional semilocal rings and their value semigroups. J. Pure Appl. Algebra **147**(3), 215–254 (2000)
2. Barucci, V., D'Anna, M., Fröberg, R.: The semigroup of values of a one-dimensional local ring with two minimal primes. Commun. Algebra **28**(8), 3607–3633 (2000)
3. Berge, C.: Hypergraphs. Ed. by North-Holland Mathematical Library, vol. 45 (1989)
4. Campillo, A., Delgado, F., Kiyek, K.: Gorenstein properties and symmetry for one-dimensional local Cohen-Macaulay rings. Manuscripta Math. **83**, 405–423 (1994)
5. Campillo, A., Delgado, F., Gusein-Zade, S.: On generators of the semigroup of a plane curve singularity. J. Lond. Math. Soc. **2**(60), 420–430 (1999)
6. Carvalho, E., Escudeiro Hernandes, M.: The semiring of values of an algebroid curve (April 2017). e-prints. arXiv
7. D'Anna, M.: Canonical module of a one-dimensional reduced local ring. Commun. Algebra **25**(09), 2939–2965 (1997)
8. D'Anna, M., García-Sánchez, P.A., Micale, V., Tozzo, L.: Good subsemigroups of $\mathbb{N}^n$. Int. J. Algebra Comput. **28**(02), 179–206 (2018)
9. D'Anna, M., Guerrieri, L., Micale, V.: The Apéry set of a good semigroup (December 2018). e-prints. arXiv
10. Delgado, F.: The semigroup of values of a curve singularity with several branches. Manuscripta Math. **59**, 347–374 (1987)
11. Delgado, F.: Gorenstein curves and symmetry of the semigroup of value. Manuscripta Math. **61**, 285–296 (1988)
12. Eiter, T., Gottlob, G.: Identifying the minimal transversals of a hypergraph and related problems. SIAM J. Comput. **24**(6), 1278–1304 (1995)
13. García, A.: Gorenstein curves and symmetry of the semigroup of value. J. Reine Angew. Math. **336**, 165–184 (1982)
14. Lipman, J.: Stable ideals and Arf ring. Am. J. Math. **93**, 649–685 (1971)
15. Murakami, K., Uno, T.: Efficient algorithms for dualizing large-scale hypergraphs. Discrete Appl. Math. **170**, 83–94 (2014)
16. Rosales, J.C., García-Sánchez, P.A., García-García, J.I., Branco, M.B.: Numerical semigroups with maximal embedding dimension. Int. J. Commutative Rings **2**, 47–53 (2003)
17. The GAP Group: GAP – Groups, Algorithms, and Programming, Version 4.10.0 (2018)
18. Waldi, R.: Wertehalbgruppe und singularität einer ebenen algebroiden kurve. Jan 1972

On Multi-Index Filtrations Associated to Weierstraß Semigroups

Julio José Moyano-Fernández

Abstract This paper is a survey on the main techniques involved in the computation of the Weierstraß semigroup at several points of curves defined over perfect fields, with special emphasis on the case of two points. Some hints about the usage of some packages of the computer algebra software SINGULAR are also given; these are however only valid for curves defined over $\mathbb{F}_p$ with p a prime number.

Keywords Algebraic curve · Adjunction theory · Normalisation · Weierstraß semigroup

1 Introduction

There are several classical problems in the theory of algebraic curves which are interesting from a computational point of view. One of them is the computation of the Weierstraß semigroup of a smooth projective algebraic curve $\widetilde{\chi}$ defined over a field $\mathbb{F}$ at a rational point P, together with a rational function $f_m \in \mathbb{F}(\widetilde{\chi})$ regular outside P and achieving a pole at P of order m, for each m in this semigroup. This problem is solved with the aid of the adjunction theory for plane curves, profusely developed by A. von Brill and M. Noether in the nineteenth century (see [3, 4, 27]) so that we assume the knowledge of a singular plane birrational model χ for the smooth curve $\widetilde{\chi}$.

Let $\mathbb{F}$ be a perfect field. For a smooth projective algebraic curve $\widetilde{\chi}$ defined over $\mathbb{F}$ and rational points $P_1, \ldots, P_r$ on $\widetilde{\chi}$, we consider the family of finitely

The author was partially supported by the Spanish Government—Ministerios de Ciencia e Innovación y de Universidades, grant PGC2018-096446-B-C22, as well as by the University Jaume I of Castellón, grant UJI-B2018-10.

J. J. Moyano-Fernández (✉)
Universitat Jaume I, IMAC–Institut de Matemàtiques i Aplicacions de Castelló, Departament de Matemàtiques, Castelló de la Plana, Spain
e-mail: moyano@uji.es

V. Barucci et al. (eds.), *Numerical Semigroups*, Springer INdAM Series 40,
https://doi.org/10.1007/978-3-030-40822-0_14

dimensional vector subspaces of $\mathbb{F}(\widetilde{\chi})$ given by the Riemann-Roch vector spaces $\mathcal{L}(\underline{m}\underline{P}) = \mathcal{L}(m_1 P_1 + m_2 P_2 + \cdots + m_r P_r)$, where $\underline{m} = (m_1, \ldots, m_r) \in \mathbb{Z}^r$. This family gives rise to a $\mathbb{Z}^r$-multi-index filtration on the $\mathbb{F}$-algebra A of the affine curve $\widetilde{\chi} \setminus \{P_1, \ldots, P_r\}$, since one has $A = \bigcup_{\underline{m} \in \mathbb{Z}^r} \mathcal{L}(\underline{m}\underline{P})$. This multi-index filtration is related to Weierstraß semigroups (see Delgado [10]) and, in case of finite fields, to the methodology for trying to improve the Goppa estimation of the minimal distance of algebraic-geometrical codes (also called Goppa codes), see for instance Carvalho and Torres [9]. A connection of that filtration with global geometrical-topological aspects in a particular case was shown by Campillo et al. [6]. Moreover, Poincaré series associated to these filtrations have been studied by the author in [25], and by Moyano-Fernández et al. [26].

Thus, a natural question is to provide a computational method in order to estimate the values of $\dim_{\mathbb{F}} \mathcal{L}(\underline{m}\underline{P}) = \ell(\underline{m}\underline{P})$ for $\underline{m} \in \mathbb{Z}^r$. More precisely, it would be convenient to estimate and compute values of type $\ell((\underline{m} + \underline{\varepsilon})\underline{P}) - \ell(\underline{m}\underline{P})$ where $\underline{\varepsilon} \in \mathbb{Z}^r$ is a vector whose components are 0 or 1. This can be done by extending the method developed by Campillo and Farrán [8] in the case $r = 1$, based on the knowledge of a plane model χ for $\widetilde{\chi}$ (with singularities) and representing the global regular differentials in terms of adjoint curves to $\widetilde{\chi}$.

This work is intended as an attempt to summarize the basic facts in the study of the theory of Weierstraß semigroups which turned out to be relevant for the Goppa codes; we will touch therefore only a few aspects of the theory rather than present a comprehensive treatise on the topic. We neither claim novelty in most of the results, which are worked out from Campillo and Farrán [8], Campillo and Castellanos [5] or Matthews [23], mainly.

The paper is organised as follows: Sect. 2 is devoted to fix the algebraic-geometrical prerequisites. Section 3 deals with the study of more specific questions concerning to our purpose, namely the adjunction theory of curves, with the remarkable Brill-Noether Theorem. In Sect. 4 we define the Weierstraß semigroup at several points and describe two methods to compute values of the form $\ell((\underline{m} + \underline{\varepsilon})\underline{P}) - \ell(\underline{m}\underline{P})$. The last section is devoted to show and explain some procedures implemented in SINGULAR [13] which are based on Sect. 4.

Notice the practical relevance of these ideas in view of the algebraic-geometric coding theory: the Weierstraß semigroup plays an important role in the decoding procedure of Feng and Rao, see e.g. Campillo and Farrán [7], or Høholdt et al. [17]. Weierstraß semigroups are also helpful in quantum coding theory, see for instance Hernando Carrillo et al. [16]. For a recent account of the theory and the bibliography we refer the reader to Martínez-Moro et al. [21, 22].

2 Terminology and Notation

Let $\mathbb{F}$ be a perfect field, and let $\overline{\mathbb{F}}$ be a fixed algebraic closure of $\mathbb{F}$. Let χ be an absolutely irreducible projective algebraic curve defined over $\mathbb{F}$. We distinguish three types of points on χ, namely the geometric points, i.e. those with coordinates on $\overline{\mathbb{F}}$; the rational points, i.e. those with coordinates on $\mathbb{F}$; and the closed points, which are residue classes of geometric points under the action of the Galois group of the field extension $\overline{\mathbb{F}}/\mathbb{F}$, namely

$$P := \{\sigma(p) : \sigma \in \mathrm{Gal}(\overline{\mathbb{F}}/\mathbb{F})\},$$

where p is a geometric point. Notice that closed points correspond one to one to points on the curve χ viewed as an $\mathbb{F}$-scheme which are closed for the Zariski topology. Every closed point has an associated residue field $\mathbb{F}'$ which is a finite extension of $\mathbb{F}$. The degree of a closed point P is defined as the cardinal of its conjugation class, which equals the degree of the extension $\mathbb{F}'/\mathbb{F}$. In particular, P is rational if and only if $\deg P = 1$.

2.1 Branches and Parametrizations

In this subsection we follow [8]. Let χ be an absolutely irreducible algebraic plane curve defined over $\mathbb{F}$, and consider a closed point P on χ, as well as the local ring $\mathcal{O} := \mathcal{O}_{\chi,P}$ at P with maximal ideal $\mathfrak{m}$; write $\overline{\mathcal{O}}$ for the semilocal ring of the normalisation of χ at P, and let $\widehat{\mathcal{O}}$ be the completion of $\mathcal{O}$ with respect to the $\mathfrak{m}$-adic topology. Each maximal ideal of $\overline{\mathcal{O}}$ (or, equivalently, every minimal prime ideal $\mathfrak{p}$ of $\widehat{\mathcal{O}}$) is said to be a *branch* of χ at P.

Let us now choose an affine chart containing P so that the curve χ has an equation $f(X, Y) = 0$, and set $A := \mathbb{F}[X, Y]/(f(X, Y))$ for the affine coordinate ring; observe that $\mathcal{O} = A_P$, therefore

$$\mathbb{F} \subseteq \mathbb{F}[X, Y]/(f(X, Y)) = A \subseteq A_P = \mathcal{O}.$$

Since $\mathbb{F}$ is perfect, we can apply Hensel's lemma to find a finite field extension $K/\mathbb{F}$ such that $K \subseteq \widehat{A_P} = \widehat{\mathcal{O}}$ is a coefficient field for $\hat{\mathcal{O}}$. Moreover, K is the integral closure of $\mathbb{F}$ in $\hat{\mathcal{O}}$. Since $\widehat{\mathcal{O}} \subseteq \widehat{\overline{\mathcal{O}}} \cong \overline{\widehat{\mathcal{O}}}$, we deduce that

$$K \subseteq \widehat{\mathcal{O}}/\mathfrak{p} \subseteq \overline{\widehat{\mathcal{O}}/\mathfrak{p}} = \widehat{\overline{\mathcal{O}}_{\mathfrak{m}}},$$

hence we can apply Hensel's lemma again to obtain a finite extension K'/K which is a coefficient field for the local ring $\widehat{\overline{\mathcal{O}}_{\mathfrak{m}}}$. Without loss of generality we can consider P as the ideal (X, Y) in $K[\![X, Y]\!]$ so that $\widehat{\mathcal{O}} \cong K[\![X, Y]\!]/(f(X, Y))$. This implies the existence of natural morphisms

$$K[\![X, Y]\!]/(f(X, Y)) \cong \widehat{\mathcal{O}} \longrightarrow \widehat{\mathcal{O}}/\mathfrak{p} \longrightarrow K'[\![t]\!] \cong \widehat{\overline{\mathcal{O}}_{\mathfrak{m}}}$$

for any local uniformizing parameter $t \in \mathfrak{m} \setminus \mathfrak{m}^2$. Notice that K can be considered isomorphic to the residue field at P. With these preliminaries as in [8], a *parametrization* of the curve χ at the point P related to the coordinates X, Y is a

K-algebra morphism $\rho : K[\![X, Y]\!] \longrightarrow K'[\![t]\!]$ which is continuous for the (X, Y)-adic and t-adic topologies and satisfies that $\mathrm{Im}(\rho) \not\subseteq K'$ and $\rho(f) = 0$. This is indeed equivalent to give formal power series $x(t), y(t) \in K'[\![t]\!]$ with $x(t) \neq 0$ or $y(t) \neq 0$ such that $f(x(t), y(t)) \equiv 0$.

Let $\rho : K[\![X, Y]\!] \to K'[\![t]\!]$ and $\sigma : K[\![X, Y]\!] \to K''[\![t]\!]$ be two parametrizations of the same rational branch. The parametrization σ is said to be *derivated* from ρ if there is a formal power series $\tau(u) \in K''[\![u]\!]$ with positive order and a continuous K-algebra morphism $\alpha : K'[\![t]\!] \to K''[\![u]\!]$ with $\alpha(t) = \tau(u)$ such that $\sigma = \alpha \circ \rho$. We write $\sigma \succ \rho$. The relation $\succ$ is a partial preorder. Two parametrizations σ and ρ are called *equivalent* if $\sigma \succ \rho$ and $\rho \succ \sigma$. Those parametrizations being minimal with respect to $\succ$ up to equivalence are called *primitive*. Equivalent primitive parametrizations are called *rational*. They always exist and are invariant under the action of the Galois group of the extension $\overline{K}/K$. Rational parametrizations are in one to one correspondence with rational branches of the curve (cf. Campillo and Castellanos [5]).

2.2 Divisors on Smooth Curves

Let us assume χ to be non-singular (or, equivalently, smooth, since $\mathbb{F}$ is perfect) for the remainder of the section. Let $\overline{\mathbb{F}}(\chi)$ be the field of rational functions of χ. Let P be a closed point on χ. The local ring $\mathcal{O}_{\chi,P}$ of χ at P with maximal ideal $\mathfrak{m}_{\chi,P}$ is therefore a discrete valuation ring with associated discrete valuation v_P. An element $f \in \mathcal{O}_{\chi,P}$ is said to vanish at P (or to have a zero at P) if $f \in \mathfrak{m}_{\chi,P}$. A rational function f such that $f \notin \mathcal{O}_{\chi,P}$ is said to have a pole at P. The order of the pole of f at P is given by $|v_P(f)|$.

A *rational divisor* D over $\mathbb{F}$ is a finite linear combination of closed points $P \in \chi$ with integer coefficients n_P, that is, $D = \sum_P n_P P$. If $n_P \geq 0$ for all P, then D is called *effective*, and we write $D \geq 0$; for two divisors D, D' we write $D \geq D'$ if the divisor $D - D'$ is effective. We define the *degree* of D as $\deg D := \sum_P n_P \deg P$, and the *support* of D as the set $\mathrm{supp}(D) = \{P \in \chi \text{ closed} \mid n_P \neq 0\}$. The set of rational divisors on $\mathbb{F}$ form an abelian group. Rational functions define *principal* divisors, namely divisors of the form

$$(f) := \sum_P \mathrm{ord}_P(f) P.$$

A rational divisor $D = \sum n_P P$ defines a $\mathbb{F}$-vector space

$$\mathcal{L}(D) = \left\{ f \in \mathbb{F}(\chi)^* \mid (f) \geq -D \right\} \cup \{0\},$$

that is, the set of rational functions f with poles only at the points P with $n_P \geq 0$ (and, furthermore, with the pole order of f at P must be less or equal than n_P), and if $n_P < 0$ such functions must have a zero at P of order greater or equal than n_P.

The dimension $\ell(D) := \dim_{\mathbb{F}} \mathcal{L}(D)$ is finite. Two elements $f, g \in \mathcal{L}(D)$ satisfy $(f) + D = (g) + D$ if and only if $f = \lambda g$, $\lambda \in \mathbb{F}$, i.e., if and only if $f = \lambda g$ for a constant $\lambda \in \mathbb{F}$. Therefore the set $|D|$ of effective divisors equivalent to D can be identified with the projective space $\mathbb{P}_{\mathcal{L}(D)}$ of dimension $\ell(D) - 1$. The set $|D|$ is called a *complete linear system* of D.

Let $\Omega_{\mathbb{F}}(\mathbb{F}(\chi))$ be the module of differentials on $\mathbb{F}(\chi)$. A differential form $\omega \in \Omega_{\mathbb{F}}(\mathbb{F}(\chi))$ defines a divisor $(\omega) := \sum_P \operatorname{ord}_P(\omega) P$, which is called canonical. A rational divisor D defines again a $\mathbb{F}$-vector space

$$\Omega(D) := \{\omega \in \Omega_{\mathbb{F}}(\mathbb{F}(\chi))^* \mid (\omega) \geq D\} \cup \{0\}$$

of finite dimension, denoted by $i(D)$. The interplay of the dimensions $\ell(D)$ and $i(D)$ is a big milestone in the theory of algebraic curves; first notice that the dimension $\ell(D)$ is bounded in the following sense:

Proposition 1 (Riemann's Inequality) *There exists a nonnegative integer g such that $\ell(D) \geq \deg D + 1 - g$ for any rational divisor D on χ.*

Definition 1 The smallest integer g satisfying the Riemann's inequality is called the *genus* of χ.

Riemann's inequality tells us that if D is a large divisor, $\mathcal{L}(D)$ is also large. But we can be a bit more precise by considering $i(D)$:

Theorem 1 (Riemann-Roch) *For D a rational divisor, $\ell(D) - i(D) = \deg D + 1 - g$.*

3 Brill-Noether Theory for Curves

This section contains a summary of the classic theory of adjunction for curves, started by Riemann [28] and developed by Brill and Noether in the nineteenth century [4].

Let P be a closed point on a plane curve χ as in the previous section. Let $\mathcal{C}_P$ be the annihilator of the $\mathcal{O}$-module $\overline{\mathcal{O}}/\mathcal{O}$, i.e.

$$\mathcal{C}_P = \mathcal{C}_{\overline{\mathcal{O}}/\mathcal{O}} = \{z \in \overline{\mathcal{O}} \mid z\overline{\mathcal{O}} \subseteq \mathcal{O}\}.$$

This set is the largest ideal in $\mathcal{O}$ which is also an ideal in $\overline{\mathcal{O}}$, and is called the *conductor ideal* of the extension $\overline{\mathcal{O}}/\mathcal{O}$. Since $\overline{\mathcal{O}}$ is a semilocal Dedekind domain with maximal ideals $\overline{\mathfrak{m}}_{Q_1}, \ldots, \overline{\mathfrak{m}}_{Q_d}$ (where Q_i denote the rational branches of χ at P), the conductor ideal has a unique factorisation

$$\mathcal{C}_P = \prod_{i=1}^{d} \overline{\mathfrak{m}}_{Q_i}^{d_{Q_i}}$$

as ideal in $\overline{O}$. The exponents d_{Q_i} can be easily computed by means of the Dedekind formula (see Zariski [30]): if $(x_i(t_i), y_i(t_i))$ is a rational parametrisation of Q_i one has

$$d_{Q_i} = \operatorname{ord}_{t_{Q_i}}\left(\frac{f_Y(X(t_{Q_i}), Y(t_{Q_i}))}{X'(t_{Q_i})}\right) = \operatorname{ord}_{t_{Q_i}}\left(\frac{f_X(X(t_{Q_i}), Y(t_{Q_i}))}{Y'(t_{Q_i})}\right). \quad (1)$$

Let $n : \widetilde{\chi} \to \chi$ be the normalisation morphism of χ. Notice that $\widetilde{\chi}$ is a nonsingular curve with $\mathbb{F}(\widetilde{\chi}) = \mathbb{F}(\chi)$. Let be $O = O_{\chi,P}$ and let $\overline{O}$ be the normalisation of O. If $Q \in n^{-1}(\{P\})$, then Q is nonsingular, hence $C_P \cdot O = \mathfrak{m}_Q^{d_Q}$ for a nonnegative integer d_Q. We define the effective divisor

$$\mathcal{A} := \sum_P \sum_{Q \in n^{-1}(\{P\})} d_Q \cdot Q$$

which is called the *adjunction divisor* of χ. Notice that $\mathcal{A}$ is a well-defined divisor on $\widetilde{\chi}$ (in fact, if P is nonsingular, there is only one $Q \in n^{-1}(\{P\})$ and in this case $d_Q = 0$). This implies in particular that the support of $\mathcal{A}$ consists of all rational branches of χ at singular points. Moreover, by setting $n_P := \dim_{\mathbb{F}} \overline{O}/C_P$ we have

$$n_P = \sum_{Q \in n^{-1}(\{P\})} d_Q$$

for every P on χ. Therefore $\deg \mathcal{A} = \sum_{P \in \chi} n_P$ (cf. Arbarello et al. [1, Appendix A]; also Tsfasman and Vlăduţ [29, 2.5.2]).

Recall that $\mathbb{F}$ is a perfect field. Let $F := F(X_0, X_1, X_2)$ be a homogeneous (absolutely irreducible) polynomial of degree d over $\mathbb{F}$ which defines the projective plane curve χ. Let $\mathcal{F}_d$ be the set of all homogeneous polynomials in three variables of degree d. Let $i : \chi \to \mathbb{P}^2_{\mathbb{F}}$ be the embedding of χ into the projective plane and $\mathbf{N} : \widetilde{\chi} \to \mathbb{P}^2_{\mathbb{F}}$ be the natural morphism given by $\mathbf{N} = i \circ n$. A rational divisor D on $\mathbb{P}^2_{\mathbb{F}}$ such that χ is not contained in $\operatorname{supp}(D)$ is called an *adjoint divisor* of χ if the pull-back divisor $\mathbf{N}^* D$ satisfies $\operatorname{supp}(\mathcal{A}) \subseteq \operatorname{supp}(\mathbf{N}^* D)$ for $\mathcal{A}$ the adjunction divisor of χ. We may consider the analogous notion for homogeneous polynomials. For $H \in \mathcal{F}_d$ with $F \nmid H$ we consider the pull-back $\mathbf{N}^* H$, which is actually the intersection divisor on $\widetilde{\chi}$ cut out by the plane curve defined by H on $\mathbb{P}^2_{\mathbb{F}}$, namely

$$\mathbf{N}^* H = \sum_{Q \in \widetilde{\chi}} r_Q \cdot Q, \quad (2)$$

with $r_Q = \operatorname{ord}_Q(h)$, with $h \in O_{\chi,n(Q)}$ a local equation of the curve defined by H at the point $n(Q)$. If H satisfies additionally that $\mathbf{N}^* D \geq \mathcal{A}$, then it will be called an *adjoint form* on χ, and the curve defined by H will be called an *adjoint curve* to χ. Notice that adjoint curves there always exist (consider e.g. the polars of the curve, cf. Brieskorn and Knörrer [2, p. 599]).

Let $d := \deg \chi$. The differentials globally defined at χ are in one to one correspondence with adjoint curves on $\widetilde{\chi}$ of degree $d-3$, as stated in the following:

Theorem 2 *Let $\mathcal{A}_n$ be the set of adjoints of degree n of the curve χ embedded in $\mathbb{P}^2_{\mathbb{F}}$, let $K_{\widetilde{\chi}}$ be a canonical divisor on $\widetilde{\chi}$ and set $d := \deg \chi$. For $n = d-3$ there is an $\mathbb{F}$-isomorphism of complete linear systems*

$$\begin{aligned} \mathcal{A}_n &\longrightarrow |K_{\widetilde{\chi}}| \\ D &\longmapsto N^*D - \mathcal{A}. \end{aligned}$$

The key idea is to realise that the map is injective since $n = d - 3 < d$; see Gorenstein [12, p. 433] or [29, 2.2.1] for further details.

In practice, we know a priori the equation of the plane curve χ (defined over a perfect field $\mathbb{F}$) given by the form $F \in \mathcal{F}_d$ and the data of a certain divisor $R = \sum_{Q'} r_{Q'} \cdot Q'$ (for finitely many points Q' on $\widetilde{\chi}$) which is effective and rational over $\mathbb{F}$, involving a finite number of rational branches Q of χ and their corresponding coefficients. Moreover, we are able to compute the adjunction divisor

$$\mathcal{A} = \sum_{Q} d_Q \cdot Q$$

of χ. Our purpose is the interpretation of the condition of being an adjoint form—called *adjoint condition*—given by (2) in terms of equations. More generally, we are interesting in finding some *adjoint* form $H \in \mathbb{F}[X_0, X_1, X_2]$ satisfying

$$\mathbf{N}^* H \geq \mathcal{A} + R. \tag{3}$$

This procedure is known as *computing adjoint forms with base conditions* (see [8], §4). Let us sketch briefly this process:

1. Choose a positive integer $\widetilde{n} \in \mathbb{N}$ in such a way that there exists an adjoint of degree $\widetilde{n}$ not being a multiple of F and satisfying (3). A bound for $\widetilde{n}$ can be found in Haché [14].
2. Take a general, arbitrary form $H \in \mathcal{F}_{\widetilde{n}}$ i.e., take a homogeneous polynomial in three variables of degree $\widetilde{n}$ leaving its coefficients as indeterminates (that is, $H(X_0, X_1, X_2) = \sum_{i+j+k=\widetilde{n}} \lambda_{i,j,k} X_0^i X_1^j X_2^k$).
3. Compute a rational primitive parametrization $\big(X(t), Y(t)\big)$ of χ at every branch involved in the support of the adjunction divisor $\mathcal{A}$ and the divisor R.
4. Get the support of the adjunction divisor $\mathcal{A}$ from the conductor ideal via the Dedekind formula (1).
5. Consider the coefficient r_Q of the divisor R at Q, and thus the local condition at Q imposed on H by (3) is given by

$$\operatorname{ord}_t h(X(t), Y(t)) \geq d_Q + r_Q, \tag{4}$$

with h the local affine equation of H at Q. The inequality (4) expresses a linear condition (given by a linear inequation) on the coefficients $\lambda_{i,j,k}$ of h.

6. The required linear equations are a consequence of the vanishing of those terms, and when Q takes all the possible values, i.e., all the possible branches of the singular points on χ and of the support of R, we get the linear equations globally imposed by the condition (3).

An easy reasoning reveals that the number of such adjoint conditions is equal to

$$\frac{1}{2}\deg \mathcal{A} + \deg R = \frac{1}{2}\sum_{P\in\chi} n_P + \deg R = \sum_{P\in\chi} \delta_P + \deg R. \tag{5}$$

Example 1 Let χ be the projective plane curve over the finite field $\mathbb{F}_2$ given by the equation $F(X,Y,Z) = X^3 - Y^2 Z$. The only singular point of χ is $P_1 = [0:0:1]$. Consider the point $P_2 = [0:1:0]$ and the effective divisor $R = 0P_1 + P_2$. The adjunction divisor of χ is $\mathcal{A} = 2P_1$. A local equation of χ with $P_1 = (0,0)$ is $f(x,y) = x^3 - y^2$. A parametrization of χ at P_1 is given by

$$X_1(t_1) = t_1^2, \quad Y_1(t_1) = t_1^3.$$

Take a form $H \in \mathcal{F}_{4-3=1}$, say $H(X,Y,Z) = aX + bY + cZ$. First we want to express the adjoint conditions in terms of the coefficients

$$\mathbf{N}^* H \geq \mathcal{A} + R = 2P_1 + P_2.$$

To this end we consider first a local equation for H at P_1, namely

$$h(x,y) = H(x,y,1) = ax + by + c.$$

Then $h(X_1(t_1), Y_1(t_1)) = h(t^2, t^3) = at_1^2 + bt_1^3 + c$. Since 2 is the coefficient for P_1 and $(X_1(t_1), Y_1(t_1))$ is a parametrization at P_1, the inequality wish to have

$$\operatorname{ord}_{t_1}(h(X_1(t_1), Y_1(t_1))) = \operatorname{ord}_{t_1}(bt_1^3 + at_1^2 + c) \geq 2$$

(continued)

Example 1 (continued)
holds if and only if $c = 0$. Thus $c = 0$ is one of the required *linear* adjoint conditions.

Next we consider a local equation for χ at P_2; this is $f'(x, z) = F(x, 1, z) = x^3 - z$, which admits a parametrization

$$X_2(t_2) = t_2, \quad Z_2(t_2) = t_2^3.$$

Consider the local equation for H at P_2

$$h'(x, z) = H(x, 1, z) = ax + b + cz.$$

The adjoint conditions imposed by $\mathbf{N}^*H \geq \mathcal{A} + R = 2P_1 + P_2$ at P_2 come from considering $h'(X_2(t_2), Z_2(t_2)) = h'(t_2, t_2^3) = at_2 + b + ct_2^3$ and they impose the conditions given by

$$\mathrm{ord}_{t_2}(h'(X_2(t_2), Z_2(t_2))) = \mathrm{ord}_{t_2}(ct_2^3 + at_2 + b) \geq 1.$$

This inequality holds whenever $b = 0$. Hence $b = 0$ is another *linear* equation to be considered in the set of adjoint conditions contained in $\mathbf{N}^*H \geq \mathcal{A} + R$. We have thus obtained two adjoint conditions, as we expected in view of (5), since $\frac{1}{2}\deg \mathcal{A} + \deg R = \frac{1}{2} \cdot 2 + 1 = 2$.

We finish this section with two remarkable results. Let χ be an absolutely irreducible projective plane curve over $\mathbb{F}$ and given by an equation $F(X_0, X_1, X_2) = 0$, where $F \in \mathcal{F}_d$. One application of the adjoint forms is the following result, due to Max Noether (of course, he did not state it in this way; our version may be found in Haché and Le Brigand [15], Theorem 4.2, and Le Brigand and Risler [20], §3.1):

Theorem 3 (Max Noether) *Let χ resp. χ' be curves as above given by homogeneous equations $F(X_0, X_1, X_2) = 0$ resp. $G(X_0, X_1, X_2) = 0$, and such that χ' does not contain χ as a component. Then, if we consider another such a curve given by $H(X_0, X_1, X_2) = 0$ with $N^*H \geq \mathcal{A} + N^*G$ (where $\mathcal{A}$ is the adjunction divisor on χ), then there exist forms A, B with coefficients in $\mathbb{F}$ such that $H = AF + BG$.*

Theorem 3 has great importance; in particular it allows us to prove the Brill-Noether theorem, which provides a basis for the vector spaces $\mathcal{L}(D)$. The reader is referred to [15, Theorem 4.4], for further details. A short remark about notation is needed: For any non effective divisor D we will write $D = D_+ - D_-$ with D_+ and D_- effective divisors of disjoint support.

Theorem 4 (Brill-Noether) *Let χ be an adjoint curve as above with normalization $\widetilde{\chi}$ and adjunction divisor $\mathcal{A}$. Let D be a divisor on $\widetilde{\chi}$ rational over $\mathbb{F}$, and consider* *a form $H_0 \in \mathcal{F}_{\tilde{n}}$ defined over $\mathbb{F}$, not divisible by F and satisfying $N^* H_0 \geq \mathcal{A} + D_+$. Then*

$$\mathcal{L}(D) = \left\{ \frac{h}{h_0} \mid H \in \mathcal{F}_{\tilde{n}}, F \nmid H \text{ and } N^* H + D \geq N^* H_0 \right\} \cup \{0\},$$

where h resp. h_0 denote the rational functions H resp. H_0 restricted on χ.

Remark 1 The choice $\tilde{n} > \max\left\{ d-1, \frac{d-3}{2} + \frac{\deg(\mathcal{A}+D_+)}{d} \right\}$ guarantees the existence of the form $H_0 \in \mathcal{F}_{\tilde{n}}$ in Theorem 4 (see Haché and Le Brigand [15] for details).

4 The Weierstraß Semigroup at Several Points

As above, let $\mathbb{F}$ be a perfect field of cardinality greater than or equal to r. Let χ be an absolutely irreducible projective algebraic plane curve defined over $\mathbb{F}$. Let $\underline{P}$ denote a set of r different points $P_1, \ldots, P_r$ on χ. Let $\tilde{\chi}$ be the normalization of χ. Consider a divisor $\underline{m}\underline{P} := m_1 P_1 + \cdots + m_r P_r$ for $m_i \in \mathbb{N}$, for all $i = 1, \ldots, r$. We will write $\underline{m} = (m_1, \ldots, m_r)$, $\varepsilon_i = (0, \ldots, 0, 1, 0, \ldots, 0)$, and $\underline{1} = (1, \ldots, 1)$.

Our purpose is to compute the dimensions of the so-called *Riemann-Roch quotients*,

$$0 \leq \dim_{\mathbb{F}} \frac{\mathcal{L}(\underline{m}\underline{P})}{\mathcal{L}((\underline{m} - \underline{1})\underline{P})} \leq r,$$

by choosing functions in $\mathcal{L}(\underline{m}\underline{P}) = \mathcal{L}(m_1 P_1 + \cdots + m_r P_r)$ but not in $\mathcal{L}((\underline{m} - \underline{1})\underline{P}) = \mathcal{L}((m_1 - 1)P_1 + \cdots + (m_r - 1)P_r)$, that is, achieving at the P_i poles of order m_i. We are going to restrict ourselves to the case $m_i \in \mathbb{N}$, for all $i = 1, \ldots, r$. These dimensions will be determined by the previous computations of the *Riemann-Roch quotients with respect to P_i*:

$$0 \leq \dim_{\mathbb{F}} \frac{\mathcal{L}(\underline{m}\underline{P})}{\mathcal{L}((\underline{m} - \varepsilon_i)\underline{P})} \leq 1,$$

where ε_i denotes the vector in $\mathbb{N}^r$ with 1 in the i-th position and 0 in the other ones. This section deals with the following issues:

- How to compute $\dim_{\mathbb{F}} \frac{\mathcal{L}(\underline{m}\underline{P})}{\mathcal{L}((\underline{m}-\varepsilon_i)\underline{P})}$ and an associated function belonging to this quotient vector space when its dimension is 1.
- How to compute $\dim_{\mathbb{F}} \frac{\mathcal{L}(\underline{m}\underline{P})}{\mathcal{L}((\underline{m}-\underline{1})\underline{P})}$ (deducing bounds).
- How to compute the Weierstraß semigroup at two points.

All the statements and proofs of this section can be found in [9, §2].

Definition 2 For $\underline{P} \in \chi$ we set

$$\Gamma_{\underline{P}} := \left\{ -(\mathrm{ord}_{P_1}(f), \ldots, \mathrm{ord}_{P_r}(f)) \mid f \in \mathbb{F}(\chi)^* \text{ regular at } \chi \setminus \underline{P} \right\}.$$

The set $\Gamma_{\underline{P}}$ is a subsemigroup of $(\mathbb{N}, +)$. Notice that, for $\underline{mP} = m_1P_1 + m_2P_2$, the fact that $f \in \mathcal{L}(\underline{mP})$ is equivalent to the fulfilling of the inequalities

$$(\star) \begin{cases} \mathrm{ord}_{P_1}(f) \geq -m_1 \\ \mathrm{ord}_{P_2}(f) \geq -m_2. \end{cases}$$

Definition 3 An element $\underline{m} \in \mathbb{N}^r$ is called a *non-gap* of $\Gamma_{\underline{P}}$ if and only if $\underline{m} \in \Gamma_{\underline{P}}$; otherwise $\underline{m}$ is called a *gap* of $\Gamma_{\underline{P}}$.

A very important characterization for the non-gaps is given by the following (see [10], p. 629):

Lemma 1 *If* $\underline{m} \in \mathbb{Z}^r$ *then one has that*

$$\underline{m} \in \Gamma_{\underline{P}} \;\; \textit{if and only if} \;\; \ell(\underline{mP}) = \ell((\underline{m} - \varepsilon_i)\underline{P}) + 1 \; \forall\, i = 1, \ldots, r.$$

For every $i = 1, \ldots, r$ and $\underline{m} = (m_1, \ldots, m_r) \in \mathbb{N}^r$, we set

$$\nabla_i(\underline{m}) := \left\{ (n_1, \ldots, n_r) \in \Gamma_{\underline{P}} \mid n_i = m_i \text{ and } n_j \leq m_j \; \forall j \neq i \right\}.$$

Then the two conditions proven to be equivalent in Lemma 1 are indeed also equivalent to $\nabla_i(\underline{m}) \neq 0$ for every $i \in \{1, \ldots, r\}$.

A gap $\underline{m}$ of $\Gamma_{\underline{P}}$ satisfying $\ell(\underline{mP}) = \ell((\underline{m} - \varepsilon_i)\underline{P})$ for all $i \in \{1, \ldots, r\}$ (or, equivalently, such that $\nabla_i(\underline{m}) = \emptyset$ for all $i \in \{1, \ldots, r\}$) is called *pure*. It is easily seen that, if $\underline{m}$ is a pure gap of $\Gamma_{\underline{P}}$, then m_i is a gap for Γ_{P_i} for every $i \in \{1, \ldots, r\}$. Furthermore, if $\underline{1} \in \Gamma_{\underline{P}}$, then $\Gamma_{\underline{P}}$ contains no pure gaps. The converse does not hold, as we will show in Example 5.

A fundamental result on Weierstraß semigroups is the following

Theorem 5 (Weierstraß Gap Theorem) *Let* $\widetilde{\chi}$ *be a curve of genus* $g \geq 1$. *Let* P *be a rational branch on* $\widetilde{\chi}$. *Then there are* g *gaps* $\gamma_1, \ldots, \gamma_g$ *of* Γ_P *such that*

$$1 = \gamma_1 < \cdots < \gamma_g \leq 2g - 1.$$

Under the above conditions we show:

Proposition 2 *Let* $\underline{m} = (m_1, \ldots, m_r) \in \mathbb{N}^r$. *If* $\underline{m}$ *is a gap of* $\Gamma_{\underline{P}}$, *then there exists a regular differential form* ω *on* $\widetilde{\chi}$ *with* $(\omega) \geq \underline{m} - \varepsilon_i$ *and a zero at* P_i *of order* $m_i - 1$ *for* some $i \in \{1, \ldots, r\}$.

Proof Set $\varphi(\underline{m}) := \ell(\underline{m}\underline{P}) - \ell((\underline{m} - \varepsilon_i)\underline{P})$ and $\psi(\underline{m}) := i((\underline{m} - \varepsilon_i)\underline{P}) - i(\underline{m}\underline{P})$; notice that $0 \le \varphi(\underline{m}), \psi(\underline{m}) \le 1$. By the Riemann-Roch theorem it is clear that

$$\ell(\underline{m}\underline{P}) - i(\underline{m}\underline{P}) = m_1 + m_2 + \cdots + m_r + 1 - g$$

$$\ell((\underline{m} - \varepsilon_i)\underline{P}) - i((\underline{m} - \varepsilon_i)\underline{P}) = m_1 + m_2 + \cdots + m_r - 1 + 1 - g.$$

By adding both equations we have that $\varphi(\underline{m}) + \psi(\underline{m}) = 1$ for every $i = 1, \ldots, r$. If $\underline{m}$ is a gap of $\Gamma_{\underline{P}}$, then $\varphi(\underline{m}) = 0$, hence $\psi(\underline{m}) = 1$, i.e. $\dim_{\mathbb{F}}\left(\frac{\Omega((\underline{m}-\varepsilon_i)\underline{P})}{\Omega(\underline{m}\underline{P})}\right) = 1$ and so there exists a regular differential form ω on $\widetilde{\chi}$ with $(\omega) \ge \underline{m} - \varepsilon_i$ and $\mathrm{ord}_{P_i}(\omega) = m_i - 1$ for *some* $i \in \{1, \ldots, r\}$.

Proposition 3 *Let χ be a plane curve of genus g, let $\underline{P}$ be a set of r closed points on χ and set $\underline{m} = (m_1, \ldots, m_r) \in \mathbb{N}^r$. If $\underline{m}$ is a gap of $\Gamma_{\underline{P}}$, then $m_1 + \cdots + m_r < 2g$.*

Proof Denote by $D_{2g,\underline{P}}$ a divisor with degree $2g$ and support $\underline{P}$, and by $D_{2g-1,\underline{P}}$ a divisor with degree $2g - 1$ and support $\underline{P}$. If $m_1 + \cdots + m_r \ge 2g - 1$ then $m_1 + \cdots + m_r \ge 0$ as a consequence of Riemann-Roch, and

$$\ell(D_{2g,\underline{P}}) = 2g + 1 - g = g + 1 \ne g = 2g - 1 + 1 - g = \ell(D_{2g-1,\underline{P}}),$$

which implies that $\underline{m} \in \Gamma_{\underline{P}}$. So, if $\underline{m} \notin \Gamma_{\underline{P}}$, then $m_1 + \cdots + m_r < 2g$.

4.1 Dimension of the Riemann-Roch Quotients with Respect to P_i and Associated Functions

We start with the computation of the dimension of the Riemann-Roch quotients associated to the points P_i.

Proposition 4 *Let $\underline{m} \in \mathbb{N}^r$ such that $\sum_{i=1}^r m_i < 2g$. For each $i \in \{1, \ldots, r\}$ we have:*

(a) $\dim_{\mathbb{F}}[\Omega((\underline{m} - \varepsilon_i)\underline{P}) \setminus \Omega(\underline{m}\underline{P})] = 1$ *if and only if there exists a homogeneous polynomial H_0 of degree $d - 3$ with $\mathbf{N}^* H_0 \ge \mathcal{A} + (\underline{m} - \varepsilon_i)\underline{P}$ such that P_i is not in the support of the effective divisor $\mathbf{N}^* H_0 - \mathcal{A} - (\underline{m} - \varepsilon_i)\underline{P}$.*

(b) There exists $\underline{m}' \ge \underline{m}$ with $\dim_{\mathbb{F}}[\Omega((\underline{m}' - \varepsilon_i)\underline{P}) \setminus \Omega(\underline{m}'\underline{P})] = 1$ *if and only if there exists a homogeneous polynomial H_0 of degree $d - 3$ such that $\mathbf{N}^* H_0 \ge \mathcal{A} + (\underline{m} - \varepsilon_i)\underline{P}$.*

Proof

(a) If $\dim_{\mathbb{F}}[\Omega((\underline{m}-\varepsilon_i)\underline{P}) \setminus \Omega(\underline{m}\underline{P})] = 1$, then this is equivalent to $\underline{m} \notin \Gamma_{\underline{P}}$ and also to the existence of an index i with $\ell(\underline{m}\underline{P}) = \ell((\underline{m}-\varepsilon_i)\underline{P})$, or, in other words, to the existence of an index i with $i((\underline{m}-\varepsilon_i)\underline{P}) = i(\underline{m}\underline{P}) + 1$; that is, there exists a homogeneous polynomial H_0 of degree $d-3$ such that $\mathbf{N}^* H_0 \ge \mathcal{A} + (\underline{m}-\varepsilon_i)\underline{P}$.

(b) If there is $\underline{m}' \geq \underline{m}$ with $\dim_{\mathbb{F}}[\Omega((\underline{m}' - \varepsilon_i)\underline{P}) \setminus \Omega(\underline{m}'\underline{P})] = 1$ then there exists an adjoint H_0 of degree $d - 3$ whose divisor is $\geq (\underline{m}' - \varepsilon_i)\underline{P}$ outside $\mathcal{A}$, i.e.,

$$\mathbf{N}^* H_0 - \mathcal{A} \geq (\underline{m}' - \varepsilon_i)\underline{P} \geq (\underline{m} - \varepsilon_i)\underline{P}.$$

Conversely, if there is H_0 of degree $d - 3$ with $\mathbf{N}^* H_0 \geq \mathcal{A} + (\underline{m} - \varepsilon_i)\underline{P}$ then there exists $\omega \neq 0$ differential form such that $(\omega) = \mathbf{N}^* H_0 - \mathcal{A} \geq (\underline{m} - \varepsilon_i)\underline{P}$. Assume that $\underline{m}' - \varepsilon_i$ are the orders of the zeros of ω at $\underline{P}$. Thus, $\underline{m}' - \varepsilon_i \geq \underline{m} - \varepsilon_i$, what implies $\underline{m}' \geq \underline{m}$ and $\omega \in \Omega((\underline{m} - \varepsilon_i)\underline{P}) \setminus \Omega(\underline{m}\underline{P})$.

The following corollary yields a way to relate the adjunction theory and the computation of the Weierstraß semigroup at several points:

Corollary 1 *Let $\underline{m} \in \mathbb{N}^r$ with $\sum_{i=1}^r m_i < 2g$. For a given form H of degree $d - 3$ and $i \in \{1, \ldots, r\}$ there exists a condition imposed by the inequality $N^* H \geq \mathcal{A} + \underline{m}\underline{P}$ at P_i which is independent of the conditions imposed by $N^* H \geq \mathcal{A} + (\underline{m} - \varepsilon_i)\underline{P}$ at P_i if and only if*

$$\dim_{\mathbb{F}} \frac{\Omega((\underline{m} - \varepsilon_i)\underline{P})}{\Omega(\underline{m}\underline{P})} = 1.$$

The second step is the computation of the rational functions associated to the nongaps of the Weierstraß semigroup at $\underline{P}$. Note that, if $\dim_{\mathbb{F}} \frac{\Omega((\underline{m}-\varepsilon_i)\underline{P})}{\Omega(\underline{m}\underline{P})} = 0$, then $\dim_{\mathbb{F}} \frac{\mathcal{L}(\underline{m}\underline{P})}{\mathcal{L}((\underline{m}-\varepsilon_i)\underline{P})} = 1$ and so there is a rational function $f_{i,\underline{m}} \in \frac{\mathcal{L}(\underline{m}\underline{P})}{\mathcal{L}((\underline{m}-\varepsilon_i)\underline{P})}$ with a pole of order m_i at P_i; this function can be computed by application of the Brill-Noether Theorem 4:

Algorithm 1 Preserving notation as above, we obtain a function $f_{i,\underline{m}} \in \frac{\mathcal{L}(\underline{m}\underline{P})}{\mathcal{L}((\underline{m}-\varepsilon_i)\underline{P})}$ with a pole of order m_i at P_i by following these steps:

1. Compute a homogeneous polynomial H_0 not divisible by F of large enough degree n in the sense of Remark 1 satisfying $\mathbf{N}^* H_0 \geq \mathcal{A} + \underline{m}\underline{P}$.
2. Calculate $R_{\underline{m}}$, which is the effective divisor such that $\mathbf{N}^* H_0 = \mathcal{A} + \underline{m}\underline{P} + R_{\underline{m}}$. Obviously $R_{\underline{m}-\varepsilon_i} = R_{\underline{m}} + P_i$.
3. Find a form $H_{\underline{m}}$ of degree n not divisible by F such that $\mathbf{N}^* H_{\underline{m}} \geq R_{\underline{m}}$ but not satisfying $\mathbf{N}^* H_{\underline{m}} \geq R_{\underline{m}-\varepsilon_i} = R_{\underline{m}} + P_i$.
4. Output: $f_{i,\underline{m}} = \frac{h_{\underline{m}}}{h_0}$, where $h_{\underline{m}}$ resp. h_0 is the restricted form on χ for $H_{\underline{m}}$ resp. for H_0.

Example 2 Let χ be the curve given by the equation $F(X, Y, Z) = X^3Z + X^4 + Y^3Z + YZ^3$ and consider the points $P_1 = [0 : 1 : 1]$ and $P_2 = [0 : 1 : 0]$ and $\underline{m} = (1, 2)$. We want to compute $\dim_{\mathbb{F}} \frac{\mathcal{L}(mP)}{\mathcal{L}((\underline{m}-\varepsilon_1)\underline{P})}$ and $\dim_{\mathbb{F}} \frac{\mathcal{L}(mP)}{\mathcal{L}(\underline{m}-\varepsilon_2)\underline{P}}$.

A local parametrization of F at P_1 is given by

$$X_1(t_1) = t_1$$
$$Y_1(t_1) = t_1^3 + t_1^4 + t_1^9 + t_1^{10} + t_1^{11} + t_1^{12} + \cdots$$

with local equation $f_1(x, y) = y^2 + y^3 + x^3 + x^4$. Analogously at P_2 we have

$$X_2(t_2) = t_2$$
$$Z_2(t_2) = t_2^4 + t_2^7 + t_2^{10} + t_2^{12} + t_2^{13} + t_2^{16} + \cdots$$

with local equation $f_2(x, z) = z + z^3 + x^3z + x^4$.

We compute the adjunction divisor, which is $\mathcal{A} = 2P_1$. Then we search for a form H of degree $d - 3 = 4 - 3 = 1$, that is, a linear form $H(X, Y, Z) = aX + bY + cZ$. The form H admits the equation $h_1(x, y) = H(X, Y - 1, 1) = ax + by + b + c$ at P_1, and H admits the equation $h_2(x, z) = H(X, 1, Z) = ax + b + cz$ at P_2; thus

$$\begin{aligned} h_1(X_1(t_1), Y_1(t_1)) &= at_1 + b(t_1^3 + t_1^4 + t_1^9 + \cdots) + b + c \\ &= (b + c) + at_1 + bt_1^3 + bt_1^4 + bt_1^9 + \cdots \end{aligned}$$

$$h_2(X_2(t_2), Z_2(t_2)) = b + at_2 + ct_2^4 + ct_2^7 + ct_2^{10} + \cdots$$

In order to compute $\dim_{\mathbb{F}} \frac{\mathcal{L}(mP)}{\mathcal{L}((\underline{m}-\varepsilon_1)\underline{P})}$ we impose the adjunction conditions at P_1, namely

$$\mathbf{N}^*H \geq \mathcal{A} + (m_1 - 1)P_1 = 2P_1 \text{ and } \mathbf{N}^*H \geq \mathcal{A} + m_1P_1 = 3P_1,$$

i.e.

$$\begin{cases} \operatorname{ord}_{t_1}(h_1(X_1(t_1), Y_1(t_1))) \geq 2 \Longrightarrow b + c = a = 0 \\ \operatorname{ord}_{t_1}(h_1(X_1(t_1), Y_1(t_1))) \geq 3 \Longrightarrow b + c = a = 0 \end{cases}$$

We see that the second equation does not add any independent condition to the first one; by Corollary 1, this means that $\dim_{\mathbb{F}} \frac{\mathcal{L}(mP)}{\mathcal{L}((\underline{m}-\varepsilon_1)\underline{P})} = 1$.

(continued)

Example 2 (continued)

We repeat this process in order to compute $\dim_{\mathbb{F}} \frac{\mathcal{L}(\underline{m}P)}{\mathcal{L}((\underline{m}-\varepsilon_2)\underline{P})}$, but imposing the adjunction conditions at P_2, which are

$$\mathbf{N}^*H \geq (m_2 - 1)P_2 = P_2 \text{ and } \mathbf{N}^*H \geq m_2P_2 = 2P_2,$$

that is,

$$\begin{cases} \operatorname{ord}_{t_2}(h_2(X_2(t_2), Z_2(t_2))) \geq 1 \Rightarrow b = 0 \\ \operatorname{ord}_{t_2}(h_2(X_2(t_2), Z_2(t_2))) \geq 2 \Rightarrow a = 0 = b. \end{cases}$$

Notice that, in this case, the adjunction divisor does not appear in the inequalities since P_2 does not belong to its support. The second system adds one independent condition to the first one, which means $\dim_{\mathbb{F}} \frac{\mathcal{L}(\underline{m}P)}{\mathcal{L}((\underline{m}-\varepsilon_2)\underline{P})} = 0$ again by Corollary 1. □

Example 3 Consider Example 2 but with $\underline{m} = (4, 6)$. As $m_1 + m_2 = 4 + 6 = 10 > 2g$, we deduce immediately that $\underline{m} \in \Gamma_{\underline{P}}$, i.e., that $\dim_{\mathbb{F}} \frac{\mathcal{L}(\underline{m}P)}{\mathcal{L}((\underline{m}-\varepsilon_i)\underline{P})} = 1$ for $i = 1, 2$. So we will look for the corresponding functions $f_{i,\underline{m}}$ with poles at P_i of order m_i for $i = 1, 2$.

First of all, we need to find an $\widetilde{n} \in \mathbb{N}$ such that $\widetilde{n} > \max\left\{3, \frac{2}{4} + \frac{12}{4}\right\} = \max\left\{3, \frac{14}{4}\right\}$. Consider for instance $\widetilde{n} = 5$.

Then we look for a form H_0 of degree $\widetilde{n} = 5$ such that $\mathbf{N}^*H_0 \geq \mathcal{A} + \underline{mP}$. In this case $\mathbf{N}^*H_0 \geq 4P_1 + 6P_2 + 2P_3$, since $\mathcal{A} = 2P_3$, with $P_3 = [0 : 0 : 1]$. After some computations we find $H_0 = X^4Z$.

In order to compute $\mathbf{N}^*H_0$, we have to calculate $\mathbf{N}^*(X)$, $\mathbf{N}^*(Y)$ and $\mathbf{N}^*(Z)$. The intersection points between $\{F = 0\}$ and $\{X = 0\}$ are $P_1 = [0 : 1 : 1]$, $P_2 = [0 : 1 : 0]$ and $P_3 = [0 : 0 : 1]$ with multiplicities 1, 1 and 2 respectively. So $\mathbf{N}^*(X) = P_1 + P_2 + 2P_3$. The intersection points between $\{F = 0\}$ and $\{Y = 0\}$ are $P_3 = [0 : 0 : 1]$ and $P_4 = [1 : 0 : 1]$ so that $\mathbf{N}^*(Y) = 3P_3 + P_4$; and the only point lying in the intersection between $\{F = 0\}$ and $\{Z = 0\}$ is $P_2 = [0 : 1 : 0]$ with multiplicity 4, therefore $\mathbf{N}^*(Z) = 4P_2$.

Hence $\mathbf{N}^*H_0 = 4\mathbf{N}^*(X) + \mathbf{N}^*(Z) = 4P_1 + 8P_2 + 8P_3$. The residue divisor $R_{\underline{m}}$ is equal to $\mathbf{N}^*H_0 - \mathcal{A} - \underline{mP} = 2P_2 + 6P_3$. Following Algorithm 1, we have to find a form H_{ε_1} such that $\mathbf{N}^*H_{\varepsilon_1} \geq R_{\underline{m}}$ but $\mathbf{N}^*H_{\varepsilon_1} \ngeq R_{\underline{m}} + P_1$. For

(continued)

Example 3 (continued)
instance we take $H_{\varepsilon_1} = Y^2Z^3$, since

$$\mathbf{N}^*H_{\varepsilon_1} = 12P_2 + 6P_3 + 2P_4 \geq 2P_2 + 6P_3$$
$$\mathbf{N}^*H_{\varepsilon_1} \ngeq P_1 + 2P_2 + 6P_3.$$

So $f_{1,\underline{m}} = \frac{Y^2Z^3}{X^4Z} = \frac{Y^2Z^2}{X^4} \in \frac{\mathcal{L}(m\underline{P})}{\mathcal{L}((\underline{m}-\varepsilon_1)\underline{P})}$.

On the other hand, we want to find a form H_{ε_2} such that $\mathbf{N}^*H_{\varepsilon_2} \geq R_{\underline{m}}$ but $\mathbf{N}^*H_{\varepsilon_2} \ngeq R_{\underline{m}} + P_2$. We can take $H_{\varepsilon_2} = X^2Y^3$, since

$$\mathbf{N}^*H_{\varepsilon_2} = 2P_1 + 2P_2 + 13P_3 + 3P_4 \geq 2P_2 + 6P_3$$
$$\mathbf{N}^*H_{\varepsilon_2} \ngeq 3P_2 + 6P_3.$$

Thus $f_{2,\underline{m}} = \frac{X^2Y^3}{X^4Z} = \frac{Y^3}{X^2Z} \in \frac{\mathcal{L}(m\underline{P})}{\mathcal{L}((\underline{m}-\varepsilon_2)\underline{P})}$. □

Algorithm 2 There is an alternative and more efficient way to compute the functions $f_{i,\underline{m}}$:

1. Take a basis of $\mathcal{L}(\underline{m}\underline{P})$, say $\{h_1, \ldots, h_s\}$.
2. Compute the pole orders at P_i, $\{-\mathrm{ord}_{P_i}(h_1), \ldots, -\mathrm{ord}_{P_i}(h_s)\}$.
3. Order these pole orders increasing, in such a way that $-\mathrm{ord}_{P_i}(h_s) = m_i$. We can assume this, as otherwise, if $-\mathrm{ord}_{P_i}(h_s) = k_i > m_i$ we can replace m_i by k_i, since $\mathcal{L}(m_1P_1 + \cdots + m_iP_i + \cdots + m_rP_r) = \mathcal{L}(m_1P_1 + \cdots + k_iP_i + \cdots + m_rP_r)$.
4. The function h_s has pole order m_i at P_i, but other functions could also have the same property. So, for any h_j satisfying $-\mathrm{ord}_{P_i}(h_j) = m_i$, there exists $\lambda_j \neq 0$ in $\mathbb{F}$ such that $h_j = \lambda_j h_s$, that is, $-\mathrm{ord}_{P_i}(h_j - \lambda_j h_s) < m_i$. So we change h_j by $g_j := h_j - \lambda_j h_s$, and $g_k := h_k$ for $k \neq j$.
5. We obtain a set of functions $\{g_1, \ldots, g_s\}$ where $g_s = f_{i,\underline{m}}$, and $g_1, \ldots, g_{s-1}$ build a basis of the vector space $\mathcal{L}((\underline{m} - \varepsilon_i)\underline{P})$.

Example 4 We present a worked example in SINGULAR for computing functions as above. First we import the library `brnoeth.lib` [11] and another virtual one—say `several.lib`—in which we have programmed

(continued)

Example 4 (continued)
the procedure `ordRF` that computes the pole orders of a rational function:

```
> LIB "brnoeth.lib";
> LIB "several.lib";
> int plevel=printlevel;
> printlevel=-1;
```

We define both the ring and the curve:

```
> ring s=2,(x,y),lp;
> list C=Adj_div(x3y+y3+x);
 ==>The genus of the curve is 3
```

The list of computed places is

```
> C=NSplaces(1,C);
> C[3];
 -->[1]:
 -->   1,1
 -->[2]:
 -->   1,2
 -->[3]:
 -->   1,3
```

The base point of the first place of degree 1 is, in homogeneous coordinates:

```
> def SS=C[5][1][1];
> setring SS;
> POINTS[1];
 -->[1]:
 -->    0
 -->[2]:
 -->    1
 -->[3]:
 -->    0
> setring s;
```

We define the divisor `G=4C[3][1]+4C[3][3]`:

```
> intvec G=4,0,4;
```

(Alternatively we can program an auxiliary procedure `zeroes` which has as output the coefficients of those points in the support of G). We need to change to the suitable ring; then, basis `LG` of $\mathcal{L}(\underline{mP})$ is supplied by the Brill-Noether algorithm:

```
> def R=C[1][2];
> setring R;
> list LG=BrillNoether(G,C);
 -->Vector basis successfully computed
> int lG=size(LG);
```

(continued)

Example 4 (continued)
The pole orders for the rational functions in LG are

```
> int j;
> intvec h;
> for (j=1;j<=lG;j=j+1){
. h[j]=ordRF(LG[j],SS,1)[1]; . }
> h;
-->0,-1,2,-2,-3,-4
```

And the desired rational function is

```
> LG[lG];
-->_[1]=xyz2+y4
-->_[2]=x4
> printlevel=plevel;
```

4.2 Computing the Weierstraß Semigroup at Two Points

In the above notation, let $\Gamma_{\underline{P}}$ resp. Γ_{P_i} be the Weierstraß semigroup at $\underline{P}$ resp. at the point P_i for $i = 1, \ldots, r$. Write $\mathbb{N}^* := \mathbb{N} \setminus \{0\}$ and $\underline{m_i} := \underline{m} - m_i\varepsilon_i$.

Proposition 5 *For* $\underline{m} \in \mathbb{N}^r$, $i \in \{1, \ldots, r\}$ *and* $\underline{m_i} \in \mathbb{N}^r \setminus \Gamma_{\underline{P}}$, *let be*

$$m := \min\Big\{n \in \mathbb{N}^* \mid \underline{m_i} + n\varepsilon_i \in \Gamma_{\underline{P}}\Big\},$$

then any vector $\underline{n} = (n_1, \ldots, n_r) \in \mathbb{N}^r$ *belongs to* $\mathbb{N}^r \setminus \Gamma_{\underline{P}}$ *whenever* $n_i = m$, *and* $n_j = m_j = 0$ *or* $n_j < m_j$ *for* $j \neq i$. *In particular, m is a gap at* P_i.

Define the usual partial order $\preceq$ over $\mathbb{N}^r$, i.e.

$$(m_1, \ldots, m_r) \preceq (n_1, \ldots, n_r) \iff m_i \leq n_i \text{ for all } i = 1, \ldots, r.$$

Proposition 6 *For* $i \in \{1, \ldots, r\}$, *let* $\underline{m} = (m_1, \ldots, m_r)$ *be a minimal element in*

$$\Big\{(n_1, \ldots, n_r) \in \Gamma_{\underline{P}} \mid n_i = m_i\Big\}$$

with respect to the partial order $\preceq$. *Assume that* $n_i > 0$ *and the existence of an index* $j \in \{1, \ldots, r\}$, $j \neq i$ *with* $m_j > 0$, *then*

- $\underline{m_i} \in \mathbb{N}^r \setminus \Gamma_{\underline{P}}$;
- $m_i = \min\{n \in \mathbb{N}^* \mid \underline{m_i} + n\varepsilon_i \in \Gamma_{\underline{P}}\}$; *in particular,* m_i *is a gap at* P_i.

Propositions 5 and 6 determine a surjective map

$$\varphi_i : \left\{\underline{m}_i \in \mathbb{N}^r \mid \underline{m}_i \in \mathbb{N}^r \setminus \Gamma_{\underline{P}}\right\} \longrightarrow \mathbb{N} \setminus \Gamma_{P_i}$$
$$\underline{m}_i \mapsto \min\left\{m \in \mathbb{N}^* \mid \underline{m}_i + m\varepsilon_i \in \Gamma_{\underline{P}}\right\}.$$

For $r = 2$ this is in fact a bijection between the set of gaps at P_1 and the set of gaps at P_2:

$$m_1 \in \mathbb{N} \setminus \Gamma_{P_1} \Leftrightarrow (m_1, 0) \in \mathbb{N}^2 \setminus \Gamma_{\underline{P}} \mapsto \beta_{m_1} := \varphi_2((m_1, 0)) \in \mathbb{N}\setminus\Gamma_{P_2}.$$

Furthermore, $m_1 = \min\left\{n \in \mathbb{N}^* \mid (n, \beta_{m_1}) \in \Gamma_{\underline{P}}\right\}$. For further details the reader is referred to Homma and Kim [18] and Kim [19], and also to Matthews [24].

For the case of two points ($r = 2$), we would like to point out some useful facts from a computational point of view:

- All the gaps at P_1 and at P_2 are also gaps at P_1, P_2.
- By Corollary 6, for any gap m_1 at P_1, one has that (m_1, β_{m_1}) are gaps at P_1, P_2 for $\beta_{m_1} = 0, 1, \ldots, \ell_{m_1}$, with $0 \leq \ell_{m_1} \leq 2g - 1$, with g the genus of the curve and ℓ_{m_1} such that $\ell_{m_1} + 1$ is a gap at P_2. Then $(m_1, \ell_{m_1} + 1) \in \Gamma_{\underline{P}}$, and we call it the *minimal (non-gap) at* m_1. We will refer to the set of the minimal non-gaps at every gap at P_1 (they will be g, since the number of gaps at P_1 is precisely g) as the set of *minimal non-gaps at* P_1.
- The gaps obtained of that form, i.e., the set

$$\left\{(m_1, \beta_{m_1}) \in \mathbb{N}^2\setminus\Gamma_{\underline{P}} \mid m_1 \in \mathbb{N}\setminus\Gamma_{P_1} \text{ and } \beta_{m_1} = 0, 1, \ldots, \ell_{m_1} \text{ with } \ell_{m_1}+1 \in \mathbb{N}\setminus\Gamma_{P_2}\right\}$$

  will be called the *set of gaps with respect to* P_1.
- Similarly, for any gap m_2 at P_2, one has that (α_{m_2}, m_2) are gaps at P_1, P_2 for $\alpha_{m_2} = 0, 1, \ldots, \ell_{m_2}$, until some $0 \leq \ell_{m_2} \leq 2g - 1$ such that ℓ_{m_2} satisfies that $\ell_{m_2} + 1$ is a gap at P_1. Then $(\ell_{m_2} + 1, m_2) \in \Gamma_{\underline{P}}$, which we will call the *minimal (non-gap) at* m_2. The set of the minimal non-gaps for every gap at P_2 will be called the set of *minimal non-gaps at* P_2. The cardinality of such a set is g, since g is the number of gaps at P_2.
- The set of gaps

$$\left\{(\alpha_{m_2}, m_2) \in \mathbb{N}^2\setminus\Gamma_{\underline{P}} \mid m_2 \in \mathbb{N}\setminus\Gamma_{P_2} \text{ and } \alpha_{m_2} = 0, 1, \ldots, \ell_{m_2} \text{ with } \ell_{m_2}+1 \in \mathbb{N}\setminus\Gamma_{P_1}\right\}$$

  is called the *set of gaps with respect to* P_2.
- The intersection between the set of gaps with respect to P_1 and with respect to P_2 is not necessarily empty. In fact, the gaps in the intersection are just the *pure gaps* at P_1, P_2.

The minimal non-gaps at P_1 and P_2 provide enough information in order to deduce the Weiestraß semigroup at P_1, P_2. Recall that we have already described algorithms to compute the dimension (and associated functions, when is possible) of the Riemann-Roch quotients $\frac{\mathcal{L}(mP)}{\mathcal{L}((\underline{m}-\varepsilon_i)P)}$ for given $\underline{m}$, and $i \in \{1, 2\}$ and two rational points P_1, P_2 on an absolutely irreducible projective algebraic plane curve χ (see Algorithms 1 and 2). An procedure to compute the set of minimal non-gaps at P_i, for $i = 1, 2$ is given by the following:

Algorithm 3 Write dim($\underline{m}$, P, C, i) for the procedure calculating the dimension of the quotient vector space $\frac{\mathcal{L}(mP)}{\mathcal{L}((\underline{m}-\varepsilon_i)\underline{P})}$:

INPUT: Points P_1, P_2, an integer $i \in \{1, 2\}$ and a curve χ.

OUTPUT: The set of minimal non-gaps at P_i.

- Let L be a empty list and g be the genus of χ;
- Let W_1 and W_2 be the lists of gaps of χ at P_1 and P_2, respectively;
- FOR $k = 1, \ldots, g$; $k = k + 1$;
  - IF $i = 1$ THEN
    - j=size of W_2;
    - WHILE $\Big(\dim((W_1[k], W_2[j]), P, \chi, i) = 1$ AND $\dim((W_1[k], W_2[j] - 1), P, \chi, i) = 1)$ OR $j = 0\Big)$ DO
      - $j = j - 1$;
    - $L = L \cup \{(W_1[k], W_2[j])\}$;
    - $W_2 = W_2 \setminus \{j\}$;
  - ELSE
    - j=size of W_1;
    - WHILE $\Big(\dim((W_1[j], W_2[k]), P, \chi, i) = 1$ AND $\dim((W_1[j] - 1, W_2[k]), P, \chi, i) = 1)$ OR $j = 0\Big)$ DO
      - $j = j - 1$;
    - $L = L \cup \{(W_1[j], W_2[k])\}$;
    - $W_1 = W_1 \setminus \{j\}$;
- RETURN(L);

Example 5 Let χ be the curve over $\mathbb{F}_2$ given by $F(X, Y, Z) = X^3Z + X^4 + Y^3Z + YZ^3$, and consider the points $P_1 = [0 : 1 : 1]$ and $P_2 = [0 : 1 : 0]$ on χ; then

$$\mathbb{N}^2 \setminus \Gamma_{\underline{P}} = \Big\{(0, 1), (0, 2), (1, 0), (1, 2), (2, 0), (2, 1)\Big\},$$

as shown in the figure below (where the black points indicate the elements of $\Gamma_{\underline{P}}$, and the white points indicate the gaps at P_1, P_2).

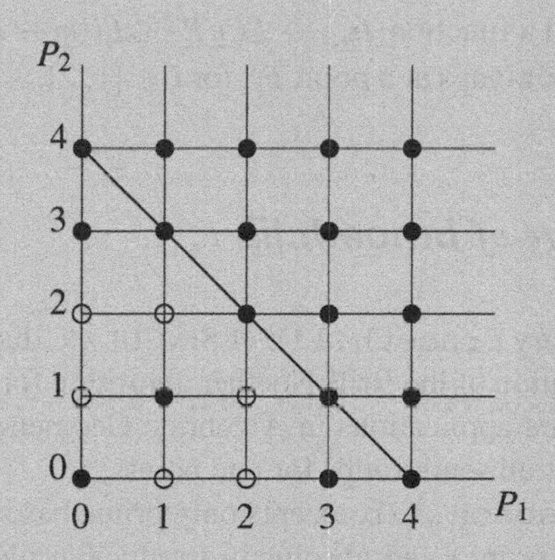

As an illustration of Corollary 6, let be $i = 1$, $\underline{m} = (m_1, m_2) = (2, 2) \in \Gamma_{\underline{P}}$ and the set $\Big\{(n_1, n_2) \in \Gamma_{\underline{P}} \mid n_1 = m_1\Big\} = \Big\{(2, n) \text{ for } n \geq 2\Big\}$. The smallest element in this set is $(2, 2)$, and

$$\underline{m}_i = \underline{m} - m_1\varepsilon_1 = (2, 2) - 2(1, 0) = (0, 2)$$

is a gap at P_1, P_2. Therefore $\min\Big\{n \in \mathbb{N}^* \mid (n, 2) \in \Gamma_{\underline{P}}\Big\} = 2 = m_1$, and $m_1 = 2$ is actually a gap at P_1.

In this example we can also see the bijection between the gaps at P_1 and the gaps at P_2: if we take $n_1 = 1$ as a gap at P_1, then $(1, 0)$ is a gap at P_1, P_2 and

$$\varphi_2((1, 0)) = \min\Big\{n \in \mathbb{N}^* \mid (1, 0)+(0, n) \in \Gamma_{\underline{P}}\Big\} = \min\Big\{n \in \mathbb{N}^* \mid (1, n) \in \Gamma_{\underline{P}}\Big\} = 1,$$

with $1 \notin \Gamma_{P_2}$. Moreover, $n_1 = 1 = \min\Big\{n \in \mathbb{N}^* \mid (n, \varphi_2((1, 0))) \in \Gamma_{\underline{P}}\Big\}$.

Consider on the other hand $p_1 = 2$ as a gap at P_1, then $\varphi_2((2, 0)) = 2$, which is a gap at P_2. Indeed $p_1 = 2 = \min\Big\{n \in \mathbb{N}^* \mid (n, \varphi_2((2, 0)) \in \Gamma_{\underline{P}}\Big\}$. The gaps at P_2 behave analogously.

5 Computational Aspects Using SINGULAR

We are interested in explaining the most important procedures implemented in SINGULAR and to give examples to show how to work with them.

More precisely, in Sect. 5.1 we give some hints of use of the library `brnoeth.lib`, since our procedures are based on most of the algorithms contained in it. In Sect. 5.2 we present the procedures which generalize the computation of the Weierstraß semigroup to the case of two points; they yield the computation of

- $\dim_{\mathbb{F}} \frac{\mathcal{L}(m\underline{P})}{\mathcal{L}((\underline{m}-\varepsilon_i)\underline{P})}$ and a function $f_{\underline{m},i} \in \mathcal{L}(m\underline{P}) \setminus \mathcal{L}((\underline{m}-\varepsilon_i)\underline{P})$ if possible;
- the set of minimal non-gaps at a point P_i, for $i \in \{1, 2\}$.

5.1 Hints of Usage of brnoeth.lib

The purpose of the library `brnoeth.lib` of SINGULAR, due to Farrán and Lossen [11], is the implementation of the Brill-Noether algorithm for solving the Riemann-Roch problem and some applications in Algebraic Geometry codes, involving the computation of Weierstraß semigroups for one point.

A first warning: `brnoeth.lib` accepts only prime base fields (i.e. finite fields of the form $\mathbb{F}_p$ for p prime) and absolutely irreducible planes curves as inputs, although the latter is not checked.

Curves are usually defined by means of polynomials in two variables, that is, by its local equation. It is possible to compute most of the data concerning the curve with the aid of the procedure `Adj_div`. We define the procedure (previously we must have defined the ring, the polynomial f and have charged the library `brnoeth.lib`):

```
> list C=Adj_div(f);
```

The output consists of a list of lists as follows:

- The first list contains the affine and the local ring.
- The second list has the degree and the genus of the curve.
- Each entry of the third list corresponds to one closed place,that is, a place and all its conjugates, which is represented by two integer, the first one the degree of the point and the second one indexing the conjugate point.
- The fourth one has the conductor of the curve.
- The fifth list consists of a list of lists, the first one, namely `C[5][d][1]` being a (local) ring over an extension of degree d and the second one (`C[5][d][2]`) containing the degrees of base points of places of degree d.

Furthermore, inside the local ring `C[5][d][1]` we can find the following lists:

- `list POINTS`: base points of the places of degree d.
- `list LOC_EQS`: local equations of the curve at the base points.
- `list BRANCHES`: Hamburger-Noether expressions of the places.
- `list PARAMETRIZATIONS`: local parametrizations of the places.

Now we explain how we can deal with the most common objects in `brnoeth.lib`. **Affine points** P are represented by a standard basis of a prime ideal, and a vector of integers containing the position of the places above P in the list supplied by `C[3]`; if the point lies at the infinity, the ideal is replaced by an homogeneous irreducible polynomial in two variables. A **place** is represented by the four lists mentioned above: a base point (`list POINTS` of homogeneous coordinates); a local equation (`list LOC_EQS`) for the curve at the base point; a Hamburger-Noether expansion of the corresponding branch (`list BRANCHES`); and a local parametrization (`list PARAMETRIZATIONS`) of that branch. A **divisor** is represented by a vector of integers, where the integer at the position i is the coefficient of the i-th place in the divisor. **Rational functions** are represented by ideals with two homogeneous generators, the first one being the numerator of the rational function, and the second one being the denominator.

Furthermore, we can compute a complete list containing all the non-singular affine (closed) places with fixed degree d just by invoking the procedure `NSplaces` in this way:

```
> C=NSplaces(1..d,C);
```

The procedure `Weierstrass`, which computes the non-gaps of the Weierstraß semigroup at one point and the associated functions with poles, is closer to our aim. It has three inputs, namely

- an *integer* indicating the rational place in which we compute the semigroup;
- an *integer* indicating how many non-gaps we want to calculate;
- the curve given in form of a *list* `C=Adj_div(f)` for some polynomial f representing the local equation of the curve at the point given in the first entry.

This procedure needs to be called from the ring `C[1][2]`. Moreover, the places must be necessarily *rational*.

5.2 Procedures Generalizing to Several Points

We present now a main procedure to compute the dimension of the so-called Riemann-Roch vector spaces of the form $\mathcal{L}(\underline{m}\underline{P}) \setminus \mathcal{L}((\underline{m}-\varepsilon_i)\underline{P})$. If this dimension equals 1, then the procedure is also able to compute a rational function belonging to the space.

The technique developed here is not the use of the adjunction theory directly, as we have developed theoretically in Sect. 4.1 (cf. Algorithm 1), because of its high computational cost, but we use a slight modification of Algorithm 2: we order the poles in a vector from the biggest one to the smallest one (in absolute value) and then we take the first entry in this vector.

```
proc RRquot (intvec m, list P, list CURVE, int chart)
"USAGE:RRquot( m, P, CURVE, ch );  m,P intvecs, CURVE a list and
ch an integer. RETURN:  an integer 0 (dimension of
L(m)\L(m-e_i)), or a list with three entries:
  @format
  RRquot[1] ideal (the associated rational function)
  RRquot[2] integer (the order of the rational function)
  RRquot[3] integer (dimension of L(m)\L(m-e_i))
  @end format
NOTE:    The procedure must be called from the ring CURVE[1][2],
         where CURVE is the output of the procedure
         @code{NSplaces}.
@*       P represents the coordinates of the place CURVE[3][P].
@*       Rational functions are represented by
         numerator/denominator
         in form of ideals with two homogeneous generators.
WARNING:  The place must be rational, i.e., necessarily
CURVE[3][P][1]=1. @* SEE ALSO: Adj_div, NSplaces, BrillNoether
EXAMPLE:  example RRquot; shows an example " {
  // computes a basis for the quotient of Riemann-Roch
  //   vector spaces L(m)\L(m-e_i)
  // where m=m_1 P_1 + ... + m_r P_r and
  //   m-e_i=m_1P_1+...+(m_i-1)P_i+...+m_r P_r,
  // a basis for the vector space L(m-e_i) and the orders of such
  //   functions, via Brill-Noether
  // returns 2 lists : the first consists of all the pole orders
  //   in increasing order and the second consists of the
  //   corresponding rational functions, where the last one is
  // the basis for the quotient vector space
  // P_1,...,P_r must be RATIONAL points on the curve.

  def BS=basering;
  def SS=CURVE[5][1][1];
  intvec posinP;
  int i,dimen;
  setring SS;
  //identify the points P in the list CURVE[3]
  int nPOINTS=size(POINTS);
  for(i=1;i<=size(m);i=i+1)
   {
     posinP[i]=isPinlist(P[i],POINTS);
   }
//in case the point P is not in the list CURVE[3]
  if (posinP==0)
   {
    ERROR("The given place is not a rational place on the curve");
   }
```

```
  setring BS;
  //define the divisor containing m in the right way
  intvec D=zeroes(m,posinP,nPOINTS);
  list Places=CURVE[3];
  intvec pl=Places[posinP[chart]];
  int dP=pl[1];
  int nP=pl[2];

  //check that the points are rational
  if (dP<>1)
  {
    ERROR("The given place is not defined over the prime field");
  }
  int auxint=0;
  ideal funcion;
  funcion[1]=1;
  funcion[2]=1;

  // Brill-Noether algorithm
  list LmP=BrillNoether(D,CURVE);
  int lmP=size(LmP);
  if (lmP==1)
  {
    dimen=0;
    return(dimen);
  }
  list ordLmP=list();
  list sortpol=list();
    for (i=1;i<=lmP;i=i+1)
    {
      ordLmP[i]=orderRF(LmP[i],SS,nP)[1];
    }
    ordLmP=extsort(ordLmP);
    if (D[posinP[chart]] <> -ordLmP[1][1])
      {
    dimen=0;
        return(dimen);
      }
    LmP=permute_L(LmP,ordLmP[2]);
    funcion=LmP[1];
    dimen=1;
    return(list(funcion,ordLmP[1][1],dimen));
}
```

Let us see an example:

```
  int plevel=printlevel;
  printlevel=-1;
  ring s=2,(x,y),lp;
  poly f=y2+y3+x3+x4;
  list C=Adj_div(f);
The genus of the curve is 2
  C=NSplaces(1,C);
  def pro_R=C[1][2];
```

```
  setring pro_R;
  intvec m=4,6;
  intvec P1=0,1,1;
  intvec P2=0,1,0;
  list P=P1,P2;
  int chart=1;
  RRquot(m,P,C,chart);
Vector basis successfully computed
-->[1]:
    _[1]=x3+yz2
    _[2]=xyz+xz2
-->[2]:
   -4
-->[3]:
   1
  printlevel=plevel;
```

This procedure needs also some auxiliar procedures, as we explain now before concluding.

As `RRquot` reads off the point through its homogeneous coordinates we need to localize that point in the list `POINTS` and make the correspondence between such a point and its position in the list of points contained in the third output of the procedure `Adj_div`. This is done by mean of a routine named `isPinlist`. Its inputs are the point P in homogeneous coordinates, that is, a vector of integers, and the list L of points from `Adj_div`. The output is an integer being zero if the point is not in the list or a positive integer indicating the position of P in L. Look at the following example:

```
  ring r=0,(x,y),ls;
  intvec P=1,0,1;
  list POINTS=list(list(1,0,1),list(1,0,0));
  isPinlist(P,POINTS);
-->1
```

We need also a procedure for ordering a list of integers. This is partially solved by the procedure `sort` from `general.lib`. But `sort` cannot order lists of negative numbers, so we have to extend this algorithm, say to `extsort`, to do so. This procedure needs to permute a vector of integers, what can be done by the procedure `perm_L` (this is actually implemented for lists of integers, see `permute_L` in `brnoeth.lib`, but not for vectors of integers). Here it is an example of usage:

```
  ring r=0,(x,y),ls;
  list L=10,9,8,0,7,1,-2,4,-6,3,0;
  extsort(L);
-->[1]:
   -6,-2,0,0,1,3,4,7,8,9,10
-->[2]:
   9,7,4,11,6,10,8,5,3,2,1
```

Finally, it is important to fix a comfortable way of reading off the data of the divisor needed in the procedure `BrillNoether`. The routine `zeroes` takes two vectors of integers `m` and `pos`, and an integer `siz` and it builds up a vector of size `siz`, with the values contained in m set in the places given by `pos` and zeroes in the other places:

```
ring r=0,(x,y),ls;
intvec m=4,6;
intvec pos=4,2;
zeroes(m,pos,5);
-->0,6,0,4,0
```

References

1. Arbarello, E., Cornalba, M., Griffiths, P.A., Harris, J.: Geometry of Algebraic Curves, vol. I. Springer, New York (1985)
2. Brieskorn, E., Knörrer, H.: Plane Algebraic Curves. Birkhäuser Verlag, Basel (1986)
3. Brill, A., Noether, M.: Ueber die algebraischen Functionen und ihre Anwendung in der Geometrie. Math. Ann. **7**, 269–310 (1874)
4. Brill, A., Noether, M.: Die Entwicklung der Theorie der algebraischen Functionen in älterer und neuerer Zeit. Jahresbericht der Deutschen Mathematiker-Vereinigung **3**, 107–566 (1892–1893)
5. Campillo, A., Castellanos, J.: Curve Singularities. An Algebraic and Geometric Approach. Hermman, Paris (2005)
6. Campillo, A., Delgado, F., Gusein-Zade, S.M.: Zeta Function at Infinity of a Plane Curve and the Ring of Functions on It. Contemporary Maths and its Applications. Vol. in honour to Pontryagin. Moscow (1999)
7. Campillo, A., Farrán, J.I.: Computing Weierstrass semigroups and the Feng-Rao distance from singular plane models. Finite Fields Appl. **6**, 71–92 (2000)
8. Campillo, A., Farrán, J.I.: Symbolic Hamburger-Noether expressions of plane curves and applications to AG-codes. Maths Comput. **71**(240), 1759–1780 (2002)
9. Carvalho, C., Torres, F.: On Goppa codes and Weierstrass gaps at several points. Des. Codes Crypt. **35**(2), 211–225 (2005)
10. Delgado de la Mata, F.: The symmetry of the Weierstrass generalized semigroups and affine embeddings. Proc. Am. Math. Soc. **108** (3), 627–631 (1990)
11. Farrán, J.I., Lossen, Ch.: `brnoeth.lib`. A SINGULAR 4-1-1 library for computing the Brill-Noether algorithm, Weierstrass semigroups and AG codes (2018)
12. Gorenstein, G.: An arithmetic theory of adjoint plane curves. Trans. Am. Math. Soc. **72**, 414–436 (1952)
13. Greuel, G.M., Pfister, G., Schönemann, H.: "SINGULAR 2.0", A computer algebra system for polynomial computations. Centre for Computer Algebra, University of Kaiserslautern, 2001
14. Haché, G.: Construction effective des codes géométriques. Ph.D. thesis, Univ. Paris 6 (1996)
15. Haché, G., Le Brigand, D.: Effective construction of Algebraic Geometry codes. IEEE Trans. Inform. Theory **41**, 1615–1628 (1995)
16. Hernando Carrillo, F.J., McGuire, G., Monserrat, F., Moyano-Fernández, J.J.: Quantum codes from a new construction of self-orthogonal algebraic geometry codes. Quantum Inf. Process. **19**, 117 (2020)
17. Høholdt, T., van Lint, J.H., Pellikaan, R.: Algebraic geometry codes. In: Pless, V.S., Huffman, W.C., Brualdi, R.A. (eds.) Handbook of Coding Theory, vol. 1, pp. 871–961. Elsevier, Amsterdam (1998)
18. Homma, H., Kim, S.G.: Goppa codes with Weierstrass pairs. J. Pure Appl. Algebra **162**, 273–290 (2001)
19. Kim, S.G.: On the index of the Weierstrass semigroup of a pair of points on a curve. Arch. Math. **62**, 73–82 (1994)

20. Le Brigand, D., Risler, J.J.: Algorithme de Brill-Noether et codes de Goppa. Bull. Soc. Math. France **116**, 231–253 (1988)
21. Martinez-Moro, E. (ed.): Algebraic Geometry Modeling in Information Theory. World Scientific, Singapore (2013)
22. Martinez-Moro, E., Munuera, C., Ruano, D. (eds.): Advances in Algebraic Geometry Codes. World Scientific, Singapore (2008)
23. Matthews, G.L.: Weierstrass pairs and minimum distance of Goppa codes. Des. Codes Crypt. **22**, 107–121 (2001)
24. Matthews, G.L.: The Weierstrass semigroup of an m-tuple of collinear points on a Hermitian curve. In: Finite fields and Applications. Lecture Notes in Computer Science 2948, pp. 12–24 (2004)
25. Moyano-Fernández, J.J.: On Weierstraß semigroups at one and two points and their corresponding Poincaré series. Abh. Math. Sem. Univ. Hambg. **81**(1), 115–127 (2011)
26. Moyano-Fernández, J.J., Tenòrio, W., Torres, F.: Generalized Weierstrass semigroups and their Poincaré series. Finite Fields Appl. (2019). arXiv:1706.03733
27. Noether, M.: Rationale Ausführung der Operationen in der Theorie der algebraischen Functionen. Math. Ann. **23**, 311–358 (1883)
28. Riemann, B.: Theorie der Abel'schen Functionen. J. Reine Angew. Math. **54**(14), 115–155 (1857)
29. Tsfasman, M.A., Vlăduţ, S.G.: Algebraic-Geometric Codes. Mathematics and its Applications, vol. 58. Kluwer Academic Publishers, Amsterdam (1991)
30. Zariski, O.: Le probleme des modules pour les branches planes. Hermann, Paris (1986)

On the Hilbert Function of Four-Generated Numerical Semigroup Rings

Anna Oneto and Grazia Tamone

Abstract In this article we study the Hilbert function H_R of one-dimensional semigroup rings $R = k[[S]]$, with embedding dimension four over an infinite field k. Let $S =< e, n_2, n_3, n_4 >$ and let $M = S \setminus \{0\}$. Consider the *Apéry set* of S with respect to the multiplicity e and its subsets $A_h = \{s \in Apéry(S) \mid s \in hM \setminus (h+1)M\}$, $h \geq 2$. Further let $D_2 \subseteq \{n_3, n_4\}$ be the set of generators with *torsion order* 1. We prove that H_R is non-decreasing at level ≤ 3 and that H_R is non decreasing in each of the following cases: if A_2 has cardinality ≤ 4, if A_3 has cardinality ≤ 3, if $A_4 = \emptyset$, if D_2 has cardinality 2, if S has multiplicity ≤ 13.

Keywords Numerical semigroup · Hilbert function · Apéry set

Mathematics Subject Classification Primary: 13H10; Secondary: 14H20

1 Introduction

The behaviour of the Hilbert function H_R of a local noetherian ring $(R, \mathbf{m}, k)$ has been studied by many authors in the last 40 years. H_R is by definition the Hilbert function of the associated graded ring $G = \oplus_{n\geq 0}(\mathbf{m}^n/\mathbf{m}^{n+1})$: $H_R(n) = dim_k(\mathbf{m}^n/\mathbf{m}^{n+1})$, $n \in \mathbb{N}$. When G is Cohen Macaulay the function H_R is non decreasing but in general H_R can be decreasing, that is $H_R(n-1) > H_R(n)$ for some value of n even if the ring R is Cohen Macaulay or Gorenstein of dimension one, see [9, 13–15, 17].

If $(R, \mathbf{m}, k)$ is a one-dimensional local Cohen-Macaulay ring with $|k| = \infty$, multiplicity e and embedding dimension v, it is known that

- *H_R is non decreasing if either* $v \leq 3$, or $v \leq e \leq v+2$, see [6, 7, 18]

A. Oneto (✉) · G. Tamone
DIMA, University of Genova, Genova, Italy
e-mail: oneto@dime.unige.it; tamone@dima.unige.it

V. Barucci et al. (eds.), *Numerical Semigroups*, Springer INdAM Series 40,
https://doi.org/10.1007/978-3-030-40822-0_15

while there are various classes of rings with $e \geq v+3$ which have H_R decreasing, see [12–14]. In the particular case $R = k[[S]]$, $S \subseteq \mathbb{N}$ numerical semigroup, we know that the function H_R is non decreasing if S verifies one of the following conditions

- *S is generated by an almost arithmetic sequence (moreover G is Cohen Macaulay if the sequence is arithmetic)* [11, 19]
- *S is minimally generated by 4 elements and either belongs to particular classes of symmetric semigroups or has Buchsbaum tangent cone* [1, 3]
- *S is minimally generated by 4 or 5 elements and* $e \leq 8$ [4, Corollary 3.14]
- *S has multiplicity* $e = v + 3 \leq 12$ [13]
- *S is constructed by "gluing" particular semigroups* [2, 10]
- *S is balanced* [3, 16].

Given a semigroup ring $R = k[[S]]$, $S \subseteq \mathbb{N}$, it is convenient to consider two families $\{C_n, n \in \mathbb{N}\}, \{D_n, n \in \mathbb{N}\}$ of subsetes of S (see Definition 2.2), which give an handy method to compute H_R, as in [3, 4, 13, 14, 16]. In this article we focus on the case $R = k[[S]]$, where $S \subseteq \mathbb{N}$ is a semigroup minimally generated by four elements: besides the sets $\{C_n, D_n\}$, the main tool we use is the Apéry set of S with respect to the multiplicity of R. Let $M = S \setminus \{0\}$ and for $s \in S$, let $ord(s) = max\{n \in \mathbb{N} \mid s \in nM\}$: we consider a partition of the Apéry set according to the $order$ of its elements as follows

$$A_h = \{s \in Apéry(S) \mid ord(s) = h\}, \ \ h \in \mathbb{N}.$$

The first case to consider is $h = 2$: since $C_2 = A_2$, a necessary condition to have H_R decreasing, is $|A_2| \geq 3$ and, when the cardinality is 3, we must have $A_2 = \{2n_p, n_p + n_q, 2n_q\}$ $(2 \leq p, q \leq 4)$, see [13, Theorem 2.1 and Theorem 2.6]. By analyzing the sets C_h, D_h, A_2 in several different subcases and by means of other technical facts, we obtain the following results:

- *The Hilbert function H_R is non decreasing at level* $\ell \leq 3$ (Theorem 7.1).
- *The Hilbert function H_R is non decreasing in each of the following cases*

  (a) $|A_2| \leq 4$ (Theorem 4.1)
  (b) $|D_2| = 2$ (Theorem 5.4)
  (c) R *has multiplicity* $e \leq 13$ (Theorem 7.2).

2 Preliminaries

Setting 2.1 In this paper $R = k[[S]]$ denotes a one-dimensional numerical semigroup ring, where k is an infinite field and $S \subseteq \mathbb{N}$ is a *numerical semigroup,* that is a submonoid of the natural numbers with finite complement to $\mathbb{N}$. Given the minimal set $\{n_1, \ldots, n_v\}$ of generators of S, we write $S = \langle n_1, n_2, \ldots, n_v \rangle$ and assume $n_1 < n_2 < \cdots < n_v$.

The ring $R = k[[t^{n_1}, \ldots, t^{n_v}]]$ has *multiplicity* $e = n_1$, *embedding dimension* v and it is the completion of the local ring $k[x_1, \ldots, x_v]_{(x_1,\ldots,x_v)}$ of the monomial affine curve C parametrized by $x_i = t^{n_i}$ $(1 \leq i \leq v)$. The maximal ideal of R is $\mathbf{m} = (t^{n_1}, \ldots, t^{n_v})$, $k = R/\mathbf{m}$, further t^e is a *superficial element of degree* 1; $dim_k(\mathbf{m}^n/\mathbf{m}^{n+1}) \leq e$ for each $n \in \mathbb{N}$ and the equality holds for each $n >> 0$; we shall denote by r the reduction exponent, i.e. the minimum integer such that $dim_k(\mathbf{m}^n/\mathbf{m}^{n+1}) = e$ $(r \leq e-1)$.

We shall denote the values of the Hilbert function H_R of R by $[a_0, a_1, .., a_p \rightarrow]$, where $a_p \rightarrow$ means that $H_R(n) = a_p$ for each $n \geq p$. Let $\nu : k((t)) \longrightarrow \mathbb{Z} \cup \{\infty\}$ be the usual valuation; then $M = S \setminus \{0\} = \nu(\mathbf{m})$ and $\nu(\mathbf{m}^h) = hM$ for each $h \geq 1$. We have: $H_R(0) = a_0 = 1$, $H_R(n) = |nM \setminus (n+1)M|$ for each $n \geq 1$.

In order to study the Hilbert function of R, we deeply use the structure of the semigroup S and mainly the subsets $A_h, C_k, D_k \subseteq S$, $h \geq 0$, $k \geq 1$. The elements in D_k are precisely the elements of order $k-1$ with $torsion\ order = 1$, as defined in [3, Page 296].

Definition 2.2 Let $S \subseteq \mathbb{N}$ be a numerical semigroup.

- The *order* of an element $g \in S$ is $ord(g) = max\,\{h \mid g \in hM\}$
- The *Apéry set of* S with respect to the multiplicity e is $Ap(S) = \{s \in S \mid s - e \notin S\}$
- A_h is the subset of the elements of order h in *Ap(S)*
- $D_1 = \emptyset$, $D_k = \{s \in S \mid ord(s) = k-1\ and\ ord(s+e) > k\}$, for each $k \geq 2$
- $D_k^h = \{s \in D_k \mid ord(s+e) = h\}$, for each $h > k$
- $C_k = \{s \in S \mid ord(s) = k\ and\ s - e \notin (k-1)M\}$, for each $k \geq 1$
- A *maximal representation* of $s \in S$ (shortly denoted by *mrp*) is any expression

$$s = \sum_{j=1}^{v} a_j n_j, \quad with\ a_j \in \mathbb{N}\ for each\ 1 \leq j \leq v\ and\ \sum_{j=1}^{v} a_j = ord(s)$$

- Given an *mrp* $s = \sum_{j=1}^{v} a_j n_j$, we say that $s' \in S$ is *induced* (by the *mrp*) if

$$s' = \sum_{j=1}^{v} b_j n_j \quad with \quad 0 \leq b_j \leq a_j\ for each\ 1 \leq j \leq v.$$

Remark 2.3 Let r be the reduction exponent of R. We have:

(1) $C_1 = \{n_2, \ldots, n_v\}$, $C_2 = A_2$, $C_h = A_h \bigcup \{\cup_k (D_k^h + e),\ 2 \leq k \leq h-1\}$, $\forall h \geq 3$
(2) $C_h = \emptyset$ for each $h \geq r+1$, $D_k = \emptyset$ for each $k \geq r$ [16, Lemmas 1.5.2–1.8.1]
(3) If $s = \sum_{j=1}^{v} a_j n_j \in C_h$, *mrp*, then $a_1 = 0$.

The following known results are crucial in the sequel.

Theorem 2.4 *Let S be be a numerical semigroup as in Setting 2.1, let H_R be the Hilbert function of $R = k[[S]]$ and let $k \geq 2$.*

(1) $H_R(k) - H_R(k-1) = |C_k| - |D_k|$ [16, Proposition 1.9.3], [3, Remark 4.1]
(2) [8, Theorem 7], [16, Theorem 1.6] *The following conditions are equivalent:*

(a) *The associated graded ring G is Cohen Macaulay*
(b) $ord(s+e) = ord(s) + 1$ *for each* $s \in S$
(c) $D_k = \emptyset$ *for each* $k \geq 2$.

(3) *If* $|D_k| \leq k+1$, *then* H_R *does not decrease at level* k [4, Corollary 3.4]
(4) *If* H_R *is decreasing, then* $|C_3| \geq 4$ [13, Proposition 1.7]
(5) *If* H_R *decreases at level* k, *then* $|C_h| \geq h+1$ *for all* $h \in [2,k]$ [4, Proposition 3.9]
In particular H_R *decreasing implies* $|A_2| \geq 3$ [4, Corollary 3.11]
(6) *If* H_R *is decreasing at level 2, then* $v \geq 6$ [4, Corollary 3.13]
(7) *If* $4 \leq v \leq 5$ *and* $e \leq 8$, *then* H_R *is non decreasing* [4, Corollary 3.14]

Proposition 2.5 *Let S be a numerical semigroup as in Setting 2.1.*

(1) *For* $s \in S$, $ord(s) = k$, *let* $s = \sum_{j=1}^{v} a_j n_j$, *mrp and let* $s' = \sum_{j\geq 1} b_j n_j$ *with* $0 \leq b_j \leq a_j$ *be an induced element. Then:*
$ord(s') = \sum_{j=1}^{v} b_j$ *and* $s \in C_k, \Longrightarrow s' \in C_h$, *where* $h = ord(s')$.
(2) *If* $n_i \in D_2$, *let* $n_i + e = \sum_2^v \lambda_j n_j$, *mrp* $ord(n_i + e) = k \geq 3$. *Then every induced element* $s = \sum_2^v \lambda'_j n_j$, *with* $\lambda'_j \in [0, \lambda_j]$ *and* $\sum_j \lambda'_j = h < k$ *belongs to* A_h.
(3) *Let* $s \in D_k$, $s + e = \sum_{j=1}^{v} \lambda_j n_j$ *with* $\sum_{j=1}^{v} \lambda_j \geq k+1$, *then* $\lambda_1 = 0$ *and if* $\sigma = \sum_{j=1}^{v} \alpha_j n_j$, $\alpha_j \in [0, \lambda_j]$, *then any mrp* $\sigma = \sum_{j=1}^{v} \gamma_j n_j$ *has* $\gamma_1 = 0$.
(4) *If* $s \in D_k$ *and* $ord(s + n_i) = k$, $(i \geq 2)$, *then* $s + n_i \in D_{k+1}$.
(5) *Let* $s_1 \neq s$, $ord(s_1) = ord(s)$, $s = \sum_{j=1}^{v} a_j n_j$, $s_1 = \sum_{j=1}^{v} b_j n_j$, *both maximal representations. Then there exist* $1 \leq i, h \leq v$ *such that* $a_i > b_i$, $a_h < b_h$.
(6) *If* $s, s_1 \in D_k$, *and* $s_1 + e = \sum_{j=2}^{v} \alpha_j n_j$, $s + e = \sum_{j=2}^{v} \lambda_j n_j$ *are both mrp, then there exist* i, h *such that* $\alpha_i > \lambda_i$, $\alpha_h < \lambda_h$.

Proof

(1) See [16, Lemma 1.11] and [13, Proposition 1.4].
(2) Clearly, $n_i + e = \sum_{j=2}^{v} \lambda_j n_j$, implies $\lambda_i = 0$ and so $a'_i = 0$; if $s \notin Ap(S)$, then $s \in M + e$ and so $n_i \in \langle n_2, \ldots, \widehat{n_i}, \ldots, n_v \rangle$ is not a minimal generator of S.
(3) If $\lambda_1 > 0$, then $s = (\lambda_1 - 1)e + \sum_{j=2}^{v} \lambda_j n_j$ would imply $ord(s) \geq k$.
$s + e = \sum_{j=1}^{v} (\lambda_j - \alpha_j) n_j + \sigma = \sum_{j=2}^{v} (\lambda_j - \alpha_j) n_j + \sum_{j=1}^{v} \gamma_j n_j$. If $\gamma_1 > 0$, then $s = (\gamma_1 - 1)e + \sum_{j=2}^{v} (\lambda_j - \alpha_j + \gamma_j) n_j$ would have order $\geq k$, since, by assumption, $\sum_{j=1}^{v} \gamma_j \geq \sum_{j=1}^{v} \alpha_j$.
(4) $ord(s + n_i + e) \geq ord(s + e) + 1 \geq k + 2$, hence $s + n_i \in D_{k+1}$.

(5–6) In fact, if $a_i \geq b_i$ for each i, then $s = s_1 + \sigma$, $ord(\sigma) \geq 1$, impossible. The second proof is similar. □

Notation 2.6 In the following we shall often use that induced elements have the properties stated in Proposition 2.5: for brevity we shall write "*induced element*" without referring to the above proposition.

Example 2.7 In general, claim (2.5.2) is false for elements $s \in Ap_k \cap D_{k+1}$, $k \geq 2$. For instance if $S =< 13, 27, 68, 150 >$, $s = 122 \in D_4 \cap A_3$, one has $s + e = 5n_2$, but $4n_2 \notin A_4$ because $4n_2 = n_2 + n_3 + e$.

Proposition 2.8 *With Setting 2.1 elements $a_1e + a_2n_2 \notin D_k$, for each $a_1, a_2 \geq 0$.*

Proof Note that $s = a_1e + a_2n_2$, or $d(s) = a_1 + a_2 = k - 1 \Longrightarrow s + e = (a_1 + 1)e + a_2n_2 < kn_2$. If $s \in D_k$, then $s + e = \lambda_2n_2 + \lambda_3n_3 + \cdots + \lambda_vn_v \geq (\lambda_2 + \ldots + \lambda_v)n_2 \geq (k + 1)n_2$, impossible. □

Proposition 2.9 *Let $s = \sum_1^v a_jn_j$, mrp, $a_1 > 0$. For $i \in [0, a_1]$ and for $p \in [2, v-1]$, let* $s_{ip} = (a_1 - i)e + a_2n_2 + \cdots + (a_p + i)n_p + \cdots + a_vn_v$, $(s = s_{0p})$:

(1) *Assume $i \geq 1$ and $s_{ip} \in D_k$, $s_{ip} + e = \sum_{j\geq 2} \lambda_{ij}n_j$ mrp. Then $\lambda_{ip} < i$ and for each $n_q > n_p$ the following conditions hold:*
$\left(1 - \sum_{j\notin\{p,q\}} \lambda_{ij}\right)n_q < \left(1 + \sum_{j\notin\{1,p,q\}}(a_j - \lambda_{ij})\right)n_q < \sum_{j\notin\{1,p,q\}}(a_j - \lambda_{ij})n_j \quad (*_i)$
(2) *If $s \in D_k$, $a_p > 0$, $s + e = \sum_{j\geq 2} \lambda_{0j}n_j$ and condition $(*_0)$ doesn't hold for s, then for all $1 \leq i \leq a_p$*
$s'_{ip} = (a_1 + i)e + a_2n_2 + \cdots + (a_p - i)n_p + \cdots + a_vn_v \notin D_k$.

Proof

(1) If $\lambda_{ip} \geq i$, the equality

$$s = (s_{ip} + e) + (i-1)e - in_p = \lambda_{i2}n_2 + \cdots + (\lambda_{ip} - i)n_p + \cdots + \lambda_{iv}n_v + (i-1)e$$

would imply $ord(s) \geq k$, against the assumption $ord(s) = k - 1$. Further we have

$$\begin{aligned}
\sigma &= \textstyle\sum_{j\neq p} \lambda_{ij}n_j = s_{ip} + e - \lambda_{ip}n_p = (a_1 - i + 1)e + (a_p + i - \lambda_{ip})n_p + \sum_{j\notin\{1,p\}} a_jn_j \\
&< \big((a_1 - i + 1) + (a_p + i - \lambda_{ip}) + a_q\big)n_q + \textstyle\sum_{j\notin\{1,p,q\}} a_jn_j = \\
&= (a_1 + a_p + a_q + 1 - \lambda_{ip})n_q + \textstyle\sum_{j\notin\{1,p,q\}} a_jn_j \Longrightarrow \\
\sigma &< (k - \textstyle\sum_{j\notin\{1,p,q\}} a_j - \lambda_{ip})n_q + \sum_{j\notin\{1,p,q\}} a_jn_j. \text{Now} \lambda_{iq} \geq k + 1 - \sum_{j\neq q} \lambda_{ij} \Longrightarrow \\
\sigma &= \textstyle\sum_{j\neq p} \lambda_{ij}n_j \geq (k + 1 - \sum_{j\neq q} \lambda_{ij})n_q + \sum_{j\notin\{p,q\}} \lambda_{ij}n_j, \text{ and so}
\end{aligned}$$

$$(k+1-\sum_{j\neq q} \lambda_{ij})n_q + \sum_{j\notin\{p,q\}} \lambda_{ij}n_j < (k - \sum_{j\notin\{1,p,q\}} a_j - \lambda_{ip})n_q + \sum_{j\notin\{1,p,q\}} a_jn_j$$

equivalently, $\left(1 + \sum_{j\notin\{1,p,q\}}(a_j - \lambda_{ij})\right)n_q < \sum_{j\notin\{1,p,q\}}(a_j - \lambda_{ij})n_j \qquad (*_i)$

(2) Consider the element $t = s'_{ip}$, then $t_{ip} = s$. If $t \in D_k$ then, by assumption, condition $(*_i)$ for t_{ip} (that is condition $(*_i)$ for s) doesn't hold. Hence by (1), $t_{ip} = s \notin D_k$, contradiction. □

3 The Case Embedding Dimension 4

From now on we restrict to the embedding dimension four case. Then H_R does not decrease at level two by (2.4.6) and H_R does not decrease if $e \leq 8$ by (2.4.7). The easy Remark 3.1 gives other conditions which assure that H_R does not decrease.

Remark 3.1

(1) Let $S = \langle e, n_2, n_3, n_4 \rangle$: recall that S is said balanced if $e + n_4 = n_2 + n_3$ in this case H_R is non decreasing by Patil and Tamone [16, Theorem 2.11].
(2) Let $S = \langle e, n_p, n_q, n_r \rangle$. If either $2n_p = e + n_q$, or $2n_q = n_p + n_r$ then respectively, $n_p - e = n_q - n_p$, and the sequence e, n_p, n_q is arithmetic or $n_q - n_r = n_q - n_p$ and the sequence n_p, n_q, n_r is arithmetic. In these cases S is generated by an almost arithmetic sequence (AAS) therefore H_R is non decreasing by Tamone [19, Corollary 2.7].

According to Theorem (2.4.5) and by the above Remark (3.1), from now on we shall consider semigroups which satisfy the following setting.

Setting 3.2 From now on we assume that $S = \langle e, n_2, n_3, n_4 \rangle$, with $e < n_2 < n_3 < n_4$ is a numerical semigroup as in (2.1), with $|A_2| \geq 3$, which is neither balanced nor AAS.

In particular, e, n_2, n_3, n_4 are such that:

- $2n_2 \neq e + n_3$, $e + n_3 \neq n_2 + n_4$, $2n_2 \neq e + n_4$, $2n_3 \neq n_2 + n_4$.
- if $n_i + n_h \notin A_2$, then $ord(n_i + n_h) \geq 3$, for each $i, h \in \{2, 3, 4\}$.

We write $S = \langle e, n_p, n_q, n_r \rangle$ when we don't require an ordered sequence of generators.

Proposition 3.3 *Let $S = \langle e, n_p, n_q, n_r \rangle$, with $n_p < n_q$. Let $s = a_1 e + a_p n_p + a_q n_q + a_r n_r \in D_k$ mrp ,with $s + e = \lambda_p n_p + \lambda_q n_q + \lambda_r n_r$.*

(1) *Assume that $(1 + a_r - \lambda_r) n_q \geq (a_r - \lambda_r) n_r$. Then $(a_1 + i)e + (a_p - i)n_p + a_q n_q + a_r n_r \notin D_k$ for each $1 \leq i \leq a_p$.*
(2) *According to the values of (p, q, r) we get*

$(p, q, r) = (2, 3, 4)$:

$$a_1 e + a_2 n_2 + a_3 n_3 \in D_k \implies (a_1 + i) e + (a_2 - i) n_2 + a_3 n_3 \notin D_k$$

$$a_1 e + a_2 n_2 + a_3 n_3 + n_4 \in D_k,\ n_4 < 2n_3 \implies (a_1 + i) e + (a_2 - i) n_2 + a_3 n_3 + n_4 \notin D_k$$

$(p, q, r) = (2, 4, 3)$:

$$a_1 e + a_2 n_2 + a_3 n_3 + a_4 n_4 \in D_k, a_3 \geq \lambda_3 - 1 \implies (a_1 + i) e + (a_2 - i) n_2 + a_3 n_3 + a_4 n_4 \notin D_k$$

$(p, q, r) = (3, 4, 2)$:

$$a_1 e + a_2 n_2 + a_3 n_3 + a_4 n_4 \in D_k,\ a_2 \geq \lambda_2 - 1 \implies (a_1 + i) e + a_2 n_2 + (a_3 - i) n_3 + a_4 n_4 \notin D_k$$

(3) *If* $ord(2n_r) \geq 3$, *in each case considered in (2), for any pair* $(b_1, b_p) \neq (a_1, a_p)$, *with* $b_1 + b_p = a_1 + a_p$, *we have* $b_1 e + b_p n_p + a_q n_q + a_r n_r \notin D_k$.
(4) *If* $ord(2n_r) \geq 3$, *then* $|D_k \cap \{a_1 e + a_p n_p + a_q n_q,\ a_1 + a_p + a_q = k-1\}| \leq k-1$.

Proof

(1) is a special case of (2.9.2) in fact by assumption we have $n_p < n_q$.
(2) It is a direct computation, since $n_p < n_q \Longrightarrow p \neq 4,\ q \neq 2$.
(3) $ord(2n_r) \geq 3 \Longrightarrow a_r, \lambda_r, \lambda_{ir} \leq 1$. By (1) we get $(a_1 + i)e + (a_p - i)n_p + a_q n_q + a_r n_r \notin D_k$ for each $1 \leq i \leq a_p$. Further in each case considered in (2) condition $(*_i)$ of (2.9) does not hold, hence by (2.9.1), $(a_1 - i)e + (a_p + i)n_p + a_q n_q + a_r n_r \notin D_k$ for each $1 \leq i \leq a_1$.
(4) Let $D'_k = D_k \cap \{a_1 e + a_p n_p + a_q n_q\}$ and let $a_1 e + a_p n_p + a_q n_q \in D_k$. By applying (2), for each (b_1, b_p) such that $b_1 + b_p = a_1 + a_p$, we get $s' = b_1 e + b_p n_p + a_q n_q \notin D_k$. Hence $|D'_k| \leq |\{0 \leq a_q \leq k-1\}|$.
If $n_p = n_2$, then $a_q \geq 1$, since $a_1 e + a_2 n_2 \notin D_h$ by (2.8).
If $n_p = n_3$, i.e., $2n_2 \notin A_2$, then $a_q = 0, s \in D'_k \Longrightarrow s + e = (a_1 + 1)e + a_3 n_3 = \lambda_2 n_2 + \lambda_3 n_3 + \lambda_4 n_4$, with $\lambda_2 \leq 1, \lambda_2 + \lambda_3 + \lambda_4 \geq k+1$: clearly this is impossible, since $(a_1 + 1)e + a_3 n_3 < n_2 + (k-1)kn_3$ and $\lambda_2 n_2 + \lambda_3 n_3 + \lambda_4 n_4 > n_2 + kn_3$. Hence $a_q \geq 1$ and the claim follows. □

Lemma 3.4 *Let* $S = \langle e, n_2, n_3, n_4 \rangle$.

(1) *Let* $n_2 + n_3 \in A_2$; *then* $n_3 \in D_2 \Longleftrightarrow n_2 + n_3 \in D_3$.
(2) *Assume* $n_2 + n_3 \notin A_2$, *i.e.*, $n_2 + n_3 = \ell e + mn_4$, *mrp with* $\ell + m \geq 3$, $m \leq 1$: *if either* $n_3 \in D_2$ *or* $n_3 + e \in D_3$, *then* $2n_3 \notin A_2$.

Proof

(1) $n_2 + n_3 \in D_3 \Longleftrightarrow (n_2 + n_3 \in A_2$ and $n_2 + n_3 + e = \lambda_2 n_2,\ \lambda_2 \geq 4) \Longleftrightarrow n_3 + e = (\lambda_2 - 1)n_2$, i.e. $n_3 \in D_2$.
(2) $n_3 \in D_2 \Longrightarrow n_3 + e = \alpha n_2 \Longrightarrow 2n_3 + e = (\alpha - 1)n_2 + n_2 + n_3 = (\alpha - 1)n_2 + \ell e + mn_4 \Longrightarrow 2n_3 = (\alpha - 1)n_2 + (\ell - 1)e + mn_4 \notin Ap(S)$.
Similarly, $n_3 + e \in D_3 \Longrightarrow 2n_3 \notin A_2$, if $m = 1$ and $2n_3 \notin Ap(S)$, if $m = 0$. □

Lemma 3.5 *Let* $S = \langle e, n_2, n_3, n_4 \rangle$. *With setting (3.2), assume* $|A_2| \geq 4$. *Then:*

(1) *At least one between* $2n_2, n_2 + n_3$ *belongs to* A_2 *and* $n_2 + n_3 \in D_3 \Longleftrightarrow n_3 \in D_2$.
(2) *If* $2n_2 \notin A_2$, *i.e.*, $2n_2 = \ell e + mn_i$, *mrp with* $i \in \{3, 4\}$, $m \leq 1$, $\ell + m \geq 3$, *then*
   (a) *If* $s = a_1 e + a_2 n_2 + a_3 n_3$ *is mrp with* $a_1 + a_2 + a_3 = k - 1$, $a_2 \in \{0, 1\}$, $k \geq 2$, *then* $s \notin D_k$. *In particular* $n_3 \notin D_2$, $n_2 + n_3, 2n_3, n_3 + e \notin D_3$
   (b) $|\{n_4 + e, n_3 + n_4\} \cap D_3| \leq 1$, $D_3 \subseteq \{n_4 + e, n_3 + n_4, n_2 + n_4, 2n_4\}$, $|D_3| \leq 3$

(c) $|D_k| \leq 2k-3$ *for each* $k \geq 3$ *and* H_R *is non-decreasing at level* ≤ 4

Proof

(1) Let $2n_2 = \ell e + mn_i$, $n_2+n_3 = \ell' e + m' n_4$, $\ell\ell' > 0$, $m, m' \in \{0,1\}$, then $n_3 - n_2 = (\ell'-\ell)e - n_i \Longrightarrow (\ell'-\ell) > 0 \Longrightarrow n_3 + n_i \notin Ap(S)$ then $|A_2| \leq 3$, against the assumption. The other cases are either similar or easy to prove.

(2.*a*) Assume $s = a_1 e + a_2 n_2 + a_3 n_3 \in D_k$, then $s + e = \lambda_2 n_2 + \lambda_3 n_3 + \lambda_4 n_4$, with $\lambda_2 + \lambda_3 + \lambda_4 \geq k+1$ $a_2, \lambda_2 \leq 1$, by assumption; impossible since

$$\left[\begin{array}{ll}\text{either } s+e = (a_1+1)e + a_2 n_2 + a_3 n_3 \leq n_2 + (k-1)n_3 \\ \text{or} \quad s+e = \lambda_2 n_2 + \lambda_3 n_3 + \lambda_4 n_4 \geq n_2 + kn_3\end{array}\right.$$

(2.*b*) It follows by (3.3.3), with $r=2$, $p=3$.

(2.*c*) By (2.*a*–*b*), we know that $2n_2 \notin A_2 \Longrightarrow |D_k| \leq |\{(k-1)n_4, (k-2)n_4 + n_3, \cdots, n_4 + (k-2)n_3\}| + |\{n_2 + (k-2)n_4, n_2 + (k-3)n_4 + n_3, \cdots, n_2 + n_4 + (k-3)n_3\}|$ hence $|D_k| \leq 2k-3$. The second claim follows by (2.4.3), since $2k-3 < k+2$ for $k \leq 4$. □

4 Cases with Cardinality of $A_2 \leq 4$

In this section, through several steps we shall prove the following theorem.

Theorem 4.1 *Assume* $S = \langle n_1 = e, n_p, n_q, n_r\rangle$ *and* $|A_2| \leq 4$, *then the Hilbert function of* $R = k[[S]]$ *is non-decreasing.*

Proof Let $|A_2| = 3$: by Oneto and Tamone [13, Prop.1.7.2, Theorem 2.1.2], if H_R decreases there exist two generators $n_p < n_q$ such that $A_2 = \{2n_p, n_p + n_q, 2n_q\}$. We show that in this case $|D_k| \leq k$ for each $k \geq 2$. In fact the possible maximal representations of the elements with *order* $k-1$ are:

$$\left[\begin{array}{l} s = (k-2)e + n_r \\ s = (k-1-a_p-a_q)e + a_p n_p + a_q n_q \end{array}\right.$$

then by (3.3.4), we have $|D_k| \leq 1 + k - 1 = k$ and the claim follows by (2.4.3).

In case $|A_2| = 4$ the proof is developed in the remaining part of this section. □

Remark 4.2 Let $S = \langle n_1 = e, n_p, n_q, n_r\rangle$ and assume $|A_2| = 4$. By Oneto and Tamone [13, Theorem 2.6] necessary conditions for the decreasing of H_R are the following four shapes of A_2:

(1) $A_2 = \{2n_p, n_2+n_3, n_2+n_4, n_3+n_4\}$, $p \in \{2,3,4\}$, $ord(2n_i) \geq 3$, *for* $i \neq p$
(2) $A_2 = \{2n_p, 2n_q, n_p+n_r, n_q+n_r\}$ $\quad ord(2n_r) \geq 3$, $ord(n_p+n_q) \geq 3$
(3) $A_2 = \{2n_p, n_p+n_q, 2n_q, n_p+n_r\}$, $\quad ord(n_i+n_r) \geq 3$, *for* $i \neq p$
(4) $A_2 = \{2n_p, 2n_q, 2n_r, n_p+n_q\}$ $\quad ord(n_i+n_r) \geq 3$, *for* $i \neq r$

Proposition 4.3 *Let $S = \langle n_1 = e, n_p, n_q, n_r\rangle$. Let $|A_2| = 4$ and assume that S verifies either case (1) or case (2) of (4.2); then H_R is non-decreasing.*

Proof In the first case we have $ord(2n_h) \geq 3$, if $h \neq p$. Then $|C_k| \leq 4$ for each $k \geq 3$:

$$C_k \subseteq \{kn_p, (k-1)n_p + n_q, (k-1)n_p + n_r, (k-2)n_p + n_q + n_r\}.$$

In the second one, any *mrp* of $s \in S$ is $s = a_1 e + a_p n_p + a_q n_q + a_r n_r$, with $a_p a_q = 0$, $a_r \leq 1$. Then:

$$C_k \subseteq \{kn_p, kn_q, (k-1)n_p + n_r, (k-1)n_q + n_r\}, \ |C_k| \leq 4 \text{ for each } k \geq 3.$$

By (2.4.5) H_R decreasing at level ℓ implies that $|C_k| \geq k+1$, $\forall k \leq \ell$, and so H_R cannot decrease at level ≥ 4. For $\ell = 3$, in case (1), if $p \in \{3, 4\}$, then $2n_2 \notin A_2$ and apply (3.5.2.*c*). If $p = 2$, by applying (3.3.4) with $p = 2, q = 3, r = 4$, $s = e+n_3$, $\sigma = n_2+n_3$ we see that $|D_3 \cap \{s, \sigma\}| \leq 1$; hence $|D_3| \leq 4$, since $D_3 \subseteq \{n_2+n_3, n_2+n_4, n_3+n_4, n_3+e, n_4+e\}$. In case (2), with $r \in \{2, 3\}$ the argument is similar to the (1) above. If $r = 4$, we have $n_2 + n_3 \notin A_2$, then by (3.4.2) we have $n_3 + e \notin D_3$ and so $|D_3| \leq 4$, since $D_3 \subseteq \{2n_3, n_2 + n_4, n_3 + n_4, n_3 + e, n_4 + e\}$. In both cases we conclude H_R is non decreasing by (2.4.3). □

Proposition 4.4 *Let $S = \langle n_1 = e, n_p, n_q, n_r\rangle$. Let $|A_2| = 4$ and assume that S verifies case (3) of (4.2). Then H_R is non-decreasing.*

Proof Let $2n_r \notin A_2$ and let $A_2 = \{2n_p, n_p+n_q, 2n_q, n_p+n_r\}$, $ord(n_q+n_r) \geq 3$.

We get $D_k = D'_k \cup D''_k$ where $\left[\begin{array}{l} D'_k = \{a_1e+a_pn_p+a_qn_q \mid a_1+a_p+a_q = k-1\} \\ D''_k = \{a_1e+a_pn_p+n_r \mid a_1+a_p = k-2\} \end{array}\right.$

We divide the proof in 6 subcases:

(3.1) $A_2 = \{2n_2, 2n_3, n_2+n_3, n_2+n_4\}$ $2n_4, n_3+n_4 \notin A_2$ $D''_k \subseteq \{a_1e+a_2n_2+n_4\}$
(3.2) $A_2 = \{2n_2, 2n_4, n_2+n_3, n_2+n_4\}$ $2n_3, n_3+n_4 \notin A_2$ $D''_k \subseteq \{a_1e+a_2n_2+n_3\}$
(3.3) $A_2 = \{2n_3, 2n_4, n_2+n_3, n_3+n_4\}$ $2n_2, n_2+n_4 \notin A_2$ $D''_k \subseteq \{a_1e+a_3n_3+n_2\}$
(3.4) $A_2 = \{2n_2, 2n_3, n_2+n_3, n_3+n_4\}$ $2n_4, n_2+n_4 \notin A_2$ $D''_k \subseteq \{a_1e+a_3n_3+n_4\}$
(3.5) $A_2 = \{2n_2, 2n_4, n_2+n_4, n_3+n_4\}$ $2n_3, n_2+n_3 \notin A_2$ $D''_k \subseteq \{a_1e+a_4n_4+n_3\}$
(3.6) $A_2 = \{2n_3, 2n_4, n_2+n_4, n_3+n_4\}$ $2n_2, n_2+n_3 \notin A_2$ *impossible by* (3.5.1)

In cases (3.1, 3.2, 3.3), by (3.3.4) we have $|D'_k| \leq k-1$ for each $k \geq 2$. We shall prove that $|D''_k| \leq 2$ and so $|D_k| \leq k+1$ and the claim follows by (2.4.3).

Case (3.1) If $s \in D''_k \Longrightarrow s + e = \lambda_2 n_2 + \lambda_3 n_3$, let $m = min\{a_2, \lambda_2\}$; then $(a_1+1)e + (a_2-m)n_2 + n_4 = (\lambda_2 - m)n_2 + \lambda_3 n_3$. If $m = a_2$, since there is an unique a_1 such that $a_1 e + n_4 \in D_{a_1+2}$ (2.5.6), we get $s = a_1e + (k-2-a_1)n_2 + n_4$ is unique; if $m = \lambda_2$, then $(a_1+1)e + (a_2-m)n_2 + n_4 = \lambda_3 n_3$ and λ_3 is unique. We deduce that $|D''_k| \leq 2$.

Case (3.2) If $s \in D''_k \Longrightarrow s+e = (a_1+1)e+a_2n_2+n_3 = \lambda_2n_2+\lambda_4n_4 \Longrightarrow \lambda_4 = 0$, otherwise $\lambda_4 > 0, \lambda_2+\lambda_4 \geq k+1 \Longrightarrow \lambda_2n_2+\lambda_4n_4 \geq kn_2+n_4$, contradiction, because $(a_1+1)e + a_2n_2 + n_3 \leq (k-1)n_2 + n_4$. Then $s+e = (a_1+1)e + a_2n_2 + n_3 = \lambda_2n_2$: clearly this means that $|D''_k| \leq 1$.

Case (3.3) In this case $D''_k = \emptyset$ by (3.5.2 a).

In cases (3.4) and (3.5) we shall construct an injective map $\Phi : D_k \longrightarrow C_k$ for each k. Let $A_2 = \{2n_2, n_2+n_p, 2n_p, n_p+n_q\}$, $ord(n_2+n_q) \geq 3$, $ord(2n_q) \geq 3$, $p, q \in \{3,4\}, p \neq q$. First we show that $(k-1)n_p+n_q \notin D^k_h+e$ for each $h \leq k-1$. In fact assume that $(k-1)n_p + n_q = s+e$, with $ord(s) \leq k-2$:

- If $s = a_1e+a_2n_2+a_pn_p \in D_h$ and $s+e = (k-1)n_p+n_q$, then $a_1+a_2 \geq 1$ (otherwise $e > n_q$) and $a_p < a_1+a_2+a_p \leq k-2$. It follows that $a_1e+a_2n_2+e = (k-1-a_3)n_p+n_q \Longrightarrow a_1e+a_2n_2 \in D_{a_1+a_2+1}$, impossible by (2.8).
- If $s = a_1e + a_pn_p + n_q \in D_h, a_1+a_p+1 \leq k-2$, $s+e = (k-1)n_p + n_q \Longrightarrow (a_1+1)e + a_pn_p = (k-1)n_p$,impossible since $(a_1+1)e + a_pn_p < (a_1+a_p+1)n_p \leq (k-2)n_p$.

Then for each h: $D^k_h + e \subseteq \{kn_2, (k-1)n_2+n_p, \ldots, kn_p\}$. Recall also that there exists at most one h_0 such that $\sigma = (h_0-2)e+n_q \in D_{h_0}$ (2.5.6).

For $s \in D_k$, *let* $s+e = \alpha_2n_2+\alpha_pn_p$, $\alpha_2+\alpha_p \geq k+1$, define a map

$$\Phi : D_k \longrightarrow C_k, \qquad \Phi(s) = \begin{cases} (k-\alpha_p)n_2+\alpha_pn_p & if\ \alpha_p < k \\ kn_p & if\ \alpha_p \geq k \end{cases}$$

Φ is injective. In fact let $s_1, s_2 \in D_k$, be such that

$$s_1+e = \alpha_2n_2+\alpha_pn_p, s_2+e = \lambda_2n_2+\lambda_pn_p$$

(*i*) If $\alpha_p, \lambda_p < k$, then $\Phi(s_1) = \Phi(s_2) \Longrightarrow (\lambda_p-\alpha_p)n_2 = (\lambda_p-\alpha_p)n_p \Longrightarrow \lambda_p = \alpha_p$. If, e.g., $\alpha_2 < \lambda_2$, then $s_2 - s_1 = (\lambda_2-\alpha_2)n_2 + (\lambda_p-\alpha_p)n_p = (\lambda_2-\alpha_2)n_2 \Longrightarrow ord(s_2) > ord(s_1)$, impossible.

(*ii*) If $\alpha_p \geq k$, $\lambda_p < k$ and $\Phi(s_1) = kn_p = \Phi(s_2) = (k-\lambda_p)n_2+\lambda_pn_p, \lambda_p < k$, then $(k-\lambda_p)n_2 = (k-\lambda_p)n_p$, hence $k = \lambda_p$, against the assumption.

(*iii*) If $\alpha_p, \lambda_p \geq k$, $\Phi(s_1) = kn_p = \Phi(s_2)$, i.e. $s_1+e = \alpha_2n_2+\alpha_pn_p, \alpha_p \geq k$, $s_2+e = \lambda_2n_2+\lambda_pn_p, \lambda_p \geq k$. Note that s_1 has *mrp* $s_1 = a_1e+a_pn_p+n_q$ (and the same holds for s_2); otherwise $s_1 = a_1e+a_2n_2+a_pn_p \Longrightarrow (a_1+1)e+a_2n_2 = \alpha_2n_2+(\alpha_p-a_p)n_p$ and so $a_1e+a_2n_2 \in D_{a_1+a_2+1}$. Then $s_1+e = (a_1+1)e+a_pn_p+n_q = \alpha_2n_2+\alpha_pn_p \Longrightarrow a_1e+n_q = \sigma \in D_{a_1+2}$; this element is unique, hence $s_1 = \sigma+(k-2-a_1)n_p = s_2$. □

The following example, as well as all the further ones are performed by using the GAP *package for numerical semigroups*, see [5].

Example 4.5

Case (1) with $n_p = n_2 = 28$. Let $S =< 13, 28, 40, 99 >$
$A_2 = \{2n_2, n_2 + n_3, n_2 + n_4, n_3 + n_4\}$ $D_2 = \{n_4\}$, $D_3 = \{n_2 + n_4, n_3 + n_4\}$, $D_4 = \{2n_2 + n_4, n_2 + n_3 + n_4\}$, $D_k = \emptyset$ for $k \geq 5$. $H_R = [4, 7, 9, 9, 11, 13 \rightarrow]$.

Case (2) with $n_p = n_2, n_q = n_4, n_r = n_3$. Let $S =< 19, 67, 81, 142 >$
$A_2 = \{2n_2, n_2 + n_3, n_3 + n_4, 2n_4\}$ $D_6 = \{2e + 3n_4 = 464\}$, $D_k = \emptyset$ for $k \neq 6$. $H_R = [4, 8, 12, 14, 16, 17, 19 \rightarrow]$

Case (3.1) Let $S =< 31, 64, 90, 96 >$ we have
$A_2 = \{2n_2, n_2+n_3, n_2+n_4, 2n_3\}$ $D_7 = \{e+5n_3\}$, $D_8 = \{e+n_2+5n_3\}$, $D_9 = \{e + 2n_2 + 5n_3\}$, $D_k = \emptyset$ for $k \notin \{7, 8, 9\}$.

Proposition 4.6 *Let* $S = \langle n_1 = e, n_p, n_q, n_r\rangle$. *Let* $|A_2| = 4$ *and assume that* S *verifies case* (4) *of* (4.2); *then* H_R *is non-decreasing.*

Proof In this case $A_2 = \{2n_p, 2n_q, 2n_r, n_p+n_q\}$, $ord(n_p+n_r) \geq 3$. Let $n_p < n_q$. If $s = a_1e + a_pn_p + a_qn_q + a_rn_r$ is *mrp*, then $a_pa_r = a_qa_r = 0$; therefore the possible *mrp* of $s \in S$ are either $s = a_1e + a_pn_p + a_qn_q$ or $s = a_1e + a_rn_r$.

Write $D_k = D'_k \cup D''_k$, where $\begin{cases} D'_k = \{s = a_1e + a_pn_p + a_qn_q \in D_k\} \\ D''_k = \{s = a_1e + a_rn_r \in D_k\} \end{cases}$

Then $D_k + e \subseteq \{\lambda_pn_p + \lambda_qn_q, \mid \lambda_p + \lambda_q \geq k+1\} \cup \{\lambda_rn_r\}$. Clearly, there exists at most one element $\sigma \in D_k$ such that $\sigma + e = \lambda_rn_r$ and this equality implies $r < 4$.

Note that $s \in D'_k$ and $s + e = \lambda_pn_p + \lambda_qn_q$, since $(a_1+1)e + a_pn_p + a_qn_q < n_p + (k-1)n_q$ and $\lambda_p + \lambda_q \geq k+1$, we get $\lambda_q \leq k-2$. Moreover, if $s, s' \in D'_k$ are such that $s + e = \lambda_pn_p + \lambda_qn_q, s' + e = \lambda'_pn_p + \lambda'_qn_q$, then $\lambda_q \neq \lambda'_q$ by (2.5.5).

Hence $(\circ)$ $|D'_k| \leq \begin{cases} k-1 \; if \; \lambda_rn_r \notin D'_k + e \\ k \quad\; if \; \lambda_rn_r \in D'_k + e \text{ (impossible if } r = 4) \end{cases}$

Now, $s \in D''_k \Longrightarrow s + e = \lambda_pn_p + \lambda_qn_q$. We have to consider the three following situations:

- $A_2 = \{2n_2, 2n_3, 2n_4, n_2 + n_4\}$, $(p, q, r) = (2, 4, 3)$. Then $D''_k = \{a_1e + a_3n_3 \in D_k\}$ and so $s \in D''_k \Longrightarrow s + e = \lambda_2n_2 + \lambda_4n_4$, where $\lambda_4 \leq k-2$ (since $n_2 < n_3 < n_4$). Then $D_k + e \subseteq \{\lambda_pn_p + \lambda_qn_q, \mid \lambda_p + \lambda_q \geq k+1, \lambda_q \leq k-2\} \cup \{\lambda_3n_3\}$, $|D_k| \leq k$.
- $A_2 = \{2n_2, 2n_3, 2n_4, n_3 + n_4\}$, $(p, q, r) = (3, 4, 2)$. Then $D''_k = \emptyset$ (2.8) and we are done.
- $A_2 = \{2n_2, 2n_3, 2n_4, n_2 + n_3\}$, $(p, q, r) = (2, 3, 4)$. By $(\circ)$, $|D'_k| \leq k-1$; further $s \in D''_k \Longrightarrow s = a_1e + a_4n_4$ then $s + e = \lambda_2n_2 + \lambda_3n_3$.

To achieve the proof we prove that

$$|\{s \in D''_k \mid s + e = \lambda_2n_2 + \lambda_3n_3, \; with \, \lambda_3 \geq k-1\}| \leq 2;$$

since there exists at most one $s \in D_k$ such that $s+e=\lambda_3 n_3$, it suffices to show that

$$\left|\{s \in D_k'' \mid s+e=\lambda_2 n_2+\lambda_3 n_3,\ \ with\ \lambda_2 \geq 1,\ \lambda_3 \geq k-1\}\right| \leq 1 \qquad (*)$$

This inequality is due to the following technical lemmas, in particular (4.8) and (4.9.*ii*). □

Lemma 4.7 *Let* $S=\langle e, n_2, n_3, n_4\rangle$ *as in (3.2), let* $A_2=\{2n_2, 2n_3, 2n_4, n_2+n_3\}$ *and let* $\begin{cases} n_2+n_4=\ell e+pn_3 : \ell+p \geq 3 \\ n_3+n_4=me+qn_2 : m+q \geq 3\end{cases} \quad (\diamond)$
Then $pq=0$, $(p,q)\neq(0,0)$; *further* $m<3 \Longrightarrow p=0$ *and* $\ell<3 \Longrightarrow q=0$.

Proof In fact $n_2+n_3+2n_4=(\ell+m)e+qn_2+pn_3 \Longrightarrow 2n_4=(\ell+m)e+(q-1)n_2+(p-1)n_3$. Since $2n_4 \in A_2$, this implies $pq=0$. □

Lemma 4.8 *Let* $S=\langle e, n_2, n_3, n_4\rangle$ *as in (3.2), let* $A_2=\{2n_2, 2n_3, 2n_4, n_2+n_3\}$ *and assume that* $(\diamond)$ *of (4.7) holds with* $m \geq 1$ *and let* $D_k''=\{s=a_1e+a_4n_4 \in D_k\}$. *Then inequality* $(*)$ *is true.*

Proof Let a_1 be the maximum value such that $s=a_1e+a_4n_4 \in D_k''$, with $\lambda_2 \geq 1, \lambda_3 \geq k-1$ and for $1 \leq i \leq a_1$, let $s_i=(a_1-i)e+(a_4+i)n_4$: we show that $s_i \notin D_k$. Hence the statement is clear. We divide the proof in 3 steps.

Step (1) Case $m \geq 2$.

$s+e+in_4=\lambda_2 n_2+\lambda_3 n_3+in_4=\lambda_2 n_2+(\lambda_3-i)n_3+i(n_3+n_4)=\lambda_2 n_2+(\lambda_3-i)n_3+i(me+qn_2)=ime+(\lambda_2+iq)n_2+(\lambda_3-i)n_3$ where $im \geq 2i \geq i+1$ and $(\lambda_3-i)>0$, since $i \leq a_1 \leq k-2$, $\lambda_3 \geq k-1$. It follows that $s_i=[(a_1+1)-(i+1)]e+(a_4+i)n_4=(im-i-1)e+(\lambda_2+iq)n_2+(\lambda_3-i)n_3$. Hence $ord(s_i) \geq im-i-1+\lambda_2+iq+\lambda_3-i \geq \lambda_2+\lambda_3-1 \geq k$, and so $s_i \notin D_k$.

Step (2) Case $m=1$, $i \leq \lambda_2$.

When $m=1$, by (4.7) we get $q \geq 2$, $p=0$, $\ell \geq 3$, $i\ell \geq 3i > i+1$. Hence $s+e+in_4=\lambda_2 n_2+\lambda_3 n_3+in_4=(\lambda_2-i)n_2+\lambda_3 n_3+i(n_2+n_4)=(\lambda_2-i)n_2+\lambda_3 n_3+i\ell e$. Then $s_i=(\lambda_2-i)n_2+\lambda_3 n_3+(i\ell-i-1)e$ and $ord(s_i) \geq \lambda_2-i+\lambda_3+i\ell-i-1 \geq k+1$, $s_i \notin D_k$

Step (3) Case $m=1$, $\lambda_2<i$.

$s+e+in_4=\lambda_2 n_2+\lambda_3 n_3+in_4=\lambda_2(n_2+n_4)+(\lambda_3+\lambda_2-i)n_3+(i-\lambda_2)(n_3+n_4)=\lambda_2\ell e+(i-\lambda_2)(e+qn_2)+(\lambda_3+\lambda_2-i)n_3=(\lambda_2\ell+i-\lambda_2)e+(i-\lambda_2)qn_2+(\lambda_3+\lambda_2-i)n_3$. Hence $s_i=(\lambda_2\ell-\lambda_2-1)e+(i-\lambda_2)qn_2+(\lambda_3+\lambda_2-i)n_3 \Longrightarrow ord(s_i) \geq \lambda_2\ell-\lambda_2-1+(i-\lambda_2)q+(\lambda_3+\lambda_2-i) \geq 3\lambda_2-\lambda_2-1+\lambda_2-i+(i-\lambda_2)2+\lambda_3 \geq i-1+\lambda_2+\lambda_3 \geq k+1$. □

Lemma 4.9 *Let* $S = \langle e, n_2, n_3, n_4\rangle$, *let* $A_2 = \{2n_2, 2n_3, 2n_4, n_2 + n_3\}$, *and assume that* ($\diamond$) *of (4.7) holds with* $m = 0$, *i.e.,* $n_2+n_4 = \ell e,\ n_3+n_4 = qn_2$ ($\ell \geq 3, q \geq 3$) :

(i) *Let* $s \in D''_k$, $s + e = \lambda_2 n_2 + \lambda_3 n_3$. *Then* $\lambda_3 \geq k - 1 \Longrightarrow \lambda_2 \leq q$.
(ii) *Inequality* $(*)$: $\left|\{s \in D''_k \mid s+e = \lambda_2 n_2 + \lambda_3 n_3, \text{ with } \lambda_2 \geq 1, \lambda_3 \geq k-1\}\right| \leq 1$ *is true.*

Proof

(i) We get $(q+1)n_2 = \ell e+n_3$. If $\lambda_2 \geq q+1$, then $s+e = (q+1)n_2+(\lambda_2-q-1)n_2+\lambda_3 n_3 = \ell e+(\lambda_2-q-1)n_2+(\lambda_3+1)n_3$ with $\ell \geq 3, \lambda_2-q-1 \geq 0$. Then $ord(s) \geq ord((\ell-1)e+(\lambda_2-q-1)n_2+(\lambda_3+1)n_3) \geq 2+\lambda_3+1 \geq k+2$, impossible.
(ii) Assume that $s = a_1e + a_4n_4$, $s' = b_1e + b_4n_4 \in D''_k$ are such that $s + e = \lambda_2 n_2 + \lambda_3 n_3, \lambda_2 \geq 1, \lambda_3 \geq k - 1, s' + e = \lambda'_2 n_2 + \lambda'_3 n_3, \lambda'_2 \geq 1, \lambda'_3 \geq k - 1$. Assume that $\lambda_3 > \lambda'_3$, then $\lambda'_2 > \lambda_2$ (2.5.6). We have $b_1e + b_4n_4 + (\lambda_3 - \lambda'_3)n_3 = a_1e + a_4n_4 + (\lambda'_2 - \lambda_2)n_2$.

Denote by $b = (\lambda'_2 - \lambda_2)$, $g = (\lambda_3 - \lambda'_3)$, $1 \leq b \leq q - 1$ and $g \geq 1$. Then:

$$b_1e + b_4n_4 + gn_3 = a_1e + a_4n_4 + bn_2 \qquad (**)$$

We consider several subcases and proceed as in the proof of Lemma 4.8.

$(g = 1)$ We have $b_1e + b_4n_4 + gn_3 = b_1e + (b_4 - 1)n_4 + (n_3 + n_4) \Longrightarrow b_1e + (b_4 - 1)n_4 + (q - b)n_2 = a_1e + a_4n_4$, which implies $ord(s) \geq k$ if $q - b \geq 2$, impossible.
If $q - b = 1$, then $b_4 > 1$, otherwise $b_1e + (q - b)n_2 \in D_k$, impossible by (2.8). Finally, $q - b = 1, b_4 \geq 2 \Longrightarrow b_1e + (b_4 - 1)n_4 + n_2 = b_1e + (b_4 - 2)n_4 + \ell e = a_1e + a_4n_4 \Longrightarrow ord(s) \geq k - 3 + \ell \geq k$, impossible.
$(2 \leq g \leq b_4)$ $b_1e + (b_4 - g)n_4 + g(n_3 + n_4) = a_1e + a_4n_4 + bn_2 \Longrightarrow$

$$b_1e + (b_4 - g)n_4 + (gq - b)n_2 = a_1e + a_4n_4 = s.$$

Then $ord(s) \geq k - 1 - g + gq - q = k + (q - 1)(g - 1) - 2 \geq k$, impossible.
$(g > b_4)$ Then $g \geq 2$, since $b_4 \geq 1$. Equality $(**) \Longrightarrow$ $b_1e+b_4n_4+gn_3 = b_1e+b_4(n_4+n_3)+(g-b_4)n_3 = b_1e+b_4qn_2+(g-b_4)n_3 = s + bn_2$ i.e. $b_1e + (b_4q - b)n_2 + (g - b_4)n_3 = s$. Then $ord(s) \geq b_1 + (b_4q - b) + (g - b_4) = b_1 + b_4(q - 1) - b + g = b_1 + b_4 + b_4(q - 2) - b + g \geq b_1 + b_4 + b_4(q - 2) - q + 1 + g = k - 1 + q - 2 - q + 1 + g = k + g - 2 \geq k$, impossible. □

Example 4.10 The semigroup $S = \langle 101, 102, 106, 302\rangle$ verifies case (4) of (4.2), (4.6). $A_2 = \{2n_2, n_2 + n_3, 2n_3, 2n_4\}$, $A_{21} = \{2210, 2214\}$, $A_h = \emptyset$ for $h \geq 22$.

$|D_i| = 1$ for $i = 8, 9, 11, 14, 17, 19, 20$, $D_i = \emptyset$ in the other cases.
$|C_1| = 3, |C_2| = 4, |C_3| = 5, |C_4| = |C_5| = 6, |C_k| \leq k$ for each $k \in [6, 22]$.

$H_R = [4,8,13,19,25,31,37,41,45,50,54,59,64,68,73,78,82,87,91,95,98,$ $100, 101 \rightarrow]$.

5 Case $D_2 = \{n_3, n_4\}$

In this section we shall prove that the Hilbert function H_R is non decreasing when the cardinality of D_2 is maximal. We begin with an useful statement regarding this case.

Proposition 5.1 *Let* $S = \langle e, n_2, n_3, n_4 \rangle$ *as in (3.2) and let* $D_2 = \{n_3, n_4\}$.

(1) *If* $s \in D_k$, $s = a_1e + a_2n_2 + a_3n_3 + a_4n_4$ *mrp, then* $a_1 = 0$.
(2) $D_k = C_{k-1} \setminus \{(k-1)n_2\}$, *for each* $k \geq 3$, $|D_3| = |A_2| - 1$.

Proof

(1) Given $s = a_1e + a_2n_2 + a_3n_3 + a_4n_4 \in D_k$, *mrp* , we have $a_1 = 0$: in fact $a_3 + a_4 \geq 1$ by (2.8) and so the assumptions $n_3, n_4 \in D_2$ imply $a_1 = 0$.

(2) If $s \in C_{k-1} \setminus \{kn_2\}$, then $a_3 + a_4 > 0$ hence $ord(s + e) > k$, i.e. $s \in D_k$. If $s \in D_k$, then either $s \in A_{k-1} \setminus \{kn_2\}$, or $s = e + s'$. In the second case, we get $ord(s') \leq k - 3$; in fact $ord(s') = k - 2$ would imply that $e + s'$ is a maximal representation of s, impossible as noted above. Hence $s \in C_{k-1}$. Further $(k-1)n_2 \notin D_k$ for any $k \geq 2$; in particular $D_3 \subseteq A_2 \setminus \{2n_2\}$. The equality $D_3 = A_2 \setminus \{2n_2\}$ follows immediately by the assumption $D_2 = \{n_3, n_4\}$, because $s = bn_2 + cn_3 + dn_4, b + c + d = 2 \in D_2, c + d \geq 1 \Longrightarrow ord(s + e) > 3$. Since $n_3 \in D_2 \Longrightarrow 2n_2 \in A_2$, by (2.5.2), (induced) we also have $|D_3| = |A_2| - 1$. □

Proposition 5.2 *Let* $S = \langle e, n_2, n_3, n_4 \rangle$ *as in (3.2) and let* $D_2 = \{n_3, n_4\}$; *then*

(1) $n_3 + e = \alpha n_2,\ \alpha \geq 3,\quad n_4 + e = \beta n_2 + \gamma n_3,\ \beta + \gamma \geq 3,\ \alpha > \beta,\ \gamma \geq 1.$
(2) $(\alpha - \beta)n_2 + n_4 = (\gamma + 1)n_3$.
(3) (*a*) $2n_3 \in A_2 \Longleftrightarrow \gamma \geq 2$.
   (*b*) $2n_3 \notin A_2 \Longrightarrow 2n_3 = (\alpha - \beta)n_2 + n_4$, *with* $\alpha \geq \beta + 2 \geq 4,\ \gamma = 1$.
(4) *If* $a_2n_2 + a_4n_4 = a_3n_3$, *then* $a_3 \geq \gamma + 1$ *and the equality holds* $\Longleftrightarrow$ *either* $a_2 = \alpha - \beta,\ a_4 = 1$, *or* $a_2 + a_3 < \alpha - \beta + 1$.

Proof

(3.*a*) If $2n_3 \in A_2$ and $\gamma = 1$, by the equality $(\alpha - \beta)n_2 + n_4 = (\gamma + 1)n_3$ in (2), we see that $2n_3 \in A_2 \Longleftrightarrow \alpha - \beta = 1$ and this would imply S is *AAS*, that contradicts assumption (3.2). Then $\gamma \geq 2$.

(3.*b*) If $2n_3 \notin A_2$, then $\gamma = 1$ by (*a*), $\alpha - \beta \geq 2$, by (2) and by assumption (3.2) and so $\alpha \geq 4$.

(4) If $a_2n_2 + a_4n_4 = a_3n_3$, with $a_3 \leq \gamma$, then $n_4 + e = \beta n_2 + a_2n_2 + a_4n_4 + (\gamma - a_3)n_3 \Longrightarrow a_4 = 0, a_2n_2 = a_3n_3$ and so $a_2 > a_3$; hence $ord(n_4 + e) \geq$

$\beta+a_2+\gamma-a_3 > \beta+\gamma$, against the assumptions. If $a_3 = \gamma+1$, (2) implies $(\alpha-\beta)n_2+n_4 = a_2n_2+a_4n_4 \Longrightarrow (a_4-1)n_4 = (\alpha-\beta-a_2)n_2$: if $a_4 = 0$, then $(a_2-\alpha+\beta)n_2 = n_4$, impossible. If $a_4 > 1$, since $n_2 < n_4$ we get $a_4-1 < \alpha-\beta-a_2$, i.e., $ord(a_2n_2+a_4n_4) < \alpha-\beta+1$. □

Proposition 5.3 *Let $S = \langle e, n_2, n_3, n_4\rangle$ as in (3.2). Assume $D_2 = \{n_3, n_4\}$, with $n_3+e = \alpha n_2$, $n_4+e = \beta n_2+\gamma n_3$ and let $s = a_2n_2+a_3n_3+a_4n_4$ mrp.*

(1) *If $s \in D_{k+1}$, then*

(*a*) $0 < a_3+(\gamma+1)a_4 < e/2$.

(*b*) $k \le e-1-a_3(\alpha-1)-a_4(\alpha\gamma+\beta-1) \le \begin{bmatrix} e-\alpha & if\ a_3>0,\ a_4=0 \\ e-\alpha-2 & if\ a_4>0 \\ e-1 & if\ a_3+a_4=0 \end{bmatrix}$

(2) (*a*) *For each $k \le e-1$ we have $kn_2 \in C_k$.*
(*b*) *For each $k \le e-\alpha$ we have $(k-1)n_2+n_3 \in D_{k+1} \subseteq C_k$.*
(*c*) $D_h \ne \emptyset \Longrightarrow h \le e-\alpha+1$ *and* $D_{e-\alpha+1} = \{(e-\alpha-1)n_2+n_3\}$.
(3) *Let $s \in D_k$ and let $s' = s+n_2$, $s'' = s+n_3$.*

(*a*) *If any mrp $s' = b_1e+b_2n_2+b_3n_3+b_4n_4$ has $b_1 = 0$ and $\alpha \ge \beta+\gamma$, then $ord(s') = k$.*
(*b*) *If $\alpha \ge \beta+\gamma$ there exists at most one element $s \in D_k$ such that $s' = s+n_2 = pe+qn_2$.*
(*c*) *If $\alpha < \beta+\gamma$ and any mrp $s'' = b_1e+b_2n_2+b_3n_3+b_4n_4$, has $b_1 = 0$, then $ord(s'') = k$.*
(*d*) *If $\alpha < \beta+\gamma$ there exist at most two elements $\sigma_1, \sigma_2 \in D_k$ such that $\sigma_i+n_3 = p_ie+q_in_2$.*

Proof

(1.*a*) Write $a_2 = k-a_3-a_4$, $n_3 = \alpha n_2-e$, $n_4 = \beta n_2+\gamma n_3-e = \beta n_2+\gamma(\alpha n_2-e)-e$: by substituting in $s = a_2n_2+a_3n_3+a_4n_4$ we get

$$(**)\quad s = [k+(\alpha-1)a_3+(\alpha\gamma+\beta-1)a_4]\,n_2-[a_3+(\gamma+1)a_4]\,e$$
$$= [k+(\alpha-3)a_3+(\alpha\gamma+\beta-2\gamma-3)a_4]\,n_2 + [a_3+(\gamma+1)a_4](2n_2-e)$$

Note that $\alpha-3 \ge 0$, $\alpha\gamma+\beta-2\gamma-3 \ge 0$ and let $a_3+(\gamma+1)a_4 = pe+q$, $p \ge 0$, $q < e$; then $[a_3+(\gamma+1)a_4](2n_2-e) = (pe+q)(2n_2-e) = [(p+1)n_2-pe-q]e+[(p-1)e+2q]n_2$. Now $p > 0$ would imply $ord(s) > k$, hence $p = 0$ and we get $[a_3+(\gamma+1)a_4](2n_2-e) = (n_2-q)e+(2q-e)n_2$; then $ord(s) = k \Longrightarrow q < e/2$.

(1.*b*) Let $h = k+(\alpha-1)a_3+(\alpha\gamma+\beta-1)a_4 = e+r \ge e$. By the above equality (**) and by (1.*a*), we would get $s = (e+r)n_2-[a_3+(\gamma+1)a_4]e = rn_2+[n_2-a_3-(\gamma+1)a_4]e$: clearly this would be a *mrp* of s and so $s \notin D_{k+1}$ by (2.8), against the assumption. Hence $h \le e-1$ and the first inequality follows. The last inequalities are immediate since $\alpha\gamma+\beta-1 \ge \alpha+1$.

(2.*a*–*b*) Since $S = \langle e, n_2, \alpha n_2 - e, \beta n_2 + \gamma n_3 - e \rangle$ we have $GCD(e, n_2) = 1$, hence

$$(***) \qquad \lambda n_2 = \mu e \Longrightarrow \lambda \geq e.$$

We prove that $s = a_2 n_2 + a_3 n_3 \in C_k$, if $a_2 + a_3 = k$, $a_3 \leq 1$. Assume that $a_2 n_2 + a_3 n_3 = b_1 e + b_2 n_2 + b_3 n_3 + b_4 n_4$, *mrp* with $b_1 + b_2 + b_3 + b_4 > a_2 + a_3$; clearly, $b_2 \leq a_2$ (since $a_3 \leq 1$) and so $b_1 > 0$, otherwise $(a_2 - b_2)n_2 + a_3 n_3 = b_3 n_3 + b_4 n_4$ implies $a_2 - b_2 + a_3 \geq b_3 + b_4$. Now $D_2 = \{n_3, n_4\} \Longrightarrow b_3 = b_4 = 0$, $a_2 n_2 + a_3 n_3 = b_1 e + b_2 n_2$, $b_1 + b_2 > a_2 + a_3$. Now let $(a_2 - b_2)n_2 + a_3 n_3 = b_1 e$.

If $a_3 = 1$, we have $(a_2 - b_2 + \alpha)n_2 = (b_1 + 1)e$ and by $(***)$, $a_2 - b_2 + \alpha \geq e$, $a_2 \geq e - \alpha$. Hence $ord(a_2 n_2 + n_3) = a_2 + 1$ for each $a_2 \leq e - \alpha - 1$, therefore $a_2 n_2 + n_3 \in D_{k+1} \subseteq C_k$ (5.1).

If $a_3 = 0$, then $a_2 - b_2 \geq e \Longrightarrow a_2 \geq e$, by $(***)$ and so $ord(kn_2) = k$, for each $k \leq e - 1$. Finally, if $k < \alpha$, $kn_2 \in A_k \subseteq C_k$. If $\alpha \leq k \leq e - 1$, then $kn_2 = (k - \alpha)n_2 + n_3 + e \in D^k_{k-\alpha+2} \subseteq C_k$ (2.3).

(2.*c*) Since $D_h \subseteq C_{h-1} \setminus \{(h-1)n_2\}$ (5.1), then $C_{h-1} \setminus \{(h-1)n_2\} \neq \emptyset$ and this claim is immediate by (1.*b*), with $k = h - 1$.

(3.*a*) Assume that $s' = b_2 n_2 + b_3 n_3 + b_4 n_4$ *mrp* with $b_2 + b_3 + b_4 \geq k + 1$. First note that $b_2 = 0$, otherwise $ord(s) \geq k$. Hence

$$(a_2+1)n_2 + a_3 n_3 + a_4 n_4 = b_3 n_3 + b_4 n_4, \text{ with } b_3 + b_4 \geq a_2 + a_3 + a_4 + 2.$$

If $b_4 \geq a_4$, $(a_2 + 1)n_2 + a_3 n_3 = b_3 n_3 + (b_4 - a_4)n_4$, then

$$(b_3 + b_4 - a_4)n_3 < (a_2 + 1)n_2 + a_3 n_3 < (a_2 + 1 + a_3)n_3$$

against the assumption $ord(s') > k$. Hence $b_4 < a_4$, $b_3 > a_3$, $(a_2+1)n_2 + (a_4 - b_4)n_4 = (b_3 - a_3)n_3$, with $a_4 - b_4 \geq 1$, $b_3 - a_3 \geq (a_2 + 1) + (a_4 - b_4) + 1$

Let $x = a_2 + 1$, $y = a_4 - b_4$, $z = b_3 - a_3$ then $xn_2 + yn_4 = zn_3$, $z \geq x + y + 2$; since $z \geq \gamma + 1$ by (5.2.4), we get $xn_2 + yn_4 = (z - \gamma - 1)n_3 + (\alpha - \beta)n_2 + n_4$, that implies $(x - 1)n_2 + (y - 1)n_4 = (z - \gamma - 1)n_3 + (\alpha - \beta - 1)n_2$. In our assumptions $ord((x - 1)n_2 + (y - 1)n_4) = x + y - 2 \Longrightarrow x + y - 2 \geq z + \alpha - \beta - \gamma - 2$, i.e., $x + y \geq z + \alpha - \beta - \gamma$.It follows: $a_2 + 1 + a_4 - b_4 \geq b_3 - a_3 + \alpha - \beta - \gamma$, $a_2 + a_3 + a_4 \geq b_3 + b_4 + \alpha - \beta - \gamma - 1 \geq a_2 + a_3 + a_4 + 2 + \alpha - \beta - \gamma - 1 \Longrightarrow \alpha \leq \beta + \gamma - 1$.

(3.*b*) Suppose $s \in D_k$ be such that $ord(s') \geq k + 1$. Then, by (3.*a*), $s' = b_1 e + b_2 n_2 + b_3 n_3 + b_4 n_4$, with $\sum_i b_i \geq k + 1$ and $b_1 > 0$, hence $b_3 = b_4 = 0$. Further $b_2 = 0$ otherwise $ord(s) \geq k$. It follows $s' = b_1 e$; then there exists at most one element such that $ord(s') \geq k + 1$.

(3.*c*) As above, let $s'' = b_2 n_2 + b_3 n_3 + b_4 n_4$, *mrp* with $b_2 + b_3 + b_4 \geq k + 1$; then, clearly $b_3 = 0$, and so $b_2 \geq a_2$, $b_4 > a_4$. Hence $(a_3 + 1)n_3 =$

$(b_2 - a_2)n_2 + (b_4 - a_4)n_4$ with $a_3 \geq \gamma$; by substituting we get

$$(a_3 - \gamma)n_3 + (\alpha - \beta)n_2 = (b_2 - a_2)n_2 + (b_4 - a_4 - 1)n_4$$

if $(\alpha - \beta) \geq (b_2 - a_2)$, then $(b_4 - a_4 - 1) < a_3 - \gamma + \alpha - \beta$, and so $b_2 + b_4 < a_2 + a_3 + a_4$ against the assumption.
If $(\alpha - \beta) < (b_2 - a_2)$, then $ord(a_3 - \gamma)n_3 = a_3 - \gamma \geq b_2 - a_2 + b_4 - a_4 - 1 - (\alpha - \beta) > b_2 + b_4 - a_2 - a_4 - 1 - \gamma \Longrightarrow a_2 + a_3 + a_4 > b_2 + b_4 - 1$, impossible.

(3.*d*) Let $\begin{cases} s + n_3 = pe + qn_2 \\ s' + n_3 = p'e + q'n_2 \end{cases}$ with $p < p'$, $q > q'$. First note that $\alpha > q > q'$: in fact $q \geq \alpha \Longrightarrow s + n_3 = (p+1)e + (q - \alpha)n_2 + n_3 \Longrightarrow (a_2 + a_3 + a_4)n_2 < a_2n_2 + a_3n_3 + a_4n_4 = (p+1)e + (q-\alpha)n_2 < (p+1+q-\alpha)n_2$ hence $a_2 + a_3 + a_4 < p + q + 1 - \alpha$, impossible since $ord(s) = a_2 + a_3 + a_4$.
$s' + n_3 = s + n_3 + (p' - p)e - (q - q')n_2 = (p' - p - 1)e + s + (\alpha - q + q')n_2$, $\alpha - q + q' = r \geq 1$:

$$s' + n_3 = s + (p' - p - 1)e + (\alpha - q + q')n_2$$
$$s' + (q - q')n_2 = s + (p' - p)e$$

Then $p' - p = 1$: in fact, $p' - p - 1 > 0 \Longrightarrow ord(s') = k - 1 \Longrightarrow a_3 = 0, a_4 > 0, s' = s - n_4 + (p' - p - 2)e + (\alpha - q' + q + \beta)n_2 + (\gamma - 1)n_3 \Longrightarrow ord(s') > k$, against the assumption $s' \in D_k$. Since $q - q' = \alpha - r$ we get:
$s' + (q - q')n_2 = s + e = a_2n_2 + a_4n_4 + e = (a_2 + \beta)n_2 + \gamma n_3 + (a_4 - 1)n_4 \Longrightarrow s' + (\alpha - r)n_2 = (a_2 + \beta)n_2 + \gamma n_3 + (a_4 - 1)n_4 \Longrightarrow s' = (a_2 + r)n_2 + a_4n_4 - (\alpha - \beta)n_2 - n_4 + \gamma n_3 = s - n_3 + rn_2$.

If $r = 1$, then $\alpha - q + q' = 1 \Longrightarrow s' = s - n_3 + n_2$, $q = \alpha - 1, q' = 0$. Since $p' - p = 1$, we see that there exists at most one pair (s, s') which verifies the above conditions. □

Theorem 5.4 *With Setting 2.1 and 3.2, assume that $D_2 = \{n_3, n_4\}$. Then H_R is non decreasing.*

Proof By (5.3.1.*b*), $s \in D_k \Longrightarrow k \leq e - a_3(\alpha - 1) - a_4(\alpha\gamma + \beta - 1) < e - (a_3 + a_4)(\alpha - 1)$
$\leq e - \alpha + 1$ and the equality holds $\Longleftrightarrow a - 3 = 1, a_4 = 0, k = e - \alpha + 1, D_k = \{(k-2)n_2 + n_3\}$: in this last case H_R is non decreasing by (2.4.3). Then assume $k \leq e - \alpha$ and let $D_k = \{s_1, \ldots, s_p\}$.

Case $\alpha \geq \beta + \gamma$. If $ord(s_i + n_2) = k$, then $s_i + n_2 \in D_{k+1} \subseteq C_k$ (5.1); further by (5.3.3.*b*) there exists at most one index i such that $ord(s_i + n_2) > k$. By (5.3.2.*a*) we know that $kn_2 \in C_k$, since $kn_2 < s_i + n_2$ for each i, we deduce that in every case $|D_k| \leq |C_k|$.

Case $\alpha < \beta + \gamma$: by (5.3.2.*c*-*d*) we deduce that $\{kn_2, (k-1)n_2 + n_3\} \subseteq C_k \setminus (D_k + n_3)$. Hence $|D_k| \leq |C_k|$ for each k. The claim follows by (2.4.1). □

6 The Hilbert Function Does Not Decrease at Level 3

We now prove, beside other facts, that the Hilbert function H_R is non decreasing at level 3. By the results of the preceding section, we can restrict to the cases S as in (3.2), with $|D_2| \leq 1, |A_2| \geq 5$. We separate the three situations $D_2 = \emptyset$, $D_2 = \{n_3\}$, $D_2 = \{n_4\}$.

Proposition 6.1 *If* $|A_2| \geq 5$ *and* $D_2 = \emptyset$, *then* $|D_3| \leq 3$, $|C_4| \leq |A_4| + 3$, H_R *is non decreasing at level 3.*

Proof

- First note that $2n_2 \notin D_3$, $n_2 + n_3 \notin D_3$, by (2.8) and (3.4.1).
  It is enough to show that $|D_3| \leq 3$: in fact $D_2 = \emptyset \Longrightarrow |C_4| = |D_2^4| + |D_3^4| + |A_4| \leq |A_4| + 3$.
  We consider separately the cases: $n_3 + e \in D_3$, $n_3 + e \notin D_3$.
- $n_3 + e \in D_3 \Longrightarrow |D_3| \leq 3$ and either $D_3 \subseteq \{n_3 + e, n_4 + e, 2n_4\}$ or $D_3 \subseteq \{n_3 + e, n_2 + n_4, 2n_4\}$, precisely

$$\begin{cases} n_3 + 2e = \alpha n_2 & \alpha \geq 4 \\ 2n_3 \notin D_3 & \\ n_3 + n_4 \notin D_3 & (n_3 + n_4 \in D_3 \Longrightarrow n_3 + n_4 + e = \beta n_2, \\ & \quad \textit{impossible since}\, n_3 + 2e = \alpha n_2) \\ |D_3 \cap \{n_4 + e,\ n_2 + n_4\}| \leq 1 & \end{cases}$$

  In fact, $2n_3 \notin D_3$, since if $2n_3 \in D_3$, then $2n_3 + e = \alpha' n_2 + n_4$, with $\alpha' < \alpha$, and so $2n_3 + 2e = \alpha' n_2 + n_4 + e = \alpha n_2 + n_3 \Longrightarrow n_4 + e = (\alpha - \alpha')n_2 + n_3$ and so $\alpha - \alpha' = 1$ because $ord(n_4 + e) = 2$. But this implies S balanced.
  Further, either $n_4 + e \notin D_3$ or $n_2 + n_4 \notin D_3$: if $n_4 + 2e = \lambda_2 n_2 + \lambda_3 n_3$ $(\lambda_2 < \alpha)$, $n_2 + n_4 + e = \beta n_3$, then $n_2 - e = (\beta - \lambda_3)n_3 - \lambda_2 n_2 \Longrightarrow (\lambda_2 + 1)n_2 = (\beta - \lambda_3)n_3 + e$
  $\Longrightarrow n_3 + 2e = (\alpha - \lambda_2 - 1)n_2 + (\lambda_2 + 1)n_2 = (\alpha - \lambda_2 - 1)n_2 + (\beta - \lambda_3)n_3 + e \Longrightarrow e \in < n_2, n_3 >$ therefore $n_3 + e \in D_3 \Longrightarrow |D_3| \leq 3$: it follows that $|C_4| \leq |A_4| + 3$.
- $n_3 + e \notin D_3$: $D_3 \subseteq \{2n_3, n_2 + n_4, n_3 + n_4, 2n_4, n_4 + e\}$. To prove the claim note that:

$$\left[\begin{array}{llll} if & n_2 + n_4 \in D_3 & then & n_2 + n_4 + e = a_3 n_3, \qquad a_3 \geq 4 \\ if & 2n_3 \in D_3 & then & \left[\begin{array}{lll} either & 2n_3 + e = b_2 n_2, & b_2 \geq 4 \\ or & 2n_3 + e = b_2 n_2 + n_4, & b_2 \geq 3 \end{array}\right. \\ if & n_3 + n_4 \in D_3 & then & n_3 + n_4 + e = c_2 n_2, \qquad c_2 \geq 4 \\ if & 2n_4 \in D_3 & then & 2n_4 + e = d_2 n_2 + d_3 n_3, \qquad d_2 + d_3 \geq 4 \end{array}\right.$$

  To show that $|D_3| \leq 3$, we divide the proof in four parts, with different results depending on $2n_3, n_3 + n_4 \in D_3$ or not.

(*a*) If $\big(2n_3 \notin D_3,\ n_3+n_4 \notin D_3\big)$, then $D_3 \subseteq \{n_2+n_4, 2n_4, n_4+e\}$.

(*b*) If $\big\{\,2n_3, n_3+n_4\,\big\} \subseteq D_3$, then $ord(n_3+n_4+e) \geq 5$ and $ord(2n_4) \geq 3$, in particular $2n_4 \notin D_3$.

In fact $n_3+n_4+e = c_2n_2 \Longrightarrow 2n_3+e = b_2n_2+n_4 \Longrightarrow b_2 < c_2$ and $n_4-n_3 = (c_2-b_2)n_2-n_4$, i.e., $2n_4 = (c_2-b_2)n_2+n_3$, then $c_2-b_2 \geq 2$ and so $2n_4 \notin A_2$, $c_2 \geq b_2+2 \geq 5$. Further $n_2+n_4 \notin D_3$, otherwise $2n_3+2e = (b_2-1)n_2+n_2+n_4+e = (b_2-1)n_2+a_3n_3$, i.e. $2e = (b_2-1)n_2+(c_3-2)n_3$, impossible. Hence $D_3 \subseteq \{2n_3, n_3+n_4, n_4+e\}$.

(*c*) If $\big(2n_3 \in D_3, n_3+n_4 \notin D_3\big)$, consider two subcases: either $2n_3+e = b_2n_2$ or $2n_3+e = b_2n_2+n_4$.

Let $2n_3+e = b_2n_2$: then either $2n_4 \notin D_3$ or $n_2+n_4 \notin D_3$.

In fact if $n_2+n_4+e = a_3n_3$, $2n_4+e = d_2n_2+d_3n_3 \Longrightarrow d_2 < b_2$ (otherwise $2n_4 = 2n_3+(d_2-b_2)n_2+d_3n_3$, impossible if $ord(2n_4) = 2$). Therefore $n_4-n_2 = d_2n_2+(d_3-a_3)n_3$ and so $d_3 < a_3$, $n_4+(a_3-d_3)n_3 = (d_2+1)n_2 \Longrightarrow 2n_3+e = (d_2+1)n_2+(b_2-d_2-1)n_2 = n_4+(a_3-d_3)n_3+(b_2-d_2-1)n_2$, impossible in every case, since $ord(n_3+e) = 2$. Then either $D_3 \subseteq \{2n_3, n_4+e, 2n_4\}$ or $D_3 \subseteq \{2n_3, n_4+e, n_2+n_4\}$.

Let $2n_3+e = b_2n_2+n_4$. In this case $n_2+n_4 \notin D_3$. In fact $n_2+n_4+e = a_3n_3 = (a_3-2)n_3+2n_3 = (a_3-2)n_3+b_2n_2+n_4-e \Longrightarrow 2e = (b_2-1)n_2+(a_3-2)n_3$, impossible. Then $D_3 \subseteq \{2n_3, n_4+e, 2n_4\}$.

(*d*) If $\big(2n_3 \notin D_3, n_3+n_4 \in D_3\big)$, then $D_3 \subseteq \{n_4+e, n_2+n_4, n_3+n_4, 2n_4\}$. We get $|D_3| \leq 3$ because either $n_2+n_4 \notin D_3$, or $2n_4 \notin D_3$. In fact $\{n_3+n_4, 2n_4\} \subseteq D_3 \Longrightarrow d_2 < c_2$ $\big($since $d_2 \geq c_2 \Longrightarrow 2n_4+e = n_3+n_4+e+(d_2-c_2)n_2+d_3n_3$, impossible$\big)$. Then $n_2+n_4+e = a_3n_3$, $n_3+n_4+e = c_2n_2$, $2n_4+e = d_2n_2+d_3n_3 \Longrightarrow$

$$(d_3+1)n_3 = (c_2-d_2)n_2+n_4 = (c_2-d_2-1)n_2+n_2+n_4 = \\ = (c_2-d_2-1)n_2+a_3n_3-e$$

this implies $d_3 > a_3$, $2n_4+e = (d_2+1)n_2+n_4+e+(d_3-a_3)n_3$, clearly impossible. □

Proposition 6.2 *If* $|A_2| \geq 5$ *and* $D_2 = \{n_3\}$, $n_3+e = \alpha n_2$, $\alpha \geq 3$, *then* $|D_3| \leq 5$ *and* $|D_3| = 5$ *implies* $|C_3| \geq 5$; *precisely*

(1) $2n_3 \notin A_2 \Longleftrightarrow 2n_3 \notin Ap$.

(2) *If* $s = a_2n_2+a_3n_3 \in D_k$, *then* $s+e = (\alpha+a_2)n_2+(a_3-1)n_3$ *is mrp*.

(3) *If* $n_3+n_4 \in D_3$, *then* $n_3+n_4+e = \alpha n_2+n_4$ *is mrp*, $n_2+n_4 \in A_2$, $2n_2+n_4 \in C_3$.

(4) *If* $n_2+n_4 \in D_3$, *then* $n_2+n_4+e = \beta n_3$, $\beta \geq 4$, $n_4+e \in D_3^h$, $h \geq 5$, $n_4+2e = (\alpha-1)n_2+(\beta-1)n_3$, *mrp*.

(5) *If* $2n_4 \in D_3$, *then* $n_4+e \notin D_3$, $n_2+n_4 \notin D_3$, $|D_3| \leq 4$ *further conditions* $\alpha = 3$ *and* $|D_3^4| = 4$ *imply* $|A_2| = 6$.

(6) *If* $2n_4 \notin D_3$, $\{n_2+n_4, n_4+e\} \subseteq D_3$, *then* $|D_3| \leq 5$ *and* $|D_3| = 5$ *implies* $\{3n_2, 2n_2+n_3, n_2+2n_3, 3n_3, 2n_2+n_4\} \subseteq C_3$, $|C_3| \geq 5 \geq |D_3|$;

if $\alpha = 3$, *then:* $|D_3^4| \leq 4$, *and* $|D_3^4| = 4$ *implies* $|A_3| \geq 4$.

(7) *If* $2n_4 \notin D_3, n_2 + n_4 \notin D_3, n_4 + e \in D_3,$ *then* $|D_3| \leq 4$;
further if $\alpha = 3$, $|D_3^4| = 4$, *then*
$$\begin{cases} n_4 + 2e = \lambda n_2 + (4-\lambda)n_3, \ \lambda \leq 2, \ \{2n_2 + n_3, 2n_2 + n_4\} \subseteq A_3, \\ \{n_2 + 2n_3, 3n_3\} \cap A_3 \neq \emptyset, \ |A_3| \geq 3 \end{cases}$$

(8) *If* $\{n_4 + e, n_2 + n_4, 2n_4\} \cap D_3 = \emptyset$, *then* $|D_3| \leq 3$.

Proof By (2.8) and the assumption, $D_3 \subseteq \{n_4 + e, n_2 + n_3, 2n_3, n_3 + n_4, n_2 + n_4, 2n_4\}$.

(1) If $D_2 = \{n_3\}$ and $2n_3 \in Ap \setminus A_2$, then $2n_3 = a_2n_2 + a_4n_4$, with $a_2 + a_4 \geq 3$, $a_2 < \alpha$, $a_4 \leq 1$. Then by the equality $n_3+\alpha n_2 = 2n_3+e = e+a_2n_2+a_4n_4$ we get $n_3 + (\alpha - a_2)n_2 = e + a_4n_4$, impossible in every case (if $a_4 = 1$, either S is balanced or $n_4 \in D_2$).

(2) Assume $s + e = \lambda_2n_2+\lambda_3n_3+\lambda_4n_4$, with $\lambda_2 + \lambda_3+\lambda_4 > \alpha + a_2 + a_3 - 1$ and consider the equalities
$$s + e = a_2n_2 + a_3n_3 + e = \lambda_2n_2+\lambda_3n_3+\lambda_4n_4 = (a_2+\alpha)n_2+(a_3 - 1)n_3$$
Then $\lambda_3 < a_3$, otherwise $a_2n_2 + e = \lambda_2n_2 + (\lambda_3-a_3)n_3 + \lambda_4n_4$, that would imply $a_2 + 1 = ord(a_2n_2 + e) \geq \lambda_2 + \lambda_3+\lambda_4 - a_3 > a_2 + 1$; hence
$$a_2n_2 + (a_3 - \lambda_3)n_3 + e = (a_2 + \alpha)n_2 + (a_3 - \lambda_3)n_3 = \lambda_2n_2 + \lambda_4n_4.$$
The last equality clearly implies $\lambda_2 > a_2 + \alpha$; hence $a_2n_2 + a_3n_3 + e = (\lambda_2 - \alpha)n_2 + (\lambda_3 + 1)n_3 + e + \lambda_4n_4$, then $a_2 + a_3 \geq \lambda_2 - \alpha + \lambda_3 + \lambda_4 + 1$, against the assumption.

(3) It follows easily since $n_4 \notin D_2$, moreover the considered elements are induced.

(4) The first equality and the fact that $ord(n_4 + e) = 2$ are clear since neither n_2, nor n_4 belong to D_2. Then $n_2 + n_4 + 2e = (\beta - 1)n_3 + n_3 + e = (\beta - 1)n_3 + \alpha n_2 \Longrightarrow n_4 + 2e = (\beta - 1)n_3 + (\alpha - 1)n_2 \Longrightarrow n_4 + e \in D_3^h$, $h \geq 5$, since $\alpha \geq 3$, $\beta \geq 4$ and this is *mrp*.

(5) $2n_4 \in D_3 \Longrightarrow 2n_4 + e = \alpha'n_2 + \gamma'n_3$, with $\alpha' + \gamma' \geq 4$; moreover $\alpha' < \alpha$, otherwise $2n_4+e = n_3+e+(\alpha'-\alpha)n_2+\gamma'n_3$, impossible by our assumptions. Now we show that $n_4 + e \notin D_3$ and so $n_2 + n_4 \notin D_3$ by (4). If $n_4 + 2e = \alpha''n_2 + \gamma''n_3$, with $\alpha'' + \gamma'' \geq 4$, then $\alpha'' < \alpha$ by (3.2) and $n_4 \notin D_2$; since $2n_4 + e = \alpha'n_2 + \gamma'n_3$, we get $n_4 - e = (\alpha' - \alpha'')n_2 + (\gamma' - \gamma'')n_3$. We get $(\alpha' - \alpha'')(\gamma' - \gamma'') < 0$, note that

- $\alpha' - \alpha'' > 0 \Longrightarrow n_4 + (\gamma'' - \gamma')n_3 = (\alpha' - \alpha'')n_2 + e \leq (\alpha - 1)n_2 + e < \alpha n_2 = n_3 + e$, impossible.
- $\alpha' - \alpha'' < 0 \Longrightarrow n_4 + (\alpha'' - \alpha')n_2 = e + (\gamma' - \gamma'')n_3 = n_3 + e + (\gamma' - \gamma'' - 1)n_3 = \alpha n_2 + (\gamma' - \gamma'' - 1)n_3$; since $\alpha'' - \alpha' < \alpha$ we would get $n_4 \in \langle n_2, n_3 \rangle$, impossible.

In case $\alpha = 3$, then $|D_3^4| = 4 \Longleftrightarrow 2n_4 \in D_3^4 \Longleftrightarrow D_3^4 = \{n_2+n_3, 2n_3, n_3+n_4, 2n_4\}$, moreover we get $D_3^4 \cup \{2n_2\} \subseteq A_2$, $|A_2| \geq 5$. Note that $|A_2| = 5$ would imply $n_2 + n_4 \notin A_2$, impossible by (3), hence $|A_2| = 6$.

(6) Assume $n_2+n_4, n_4+e \in D_3$, $2n_4 \notin D_3$, with $n_2+n_4+e = \beta n_3$, $n_4+2e = (\alpha-1)n_2 + (\beta-1)n_3$ (4-*b*). Clearly, $2n_2, 2n_4 \notin D_3 \Longrightarrow |D_3| \leq 5$.

Since $n_3 + n_4 + e = \alpha n_2 + n_4$, $2n_3 + e = \alpha n_2 + n_3$ are *mrp*, by (2-3) and (2.5) we obtain $\{3n_2, 2n_2+n_3, n_2+2n_3, 3n_3, 2n_2+n_4\} \subseteq C_3$. These five induced elements are distinct: for this it is enough to note that $3n_3 = 2n_2 + n_4 \Longrightarrow n_2+n_4+e = 2n_2+n_4+(\beta-3)n_3$, impossible.

If $\alpha = 3$, $|D_3^4| = 4 \Longleftrightarrow D_3^4 = \{n_2+n_3, 2n_3, n_2+n_4, n_3+n_4\} \Longleftrightarrow n_2+n_4+e = 4n_3$. Since $D_3+e = \{4n_2, 3n_2+n_3, 4n_3, 3n_2+n_4, 2n_2+2n_3 (= n_4+2e)\} \subseteq C_4$ (all *mrp*) we get $\{2n_2+n_3, 3n_3, n_2+2n_3, 2n_2+n_4\} \subseteq C_3 \setminus \{D_2^3+e\} = A_3$, therefore $|A_3| \geq 3$.

(7) By assumption we have $D_3 \subseteq \{n_2+n_3, 2n_3, n_3+n_4, n_4+e\}$ and so $|D_3^4| \leq |D_3| \leq 4$.

In case $\alpha = 3$, $|D_3^4| = 4$, then $ord(n_4+2e) = 4$, $n_4 + 2e = \lambda n_2 + (4-\lambda)n_3$, $0 \leq \lambda \leq \alpha - 1 = 2$ and so $\{n_2+2n_3, 3n_3\} \cap A_3 \neq \emptyset$; further, as above one can prove that $\{2n_2+n_3, 2n_2+n_4\} \subseteq A_3$, hence $|A_3| \geq 3$.

(8) is obvious. □

Proposition 6.3 *If $|A_2| \geq 5$ and $D_2 = \{n_3\}$, let $n_3 + e = \alpha n_2$, $\alpha \geq 3$. Then*

(1) *H_R is non decreasing at level 3.*
(2) *If $\alpha = 3$, then $|D_3^4| \leq 4$, $|C_4| \leq |A_4| + 4$.*
(3) *If $\alpha \geq 4$, then $|D_3^4| \leq 1$, $|C_4| \leq |A_4| + 2$.*

Proof Claim (1) follows directly by the above lemma 6.2, by using (1) and (3) of (2.4): in fact either $|D_3| \leq 4$, or $|D_3| = 5$, $|C_3| \geq 5$.

(2) and (3) are immediate by (6.2), since: $C_4 = A_4 \cup D_3^4 \cup D_2^4$ (2.3), $|D_2^4| \leq 1$ and $\alpha = 3 \Longrightarrow D_2^4 = \emptyset$. □

Example 6.4 The semigroup $S = \langle 15, 16, 33, 101\rangle$ verifies $A_2 = \{2n_2, n_2+n_3, 2n_3, n_2+n_4, n_3+n_4\}$, $|A_2| = 5$, $D_2 = \{n_3\}$, $|D_3| = 5$; $|D_2| = 1, |C_2| = 5$, $|D_3| = |C_3| = 5$, $|D_4| = 4, |C_4| = 6$, $|D_5| = |C_5| = 5, \ldots, |D_{11}| = |C_{11}| = 2$: Hilbert function =[4, 8, 8, 10, 10, 11, 11, 12, 12, 13, 13, 14, 14, 15 →].

Proposition 6.5 *If $|A_2| \geq 5$ and $D_2 = \{n_4\}$, then H_R is non decreasing at level 3. More precisely:*

(1) $|D_3| \leq 5$, $|C_4| \leq |A_4| + 4$ *and if* $|A_2| = 5$, *then* $|D_3| \leq 4, |D_3^4| \leq 3$.
(2) *If* $|D_3| = 5$, *then* $|A_2| = 6$, $|C_3| \geq 5$, $|A_3| \geq 4$, $|D_3^4| \leq 4$.

Proof Let $n_4 + e = \beta n_2 + \gamma n_3$ (*mrp*) with $\beta + \gamma \geq 3$.

(1) $n_2 + n_3 \notin D_3$ by (3.4), $D_3 \subseteq A_2 \cup \{n_3 + e\} \setminus \{2n_2, n_2+n_3\} \subseteq \{n_3+e, 2n_3, n_2+n_4, n_3+n_4, 2n_4\}$. This proves that $|D_3| \leq 5$.

The inequality $|C_4| \leq |A_4| + 4$ will follow during the proof. Note that

(a) $n_3 + e \in D_3 \Longrightarrow n_3 + 2e = \alpha n_2,\ mrp,\ \alpha \geq 4,\ \alpha > \beta,\ \gamma > 0,\ 2n_2 \in A_2,$
$\beta = 0 \Longrightarrow 2n_3 \in A_2, \quad \beta \geq 1 \Longrightarrow n_2 + n_3 \in A_2$

(b) $2n_3 \in D_3 \quad \Longrightarrow 2n_3 + e = \alpha' n_2 + \delta n_4,\ mrp,\ \alpha' + \delta \geq 4,\ \alpha' \geq 3,\ \delta \leq 1.$

Now consider the following subcases:

- $\{n_3 + e, 2n_3\} \subseteq D_3$:

$$\begin{cases} \delta\gamma = 1,\ \alpha = \alpha' + \beta \geq 5,\ \beta \geq 2, & \\ n_4 + e = \beta n_2 + n_3, & 2n_2, n_2 + n_3, 2n_3 \in A_2,\ n_2 + n_4 \in D_3 \\ n_3 + 2e = \alpha n_2, & 3n_2 \in A_3 \\ 2n_3 + e = (\alpha - \beta)n_2 + n_4, & 2n_2 + n_4 \in C_3 \setminus (D_2 + e) \subseteq A_3 \\ n_2 + n_4 + e = (\beta+1)n_2 + n_3, & 2n_2 + n_3 \in C_3 \\ n_3 + n_4 + e = \beta n_2 + 2n_3, & n_3 + n_4 \in A_2 \Longrightarrow n_2 + 2n_3 \in A_3 \\ 2n_4 + e = \beta n_2 + n_3 + n_4, & n_3 + n_4, 2n_4 \in A_2 \Longrightarrow n_2 + n_3 + n_4 \in A_3 \end{cases}$$

By (a) and (b) it follows that $2n_3 + e = \alpha' n_2 + n_4, \delta = 1$. Therefore $\alpha n_2 + n_3 = 2n_3 + 2e = \alpha' n_2 + n_4 + e = (\alpha' + \beta)n_2 + \gamma n_3$ and so $(\alpha - \alpha' - \beta)n_2 = (\gamma - 1)n_3$; since $ord((\gamma - 1)n_3) = \gamma - 1$, from the mrp of $n_4 + e$ we get $\alpha - \alpha' - \beta = \gamma - 1 = 0$. Then $\beta \geq 2,\ \alpha = \alpha' + \beta \geq 5$. Hence $\{2n_2, n_2 + n_3, n_2 + n_4, 2n_3\} \subseteq A_2,\ n_2 + n_4 \in D_3$, since the first 3 elements are induced and $2n_3 \in D_3 \Longrightarrow 2n_3 \in A_2$ by the assumption (3.2). Moreover they are distinct by (3.2) and so $|A_2| = 5 \Longrightarrow |A_2 \cap \{n_3 + n_4, 2n_4\}| \leq 1$. Clearly $n_2 + n_4 + e = (\beta+1)n_2 + n_3$ is mrp, then $2n_2 + n_3 \in C_3$ (induced element); further $2n_2 + n_4 \in A_3$ since it is induced and different from $n_4 + e$. Also, $|D_3^4| \leq 4$, since $n_3 + e \notin D_3^4$ and $|D_3^4| = 4 \Longrightarrow \beta = 2$ and so $\beta + \gamma = 3,\ |D_2^4| = 0,\ |C_4| \leq |A_4| + 4$. We conclude that

(c) $\{n_3 + e, 2n_3\} \subseteq D_3 : \Longrightarrow |D_3^4| \leq 4,\ |C_4| \leq |A_4| + 4$
(d) $|A_2| = 5 \Longrightarrow |D_3| \leq 4$ and $|D_3^4| \leq 3$.

If $n_3 + n_4 \in A_2$ one can easily see that the representation $n_3 + n_4 + e = 2n_3 + (\alpha - \alpha')n_2$ is mrp and so we get the induced element $n_2 + 2n_3 \in C_3 \cap A_3$.

Finally, if $n_3 + n_4, 2n_4 \in A_2$, then $ord(n_2 + n_3 + n_4) = 3$. In fact $ord(n_2 + n_3 + n_4) > 3$ implies $n_2 + n_3 + n_4 = \mu e$, impossible since the equality $2n_4 + e = (\alpha - \alpha')n_2 + n_3 + n_4$ would imply $ord(2n_4) > 3$.

- $2n_3 \notin D_3,\ n_3 + e \in D_3$:

$D_3 \subseteq \{n_3 + e\} \cup \{n_2 + n_4, n_3 + n_4, 2n_4\}$. We get $|D_3^4| \leq 3$; in fact

$2n_3 \in A_2 \setminus D_3 \Longrightarrow n_2 + n_3 \in A_2$ and so, $2n_2 \in A_2$,
$|A_2| = 5 \Longrightarrow |D_3| \leq 3$.

$2n_3 \notin A_2 \Longrightarrow ord(2n_3) \geq 3$, by (3.2), hence $\gamma = 1$, by (a),
$\beta \geq 2$, and so,
$n_3 + n_4 \in D_3, n_3 + n_4 + e = \beta n_2 + 2n_3 \Longrightarrow$
$n_3 + n_4 \in D_3^{h \geq 5}$.

- $2n_3 \in D_3, n_3 + e \notin D_3$:

then $\left[\begin{array}{ll} \text{either } \delta = 1, \alpha' < \beta, \gamma = 0, 2n_4 \notin A_2, |D_3| \leq 3, |A_2| = 5 \\ \text{or} \quad \delta = 0, \alpha' > \beta, \gamma > 0, |A_2| = 5 \Longrightarrow |D_3| \leq 3 \end{array}\right.$

If $\delta = 1 \Longrightarrow \gamma = 0$; in fact $2n_3 + 2e = \alpha' n_2 + n_4 + e = (\alpha' + \beta)n_2 + \gamma n_3 \Longrightarrow \gamma \leq 1$; further $\gamma = 1 \Longrightarrow n_3 + 2e = (\alpha' + \beta)n_2 \Longrightarrow n_3 + e \in D_3$ (since $n_3 \notin D_3 \Longrightarrow ord(n_3 + e) = 2$), against the assumption. Hence $n_4 + e = \beta n_2, 2n_3 + e = \alpha' n_2 + n_4$; these equalities by substitution imply $2n_3 = (\alpha' - \beta)n_2 + 2n_4$, $(\alpha' - \beta) < 0$, $2n_4 \notin A_2$

If $\delta = 0 \Longrightarrow 2n_3 + e = \alpha' n_2$, $n_4 + e = \beta n_2 + \gamma n_3$, $\gamma > 0, \alpha' > \beta$. Then $2n_2, n_2 + n_3, 2n_3 \in A_2$ and so $A_2 = 5 \Longrightarrow \{n_2 + n_4, n_3 + n_4, 2n_4\} \cap A_2 \leq 2 \Longrightarrow |D_3| \leq 3$. In any case $|C_4| \leq |A_4| + 4$. This concludes the proof of (1).

(2) Assume $|D_3| = 5$, i.e. $D_3 = \{n_3 + e, 2n_3, n_2 + n_4, n_3 + n_4, 2n_4\}$, since $n_2 + n_3 \notin D_3$. Then by (*c*) and (*d*) we see that $|A_2| = 6$ and $|D_3^4| \leq 4$. Further these conditions imply $\{3n_2, 2n_2 + n_4, n_2 + 2n_3, n_2 + n_3 + n_4\} \subseteq A_3$ and $\{3n_2, 2n_2 + n_4, n_2 + 2n_3, n_2 + n_3 + n_4, 2n_2 + n_3\} \subseteq C_3$.

Finally, by (2.4.1–3) and by (1) and (2) we see that H_R is non decreasing at level 3. □

Example 6.6 The semigroup $S = \langle 15, 16, 50, 67 \rangle$ has $D_2 = \{n_4\}, |A_2| = 6, |D_3| = 5$. Hilbert function: [4, 9, 9, 10, 10, 11, 11, 12, 12, 13, 13, 13, 14, 15].

7 Conclusions

In this section we summarize all the proved results in the following theorem and we prove that the Hilbert function is non decreasing if the multiplicity of R is ≤ 13.

Theorem 7.1 *Let S be a numerical semigroup minimally generated by e, n_2, n_3, n_4 and let $R = k[[S]]$. Then:*

(1) *The Hilbert function H_R is non decreasing at level ≤ 3*
(2) *The Hilbert function H_R is non decreasing if the semigroup S verifies one of the following cases*

(a) $|A_2| \leq 4$ (4.1)
(b) $|D_2| = 2$ (5.4)
(c) $|A_3| \leq 2$.
(d) $|A_3| + |D_2^3| \leq 3$
(e) $A_4 = \emptyset$

Proof By Remark 3.1 it is enough to restrict to Setting 3.2. Further, by (2.*b*) we can assume $|D_2| \leq 1$; in particular, (2.*c*) is immediate from (2.*b*) and (2.*d*).

(1) is a corollary of Theorem (5.4) and Propositions (6.3), (6.5), (6.1).

(2.*d*–*e*) Recall that $C_3 = A_3 \cup D_2^3$ (2.3.1). If $A_4 = \emptyset$, by (1) it is enough to show that H_R is non-decreasing at level ≥ 4. By (6.3), (6.5.1), (6.1), we have

$|C_4| \le 4$. Then in both cases apply (2.4.5): H_R decreasing at level $\ell \ge 3$ would imply $|C_k| \ge k+1$ for each $2 \le k \le \ell$. □

Let $R' = R/t^e R$: the Hilbert function of R' is

$$H_{R'} = \left[1 = |A_0|,\ v-1 = |A_1|,\ |A_2|,\ \cdots,\ |A_d|\right]$$

where $d = max\{ord(\sigma)\,|\,\sigma \in Ap(S)\}$, so that $A_{d+1} = \emptyset$, see e.g., [16, Lemma 1.3]. The case $e \le 13$ is mostly a corollary of the above results:

Theorem 7.2 *With Setting 2.1, let S be a numerical semigroup minimally generated by* e, n_2, n_3, n_4 *and let* $R = k[[S]]$. *If the multiplicity e of R is* ≤ 13, *then* H_R *is non-decreasing.*

Proof We already know that for $e \le 8$ the function H_R is non-decreasing as proved in [4, Corollary 3.14]. We classify the remaining cases according to the Hilbert function of $R' = R/t^e R$:

$$H_{R'} = \left[1,\ 3,\ |A_2|, \cdots, |A_d|\right]$$

where d is such that $A_{d+1} = \emptyset$, $\sum_2^d |A_k| = e-4$, and according to the admissibility theorem of Macaulay. By (7.1), it is enough to show that H_R doesn't decrease at level $\ell \ge 4$, when $|A_2| \ge 5$, $|A_3| \ge 3$ and $|D_2| \le 1$, under the assumption (3.2).

We have $|A_3| \le e - 4 - |A_2| \le e - 9$; then, if either $e \le 11$, or $e = 12$ and $|A_2| = 6$, we get $|A_3| \le 2$ and apply Theorem (7.1.2.*d*). Now the remaining cases are:

$$\begin{array}{llll}
(1) & e = 12, & H_{R'} = [1, 3, 5, 3, 0], & |D_2^3| = 1 \\
(2) & e = 13, & H_{R'} = [1, 3, 5, 4, 0], & |D_2^3| \le 1 \\
(3) & e = 13, & H_{R'} = [1, 3, 6, 3, 0], & |D_2^3| = 1 \\
(4) & e = 13, & H_{R'} = [1, 3, 5, 3, 1, 0], & |D_2^3| = 1
\end{array}$$

The claim in the first 3 ones is clear by (7.1.2.*e*).

In case (4), in most situations $|C_4| = |A_4| + |D_3^4| + |D_2^4| \le 4$ and it suffices to use (2.4.5):

- if $D_2 = \emptyset$, then $|C_4| \le |A_4| + 3 = 4$, by (6.1);
- if $D_2 = \{n_4\}$, then $|D_3^4| \le 3$ and $|D_2^4| = 0$, $|C_4| \le 1 + 3 = 4$, by (6.5.1);
- if $D_2 = \{n_3\}$, $n_3 + e = \alpha n_2$, with $\alpha \ge 4$, then $|D_3^4| \le 3$ and $|D_2^4| = 0$, $|C_4| \le 1 + 3 = 4$, by (6.3.3). It remains the case $D_2 = \{n_3\}$, with $\alpha = 3$: by (6.2) the unique possible "decreasing" situation is *case* (6.2.7), with

$$|D_3^4| = 4,\ n_4 + 2e = \lambda n_2 + (4-\lambda) n_3,\ 0 \le \lambda \le 2, \quad |A_3| = 3 \qquad (\diamond)$$

where either $A_3 = \{2n_2 + n_3, 2n_2 + n_4, 3n_3\}$, or $A_3 = \{2n_2 + n_3, 2n_2 + n_4, n_2 + 2n_3\}$. Then it is clear that $n_2 + n_3 + n_4 \notin A_3$, hence $n_2 + n_3 + n_4 = \mu e$, $\mu \geq 4$, since in this case $\{n_2 + n_3, n_2 + n_4, n_3 + n_4\} \subseteq A_2$:

$$\begin{aligned} \lambda = 0 &\Longrightarrow (\mu + 2)e = n_2 + n_3 + n_4 + 2e = n_2 + 5n_3 = n_2 + 5(3n_2 - e) \\ &\Longrightarrow (\mu + 7)13 = 16n_2 \\ \lambda = 1 &\Longrightarrow (\mu + 2)e = n_2 + n_3 + n_4 + 2e = 2n_2 + 4n_3 = 2n_2 + 4(3n_2 - e) \\ &\Longrightarrow (\mu + 6)13 = 14n_2 \end{aligned}$$

similarly, $\lambda = 2 \Longrightarrow (\mu + 5)13 = 12n_2$. For each value of λ we see that $e = 13$ must divide n_2, impossible; this proves that case ($\diamond$) cannot exist. Hence H_R is non decreasing also under the assumption (4). □

Acknowledgements This work is supported by the National Group of Algebraic and Geometric Structures and their Applications (GNSAGA)-INDAM—Italy.

References

1. Arslan, F., Mete P.: Hilbert functions of Gorenstein monomial curves. Proc. Am. Math. Soc. **135**(7), 1993–2002 (2007)
2. Arslan, F., Mete, P., Sahin, M.: Gluing and Hilbert functions of monomial curves. Proc. Am. Math. Soc. **137**, 2225–2232 (2009)
3. Cortadellas Benitez, T., Jafari, R., Zarzuela Armengou, S.: On the Apéry set of monomial curves. Semigroup Forum **86**(2), 289–320 (2013)
4. D'Anna, M., Di Marca, M., Micale, V.: On the Hilbert function of the tangent cone of a monomial curve. Semigroup Forum **91**(3), 718–730 (2015)
5. Delgado, M., García-Sánchez, P.A., Morais, J.: NumericalSgps. A GAP package for numerical semigroups, Version 1.1.5 (2017). (Refereed GAP package, https://gap-packages.github.io/numericalsgps)
6. Elias, J.: The conjecture of Sally on the Hilbert function for curve singularities. J. Algebra **160**(1), 42–49 (1993)
7. Elias, J.: On the depth of the tangent cone and the growth of the Hilbert function. Trans. Am. Math. Soc. **351**(10), 4027–4042 (1999)
8. García, A.: Cohen-Macaulayness of the associated graded ring of a semigroup ring. Commun. Algebra **10**(4), 393–415 (1982)
9. Gupta, S.K., Roberts, L.G.: Cartesian squares of ordinary singularities of curves. Commun. Algebra **11**(2), 127–182 (1983)
10. Jafari, R., Zarzuela Armengou, S.: On monomial curves obtained by gluing. Semigroup Forum **88**(2), 397–416 (2014)
11. Molinelli, S., Tamone, G.: On the Hilbert function of certain rings of monomial curves. J. Pure Appl. Algebra **101**(2), 191–206 (1995)
12. Molinelli, S., Tamone, G.: On the Hilbert function of certain non Cohen-Macaulay one-dimensional rings. Rocky Mountain J. Math. **29**(1), 271–300 (1999)
13. Oneto, A., Tamone, G.: On semigroup rings with decreasing Hilbert function. J. Algebra **489**, 373–398 (2017)
14. Oneto, A., Strazzanti, F., Tamone, G.: One-dimensional Gorenstein local rings with decreasing Hilbert function. J. Algebra **489**, 91–114 (2017)

15. Orecchia, F.: One dimensional local rings with reduced associated ring and their Hilbert function. Manuscripta Math. **32**(3–4), 391–405 (1980)
16. Patil, D.P., Tamone, G.: CM defect and Hilbert functions of monomial curves. J. Pure Appl. Algebra **215**, 1539–1551 (2011)
17. Roberts, L.G.: Ordinary singularities with decreasing Hilbert function. Can. J. Math. **34**, 169–180 (1982)
18. Rossi, M.E., Valla, G.: Cohen-Macaulay local rings of embedding dimension $e + d - 3$. Proc. Lond. Math. Soc. **80**, 107–126 (2000)
19. Tamone, G.: On the Hilbert function of some non Cohen-Macaulay graded rings. Commun. Algebra **26**(12), 4221–4231 (1998)

Lattice Ideals, Semigroups and Toric Codes

Mesut Şahin

Abstract Let X be a complete n-dimensional simplicial toric variety over a finite field with homogeneous coordinate ring S. In this survey, we review algebraic methods for studying evaluation codes defined on subsets of the algebraic torus T_X. The key object is the vanishing ideal of the subset and its Hilbert function. We also explore the nice correspondence between subgroups of the group T_X and lattice ideals as their vanishing ideals. We present recent results for obtaining a basis for the lattice and for computing a minimal generating set of its ideal.

Keywords Toric code · Toric variety · Linear code · Lattice ideal · Affine semigroup · Numerical semigroup

1 Introduction

Let $\mathbb{K} = \mathbb{F}_q$ be a finite field with q elements. A complete simplicial toric variety X of dimension n over $\mathbb{K}$ corresponds to a fan Σ whose cones fill up the space $\mathbb{R}^n$, see the wonderful book [4] by Cox, Little and Schenk for the general theory of toric varieties. Standard multiplication action of the algebraic torus $T_X \cong (\mathbb{K}^*)^n$ on itself extends to an action on X and orbits of this algebraic action correspond to cones in Σ. One dimensional cones of Σ, generated by primitive lattice vectors $\mathbf{v}_1, \dots, \mathbf{v}_r \in \mathbb{Z}^n$, are denoted by $\rho_1, \dots, \rho_r$. We use $[m]$ to denote the set $\{1, \dots, m\}$ for any positive integer $m \geq 1$ and $x^{\mathbf{u}}$ to denote the Laurent monomial $\mathbf{x}^{\mathbf{u}} = x_1^{u_1} \dots x_m^{u_m}$ for

The author is supported by TÜBİTAK Project No. 114F094.

M. Şahin (✉)
Hacettepe University, Ankara, Turkey
e-mail: mesut.sahin@hacettepe.edu.tr

V. Barucci et al. (eds.), *Numerical Semigroups*, Springer INdAM Series 40,
https://doi.org/10.1007/978-3-030-40822-0_16

a vector $\mathbf{u} \in \mathbb{Z}^m$. If ϕ is the matrix with rows $\mathbf{v}_1, \ldots, \mathbf{v}_r$ then we have the following short exact sequence:

$$\mathfrak{P} : 0 \longrightarrow \mathbb{Z}^n \xrightarrow{\phi} \mathbb{Z}^r \xrightarrow{\beta} \mathcal{A} \longrightarrow 0 ,$$

where $\mathcal{A}$ is isomorhic to the class group of X. Smooth X with an n-dimensional cone $\sigma \in \Sigma$ has a torsion free class group by [4, Proposition 4.2.5]. So, we assume that $\mathcal{A} \cong \mathbb{Z}^d$, where $d := r - n$. Applying the contravariant functor $Hom(-, \mathbb{K}^*)$ to $\mathfrak{P}$ gives the following short exact sequence:

$$\mathfrak{P}^* : 1 \longrightarrow G \xrightarrow{i} (\mathbb{K}^*)^r \xrightarrow{\pi} (\mathbb{K}^*)^n \longrightarrow 1 ,$$

where $\pi : (t_1, \ldots, t_r) \mapsto (\mathbf{t}^{\mathbf{u}_1}, \ldots, \mathbf{t}^{\mathbf{u}_n})$, with $\mathbf{u}_1, \ldots, \mathbf{u}_n$ being the columns of ϕ.

Our algebraic object of interest is the polynomial ring

$$S = \mathbb{K}[x_1, \ldots, x_r] = \bigoplus_{\alpha \in \mathcal{A}} S_{\alpha}$$

multigraded by $\mathcal{A}$ via $\deg(x_j) := \beta_j := \beta(e_j)$, where e_j is the standard basis element of $\mathbb{Z}^r$ for each $j \in [r]$. The irrelevant ideal of S is the monomial ideal $B = \langle x^{\hat{\sigma}} : \sigma \in \Sigma \rangle$, where $x^{\hat{\sigma}} = \prod_{\rho_i \notin \sigma} x_i$. The subgroup $G \subset (\mathbb{K}^*)^r$ acts naturally on $\mathbb{K}^r \setminus V(B)$ and the resulting orbits $[P] := G \cdot P$ make into a variety isomorphic to X by a fundamental result of Cox [3]. This is because G is reductive when $\mathcal{A}$ is torsion free, and thus arguments of [3] applies even thought $\mathbb{K}$ is a finite field. Thus, any representative of the orbit $[P]$ is said to be a "homogeneous coordinate" for a point of X. Due to this relation, S is also known as the Cox ring of X. The multigraded polynomials of S are supported in the semigroup $\mathbb{N}\beta$ generated by $\beta_1, \ldots, \beta_r$, i.e. $\dim_{\mathbb{K}} S_{\alpha} = 0$ when $\alpha \notin \mathbb{N}\beta$. Since X is complete, $\dim_{\mathbb{K}} S_{\alpha}$ is finite for all $\alpha \in \mathbb{N}\beta$.

Next, we recall linear codes obtained by evaluating homogeneous polynomials on arbitrary subsets $Y = \{[P_1], \ldots, [P_N]\}$ of X. For a fix degree $\alpha \in \mathbb{N}\beta$, we define the *evaluation map*

$$\mathrm{ev} : S_{\alpha} \to \mathbb{F}_q^N, \quad F \mapsto (F(P_1), \ldots, F(P_N)) . \tag{1}$$

The image $C_{\alpha,Y} = \mathrm{ev}_Y(S_{\alpha})$ of this linear map is a subspace of the $\mathbb{F}_q$-vector space $\mathbb{F}_q^N$ and is called a *generalized toric code*. There are three basic parameters of a linear code. The length N is the number of points in Y, the dimension k is the usual dimension of $C_{\alpha,Y}$ as a vector space over $\mathbb{F}_q$, and the minimum distance $d = d(C_{\alpha,Y})$ is the minimum of weights of nonzero vectors in $C_{\alpha,Y}$, where weight of a vector is the number of nonzero components. For more details on error-correcting linear codes, consult e.g. [11, 20].

Toric codes, which are the evaluation codes obtained from the torus $Y = T_X$ were studied first by Hansen around 1998 and have been studied until now by many others, see [12] for a nice geometric introduction to the topic. See also [9] for a more recent geometric approach. They are used many times in literature for obtaining examples of linear codes with the parameters better than existing ones, see [2].

In this survey, we restrict ourselves to reviewing combinatorial commutative algebraic methods for studying evaluation codes defined on subsets of the algebraic torus T_X. The key object is the vanishing ideal of the subset and its Hilbert function. We also explore the nice correspondence between subgroups of the group T_X and lattice ideals as their vanishing ideals. We present recent results for obtaining a basis for the lattice and for computing a minimal generating set of its ideal.

2 The Algebraic Approach

In this section, we review methods offered by combinatorial commutative algebra for computing main parameters of codes obtained from subsets of a toric variety. *The essense of this approach is to compute or estimate parameters without constructing the code using algebraic and geometric properties of the subset and of the toric variety.* We refer the reader to an excellent book [13] written by Miller and Sturmfels for many aspects of commutative algebra related to combinatorics and geometry. Before we go further, let us present two important examples of toric varieties we use to illustrate the main results.

The first one is the class of *weighted projective spaces* closely related to the well known class of *numerical semigroups.*

Example 2.1 Let $\beta = [w_1 \cdots w_r]$ be the matrix representing the map β in the first exact sequence $\mathfrak{P}$, where $w_1, \ldots, w_r$ are relatively prime positive integers. We grade the polynomial ring $S = \mathbb{K}[x_1, \ldots, x_r]$ by assigning $\deg(x_i) = w_i$, for $i \in [r]$. The semigroup $\mathbb{N}\beta$ is generated, not necessarily minimally, by $w_1, \ldots, w_r$. There are only finitely many positive integers not belonging to $\mathbb{N}\beta$, known as the gaps, and thus $N\beta$ is a numerical semigroup. Let L_β be the subgroup of $\mathbb{Z}^r$ whose elements are sent to zero under β. Since $\mathbb{Z}\beta = \mathbb{Z}$ is free, the lattice ideal

$$I_{L_\beta} := \langle \mathbf{x}^{\mathbf{u}} - \mathbf{x}^{\mathbf{v}} \, : \, \mathbf{u}, \mathbf{v} \in \mathbb{N}^r \text{ and } \beta\mathbf{u} = \beta\mathbf{v} \rangle$$

is the toric ideal of the semigroup $\mathbb{N}\beta$ whose zero locus is the monomial curve:

$$V(I_{L_\beta}) = \{(t^{w_1}, \ldots, t^{w_r}) \, : \, t \in \mathbb{K}\}.$$

By the dual short exact sequence $\mathfrak{P}^*$, the group $G = V(I_{L_\beta}) \cap (\mathbb{K}^*)^r$ is the zero locus in $(\mathbb{K}^*)^r$ of this monomial curve, and thus, $G = \{(t^{w_1}, \ldots, t^{w_r}) \, : \, t \in \mathbb{K}^*\}$. The irrelevant ideal is $B = \langle x_1, \ldots, x_r \rangle$ with a zero set $V(B) = (0, \ldots, 0)$. So, points of the weighted projective space $X = \mathbb{P}(w_1, \ldots, w_r)$ are the following equivalence classes:

$$[x_1 : \cdots : x_r] = \{(t^{w_1}x_1, \ldots, t^{w_r}x_r) \, : \, t \in \mathbb{K}^*\}, \text{ where } (x_1, \ldots, x_r) \neq (0, \ldots, 0).$$

The toric variety $X = \mathbb{P}(w_1, \ldots, w_r)$ is smooth if and only if it is the usual projective space $\mathbb{P}^{r-1}$, i.e., $w_1 = \cdots = w_r = 1$. □

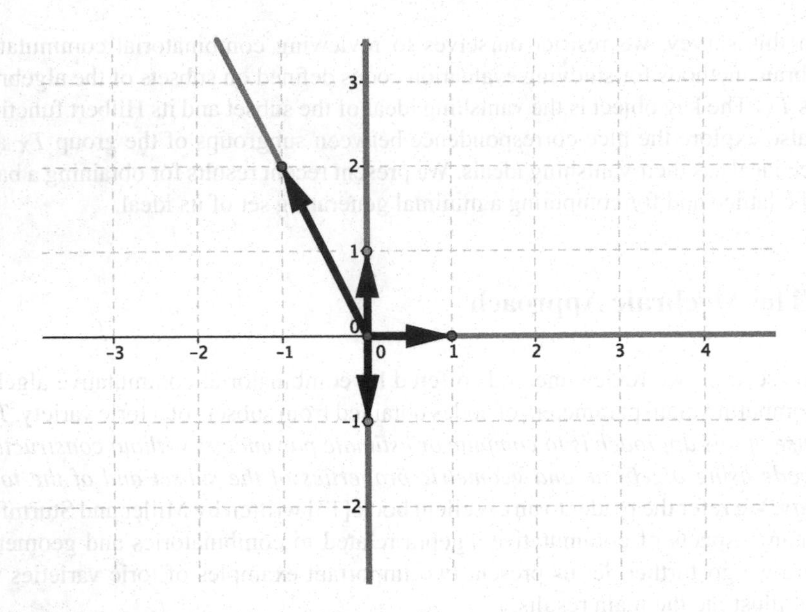

Fig. 1 The fan for the Hirzebruch surface

Our second example is a smooth surface that will be running throughout.

Example 2.2 Let $X = H_2$ be the Hirzebruch surface corresponding to a fan in $\mathbb{R}^2$ generated by $\mathbf{v}_1 = (1, 0)$, $\mathbf{v}_2 = (0, 1)$, $\mathbf{v}_3 = (-1, 2)$, and $\mathbf{v}_4 = (0, -1)$ (Fig. 1).

So, H_2 has the following exact sequence:

$$\mathfrak{P} : 0 \longrightarrow \mathbb{Z}^2 \xrightarrow{\phi} \mathbb{Z}^4 \xrightarrow{\beta} \mathbb{Z}^2 \longrightarrow 0\,,$$

where

$$\phi = \begin{bmatrix} 1 & 0 & -1 & 0 \\ 0 & 1 & 2 & -1 \end{bmatrix}^T \quad \text{and} \quad \beta = \begin{bmatrix} 1 & 0 & 1 & 2 \\ 0 & 1 & 0 & 1 \end{bmatrix}.$$

The ring $S = \mathbb{K}[x, y, z, w]$ is multigraded via

$$\deg(x) = \deg(z) = (1, 0),\ \deg(y) = (0, 1) \text{ and } \deg(w) = (2, 1).$$

- The irrelevant ideal is $B = \langle xy, yz, zw, wx \rangle$ with the following zero set

$$V(B) = V(x, z) \cup V(y, w).$$

- $X = H_2 = (\mathbb{K}^4 \setminus V(B))/G$, where

$$G = \{(x, y, z, w) \in (\mathbb{K}^*)^4 \,|\, xz^{-1} = yz^2w^{-1} = 1\} = \{(x, y, x, x^2y) \,|\, x, y \in \mathbb{K}^*\}.$$

- A point of X is an orbit such as the following one:

$$[1:1:0:0] := G \cdot (1,1,0,0) = \{(x, y, 0, 0) \mid x, y \in \mathbb{K}^*\}.$$

The key algebraic tool in the course of studying codes on toric varieties is the multigraded Hilbert function.

Definition 2.3 Let $I(Y)$ be the multigraded vanishing ideal of $Y \subseteq X$, i.e. it is the ideal generated by homogeneous polynomials vanishing on Y. Then, the multigraded Hilbert function and series are defined respectively by

$$H_Y(\boldsymbol{\alpha}) = \dim_{\mathbb{K}} S_{\boldsymbol{\alpha}} - \dim_{\mathbb{K}} I_{\boldsymbol{\alpha}}(Y) \quad \text{and} \quad HS_Y(\mathbf{t}) = \sum_{\boldsymbol{\alpha} \in \mathcal{A}} H_Y(\boldsymbol{\alpha})\mathbf{t}^{\boldsymbol{\alpha}}.$$

Since the kernel of the evaluation map is the subspace $I_{\boldsymbol{\alpha}}(Y)$ of $S_{\boldsymbol{\alpha}}$ which is spanned by the homogeneous polynomials vanishing at all the points of Y, dimension of the image is nothing but the value of the Hilbert function of Y at $\boldsymbol{\alpha}$.

Proposition 2.4 *The dimension of* $C_{\boldsymbol{\alpha},Y}$ *equals* $H_Y(\boldsymbol{\alpha})$. □

So, it is desirable to investigate properties of the Hilbert function to understand codes in question better. We use the partial ordering $\preceq$ below, where $\boldsymbol{\alpha} \preceq \boldsymbol{\alpha}'$ if $\boldsymbol{\alpha}' - \boldsymbol{\alpha} \in \mathbb{N}\beta$.

Theorem 2.5 ([17, Theorem 3.7]) *The multigraded Hilbert function H_Y of $Y \subseteq X$ has the following properties.*

(i) If $\boldsymbol{\alpha} - \boldsymbol{\alpha}_i \notin \mathbb{N}\beta$, for every degree $\boldsymbol{\alpha}_i$ of minimal generators of $I(Y)$, then $H_Y(\alpha) = \dim_{\mathbb{K}} S_{\boldsymbol{\alpha}}$,
(ii) If there is a non-zerodivisor in $S/I(Y)$ of degree β_j, for each $j \in [r]$, then it is non-decreasing: $H_Y(\boldsymbol{\alpha}) \leq H_Y(\boldsymbol{\alpha}')$ for all $\boldsymbol{\alpha} \preceq \boldsymbol{\alpha}'$.
(iii) $H_Y(\boldsymbol{\alpha}) \leq |Y|$, for all $\boldsymbol{\alpha} \in \mathbb{N}\beta$.

□

Remark 2.6 The Hilbert function is non-decreasing especially for subsets $Y \subseteq T_X$ since there exists a non-zerodivisor in $S/I(Y)$ of any degree $\alpha \in \mathbb{N}\beta$ by [17, Lemma 3.6]. □

The fan Σ of X determines an important subsemigroup $\mathcal{K}$ of the semigroup $\mathbb{N}\beta$, containing some what more important degrees which we isolate here.

Definition 2.7 Let $\mathbb{N}\hat{\sigma}$ be the semigroup generated by the subset $\{\beta_j : \rho_j \notin \sigma\}$ for a cone $\sigma \in \Sigma$. Then,

$$\mathcal{K} = \bigcap_{\sigma \in \Sigma} \mathbb{N}\hat{\sigma}.$$

Remark 2.8 (Algebraic Importance of $\mathcal{K}$) If X is smooth, then by Remark 2.5 in [10], there exists a non-zerodivisor in $S/I(Y)$ of degree $\boldsymbol{\alpha} \in \mathcal{K}$. □

Remark 2.9 (Geometric Importance of $\mathcal{K}$) The class group $\mathrm{Cl}(X)$ is a group of Weil divisors on X modulo linear equivalence. The isomorphism $\mathcal{A} \cong \mathrm{Cl}(X)$ allows us to speak of degrees $\alpha \in \mathcal{A}$ lying in $\mathrm{Cl}(X)$. Furthermore, a Cartier divisor D on X is said to be *semi-ample* (or basepoint free as in [4]) if the corresponding line bundle $O(D)$ is generated by global sections. Since the property of being ample (resp. semi-ample) is preserved under linear equivalence we may speak of ample (resp. semi-ample) degrees α in $\mathcal{A}$. Geometrically, $\mathbb{N}\beta$ corresponds to the subset of $\mathrm{Cl}(X)$ containing the classes of effective Weil divisors on X and $\mathcal{K}$ corresponds to the subset containing the classes of *numerically* effective line bundles on X. By [4, Theorem 6.3.12], $\mathcal{K}$ is the set of semi-ample degrees in $\mathbb{N}\beta \subseteq \mathcal{A}$. □

Remark 2.10 When $X = \mathbb{P}^{n_1} \times \cdots \times \mathbb{P}^{n_r}$, the degree of each of the variables $x_{j,0}, x_{j,1}, \ldots, x_{j,n_j}$ is the standard basis vector e_j, for each $j \in [r]$. In addition to the properties above, Sidman and Van Tuyl showed in [18, Proposition 1.9] that the Hilbert function stabilizes in the direction of e_j, that is, $H_Y(\boldsymbol{\alpha}+e_j) = H_Y(\boldsymbol{\alpha}+ke_j)$, for any positive integer k, if $H_Y(\alpha) = H_Y(\boldsymbol{\alpha} + e_j)$ for some $j \in [r]$ and for some $\boldsymbol{\alpha} \in \mathbb{N}^r$. □

There are infinitely many elements of $\mathbb{N}\beta$ yielding evaluation codes on the same subset Y. Some of them have the same parameters. In the course of searching codes with good parameters, one wants to avoid dealing with each of these "equivalent codes" separately. Here is the precise definition of "equivalence" we need:

Definition 2.11 Let C_1 and C_2 be two subspaces of $\mathbb{F}_q^N$. We say that the linear codes C_1 and C_2 are "monomially equivalent" if there are non-zero scalars $t_1, \ldots, t_N \in \mathbb{F}_q^*$ and a permutation $\mathcal{P}$ of $\{1, \ldots, N\}$ such that $(c_1, \ldots, c_N) \in C_1$ if and only if $(t_1 c_{\mathcal{P}(1)}, \ldots, t_N c_{\mathcal{P}(N)}) \in C_2$. □

The stabilization property of the Hilbert function is very important to detect equivalent codes as the following reveals.

Proposition 2.12 *If there is a non-zerodivisor in $S/I(Y)$ of degree $\alpha_0 \in \mathbb{N}\beta$ and $H_Y(\boldsymbol{\alpha}) = H_Y(\boldsymbol{\alpha}+\boldsymbol{\alpha}_0)$ then the codes $C_{\boldsymbol{\alpha},Y}$ and $C_{\boldsymbol{\alpha}+\boldsymbol{\alpha}_0,Y}$ are monomially equivalent. Therefore, there are only finitely many non-equivalent codes $C_{\boldsymbol{\alpha},Y}$, in particular, for subsets Y lying inside the torus T_X.* □

Proof The proof of [17, Proposition 4.3] essentially works fine under the hypothesis here. □

Problem 2.13 Let X be a toric variety and $H_Y(\boldsymbol{\alpha}) = H_Y(\boldsymbol{\alpha} + \beta_j)$ for some $j \in [r]$ and $\boldsymbol{\alpha} \in \mathbb{N}\beta$. Is it true that the Hilbert function stabilizes in the direction of β_j? □

It would be nice to eliminate trivial codes also. Due to the Singleton bound, the minimum distance satisfies $d \leq N + 1 - k$, where $N = |Y|$ is the length and

$k = H_Y(\boldsymbol{\alpha})$ is the dimension. A code is *trivial* if the dimension takes the maximum possible value; namely $k = N$, as in this case $d = 1$. This motivates the following

Definition 2.14 The multigraded regularity of Y, denoted by $\text{reg}(Y)$, is the set of degrees $\boldsymbol{\alpha} \in \mathbb{N}\beta$ for which $H_Y(\boldsymbol{\alpha}) = |Y|$, the length of $C_{\boldsymbol{\alpha},Y}$. □

Remark 2.15 The subset Y may be given implicitly as in Definition 3.1. In order to construct a code of prescribed length, one needs to know the length $|Y|$ beforehand. The Hilbert function gives not only the dimension of the code but also the length even before constructing the code, since for any $\boldsymbol{\alpha} \in \text{reg}(Y)$, we have $H_Y(\boldsymbol{\alpha}) = |Y|$. In order for this to work efficiently, we need to know at least one element from the set $\text{reg}(Y)$. □

In general, the set $\text{reg}(Y)$ is not determined by a number but we have good fortune in some nice cases as we demonstrate next.

Proposition 2.16 ([17, Proposition 3.12]) *Let $X = \mathbb{P}(w_1, \ldots, w_r)$ be a weighted projective space and Y be a subset such that $S/I(Y)$ has a non-zerodivisor of degree 1. Then, there is an integer a_Y satisfying*

$$\text{reg}(Y) = 1 + a_Y + \mathbb{N}.$$

Moreover, a_Y equals the degree of the rational function corresponding to the Hilbert series of $S/I(Y)$. This is valid, in particular, when $Y \subseteq T_X$ and $w_1 = 1$. □

Problem 2.17 Find the invariant a_Y for a given $Y \subseteq T_X$ for the weighted projective space $X = \mathbb{P}(w_1, \ldots, w_r)$ such that $1 + a_Y + \mathbb{N} \subseteq \text{reg}(Y)$. □

In the particular case of the torus $Y = T_X$, for $X = \mathbb{P}(w_1, \ldots, w_r)$, we have the following nice result relating a_Y with a famous invariant $g(W)$ of the semigroup W generated by $w_1, \ldots, w_r$, where $g(W)$ is the largest integer not belonging to W.

Corollary 2.18 ([5, Corollary 3.9]) *If $Y = T_X$ and $g(W)$ is the Frobeneous number of W, then*

$$a_Y = (q-2)[w_1 + \cdots + w_r + g(W)] + g(W).$$

When $\text{reg}(Y)$ can not be determined exactly, it may be sufficient to give bounds on it. A subset Y is a *complete intersection* if $I(Y)$ is generated by a regular sequence of homogeneous polynomials $F_1, \ldots, F_n \in S$ where n is the dimension of X. The following lower bound is given in [17, Theorem 3.16].

Theorem 2.19 *Let $Y \subseteq X$ be a complete intersection of n hypersurfaces of degrees $\boldsymbol{\alpha}_1, \ldots, \boldsymbol{\alpha}_n$ in $\mathcal{K}$. If there is a non-zerodivisor in $S/I(Y)$ of degree β_j, for each $j \in [r]$, then,*

$$\boldsymbol{\alpha}_1 + \cdots + \boldsymbol{\alpha}_n + \mathbb{N}\beta \subseteq \text{reg}(Y).$$

Another general bound for the set reg(Y) following from [10, Proposition 2.10] is given as

Theorem 2.20 *Let X be smooth and $Y \subseteq X$. If the free modules $\mathcal{F}_i$ in a minimal free resolution of $I(Y)$ are of the following form*

$$\mathcal{F}_i = \bigoplus_{j=1}^{\beta_i} S[-\boldsymbol{\alpha}_{i,j}] \quad \text{then} \quad \bigcap_{i,j}(\boldsymbol{\alpha}_{i,j} + \mathcal{K}) \subseteq \text{reg}(Y).$$

Thus, if Y is a complete intersection of n hypersurfaces of degrees $\boldsymbol{\alpha}_1, \ldots, \boldsymbol{\alpha}_n$ in $\mathcal{K}$, the intersection becomes $\boldsymbol{\alpha}_1 + \cdots + \boldsymbol{\alpha}_n + \mathcal{K} \subseteq \boldsymbol{\alpha}_1 + \cdots + \boldsymbol{\alpha}_n + \mathbb{N}\beta \subseteq \text{reg}(Y)$. □

Problem 2.21 Find a non-empty subset of reg(Y) for an arbitrary $Y \subseteq X$ and for any toric variety X.

Having seen non-decreasing property of the dimension of the codes as the degree increases, we may wonder about the behavior of the minimum distance. As we discuss next, the minimum distance does not increase in any direction.

Proposition 2.22 *If there is a non-zerodivisor in $S/I(Y)$ of degree β_j, for each $j \in [r]$, the minimum distance is non-increasing in the sense that $d(\mathcal{C}_{\boldsymbol{\alpha},Y}) \geq d(\mathcal{C}_{\boldsymbol{\alpha}',Y})$ for all $\boldsymbol{\alpha} \preceq \boldsymbol{\alpha}'$.* □

Proof Let $F \in S_{\boldsymbol{\alpha}}$ be a polynomial with image $\text{ev}(F) = (F(P_1), \ldots, F(P_N))$ having weight $d = d(\mathcal{C}_{\boldsymbol{\alpha},Y})$. This means that F has exactly $N - d$ roots among the elements of Y and $F \notin I(Y)$. If the polynomial G_j is a non-zerodivisor in $S/I(Y)$ of degree β_j, then G_jF has at least $N - d$ roots and $G_jF \notin I(Y)$. It follows that $\text{ev}(G_jF) = (G_j(P_1)F(P_1), \ldots, G_j(P_N)F(P_N))$ lies in $\mathcal{C}_{\alpha+\beta_j,Y} \setminus \{0\}$ and has weight at most d. Thus, $d \geq d(\mathcal{C}_{\boldsymbol{\alpha}+\beta_j,Y})$. As these are true for every $j \in [r]$, the claim follows. □

As for the minimum distance, we have the following lower bound provided by Soprunov and stated here using the language of this paper, see [19, Theorem 3.2].

Theorem 2.23 *Let $Y \subseteq T_X$ be a reduced complete intersection of hypersurfaces of degrees $\boldsymbol{\alpha}_1, \ldots, \boldsymbol{\alpha}_n \in \mathcal{K}$ and denote $\alpha_Y = \boldsymbol{\alpha}_1 + \cdots + \boldsymbol{\alpha}_n - \beta_1 - \cdots - \beta_r$. If any subset $Z \subset Y$ of size m has $H_Z(\boldsymbol{\alpha}') = m$ for some $\boldsymbol{\alpha}' \in \mathbb{N}\beta$ with $\boldsymbol{\alpha} + \boldsymbol{\alpha}' \preceq \alpha_Y$, then $d(\mathcal{C}_{\boldsymbol{\alpha},Y}) \geq m + 1$.* □

Proof We outline how this version follows from the original one by relating notations of two papers. First, we introduce local coordinates for the points in the torus T_X by:

$$t_1 = \mathbf{x}^{\mathbf{u}_1}, \ldots, t_n = \mathbf{x}^{\mathbf{u}_n}. \tag{2}$$

So, the coordinate ring $\mathbb{K}[T_X]$ of T_X can be identified with the ring of Laurent polynomials $\mathbb{K}[t_1^{\pm 1}, \ldots, t_n^{\pm 1}]$. For any subset $A \subseteq \mathbb{R}^n$, we denote by $\mathcal{L}(A)$ the vector space over $\mathbb{K}$ spanned by monomials $\mathbf{t}^{\mathfrak{m}} = t_1^{m_1} \cdots t_n^{m_n}$, where $m_1, \ldots, m_n$ are coordinates of $\mathfrak{m} \in A \cap \mathbb{Z}^n$.

Recall that every $\mathbf{a} = (a_1, \ldots, a_r) \in \beta^{-1}(\boldsymbol{\alpha})$ defines a rational polytope

$$P_{\mathbf{a}} := \{\mathbf{u} \in \mathbb{R}^n : \langle \mathbf{u}, \mathbf{v}_j \rangle \geq -a_j, \quad \forall j \in [r]\}.$$

Note that different elements in $\beta^{-1}(\boldsymbol{\alpha})$ will have the same polytope up to a lattice translation. We denote by $P_{\boldsymbol{\alpha}}$ the class of these polytopes and abuse notation using $P_{\boldsymbol{\alpha}}$ instead of one $P_{\mathbf{a}}$ expecting that the context will clarify which one is meant. This leads to the vector space isomorphism $\mathbf{t}^{\mathfrak{m}} \to \mathbf{x}^{\phi(\mathfrak{m})+\mathbf{a}}$ from $\mathcal{L}(P_{\mathbf{a}} \cap \mathbb{Z}^n)$ to $S_{\boldsymbol{\alpha}}$, for any $\mathbf{a} \in \beta^{-1}(\boldsymbol{\alpha})$.

One can prove that the assumptions (1)–(3) in [19] are satisfied since $Y \subseteq T_X$ is a reduced complete intersection of hypersurfaces of degrees $\boldsymbol{\alpha}_1, \ldots, \boldsymbol{\alpha}_n \in \mathcal{K}$, see e.g. the proof of Proposition 4.2 in the first arxiv version of [17]. Let $P = P_{\alpha_1} + \cdots + P_{\boldsymbol{\alpha}_n}$ be the Minkowski sum of polytopes corresponding to $\boldsymbol{\alpha}_1, \ldots, \boldsymbol{\alpha}_n$. Then, $S_{\boldsymbol{\alpha}_Y}$ is isomorphic to $\mathcal{L}(P^\circ \cap \mathbb{Z}^n)$ for the interior P° of P and $\boldsymbol{\alpha}_Y = \alpha_1 + \cdots + \boldsymbol{\alpha}_n - \beta_1 - \cdots - \beta_r$. So, if $\boldsymbol{\alpha} + \boldsymbol{\alpha}' \preceq \boldsymbol{\alpha}_Y$, then for $A = P_{\boldsymbol{\alpha}}$ and $B = P_{\boldsymbol{\alpha}'}$, we have $A + B \subseteq P_{\alpha+\alpha'} \subseteq P^\circ$. Finally, if any subset $Z \subset Y$ of size m has $H_Z(\boldsymbol{\alpha}') = m$, then the linear map $\mathrm{ev}_Z : S_{\boldsymbol{\alpha}'} \to \mathbb{K}^m$ obtained by evaluation of polynomials in $S_{\boldsymbol{\alpha}'}$ at the points in Z is surjective, since $S_{\boldsymbol{\alpha}'}/\ker(\mathrm{ev}_Z) \cong \mathbb{K}^m$. This means that the map $\mathrm{ev}_Z : \mathcal{L}(B) \to \mathbb{K}^m$ is surjective. Therefore, hypotheses of [19, Theorem 3.2] are satisfied, completing the proof. □

Complete intersections Y satisfying the extra condition that $I(Z) \cap S_{\boldsymbol{\alpha}'} = 0$ for any subset $Z \subset Y$ of size $m = \dim_{\mathbb{K}} S_{\boldsymbol{\alpha}'}$, have the following bigger lower bound given again by Soprunov, see [19, Theorem 3.9].

Theorem 2.24 *Let $Y \subseteq T_X$ be a reduced complete intersection of hypersurfaces of degrees $\boldsymbol{\alpha}_1, \ldots, \boldsymbol{\alpha}_n \in \mathcal{K}$ and $\alpha_Y = \boldsymbol{\alpha}_1 + \cdots + \boldsymbol{\alpha}_n - \beta_1 - \cdots - \beta_r$. If any subset $Z \subset Y$ of size $m = \dim_{\mathbb{K}} S_{\boldsymbol{\alpha}'}$ has $H_Z(\boldsymbol{\alpha}') = m$ for some $\boldsymbol{\alpha}' \in \mathbb{N}\beta$ with $\boldsymbol{\alpha} + k\boldsymbol{\alpha}' \preceq \boldsymbol{\alpha}_Y$, then $d(C_{\boldsymbol{\alpha},Y}) \geq k(m-1) + 2$.* □

Proof Relying on the notations of the previous proof, let us reveal why hypotheses of [19, Theorem 3.9] follows from our assumptions. As before the assumptions (1)–(3) are satisfied. If any subset $Z \subset Y$ of size m has $H_Z(\boldsymbol{\alpha}') = m$, then $S_{\boldsymbol{\alpha}'}/\ker(\mathrm{ev}_Z) \cong \mathbb{K}^m$. If this happens for $m = \dim_{\mathbb{K}} S_{\boldsymbol{\alpha}'}$, then $\ker(\mathrm{ev}_Z) = 0$ meaning that the evaluation map is an isomorphism. This implies that the assumption (4) is satisfied for the polytope $Q = P_{\boldsymbol{\alpha}'}$. When $\boldsymbol{\alpha} + k\boldsymbol{\alpha}' \preceq \boldsymbol{\alpha}_Y$, we have the inclusion $A + kQ = P_{\boldsymbol{\alpha}} + kP_{\boldsymbol{\alpha}'} \subseteq P_{\boldsymbol{\alpha}+k\boldsymbol{\alpha}'} \subseteq P^\circ$, as required. □

Problem 2.25 Find bounds on the minimum distance for an arbitrary $Y \subseteq X$ and for any toric variety X.

3 Lattice Ideals and Subgroups of the Torus T_X

The previous section motivates studying vanishing ideals of subsets Y of the torus as in this case hypothesis about the existence of non-zerodivisors hold automatically. So as to take full advantage of the algebraic approach offered by combinatorial commutative algebra, it is a good idea to focus on subgroups of T_X. In fact, in the particular case of $X = \mathbb{P}(w_1, \dots, w_r)$ and $Y = T_X$, the ideal $I(Y)$ has shown to have a special form in [5]. Namely, it is a special lattice ideal related to the defining ideal of the semigroup generated by the degrees $\beta_j := \deg(x_j) = w_j$, for $j = 1, \dots, r$. The main result of this section is from [16] and uncovers the relation between lattice ideals and subgroups of T_X.

A lattice L is a subgroup of $\mathbb{Z}^r$. We can write a vector in $\mathbb{Z}^r$ as a difference $\mathfrak{m} = \mathfrak{m}^+ - \mathfrak{m}^-$, of two vectors $\mathfrak{m}^+, \mathfrak{m}^- \in \mathbb{N}^r$. If we let $F_\mathfrak{m} = \mathbf{x}^{\mathfrak{m}^+} - \mathbf{x}^{\mathfrak{m}^-}$, the binomial ideal I_L generated by these special binomials $F_\mathfrak{m}$ arising from the lattice $L \subset \mathbb{Z}^r$ is called the *lattice ideal* of L. In short, $I_L = \langle F_\mathfrak{m} \mid \mathfrak{m} \in L \rangle$.

Recall that a point of the toric variety X is an orbit of a point P from the affine space $\mathbb{K}^r$ and so is denoted by $[P] := G \cdot P = [p_1 : \dots : p_r]$. We use [1] shortly to mean the point $[1 : \dots : 1]$.

Definition 3.1 Every matrix $Q = [\mathbf{q}_1 \mathbf{q}_2 \cdots \mathbf{q}_r] \in M_{s \times r}(\mathbb{Z})$ defines a subgroup

$$Y_Q = \{[\mathbf{t}^{\mathbf{q}_1} : \dots : \mathbf{t}^{\mathbf{q}_r}] | \mathbf{t} \in (\mathbb{K}^*)^s\} \subseteq T_X$$

of the torus T_X, which we call the toric set parameterized by Q.

In [15], the vanishing ideals of these toric sets parameterised by monomials are shown to be lattice ideals of dimension 1, when the toric variety is a projective space, i.e., $X = \mathbb{P}(w_1, \dots, w_r)$ with $w_1 = \dots = w_r = 1$.

Definition 3.2 Given an $s \times r$ integer matrix B, let $L_B = \mathbb{Z}^r \cap \ker B$ be the sublattice of $\mathbb{Z}^r$ determined by B. A lattice L is called **homogeneous** if $L \subseteq L_\beta$, where β is the matrix representing the second map in the first short exact sequence $\mathfrak{P}$. □

The following fact given for the first time in [16, Proposition 2.3] justifies our choice of terminology.

Proposition 3.3 *L is homogeneous if and only if I_L is homogeneous.* □

For a homogeneous ideal J of S, let

$$V_X(J) := \{[P] \in X : F(P) = 0, \text{ for all homogeneous } F \in J\}.$$

Summarizing some of the results of [16], we get the following nice relations:

Theorem 3.4 *The following are equivalent:*

(i) Y is a subgroup of T_X,
(ii) $Y \subseteq T_X$ and $I(Y)$ is a radical lattice ideal of dimension $r - n$,
(iii) $Y = Y_Q$ for a square Q.

□

Proof (i) $\Longrightarrow$ (ii) follows from Theorem 2.9 and Theorem 5.1 in [16].
(ii) $\Longrightarrow$ (iii): If $I(Y) = I_L$ then $V_X(I_L) \cap T_X = Y_Q$ for a square Q, by Proposition 3.4 of [16]. Since $Y \subseteq T_X$ and $Y = V_X(I(Y))$, the claim follows.
(iii) $\Longrightarrow$ (i) is clear. □

The last result suggests that for studying subgroups of the torus T_X it is sufficient to focus on subgroups of the form Y_Q for a square matrix Q.

3.1 Degenerate Tori

In this section, we pay attention to subgroups Y_A of T_X parameterized by diagonal matrices $A = \text{diag}(a_1, \ldots, a_r)$.

Definition 3.5 The subset $Y_A = \{[t_1^{a_1} : \cdots : t_r^{a_r}] : t_i \in \mathbb{K}^*\}$ of the torus T_X is called a **degenerate torus.** □

If $\mathbb{K}^* = \langle \eta \rangle$, every $t_i \in \mathbb{K}^*$ is of the form $t_i = \eta^{s_i}$, for some $0 \leq s_i \leq q-2$. Let $d_i = \text{ord}(\eta^{a_i})$ and $D = \text{diag}(d_1, \ldots, d_r)$ be the matrix defining an automorphism of $\mathbb{Z}^r$. Then, the ideal $I(Y_A)$ is determined by D and β in this case.

Theorem 3.6 ([16, Theorem 4.5]) *If $Y = Y_A$ then $I(Y) = I_L$ for $L = D(L_{\beta D})$.* □

Evaluation codes on complete intersections have been studied before in literature, e.g. [6–8]. Following these results, we study vanishing ideals of special subsets of the torus T_X and characterize when they are complete intersections using mixed dominating matrices we define now.

Definition 3.7 If each column of a matrix has both a positive and a negative entry we say that it is *mixed*. If it does not have a square mixed submatrix, then it is called *dominating*. □

Theorem 3.8 ([14, Theorem 3.9]) *Let $L \subset \mathbb{Z}^r$ be a nonzero lattice with $L \cap \mathbb{N}^r = 0$. Then I_L is complete intersection iff L has a basis $\mathfrak{m}_1, \ldots, \mathfrak{m}_n$ such that the matrix $[\mathfrak{m}_1 \cdots \mathfrak{m}_n]$ is mixed dominating. In the affirmative case, we have*

$$I_L = \langle x^{m_1^+} - x^{m_1^-}, \ldots, x^{m_n^+} - x^{m_n^-} \rangle.$$

□

Using Theorem 3.8, we prove the following.

Proposition 3.9 ([16, Proposition 4.12]) *$I(Y_A)$ is a complete intersection iff so is the toric ideal $I_{L_{\beta D}}$. A minimal generating system of binomials for $I(Y_A)$ is obtained from that of $I_{L_{\beta D}}$ by replacing x_i with $x_i^{d_i}$.* □

Corollary 3.10 ([16, Corollary 4.14]) *We have the following:*
(i) if $Y = \{[1]\}$ then $I(Y) = I_{L_\beta}$,
(ii) if $Y = T_X$ then $I(Y) = I_L$, for $L = (q-1)L_\beta$,

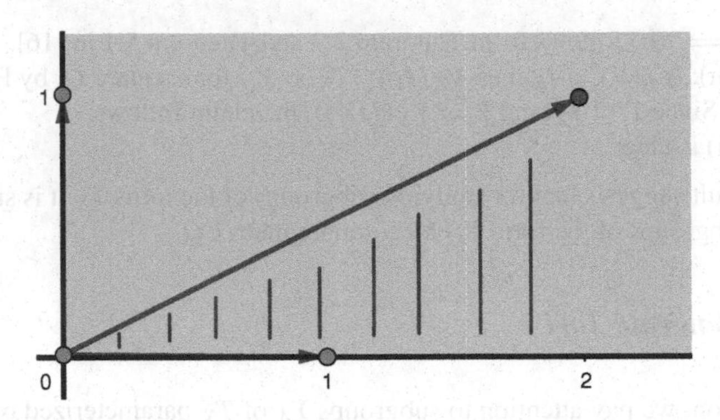

Fig. 2 The subsemigroup $\mathcal{K}$

(iii) $I(T_X)$ *is a complete intersection iff so is the toric ideal* I_{L_β}*, which is independent of* q.

□

Using the matrix ϕ defined by the fan Σ and the result presented in this section one can easily check whether the vanishing ideal of T_X is a complete intersection.

Example 3.11 Consider the Hirzebruch surface $X = H_\ell$, whose fan gives the following mixed dominating matrix

$$\phi = \begin{bmatrix} 1 & 0 & -1 & 0 \\ 0 & 1 & \ell & -1 \end{bmatrix}^T.$$

- So, $I_L = \langle x_1 - x_3, x_2 x_3{}^{\ell} - x_4 \rangle$ is a complete intersection.
- Thus, tori T_X are all complete intersections for every q and ℓ:

$$I(T_X) = \langle x_1{}^{q-1} - x_3{}^{q-1}, x_2{}^{q-1} x_3{}^{\ell(q-1)} - x_4{}^{q-1} \rangle.$$

- So, $\alpha_1 = (q-1, 0)$ and $\alpha_2 = (2q-2, q-1)$ are in $\mathcal{K}$ bounding the multigraded regularity, see Fig. 2.
- Toric codes are trivial after degree $\alpha_1 + \alpha_2 = (3q-3, q-1)$, see Fig. 3.

□

4 Vanishing Ideals of Subsgroups of T_X

Studying evaluation codes defined on subgroups of the torus is reduced to an investigation about parameterized subgroups Y_Q by the virtue of Theorem 3.4. In this section we give algorithms for determining a generating set of the lattice ideal $I(Y_Q)$ and for computing directly $|Y_Q|$, see [1]. These generalize the corresponding

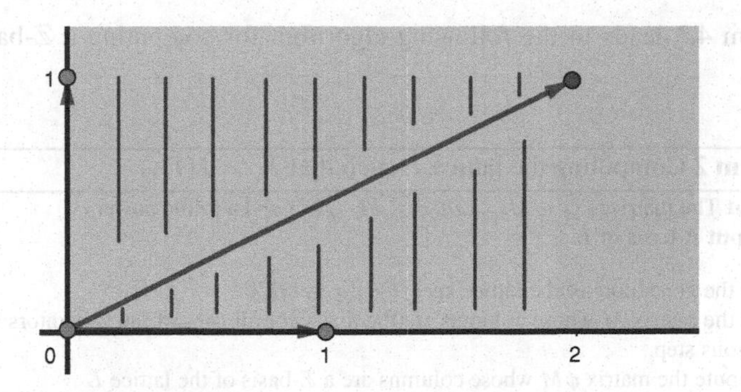

Fig. 3 The semigroup $\mathbb{N}\beta$

results of [15] for the case $X = \mathbb{P}^n$ of projective space, and those of [5] for the case $X = \mathbb{P}(w_1, \ldots, w_r)$ of weighted projective spaces and $Y_Q = T_X$.

The first description of the vanishing ideal is via Elimination theory.

Theorem 4.1 ([1, Theorem 2.3]) *Let* $R = S[y_1, \ldots, y_s, z_1, \ldots, z_d, w]$ *be an extension of* $S = \mathbb{K}[x_1, \ldots, x_r]$. *Then* $I(Y_Q) = J \cap S$, *where*

$$J = \langle \{x_j \mathbf{y}^{\mathbf{q}_j^-} \mathbf{z}^{\beta_j^-} - \mathbf{y}^{\mathbf{q}_j^+} \mathbf{z}^{\beta_j^+}\}_{j=1}^r \cup \{y_i^{q-1} - 1\}_{i=1}^s, w\mathbf{y}^{\mathbf{q}_1^-} \mathbf{z}^{\beta_1^-} \cdots \mathbf{y}^{\mathbf{q}_r^-} \mathbf{z}^{\beta_r^-} - 1\rangle.$$

Theorem 4.1 gives rise to the following algorithm for computing the binomial generators of $I(Y_Q)$.

Algorithm 1 Computing the generators of vanishing ideal $I(Y_Q)$

Input The matrices $Q \in M_{s\times r}(\mathbb{Z})$, $\beta \in M_{d\times r}(\mathbb{Z})$ and a prime power q.
Output The generators of $I(Y_Q)$.

1: Write the ideal J of R using Theorem 4.1.
2: Find the Gröbner basis G of J using the lexicographic monomial ordering with $w > z_1 > \cdots > z_d > y_1 > \cdots > y_s > x_1 > \cdots > x_r$.
3: Find $G \cap S$ so that $I(Y_Q) = \langle G \cap S\rangle$.

Example 4.2 Consider the Hirzebruch surface $X = H_2$ over $\mathbb{F}_{11}$ and the toric set parameterized by $Q = [1\ 2\ 3\ 4]$, that is, $Y_Q = \{[t : t^2 : t^3 : t^4] | t \in \mathbb{F}_{11}^*\}$. We compute the generators of $I(Y_Q)$ and obtain $I(Y_Q) = \langle x_1^5 - x_3^5, x_1^2 x_2 - x_4\rangle$, see [1] for the code used. □

Here is another description of the vanishing ideal relying on the underlying lattice.

Theorem 4.3 ([1, Theorem 3.4]) *Let* $\pi_s : \mathbb{Z}^{n+s} \to \mathbb{Z}^n$ *be the projection map sending* $(c_1, \ldots, c_n, c_{n+1}, \ldots, c_{n+s})$ *to* $(c_1, \ldots, c_n)$. *Then, the ideal* $I(Y_Q) = I_L$, *for the lattice* $L = \{\phi\mathbf{c} : \mathbf{c} \in \pi_s\,(\ker_{\mathbb{Z}}[Q\phi|(q-1)I_s])\}$. □

Theorem 4.3 leads to the following algorithm for computing a $\mathbb{Z}$-basis of the lattice L.

Algorithm 2 Computing the lattice L such that $I_L = I(Y_Q)$

Input The matrices $Q \in M_{s\times r}(\mathbb{Z})$, $\phi \in M_{r\times n}(\mathbb{Z})$ and a prime power q.
Output A basis of L.

1: Find the generators of the lattice $\ker_{\mathbb{Z}}[Q\phi|(q-1)I_s]$.
2: Find the matrix M whose columns are the first s coordinates of the generators found in the previous step.
3: Compute the matrix ϕM whose columns are a $\mathbb{Z}$-basis of the lattice L

Example 4.4 Let $X = \mathcal{H}_2$ over $\mathbb{F}_{11}$ and $Q = [1\ 2\ 3\ 4]$. Then, using this algorithm we get the lattice L whose basis vectors appear as the columns of the matrix below:

$$ML = \begin{bmatrix} 2 & 1 & 0 & -1 \\ -5 & 0 & 5 & 0 \end{bmatrix}^T.$$

The lattice ideal I_L is the saturation of the lattice basis ideal $\langle x_1^5 - x_3^5, x_1^2 x_2 - x_4\rangle$. Since ML is mixed dominating, $I(Y_Q) = I_L$ is complete intersection meaning that it is already saturated. Therefore, we get $I_L = \langle x_1^5 - x_3^5, x_1^2 x_2 - x_4\rangle$. If we compute Hilbert function of Y_Q at some degrees, we get the following table:

```
{1, 2, 3, 4, 5, 5, 5, 5}
{1, 2, 3, 4, 5, 5, 5, 5}
{1, 2, 3, 4, 5, 5, 5, 5}.
```

The table represents values of the Hilbert function H_{Y_Q} at elements $\boldsymbol{\alpha} \in \mathbb{N}\beta = \mathbb{N}^2$. It should be thought of as the usual first quadrant as in Fig. 3: the left-bottom value 1 corresponds to the origin. The others from the bottom to the top should be regarded as $H_{Y_Q}(\boldsymbol{\alpha})$ for $\boldsymbol{\alpha} \in \{(0,1),(0,2)\}$. Similarly, values $1, 2, 3$ at the bottom correspond to elements $\boldsymbol{\alpha} \in \{(0,0),(1,0),(2,0)\}$. In order to understand the cardinality $|Y_Q|$ and the set $\operatorname{reg}(Y_Q)$, we need to know more about the behaviour of the Hilbert function. As Y_Q lies in the torus T_X, variables are non-zerodivisor, so Hilbert function is non-decreasing in the direction of $\beta_1 = (1,0)$ and $\beta_2 = (0,1)$. Apriori, we do not know if H_{Y_Q} stabilizes after it repeats itself once. So, we can not determine the rest of the table. In other words, it does not follow from $H_{Y_Q}(4,0) = 5 = H_{Y_Q}(5,0)$ that $H_{Y_Q}(a,0) = 5$ for all $a > 5$ as in the case of $X = \mathbb{P}^{n_1} \times \cdots \times \mathbb{P}^{n_r}$. Since $X = \mathcal{H}_2$ is smooth and Y_Q is a complete intersection of hypersurfaces of degrees $\alpha_1 = (2,1)$ and $\alpha_2 = (5,0)$, we can use either of the results Theorem 2.19 or Theorem 2.20 to deduce that $\boldsymbol{\alpha}_1 + \boldsymbol{\alpha}_2 = (7,1) \in \operatorname{reg}(Y_Q)$. Therefore, $|Y_Q| = H_{Y_Q}(7,1) = 5$ and thus $\operatorname{reg}(Y_Q) = (4,0) + \mathbb{N}^2$.

Similarly, we can not make sure apriori that the rows of the table repeat itself forever. If we look at this example more closely, we realize the following, for every $b \in \mathbb{N}$:

$$S_{(0,b)} = \{x_2^b\}, \quad S_{(1,b)} = \{x_1x_2^b, x_1x_2^b\}, \quad S_{(2,b)} = \{x_1^2x_2^b, x_1x_3x_2^b, x_3^2x_2^b, x_2^{b-1}x_4\}$$

$$S_{(3,b)} = \{x_1^3x_2^b, x_1^2x_3x_2^b, x_1x_3^2x_2^b, x_3^3x_2^b, x_1x_2^{b-1}x_4, x_3x_2^{b-1}x_4\}.$$

Since $x_1^2x_2 - x_4 \in I(Y_Q)$, it follows that $x_1^2x_2^b + I(Y_Q) = x_2^{b-1}x_4 + I(Y_Q)$ in the quotient ring $S/I(Y_Q)$ and thus for every $b \in \mathbb{N}$, we have

$$H_{Y_Q}(0,b) = 1, \quad H_{Y_Q}(1,b) = 2, \quad H_{Y_Q}(2,b) = 3, \quad H_{Y_Q}(3,b) = 4.$$

Hence, the only non-equivalent and non-trivial codes are the generalized toric codes C_{α,Y_Q} for the degrees $\alpha \in \{(1,0),(2,0),(3,0)\}$. □

4.1 The Length of the Code C_{α,Y_Q}

In this section, we give an algorithm for computing the length of the code C_{α,Y_Q} directly using the parameterization of Y_Q. It is clear that T_X and Y_Q are groups under the componentwise multiplication and that the map

$$\varphi_Q : (\mathbb{K}^*)^s \to Y_Q, \quad \mathbf{t} \to [\mathbf{t}^{\mathbf{q}_1} : \cdots : \mathbf{t}^{\mathbf{q}_r}]$$

is a group epimorphism. It follows that $Y_Q \cong (\mathbb{K}^*)^s/\ker(\varphi_Q)$ and thus,

$$|Y_Q| = |(\mathbb{K}^*)^s|/|\ker(\varphi_Q)| = (q-1)^s/|\ker(\varphi_Q)|.$$

Hence, computation of the length of the code C_{α,Y_Q} is reduced to determining the number $|\ker(\varphi_Q)|$.

Proposition 4.5 ([1, Proposition 5.1]) *Let* $H = \{1,\ldots,q-1\} \times \cdots \times \{1,\ldots,q-1\} \subset \mathbb{Z}^s$ *and* η *be a generator of* $\mathbb{K}^*$. *If* $\mathcal{P} = \{\mathbf{h} \in H | \mathbf{h}Q\phi \equiv 0 \bmod q-1\}$, *then we have* $\ker(\varphi_Q) = \{(\eta^{h_1},\ldots,\eta^{h_s}) | \mathbf{h} = (h_1,\ldots,h_s) \in \mathcal{P}\}$. *Therefore,* $|\ker(\varphi_Q)| = |\mathcal{P}|$. □

We can compute $k = |\ker(\varphi_Q)| = |\mathcal{P}|$ and thereby the length of the code easily using the following:

Procedure 4.6 The following code in `Macaulay2` computes length of C_{α,Y_Q}.

```
i2 : r=numRows Phi;s=numRows Q;n=numColumns Phi;
i3 : L=for i from 1 to q-1 list i;
i4 : L= set L;L=L^**(s);L=toList L;
i5 : k=0;
```

```
i6 : scan(L,i->if ((matrix{{i}}*Q*Phi)%(map((ZZ)^1,n,(i,j)>
(q-1))))==(matrix mutableMatrix(ZZ,1,n)) then k=k+1)
i7: length=((q-1)^s)/k
```

Example 4.7 We can calculate the length of the code corresponding to the Example 4.4 directly using the Procedure 4.6 with the following input:

```
i1 : q=11;Phi=matrix{{1,0},{0,1},{-1,2},{0,-1}};
Q=matrix {{1,2,3,4}};
```

We finish with the following example to illustrate the advantage of computing length beforehand in order to determine $\text{reg}(Y_Q)$.

Example 4.8 We can calculate the cardinality $|Y_Q|$ of a subset $Y_Q \subseteq \mathbb{P}(3, 4, 5, 6)$ directly using the Procedure 4.6 with the following input:

```
i1 : q=11; Phi= syz matrix {3,4,5,6};
Q=matrix{{3,2,1,4},{1,2,3,4},{3,4,5,6}};
```

This reveals that $|Y_Q| = 25$. In order to determine $\text{reg}(Y_Q)$, we compute the vanishing ideal $I(Y_Q)$ and the first 60 values of its Hilbert function:

```
                2         8       2   5        7      3 5    10      5
o62 = ideal (x    - x x , x   - x x x , x x  - x x , x     - x )
                2     1 3   3     1 2 4   2 3    1 4   1       4

i65 : apply(60,i-> hilbertFunction({i},IYQ))

o65 = {1, 0, 0, 1, 1, 1, 2, 1, 1, 3, 3, 2, 4, 3, 3, 6, 5, 4,
7, 6, 6, 9, 8, 7, 11, 10, 9, 13, 12, 11, 15, 14, 13, 17, 16,
15, 19, 18, 17, 20, 19, 19, 22, 21, 20, 22, 22, 22, 24, 23,
22, 24, 24, 24, 25, 24, 24, 25, 25, 25}.
```

Since the ideal is not a complete intersection we currently do not know anything about the regularity set. As the length is found 25 beforehand, it is certain that

$$\text{reg}(Y_Q) = \{d \in \mathbb{N}\beta \; : \; H_{Y_Q}(d) = 25\}.$$

It is not clear if the Hilbert function stabilizes after these 60 values. Using its non-decreasing behavior, we can say that $H_{Y_Q}(d) \leq H_{Y_Q}(d + w_i)$, for $w_i \in \{3, 4, 5, 6\}$ and for all $d > 0$. Since $H_{Y_Q}(57) = 25$, it follows that $H_{Y_Q}(d) = 25$, for all degrees $d = 60, 61, 62$ and 63. Thus, $H_{Y_Q}(d) = 25$ forever, giving a lower bound for the regularity $57 + \mathbb{N} \subseteq \text{reg}(Y_Q)$. Indeed, we have $\text{reg}(Y_Q) \setminus (57 + \mathbb{N}) = \{54\}$. This implies that for $a_{Y_Q} = 56$ we do not have equality $\text{reg}(Y_Q) = a_{Y_Q} + 1 + \mathbb{N}$, as in Proposition 2.16.

Note that the degree of the following rational function representing the Hilbert series is 56:

```
i66 : hilbertSeries IYQ

              8    30    38    39    40    43    44    69    70    73    74
       1- T - T  + T  - T  - T  + T  + T  + T  + T  - T  - T
o66 = ------------------------------------------------------------
                       6       5       4       3
              (1 - T )(1 - T )(1 - T )(1 - T )
```

□

Acknowledgement We thank the editors and an anonymous referee for their comments which improved the presentation.

References

1. Baran, E., Şahin, M.: Vanishing ideals of parameterized toric codes. http://www.arxiv.org/abs/1802.04083
2. Brown, G., Kasprzyk, A.M.: Seven new champion linear codes. LMS J. Comput. Math. **16**, 109–117 (2013)
3. Cox, D.A.: The homogeneous coordinate ring of a toric variety. J. Algebraic Geom. **4**, 17–50 (1995)
4. Cox, D.A., Little, J., Schenck, H.: Toric Varieties. Graduate Studies in Mathematics, vol. 124. AMS, Providence (2011)
5. Dias, E., Neves, J.: Codes over a weighted torus. Finite Fields Appl. **33**, 66–79 (2015)
6. Duursma, I., Rentería, C., Tapia-Recillas, H.: Reed-Muller codes on complete intersections. Appl. Algebra Eng. Commun. Comput. **11**, 455–462 (2001)
7. Gold, L., Little, J., Schenck, H.: Cayley–Bacharach and evaluation codes on complete intersections. J. Pure Appl. Algebra **196**(1), 91–99 (2005)
8. Hansen, J.: Linkage and codes on complete intersections. Appl. Algebra Eng. Commun. Comput. **14**, 175–185 (2003)
9. Little, J.B.: Toric codes and finite geometries. Finite Fields Appl. **45**, 203–216 (2017)
10. Maclagan, D., Smith, G.: Uniform bounds on multigraded regularity. J. Algebraic Geom. **14**(1), 137–164 (2005)
11. MacWilliams, F.J., Sloane, N.J.A.: The Theory of Error-Correcting Codes. North-Holland, Amsterdam (1978)
12. Martínez-Moro, E., Ruano, D.: Toric Codes. Advances in Algebraic Geometry Codes. Coding Theory and Cryptology, vol. 5, pp. 295–322. World Scientific, Hackensack (2008)
13. Miller, E., Sturmfels, B.: Combinatorial Commutative Algebra. Graduate Texts in Mathematics, vol. 227. Springer, New York (2005)
14. Morales, M., Thoma, A.: Complete Intersection Lattice Ideals. J. Algebra **284**, 755–770 (2005)
15. Rentería, C., Simis, A., Villarreal, R.H.: Algebraic methods for parameterized codes and invariants of vanishing ideals over finite fields. Finite Fields Appl. **17**(1), 81–104 (2011)
16. Şahin, M.: Toric Codes and Lattice Ideals. Finite Fields Appl. **52**, 243–260 (2018)
17. Şahin, M., Soprunov, I.: Multigraded Hilbert functions and toric complete intersection codes. J. Algebra **459**, 446–467 (2016)

18. Sidman, J., Van Tuyl, A.: Multigraded regularity: syzygies and fat points. Beiträge Algebra Geom. **47**(1), 67–87 (2006)
19. Soprunov, I.: Toric complete intersection codes. J. Symb. Comput. **50**, 374–385 (2013)
20. J.H. van Lint, Introduction to Coding Theory, Graduate Texts in Mathematics, vol. 86. Springer, Heidelberg (1982)

The Number of Star Operations on Numerical Semigroups and on Related Integral Domains

Dario Spirito

Abstract We study the cardinality of the set Star(S) of star operations on a numerical semigroup S; in particular, we study ways to estimate Star(S) and to bound the number of nonsymmetric numerical semigroups such that $|\text{Star}(S)| \leq n$. We also study this problem in the setting of analytically irreducible, residually rational rings whose integral closure is a fixed discrete valuation ring.

Keywords Numerical semigroups · Star operations · Residually rational rings

2020 MSC Classification 20M12, 13G05

1 Introduction

A *star operation* on an integral domain D is a particular closure operation on the set of fractional ideals of D; this notion was defined to generalize the *divisorial closure* [4, 13] and has been further generalized to the notion of *semistar operation* [17]. Star operations have also been defined on cancellative semigroups in order to obtain semigroup-theoretic analogues of some ring-theoretic (multiplicative) definitions [11]. A classical result characterizes the Noetherian domains D in which every ideal is divisorial or, equivalently, which Noetherian domains admit only one star operation: this happens if and only if D is Gorenstein of dimension one [2]. Recently, this result has been a starting point of a deeper investigation on the cardinality of the set Star(D) of the star operations on D, obtaining a precise counting for h-local Prüfer domains [7] (and, more generally, an algorithm to calculate their number for semilocal Prüfer domains [23]), some pseudo-valuation domains [18, 26] and some Noetherian one-dimensional domains [6, 8, 25]. In particular, for Noetherian

D. Spirito (✉)
Dipartimento di Matematica e Fisica, Università degli Studi "Roma Tre", Roma, Italy
e-mail: spirito@mat.uniroma3.it

V. Barucci et al. (eds.), *Numerical Semigroups*, Springer INdAM Series 40, https://doi.org/10.1007/978-3-030-40822-0_17

domains, a rich source of examples are *numerical semigroup rings*, that is, rings in the form $K[[S]] := K[[X^s \mid s \in S]]$, where K is a field and S is a numerical semigroup.

Inspired by this example, the study of star operations on numerical semigroups (and, in particular, of their cardinality) was initiated in [21]. In particular, the main problem that was tackled was the following: given a (fixed) integer n, how many numerical semigroups have exactly n star operations? By estimating the cardinality of $\mathrm{Star}(S)$, it was shown that this number is always finite, and that the same holds for residually rational rings (see Sect. 10 for a precise statement). Subsequently, in [27], better estimates on $|\mathrm{Star}(S)|$ allowed to give a much better bound the number of semigroups with at most n star operations, while in [22] the set $\mathrm{Star}(S)$ was described in a very precise way when the semigroup S has multiplicity 3.

In this paper, we give a unified treatment of the study of $\mathrm{Star}(S)$, surveying the main results of [21, 22, 27] and [24] and deepening them. In particular, we give a rather precise asymptotic expression for the number of semigroups of multiplicity 3 with less than n star operations (Theorem 6.4), an $O(n^\epsilon)$ bound for the semigroups of prime multiplicity (Theorem 7.4), we list all nonsymmetric numerical semigroups with 150 or less star operations (Table 4), and prove an explicit bound for residually rational rings (Theorem 10.5).

The structure of the paper is as follows: Sects. 2 and 3 present basic material; Sects. 4 and 5 present estimates already present in [21] and [27]; Sect. 6 deepens the analysis of [22] on semigroups of multiplicity 3; Sect. 7 studies the case where the multiplicity is prime (and bigger than 3); Sect. 8 introduces the concept of *linear families* (one example of which was analyzed in [24]); Sect. 9 is devoted to algorithms to calculate $|\mathrm{Star}(S)|$ and to determine all the nonsymmetric semigroups with at most n star operations; Sect. 10 studies the domain case, and contains analogues of the results of Sect. 4 for residually rational domains.

2 Notation

For all unreferenced results on numerical semigroups we refer the reader to [19].

A *numerical semigroup* is a set $S \subseteq \mathbb{N}$ that contains 0, is closed by addition and such that $\mathbb{N} \setminus S$ is finite. If $a_1, \dots, a_n$ are coprime positive integers, the numerical semigroup *generated* by $a_1, \dots, a_n$ is $\langle a_1, \dots, a_n\rangle := \left\{\sum_{i=1}^n t_i a_i \mid t_i \in \mathbb{N}\right\}$. The notation $S = \{0, b_1, \dots, b_n, \rightarrow\}$ indicates that S is the set containing $0, b_1, \dots, b_n$ and all integers bigger than b_n.

To any numerical semigroup S are associated some natural numbers:

- the *genus* of S is $g(S) := |\mathbb{N} \setminus S|$;
- the *Frobenius number* of S is $F(S) := \sup(\mathbb{N} \setminus S)$;
- the *multiplicity* of S is $m(S) := \inf(S \setminus \{0\})$.

The *Apéry set* of S with respect to $n \in S$ is the set $\mathrm{Ap}(S, n) := \{x \in S \mid x - n \notin S\}$. Without specifications, the *Apéry set* of S is the Apéry set with respect to the multiplicity; we write $\mathrm{Ap}(S) := \mathrm{Ap}(S, m(S))$.

A *hole* of S is an integer $x \in \mathbb{N} \setminus S$ such that $F(S) - x \notin S$. A semigroup S is *symmetric* if it has no holes, while it is *pseudosymmetric* if $g(S)$ is even and $g(S)/2$ is its only hole. Setting $T(S) := \{x \in \mathbb{N} \setminus S \mid x + (S \setminus \{0\}) \subseteq S\}$, we also have that S is symmetric if and only if $T(S) = \{F(S)\}$ [19, Corollary 4.11].

An *integral ideal* of S is a nonempty subset $I \subseteq S$ such that $I + S \subseteq I$, i.e., such that $i + s \in I$ for all $i \in I$, $s \in S$. A *fractional ideal* of S is a subset $I \subseteq \mathbb{Z}$ such that $d + I$ is an integral ideal for some $d \in \mathbb{Z}$, or equivalently an $I \subsetneq \mathbb{Z}$ such that $I + S \subseteq I$. We shall use the term "ideal" as a shorthand for "fractional ideal".

If $\{I_\alpha\}_{\alpha \in A}$ is a family of ideals, then its intersection (if nonempty) is an ideal, while its union is an ideal if and only if there is a $d \in \mathbb{Z}$ such that $d < i$ for all i in the union. If I, J are ideals, the set $(I - J) := \{x \in \mathbb{Z} \mid x + J \subseteq I\}$ is still an ideal of S.

We denote by $\mathcal{F}(S)$ the set of fractional ideals of S, and by $\mathcal{F}_0(S)$ the set of fractional ideals contained between S and $\mathbb{N}$; equivalently, $\mathcal{F}_0(S) = \{I \in \mathcal{F}(S) \mid 0 = \inf(I)\}$. For every ideal I, there is a unique d such that $-d + I \in \mathcal{F}_0(S)$ (namely, $d = \inf(I)$).

If a, b are integers, we use (a, b) to indicate their greatest common divisor. If f, g are functions of n, we use $f = O(g)$ to mean that there is a constant C such that $f(n) \leq C \cdot g(n)$ for all $n \geq 0$.

3 Star Operations

Definition 3.1 ([21, Definition 3.1]) Let S be a numerical semigroup. A *star operation* on S is a map $* : \mathcal{F}(S) \longrightarrow \mathcal{F}(S)$, $I \mapsto I^*$, that satisfies the following properties:

- $*$ is *extensive*: $I \subseteq I^*$ for every $I \in \mathcal{F}(S)$;
- $*$ is *order-preserving*: if $I, J \in S$ and $I \subseteq J$, then $I^* \subseteq J^*$;
- $*$ is *idempotent*: $(I^*)^* = I^*$ for every $I \in \mathcal{F}(S)$;
- $*$ fixes S, that is, $S^* = S$;
- $*$ is *translation-invariant*: $d + I^* = (d + I)^*$ for every $I \in \mathcal{F}(S)$ and every $d \in \mathbb{Z}$.

We denote by $\mathrm{Star}(S)$ the set of star operations on S.

If $I = I^*$, we say that I is $*$-*closed*; we denote the set of $*$-closed ideals by $\mathcal{F}^*(S)$.

The set $\mathrm{Star}(S)$ can be endowed with a natural partial order: we say that $*_1 \leq *_2$ if $I^{*_1} \subseteq I^{*_2}$ for every ideal I, or equivalently if $\mathcal{F}^{*_2}(S) \subseteq \mathcal{F}^{*_1}(S)$. Under this order, $\mathrm{Star}(S)$ is a complete lattice: its minimum is the identity, while its maximum is the v-*operation* (or *divisorial closure*) $v : I \mapsto (S - (S - I))$.

Since $\mathbb{N}$ is v-closed, any star operation restricts to a map $*_0 : \mathcal{F}_0(S) \longrightarrow \mathcal{F}_0(S)$; furthermore, $*_0$ uniquely determines $*$ (since every ideal can be translated into $\mathcal{F}_0(S)$). We define $\mathcal{G}_0(S) := \mathcal{F}_0(S) \setminus \mathcal{F}^v(S)$, that is, $\mathcal{G}_0(S)$ is the set of ideals I of S such that $0 = \inf I$ and $I \neq I^v$.

Since $\mathcal{F}_0(S)$ is finite, Star(S) is a finite set for all numerical semigroups S [21, Proposition 3.2]. Furthermore, $|\mathrm{Star}(S)| = 1$ if and only if v is the identity, which happens if and only if S is symmetric [1, Proposition I.1.15].

4 Estimates Through the Genus

Our main interest in this paper will be the function $\Xi(n)$ that associates to every integer $n > 1$ the number of numerical semigroups S such that $2 \leq |\mathrm{Star}(S)| \leq n$. More generally, if $\mathcal{S}$ is a set of numerical semigroups, we define $\Xi_{\mathcal{S}}(n)$ as the number of semigroups $S \in \mathcal{S}$ such that $2 \leq |\mathrm{Star}(S)| \leq n$. We will mainly be interested in the asymptotic growth and in asymptotic bounds of Ξ and $\Xi_{\mathcal{S}}$, for some distinguished sets $\mathcal{S}$ of semigroups.

It is very difficult to determine precisely the number of star operations on a numerical semigroup S, while it is easier to find estimates for $|\mathrm{Star}(S)|$: for this reason, we work with Ξ instead of the function that counts the number of semigroups with exactly n star operations. Most of the bounds proven in the paper will be obtained in a two-step process:

1. find a function ϕ (depending on some of the invariants of S) such that $|\mathrm{Star}(S)| \geq \phi(S)$ for all $S \in \mathcal{S}$;
2. estimate the number of $S \in \mathcal{S}$ satisfying $\phi(S) \leq n$.

In this way, we obtain an estimate on the number of semigroups $S \in \mathcal{S}$ satisfying $|\mathrm{Star}(S)| \leq n$: indeed, if $|\mathrm{Star}(S)| \leq n$ then we must also have $\phi(S) \leq n$.

The first important result is to prove that Ξ is actually well-defined, that is, that there are only a finite number of numerical semigroups satisfying $2 \leq |\mathrm{Star}(S)| \leq n$. To do so, the first estimate involves the genus of S.

Theorem 4.1 ([27, Proposition 8.1]) *Let S be a nonsymmetric numerical semigroup. Then, $|\mathrm{Star}(S)| \geq g(S) + 1$.*

Sketch of Proof For every ideal $I \in \mathcal{G}_0(S)$, we define $*_I$ as the largest star operation $*$ such that $I = I^*$; equivalently, $*_I$ is the map such that

$$J^{*_I} = J^v \cap (I - (I - J))$$

for every ideal J [21, Proposition 3.6]. Then, $*_I = *_J$ if and only if $I = J$ [21, Theorem 3.8]. Since S is nonsymmetric, there is a $\tau \in T(S) \setminus \{F(S)\}$ [19, Corollary 4.11]; let $\lambda := \min\{\tau, F(S) - \tau\}$. If $x \in \mathbb{N} \setminus S$, let $M_x := \{z \in \mathbb{N} \mid x - z \notin S\}$; then, M_x is an ideal (which is not always divisorial). We associate to each $x \in \mathbb{N} \setminus S$ an ideal I_x:

- if $x < \lambda$ and $\lambda - x \notin S$, then $I_x := S \cup \{z \in \mathbb{N} \mid z > x, z \in M_\lambda\}$;
- if $x < \lambda$ and $\lambda - x \in S$, then $I_x := S \cup \{z \in \mathbb{N} \mid z > g - (\lambda - x)\}$;
- if $x \geq \lambda$, then $I_x := M_x$.

Then, $I_x \neq I_y$ if $x \neq y$. Each I_x is not divisorial: in the first case because $\sup(\mathbb{N} \setminus S) = \lambda \notin S$ and by [21, Lemma 4.7], in the second case because $\tau \in I_x^v \setminus I^v$ (and by [21, Proposition 3.11]), in the third case by [21, Lemma 4.8]. Hence, they generate $g(S)$ different star operations, all different from the divisorial closure. Thus, $|\mathrm{Star}(S)| \geq g(S) + 1$. □

We now translate this estimate to a bound on Ξ.

Theorem 4.2 ([27, Section 8]) *Preserve the notation above.*

(a) $\Xi(n) < \infty$ *for every* $n > 1$.
(b) If $\varphi := \frac{\sqrt{5}+1}{2}$ *is the golden ratio, then*

$$\Xi(n) = O(\varphi^n) = O(\exp(n \log \varphi)).$$

Proof By [32], the number of numerical semigroups of genus at most n is $O(\varphi^n)$. The claim follows from Theorem 4.1. □

5 Estimates Through the Multiplicity

The proof of Theorem 4.1 involves star operations generated by a single ideal (called *principal* star operations). In general, not all star operations have this form; to work more generally we define, given $\Delta \subseteq \mathcal{G}_0(S)$, the star operation *induced by* Δ as

$$*_\Delta := \inf\{*_I \mid I \in \Delta\}.$$

Every star operation can be represented in this form [27, Section 3], but in general we may have $*_\Delta = *_\Lambda$ even if $\Delta \neq \Lambda$. To obtain better estimates, we want to identify special subsets of $\mathcal{G}_0(S)$ that induce pairwise different star operations. We introduce the following definitions.

Definition 5.1 ([27, Definition 3.1]) The $*$*-order* on $\mathcal{G}_0(S)$ is the partial order $\leq_*$ defined by $I \leq_* J$ if and only if $*_I \geq *_J$; equivalently, $I \leq_* J$ if I is $*_J$-closed.

The fact that, for $I, J \in \mathcal{G}_0(S)$, $*_I = *_J$ if and only if $I = J$ guarantees that the $*$-order is really a partial order (see [21, Corollary 3.9] or the proof of Theorem 4.1); on the other hand, the same relation defined on the whole $\mathcal{F}(S)$ is only a preorder (see the discussion after [27, Definition 3.1]).

Definition 5.2 Let $(\mathcal{P}, \leq)$ be a partially ordered set. An *antichain* of $\mathcal{P}$ is a subset of pairwise noncomparable elements.

Definition 5.3 Let $a \in \mathbb{N} \setminus S$. Then, $\mathcal{Q}_a$ is the set of ideals $I \in \mathcal{G}_0(S)$ such that $a = \sup(\mathbb{N} \setminus I)$ and such that $a \in I^v$.

The set $\mathcal{Q}_a$ is nonempty if and only if M_a is nondivisorial (in which case $M_a \in \mathcal{Q}_a$) [27, Proposition 5.2].

Proposition 5.4 ([27, Proposition 5.11]) *Let $a, b \in \mathbb{N} \setminus S$, and let $\Delta \subseteq \mathcal{Q}_a$, $\Lambda \subseteq \mathcal{Q}_b$ be two nonempty sets of ideals that are antichains with respect to set inclusion. If $\Delta \neq \Lambda$, then $*_\Delta \neq *_\Lambda$.*

Given $\mathcal{P} \subseteq \mathcal{G}_0(S)$, we denote by $\omega_i(\mathcal{P})$ the number of antichains of $\mathcal{P}$ with respect to set inclusion.

Corollary 5.5 ([27, Corollary 5.12]) *For every numerical semigroup S, we have*

$$|\mathrm{Star}(S)| \geq 1 + \sum_{a \in \mathbb{N} \setminus S} (\omega_i(\mathcal{Q}_a) - 1).$$

Remark 5.6 In [27], the notation $\omega(\mathcal{P})$ was used for the number of antichain of $\mathcal{P}$ with respect to the $*$-order, and $\omega_i(\mathcal{P})$, with "i" standing for "inclusion", was used to distinguish the two quantities. In this paper we do not use directly the antichains in the $*$-order, but we preserve the notation $\omega_i(\mathcal{P})$ for the sake of consistency.

Corollary 5.5 allows a relatively quick estimate of $\mathrm{Star}(S)$ when S is a fixed semigroup, since finding $\mathcal{Q}_a$ and counting the antichains with respect to inclusion is much quicker than determining and comparing star operations. From a theoretical point of view, it can be used through the following construction.

Suppose a is a hole of S. Let $J := S \cup \{x \in \mathbb{N} \mid x > a\}$, and let $Z(a) := \{a - m + 1, \ldots, a - 1\} \setminus S$. For every $A \subseteq Z(a)$, the set $I_A := J \cup A$ is an ideal of S, and it belongs to $\mathcal{Q}_a$ since $g - a \notin S$ [21, Lemma 4.7]. Furthermore, $I_A \subseteq I_B$ if and only if $A \subseteq B$; hence, the set of the I_A (under the containment order) is isomorphic to the power set of $Z(a)$. The number of antichains of the power set of a set with n elements is called the n-th *Dedekind number*, and we denote it by $\omega(n)$. The sequence $\{\omega(n)\}$ grows extremely quickly (as an exponential of an exponential), and for this reason it is known only up to $n = 8$ [12, 31].

A similar construction can be done if $a < m(S)$ is not a hole, but there is a hole $b < a$; in this case, we consider $Z(a) = \{1, \ldots, a - 2\}$, and the best estimate is obtained with $a = m(S) - 1$. Using these constructions (and some variants), we can prove the following.

Proposition 5.7 ([27, Propositions 5.19 and 5.21]) *Let S be a nonsymmetric numerical semigroup, and let $\nu(S) := \left\lceil \frac{m(S)-1}{2} \right\rceil$. Let $a \in \mathbb{N} \setminus S$.*

(a) If $m(S) < a \leq g/2$ and $g - a \notin S$ then $\omega_i(\mathcal{Q}_a) \geq \omega(\nu(S))$.
(b) If $2m(S) < a \leq g/2$ and $g - a \notin S$ then $\omega_i(\mathcal{Q}_a) \geq 2\omega(\nu(S)) - 2$.
(c) If $a < m(S)$ and $g - a \notin S$ then $\omega_i(\mathcal{Q}_a) \geq \omega(a - 1)$.
(d) If $a < m(S)$ and there is a hole $b < a$ of S, then $\omega_i(\mathcal{Q}_a) \geq \omega(a - 2)$.

In particular, $|\mathrm{Star}(S)| \geq \omega(\nu(S))$.

As in Sect. 4, we can use the last estimate to obtain a bound on Ξ.

Theorem 5.8 ([27, Theorem 8.4]) *For every $\epsilon > 0$,*

$$\Xi(n) = O\left[\exp\left(\left(\frac{2}{\log 2} + \epsilon\right)\log(n)\log\log(n)\right)\right].$$

Sketch of Proof Let $A_\epsilon := \frac{2}{\log 2} + \epsilon$. Using Proposition 5.7 and the estimates in [12], we have that if $|\mathrm{Star}(S)| \leq n$ then (for any $\epsilon' > 0$ and $n \geq n_0(\epsilon')$)

$$n \geq \omega(\nu(S)) \geq 2^{\binom{\nu(S)}{\lceil \nu(S)/2 \rceil}} \geq 2^{2^{(1-\epsilon')\nu(S)}}$$

when $\nu(S)$ is large. Writing it as a function of $m(S)$, we get $m(S) \leq A_\epsilon \log\log n$.

Let $\Xi_\mu(n)$ be the number of nonsymmetric numerical semigroups of multiplicity μ with at most n star operations: then, using Theorem 4.1, $\Xi_\mu(n)$ is at most equal to the number of numerical semigroups of multiplicity μ of genus $\leq n$, which is at most $(n-1)^{\mu-1}$. It follows that

$$\Xi(n) \leq \sum_{\mu=3}^{A_\epsilon \log\log n} (n-1)^{\mu-1} \leq n^{A_\epsilon \log\log(n)} \leq \exp(A_\epsilon \log(n)\log\log(n)),$$

as claimed. □

6 Multiplicity 3

In the last passage of the proof of Theorem 5.8, we needed to estimate the function $\Xi_\mu(n)$ counting the nonsymmetric numerical semigroups of multiplicity μ with at most n star operations. While a very crude bound was enough to obtain the theorem, it is reasonable to ask for more precise estimates: in this section we analyze the case of multiplicity 3, while in the next one we study the case where $m(S) > 3$ is prime.

The case of numerical semigroups of multiplicity 3 can be analyzed very thoroughly, obtaining a complete solution to the problem of finding the set of star operations on S.

Theorem 6.1 *Let $S := \langle 3, 3\alpha+1, 3\beta+2\rangle$ be a numerical semigroup of multiplicity 3, where* $\mathrm{Ap}(S) = \{3, 3\alpha+1, 3\beta+2\}$.

(a) [22, Theorem 7.4] $(\mathcal{G}_0(S), \leq_*)$ is order-isomorphic to the direct product $\{1, \dots, 2\alpha-\beta\} \times \{1, \dots, 2\beta-\alpha+1\}$.
(b) [22, Corollary 6.5] $\mathrm{Star}(S)$ *is order-isomorphic to the set of antichains of $(\mathcal{G}_0(S), \leq_*)$.*
(c) [22, Theorem 7.6] $|\mathrm{Star}(S)| = \binom{\alpha+\beta+1}{2\alpha-\beta} = \binom{\alpha+\beta+1}{2\beta-\alpha+1} = \binom{g(S)+1}{F(S)-g(S)+2}$.

Using Proposition 5.4, we can also improve [22, Proposition 7.8].

Proposition 6.2 *Let S be a nonsymmetric numerical semigroup. Then, the following are equivalent:*

(i) S *is a pseudosymmetric semigroup of multiplicity 3;*
(ii) $(\mathcal{G}_0(S), \leq_*)$ *is linearly ordered;*
(iii) $\mathrm{Star}(S)$ *is linearly ordered.*

Proof If $m(S) = 3$, the result is exactly [22, Proposition 7.8]. Suppose thus $m(S) > 3$; we need to show that $(\mathcal{G}_0(S), \leq_*)$ is not linearly ordered, and to do so it is enough (by Proposition 5.4) to find two ideals J_1, J_2 in some $\mathcal{Q}_a$ that are not comparable. Let τ be a hole of S such that $\tau \leq g/2$ (it exists because S is not symmetric). We distinguish several cases.

If $\tau \geq 3$, then by [21, Lemma 4.13] we can find $a_1, a_2 \in (\{\tau - m + 1, \ldots, \tau - 1\} \cap \mathbb{N}) \setminus S$; then, we set $J_i := S \cup \{x \in \mathbb{N} \mid x > \tau\} \cup \{a_i\}$.

If $\tau < 3$ and $m(S) > 4$, consider $b := 4$: then, the set $\{1, 2, 3\} \setminus \{3 - \tau\}$ contains two different elements, say x_1 and x_2, and we take $J_i := S \cup \{x \in \mathbb{N} \mid x > 3\} \cup \{3 - \tau, x_i\}$ (they belong to $\mathcal{Q}_3$ by the proof of [27, Proposition 5.20]).

Suppose $m(S) = 4$ and $\tau \leq 2$. If $\tau = 1$ then one between $g := g(S)$ and $g - 1$ is even; call it e. Then, $e/2$ is a hole of S which is not bigger then $g/2$; in particular, if $\frac{e}{2} \geq 3$ we are in the case above. If $\frac{e}{2} \leq 2$, then $g \leq 5$, and so either $g = 3$ or $g = 5$. In the latter case we would have $g - 1 = 4 \notin S$, a contradiction; in the former case, $S = \langle 4, 5, 6, 7\rangle$, and by direct inspection $\mathcal{G}_0(S)$ is not linearly ordered (see [27, Example 5.21]).

If $\tau = 2$, consider $J_1 := S \cup \{g - 2\}$ and $J_2 := S \cup (2 + S)$. Then, both are elements of $\mathcal{Q}_g$, and $g - 2 \notin J_2$ (otherwise $g - 2 - 2 = g - 4 = g - m \in S$, which is absurd); furthermore, $J_1 \neq J_2$ since otherwise $2 = g - 2$, i.e., $g = 4$, a contradiction, and so they are noncomparable.

Therefore, if $m(S) > 3$ the $*$-order on $\mathcal{G}_0(S)$ is not total, as claimed. □

We now want to use Theorem 6.1 to calculate $\Xi_3(n)$. The idea is to divide the set of semigroups of multiplicity 3 in sets defined by the relation $2\alpha - \beta = k$ (if $\alpha \leq \beta$) or $2\beta - \alpha + 1 = k$ (if $\alpha > \beta$), and then estimate $\Xi_S(n)$ for each of these families.

Lemma 6.3 *Let k, n be integers, and define*

$$p_{k,n}(X) := \frac{X(X-1)\cdots(X-k+1)}{k!} - n.$$

Then:

(a) $p_{k,n}$ *has a unique zero* $x_{k,n}$ *that satisfies* $x_{k,n} > k - 1$*;*
(b) *for all k, there is a $n_0(k)$ such that, for all $n \geq n_0(k)$,*

$$(k!n)^{1/k} - 1 < x_{k,n} < (k!n)^{1/k} + k - 1.$$

Proof

(a) Let $\widetilde{p}_{k,n}(X) := p_{k,n}(X + k - 1) = \frac{X(X+1)\cdots(X+k-1)}{k!} - n$: then, $\widetilde{p}_{k,n}$ is a polynomial whose coefficients are all positive, and thus $\widetilde{p}_{k,n}$ is increasing for $X > 0$, i.e., $p_{k,n}$ is increasing for $X > k - 1$. Furthermore, $p_{k,n}(k - 1) = \widetilde{p}_n(0) = -n$, and thus $p_{k,n}$ has a unique zero $x_{k,n} > k - 1$.

(b) We have

$$p_{k,n}((k!n)^{1/k} + k - 1) = \widetilde{p}_{k,n}((k!n)^{1/k}) > \frac{((k!n)^{1/k})^k}{k!} - n = n - n = 0,$$

and thus $x_{k,n} < (k!n)^{1/k}+k-1$. On the other hand, write $k!\widetilde{p}_{k,n}(X) = \sum_{t=0}^{k} \lambda_t X^t$: then, $\lambda_k = 1$ and $\lambda_0 = -k!n$. We have

$$\lambda_t \cdot ((k!n)^{1/k} - k)^t = \lambda_t \sum_{i=0}^{t} \binom{t}{i} (-1)^{t-i} (k!n)^{i/k} k^{(t-i)/k}.$$

Adding all these terms, we see that $k!\widetilde{p}_{k,n}(X)$ is a sum of monomials (with fractional exponent) in n. The maximal exponent is 1, which appears twice: for $t = k = i$ and for $t = 0$. The former is equal to $k!n$ and the latter to $-k!n$, and so their sum is zero. The next term is the one with exponent $(k - 1)/k$, and again we have two monomials: for $t = k$ and $i = 1$ and for $t = k - 1 = i$. Hence, the leading term of $k!\widetilde{p}_{k,n}((k!n)^{1/k} - k)$, as a function of n, is

$$-\binom{k}{1}(k!n)^{(k-1)/k} \cdot k + \lambda_{k-1}(k!n)^{(k-1)/k} = k!^{(k-1)/k}(-k^2 + \lambda_{k-1})n^{(k-1)/k}.$$

We have $\lambda_{k-1} = 1+2+\cdots+k-1 = \frac{k(k-1)}{2}$; hence, the sign of $k!\widetilde{p}_n((k!n)^{1/k} - k)$ is equal to the sign of

$$-k^2 + \lambda_{k-1} = -k^2 + \frac{k(k-1)}{2} = -\frac{k^2 + k}{2} < 0.$$

Therefore, for large n we have $x_{k,n} > (k!n)^{1/k} - k + (k - 1) = (k!n)^{1/k} - 1$, as claimed. □

Theorem 6.4 *For every integer $t > 1$, we have*

$$\Xi_3(n) = \frac{2}{3}\left(\sum_{k=1}^{t-1}(k!)^{1/k} \cdot n^{1/k}\right) + O(n^{1/t}\log^2 n).$$

Proof Given a numerical semigroup $S = \langle 3, 3\alpha + 1, 3\beta + 2\rangle$ of multiplicity 3, let $p(S) := \alpha + \beta + 1$ and $q(S) := 2\alpha - \beta$. Then, $p(S) + q(S) = 3\alpha + 1$; we have

$p(S) > q(S)$ for all nonsymmetric semigroups, and furthermore $p(S) \neq 2q(S)$ for all S, which means that $p(S) < 2q(S)$ or $p(S) > 2q(S)$.

Given an integer $k \geq 1$, define the following sets: $\mathcal{S}_k$ is the set of numerical semigroups with $p(S) < 2q(S)$ and $q(S) = k$, while $\mathcal{S}_{-k}$ is the set of semigroups with $p(S) > 2q(S)$ and $p(S) - q(S) = k$. Then, each nonsymmetric semigroup belongs to exactly one $\mathcal{S}_k$ or $\mathcal{S}_{-k}$, and thus

$$\Xi_3(n) = \sum_{k\geq 1} \Xi_{\mathcal{S}_k}(n) + \Xi_{\mathcal{S}_{-k}}(n).$$

We claim that $\Xi_{\mathcal{S}_k}(n) = (k!)^{1/k} \cdot n^{1/k} + O(1)$ for each k.

Indeed, $\Xi_{\mathcal{S}_k}(n)$ is equal to the number of integer solutions of the system

$$\begin{cases} \binom{X}{k} \leq n \\ X + k \equiv 1 \bmod 3 \\ X \geq 2k \end{cases}$$

In the notation of Lemma 6.3, the first equation is exactly $p_{k,n}(X) \leq 0$; hence, the number of solutions is $\frac{1}{3}(x_{k,n} - 2k) + \epsilon$ for some $|\epsilon| \leq 1$ (depending on k and n). For large n, using Lemma 6.3(b) this is equal to

$$\frac{1}{3}k!^{1/k}n^{1/k} - \frac{2}{3}k + O(1) = \frac{1}{3}k!^{1/k}n^{1/k} + O(1)$$

for k fixed, as claimed. A completely analogous reasoning holds for $\mathcal{S}_{-k}$, since also $\binom{X}{X-k} = p_{k,n}(X)$.

Take any integer t and let $\mathcal{S} := \bigcup_{k<t} \mathcal{S}_k \cup \mathcal{S}_{-k}$. Then,

$$\begin{aligned}\Xi_{\mathcal{S}}(n) &= \sum_{i=1}^{t-1} \Xi_{\mathcal{S}_k}(n) + \Xi_{\mathcal{S}_{-k}}(n) = \sum_{i=1}^{t-1}\left(\frac{2}{3}k!^{1/k}n^{1/k} + O(1)\right) \\ &= \frac{2}{3}\left(\sum_{i=1}^{t-1} k!^{1/k}n^{1/k}\right) + O(t).\end{aligned}$$

Let $\mathcal{S}'$ be the complement of $\mathcal{S}$ in the set of all numerical semigroups of multiplicity 3, and consider $\Xi_{\mathcal{S}'}(n)$. Let $G_r(n)$ be the number of binomial coefficients $\binom{a}{b}$ such that $\binom{a}{b} \leq n$, $b \geq t$ and $a \geq 2b$; then, since a binomial coefficient arises from at most one semigroup, we have

$$\Xi_{\mathcal{S}'}(n) \leq 2\sum_{r=t}^{\infty} G_r(n). \tag{1}$$

If $k > \log_2(n)$, then

$$\binom{2k}{k} \geq \frac{4^{\log_2(n)}}{\sqrt{4\log_2(n)}} \geq \frac{n^2}{\sqrt{4\log_2(n)}} > n$$

for large n. Thus, it is enough to consider the sum in (1) only for k going from t to $\log_2(n)$.

By Lemma 6.3, if $\binom{a}{t} \geq n$ then $a \leq (k!n)^{1/k}$; hence, $G_k(n) \leq (k!n)^{1/k}$ and

$$\Xi_{S'}(n) \leq 2 \sum_{k=t}^{\log_2(n)} (k!n)^{1/k} \leq 2n^{1/t} \sum_{k=t}^{\log_2 n} (k!)^{1/k} = O(n^{1/t} \log^2 n).$$

since $(k!)^{1/k} \leq k$. The claim is proved. □

Note that we *cannot* write Ξ_3 as the series

$$\Xi_3(n) = \frac{2}{3} \sum_{k=1}^{\infty} (k!)^{1/k} \cdot n^{1/k},$$

because at fixed n the terms have limit 1, and so the series does not converge. When n is fixed, a good approximation for $\Xi_3(n)$ is obtained stopping the series at $k = \log_2(n)$; an even better approximation can be obtained stopping it at $k = \frac{1}{2}(\log_2 n + \log_2 \log_2 n)$, since also for this value we have $\binom{2k}{k} > n$.

7 Prime Multiplicity

The formula for $|\mathrm{Star}(S)|$ in the previous section was based on an explicit (and very regular) description of $\mathcal{G}_0(S)$. For semigroups of bigger multiplicity, both listing all non-divisorial ideals and understanding the $*$-order becomes much more complicated (see the examples in [24]), and so we need to rely on estimates. In this section, we shall obtain good estimates for some particular classes of semigroups.

The main idea is to generalize the reasoning used to obtain the estimate $|\mathrm{Star}(S)| \geq \omega(\nu(S))$ by considering not only the elements $b \in \{a - m(S) + 1, \ldots, a - 1\} \setminus S$, but also the integers in the form $b - km$.

Theorem 7.1 *Let S be a nonsymmetric numerical semigroup of multiplicity m, and let $a \in \mathbb{N} \setminus S$ be a hole of S. Suppose that there are $b_1, b_2 \in (a - m, a) \cap \mathbb{N}$ and $\sigma \in \mathbb{N}$ such that:*

- *$b_1, b_2 \notin S$;*
- *for $c \in \{a - b_1, a - b_2, |b_1 - b_2|\}$, the element $a_c \in \mathrm{Ap}(S, m)$ congruent to c modulo m satisfies $a_c \geq \sigma m$.*

Then, $|\mathrm{Star}(S)| \geq \binom{2\sigma}{\sigma}$.

Proof For $0 \leq j, k < \sigma$, let $I(j, k)$ be the ideal

$$I(j, k) := S \cup \{x \in \mathbb{N} \mid x > a\} \cup (b_1 - jm + S) \cup (b_2 - km + S).$$

We first prove that $\max(\mathbb{N} \setminus I(j, k)) = a$. Clearly, every element larger than a is in $I(j, k)$. On the other hand, $a \notin S$, while $a \in b_1 - jm + S$ is equivalent to $a-(b_1-jm) \in S$, and the latter is impossible since $a-(b_1-jm) = (a-b_1)+jm < \sigma m$; hence, $a \notin b_1 - jm + S$, and in the same way $a \notin b_2 - km + S$.

Furthermore, $b_1 - jm - m \notin I(j, k)$: the only possibility would be $b_1 - jm - m \in b_2 - km + S$, but his would imply

$$b_1 - jm - m - (b_2 - km) = b_1 - b_2 + (k - j - 1)m \in S,$$

which is impossible since $b_1 - b_2 + (k - j - 1)m < \sigma m$. Hence, the Apéry set of $I(j, k)$ contains a, $b_1 - jm$ and $b_2 - km$; in particular, these ideals all distinct.

Since a is a hole of S, all the $I(j, k)$ belong to $\mathcal{Q}_a$, and by Proposition 5.4 every nonempty antichain with respect to containment induces a different star operation on S. Under the containment order, the set of the $I(j, k)$ is isomorphic to the direct product $\{1, \ldots, \sigma\} \times \{1, \ldots, \sigma\}$; by [22, Lemma 7.5], the latter set has $\binom{2\sigma}{\sigma}$ antichains. The claim now follows from Corollary 5.5. □

When instead of b_1 and b_2 we have z elements, say $b_1, \ldots, b_z$, in $(a - m, a) \cap \mathbb{N}$ but out of S, the same reasoning (with the natural modifications to the hypothesis) can be applied, considering the set containing the ideals in the form

$$I(j_1, \ldots, j_z) := S \cup \{x \in \mathbb{N} \mid x > a\} \cup \bigcup_{i=1}^{z} (b_i - j_i m + S),$$

which will be isomorphic to $\{1, \ldots, \sigma\}^z$. Numerically, this version gives a much better bound on $|\text{Star}(S)|$, although there isn't a simple formula to express it; however, the version of the theorem with only b_1 and b_2 will suffice for our purpose.

Lemma 7.2 *If a is a hole of a numerical semigroup S and $a + m(S) \notin S$, then $a + m(S)$ is a hole of S.*

Proof Immediate from the fact that $F(S) - (a + m(S)) = (F(S) - a) - m(S)$ can't belong to S if $F(S) - a \notin S$. □

Lemma 7.3 *Let S be a numerical semigroup with multiplicity m, and let $a \in \mathrm{Ap}(S, m)$. If $(a, m) | (F(S), m)$, then*

$$a \geq \frac{F(S) + m}{m - 1}$$

Proof Suppose first that $(a, m) = 1$: then, $S' := \langle m, a \rangle$ is a numerical semigroup, and $F(S) \leq F(S')$. However, $F(S') = am - a - m = a(m-1) - m$; solving for a we have our claim.

If $(a, m) =: d > 1$, we consider the semigroup $S' := S/d := \{x/d \mid x \in S \cap d\mathbb{N}\}$: then, since d divides m and $F(S)$, we have $m(S') = m(S)/d$, $F(S') = F(S)/d$ and $a/d \in S'$. By the previous part of the proof,

$$\frac{a}{d} \geq \frac{F(S') + m(S')}{m(S') - 1} = \frac{F(S) + m(S)}{d} \frac{d}{m(S) + d} = \frac{F(S) + m}{m - d} \geq \frac{F(S) + m}{m - 1},$$

and the claim is proved. □

Theorem 7.4 *Let $m > 3$ be a prime number. Then, for every $\epsilon > 0$,*

$$\Xi_m(n) = O(\log^{m-1} n) = O(n^\epsilon).$$

Proof There are only finitely many numerical semigroups of multiplicity m satisfying $F(S) < km$, for every $k \in \mathbb{N}$; hence, we can ignore them and only consider (nonsymmetric) semigroups satisfying $F(S) > m^3$.

Fix such a semigroup S, and let a be a hole of S satisfying $a \leq F(S)/2$. Applying Lemma 7.2, we see that, for any $k \in \mathbb{N}$, the element $a + km$ is either a holc of S or belongs to S; let h be the largest of such holes that is also smaller or equal than $F(S)/2$. By Lemma 7.3, and since $m > 3$, we must have $h \geq \frac{F(S)+m}{m-1} - m \geq \frac{F(S)-m^2}{m-1}$. Note that, since $F(S) > m^3$, we have $h > m$.

By [21, Lemma 4.13], since $m < h \leq F(S)/2$, there are two elements $b_1, b_2 \in (a - m, m) \setminus S$; taking $\sigma := \left\lfloor \frac{1}{m} \frac{F(S)+m}{m-1} \right\rfloor$, we can apply Theorem 7.1, obtaining $|\mathrm{Star}(S)| \geq \binom{2\sigma}{\sigma}$. Now

$$\left\lfloor \frac{1}{m} \frac{F(S) + m}{m - 1} \right\rfloor \geq \frac{1}{m} \frac{F(S) + m}{m - 1} - 1 = \frac{F(S)}{m(m-1)} + \frac{1}{m - 1} - 1 \geq \frac{F(S)}{m^2}$$

using $F(S) > m^3$. Setting $\sigma' := \left\lceil \frac{F(S)}{m^2} \right\rceil$, for these semigroups we have

$$|\mathrm{Star}(S)| \geq \binom{2\sigma'}{\sigma'} \geq \frac{2^{2\sigma'-1}}{\sqrt{\sigma'}} \geq 2^{\sigma'}.$$

If $|\mathrm{Star}(S)| \leq n$, this means that $\sigma' \leq \log_2 n$, i.e.,

$$\frac{F(S)}{m^2} < \log_2 n \Longrightarrow F(S) < m^2 \log_2 n.$$

Therefore,

$$\Xi_m(n) \le C+(m^2\log_2 n)^{m-1} = C+m^{2(m-1)}(\log_2 n)^{m-1} = O(\log^{m-1} n) = O(n^\epsilon)$$

for every $\epsilon > 0$. □

Corollary 7.5 *Let $\mathcal{S}$ be the set of all numerical semigroups whose multiplicity is a prime number > 3. Then, for every $\epsilon > 0$, we have*

$$\Xi_{\mathcal{S}}(n) = O(n^\epsilon).$$

Proof By [27, Proposition 8.2], we need to consider only semigroups with multiplicity up to $A_\epsilon \log\log n$, where $A_\epsilon := \frac{2}{\log 2} + \epsilon$.

There are at most $(m^2)^{m-1} = m^{2(m-1)}$ semigroups of multiplicity m with $F(S) < m^3$; hence, by the proof of the previous theorem we have

$$\Xi_m(n) \le m^{2(m-1)} + \frac{2}{\log 2} m^{2(m-1)} \log^{m-1} n \le \frac{4}{\log 2} \log^{m+2} n$$

for large n, since $m^{2(m-1)} \le (A_\epsilon \log\log n)^{2A_\epsilon \log\log n} \le \log^3 n$. Therefore,

$$\Xi_{\mathcal{S}}(n) = \sum_{m>3 \text{ prime}} \Xi_m(n) = \sum_{\substack{m=5\\ m \text{ prime}}}^{A_\epsilon \log\log n} \Xi_m(n) \le (A_\epsilon \log\log n) \cdot \frac{4}{\log 2} (\log n)^{A_\epsilon \log\log n},$$

which is $O(n^\epsilon)$. The claim is proved. □

The proof above is based on the fact that if $m(S)$ is prime then no generator of S can be too small. The same happens if we consider only the elements of the Apéry set that are coprime with $m(S)$; however, in this case, we also need to find a large hole. If $F(S)$ is even, one easy solution is using $F(S)/2$.

Theorem 7.6 *Let $\mathcal{S}$ be the set of numerical semigroups of multiplicity $m \ge 4$ such that $3 \nmid m$ and $F(S) \equiv 0 \bmod 2$. Then, for every $\epsilon > 0$,*

$$\Xi_{\mathcal{S}}(n) = O(n^\epsilon).$$

Proof Let $\mathcal{S}_m$ be the set of numerical semigroup with (fixed) multiplicity m satisfying $F(S) \equiv 0 \bmod 2$; for large n, by the proof of Theorem 5.8 we have $\Xi_{\mathcal{S}_m}(n) = 0$ if $m > 2\log\log n$.

As in the previous proof, there are at most m^{2m} semigroups S of multiplicity m with $F(S) \le 2m^2$.

Fix a semigroup S such that $F(S) > 2m^2$, and let $\tau := F(S)/2$: then, τ is a hole of S and, since $F(S) > 2m^2$, we have $\tau > m^2$. Consider the elements $\tau - 2$ and $\tau - 1$.

If $\tau_1, \tau_2 \notin S$, then we can apply Theorem 7.1 with $b_1 = \tau - 2$, $b_2 = \tau - 1$ and $\sigma = \left\lfloor \frac{F(S)}{m^2} \right\rfloor$, applying Lemma 7.3 (since both $(1, m)$ and $(2, m)$ divide $(m, F(S))$).

If $\tau_1, \tau_2 \in S$, then $\tau + 1$ and $\tau + 2$ cannot belong to S (otherwise $\tau - 1 + \tau + 1 = 2\tau = F(S) \in S$, a contradiction, and analogously for $\tau - 2$). Hence, we can apply Theorem 7.1 with $b_1 = \tau - m + 2$, $b_2 = \tau - m + 1$ and $\sigma = \left\lfloor \frac{F(S)}{m^2} \right\rfloor$.

Suppose that $\tau - 2 \in S$ while $\tau - 1 \notin S$. As before, $\tau + 2 \notin S$, and we take $b_1 := \tau - m + 2$ and $b_2 := \tau - 1$. Then, $b_2 - b_1 = m - 3$, and so $(m, m - 3) = 1$ (since $3 \nmid m$). Using Lemma 7.3 we can apply Theorem 7.1 with $\sigma = \left\lfloor \frac{F(S)}{m^2} \right\rfloor$. Analogously, if $\tau - 2 \notin S$ and $\tau - 1 \in S$ we use $b_1 := \tau - m + 1$ and $b_2 := \tau - 2$.

In all cases, we have $|\mathrm{Star}(S)| \geq \binom{2\sigma}{\sigma} \geq 2^\sigma$. Hence, for large n, is $S \in \mathcal{S}_m$ satisfies $|\mathrm{Star}(S)| \geq n$ we must have $F(S) < m^2 \log_2 n$; as in the proof of Theorem 7.4 it follows that

$$\Xi_{\mathcal{S}_m}(n) \leq m^{2m} + \frac{2}{\log 2} \log^{m+2} n$$

for large n, and summing on m we have

$$\Xi_{\mathcal{S}}(n) \leq (2 \log\log n)^{4 \log\log n + 1} + \frac{2}{\log 2} (\log n)^{A_\epsilon \log\log n} = O(n^\epsilon)$$

for every $\epsilon > 0$. □

Proposition 7.7 *Let $\mathcal{S}$ be the set of numerical semigroups of multiplicity $m \geq 4$ such that $F \equiv 0 \bmod 6$. Then, for every $\epsilon > 0$,*

$$\Xi_{\mathcal{S}}(n) = O(n^\epsilon).$$

Proof The proof is entirely analogous to the proof of Theorem 7.6. □

An interesting point to note is that, if we are interested in an asymptotic bound or expression for $\Xi(n)$, the families considered in Theorems 7.4 and 7.6 or in Proposition 7.7 give a contribution of a lower order than Ξ_3 (for which Theorem 6.4 gives a linear term); hence, these families are irrelevant when considering (the dominant term of) the asymptotic growth for Ξ.

8 Linear Families

In the previous section, Theorem 7.1 has been applied on families where, while the Frobenius number increases, also the generators (or at least some of them) increase; this is then used to prove an exponential bound on $|\mathrm{Star}(S)|$, which in turn gives a bound of type $O(n^\epsilon)$ on $\Xi_{\mathcal{S}}$. In general, however, it is possible to have a family of semigroups where the Frobenius number increases, while some generators remain fixed.

Let S be a numerical semigroup and $d > 1$ be an integer dividing $m(S)$. Let $\{b_1, \dots, b_s\}$ be integers such that $b_i \geq d \cdot (F(S) + m(S))$ and such that each b_i is coprime with $m(S)$. Then, $T := \langle dS, b_1, \dots, b_s \rangle$ is a numerical semigroup. We can divide the Apéry set of T into two parts, $d\mathrm{Ap}(S)$ and a set $A := \{a_1, \dots, a_t\}$ where each a_i is bigger than every element of $d\mathrm{Ap}(S)$.

For every $k \geq 0$, let now $T_k := \langle dS, A + kd \rangle$; then, T_k is still a numerical semigroup, and $T_k = dS \cup (A + kd + m(T)\mathbb{N})$. Considering the family $\{T_k\}_{k \geq 0}$, this means that one part of the semigroup remains fixed for every member of the family, while another part gets smaller and smaller.

We call a family $\mathcal{T} := \{T_k\}_{k \geq 1}$ constructed in this way the *linear family* constructed from S, d and $\{b_1, \dots, b_s\}$.

In particular, we have $F(T_k) = F(T) + kd$; furthermore, if $x \in \mathbb{N} \setminus S$ and $x + m(S) \in dS$, then $F(T) - x \in T$ if and only if $F(T_k) - x = F(T) + kd - x \in T_k$. Suppose now that T has only two holes, x and $F(T) - x$, and suppose that $x + m(S) \in dS$. Then, the only holes of T_k will be x and $F(T) + kd - x$; in particular, the method applied in the previous section using Theorem 7.1 can fail badly, in the sense that the integer σ will be the same for all members of the family. In particular, the bound on $|\mathrm{Star}(S)|$ does not increase with k.

Example 8.1 Start from $S = \langle 2, 3 \rangle$ and take $d = 2$. Then, $d(F(S) + m(S)) = 6$, so we can take $\{b_1, b_2\} = \{9, 11\}$. Hence, $T := \langle 4, 6, 9, 11 \rangle$, while $T_k := \langle 4, 6, 9 + 2k, 11 + 2k \rangle$. The only holes of T are 2 and 7, so the holes of T_k are 2 and $7 + 2k$. For the hole $a = 2$, the only possible σ is 0, while for the hole $a = 7 + 2k$ the set $\{a - m + 1, \dots, a - 1\}$ contains a unique element out of S, namely $a - m + 2 = 5 + 2k$, and thus Theorem 7.1 cannot even be applied to $7 + 2k$.

The only estimate we have is thus Theorem 4.1, which gives $|\mathrm{Star}(T_k)| \geq g(T_k) + 1 = k + 5$ and corresponds to a bound $\Xi_{\mathcal{T}}(n) \leq n - 4$, where $\mathcal{T} := \{T_k\}_{k \geq 1}$.

For this particular family, [24, Proposition 5.8] gives the *upper* bound $|\mathrm{Star}(T_k)| \leq 65 + 30k$, which in particular implies $\Xi_{\mathcal{T}}(n) \geq \frac{1}{30}n - \frac{65}{30}$.

A calculation of $|\mathrm{Star}(T_k)|$ for low k suggests that the behavior of $|\mathrm{Star}(T_k)|$ is linear in k; more precisely, that $|\mathrm{Star}(T_k)| = 51 + 20k$, and thus that $\Xi_{\mathcal{T}}(n) = \frac{1}{20}n - \frac{31}{20} = \frac{1}{20}(n - 31)$.

In general, there will be linear families for which $|\mathrm{Star}(T_k)|$ does not exhibit a linear behavior: for example, if $m(S)$ is odd and coprime with 3 (and so d must be odd too) then $F(T_k)$ will be alternatively even and odd, and so for at least one half of the semigroups of the family we can apply Theorem 7.6; the same happens if T has holes that are bigger than the elements of $d\mathrm{Ap}(S)$.

On the other hand, if the behavior of $|\mathrm{Star}(T_k)|$ *is* linear (as it seems to happen in the example), then the contribution of $\Xi_{\mathcal{T}}$ to Ξ has the same asymptotic growth as Ξ_3, contrary to what happens for the families of Sect. 7. In particular, the overall contribution of these families will depend also on the precise value of the linear bounds on $\Xi_{\mathcal{T}}$, which seem difficult to calculate theoretically for all families.

In Table 1, we list the precise value of $|\mathrm{Star}(T_k)|$ for a few families obtained with the above construction and for which the sequence $\{|\mathrm{Star}(T_k)|\}$ exhibits (experimentally) a linear behavior.

Table 1 Linear behavior of $|\mathrm{Star}(S)|$

S	d	$\{b_1, \ldots, b_s\}$	T_k	$\lvert\mathrm{Star}(T_k)\rvert$	Range checked
$\langle 2,3\rangle$	2	$\{9, 11\}$	$\langle 4, 6, 9+2k, 11+2k\rangle$	$51+20k$	$0 \leq k \leq 20$
$\langle 2,5\rangle$	2	$\{15, 21\}$	$\langle 4, 10, 15+2k, 21+2k\rangle$	$1368+400k$	$0 \leq k \leq 15$
$\langle 2,7\rangle$	2	$\{21, 23\}$	$\langle 4, 14, 21+2k, 23+2k\rangle$	$29{,}800+6800k$	$0 \leq k \leq 4$

9 Algorithms and Explicit Data

A star operation $*$ is uniquely determined by its restriction $* : \mathcal{F}_0(S) \longrightarrow \mathcal{F}_0(S)$. Since $\mathcal{F}_0(S)$ is a finite set that can be computed explicitly, the set of star operations (and, in particular, its cardinality) can be determined just by listing all maps from $\mathcal{F}_0(S)$ to itself and checking which ones satisfy the properties of a star operation.

An easier way to work algorithmically is to consider the set of closed ideals. Indeed, a star operation $*$ is also uniquely determined by the set $\mathcal{F}_0^*(S) := \{I \in \mathcal{F}_0(S) \mid I = I^*\}$; furthermore, a set $\Delta \subseteq \mathcal{F}_0(S)$ is equal to $\mathcal{F}_0^*(S)$ for some $*$ if and only if it satisfies the following conditions [21, Lemma 3.3]:

- $S \in \Delta$;
- if $I, J \in \Delta$, then $I \cap J \in \Delta$;
- if $I \in \Delta$ and $k \in I$, then $(-k + I) \cap \mathbb{N} \in \Delta$.

In particular, since every star operation is smaller than the divisorial closure, Δ must also contain the set $\mathcal{F}_0^v(S) = \{I \in \mathcal{F}_0(S) \mid I = I^v\}$.

Hence, we can write $\mathcal{F}_0^*(S) = \mathcal{F}_0^v(S) \cup \mathcal{G}_0^*(S)$, where $\mathcal{G}_0^*(S) := \mathcal{G}_0(S) \cap \mathcal{F}_0^*(S)$. By definition, $\mathcal{G}_0^*(S)$ must be downward closed in the $*$-order: thus, we need only to check the subsets of $\mathcal{G}_0(S)$ that are downward closed, and these can be constructed recursively (either directly or by constructing the antichains Θ of $\mathcal{G}_0(S)$ and then considering the sets $\Theta^{\downarrow} := \{J \mid J \leq_* I \text{ for some } I \in \Theta\}$). Furthermore, for any ideal I, the ideals $I \cap J$ (for J divisorial) and $(-k + I) \cap \mathbb{N}$ (for $k \in I$) are always smaller than I in the $*$-order, and thus they do not need to be checked.

Therefore, we can write the following algorithm to calculate the cardinality of $\mathrm{Star}(S)$.

1. Find all ideals in $\mathcal{F}_0(S)$:
   (a) find $\mathrm{Ap}(S) = \{0 = a_0, a_1, \ldots, a_{m-1}\}$, where $m = m(S)$ and $a_i \equiv i \bmod m$;
   (b) for each $1 \leq i \leq m-1$, let $b_i := \lfloor a_i/m \rfloor$;
   (c) for each vector $\mathbf{v} := [c_1, \ldots, c_{m-1}]$ such that $0 \leq c_i \leq b_i$ for all i, consider the set $I(\mathbf{v}) := S \cup \bigcup_i (c_i + m\mathbb{N})$;
   (d) if $I(\mathbf{v})$ is an ideal, store it into $\mathcal{F}_0(S)$.
2. Divide $\mathcal{F}_0(S)$ into $\mathcal{F}_0^v(S)$ and $\mathcal{G}_0(S)$ by checking whether $I = I^v$ or $I \neq I^v$ for all $I \in \mathcal{F}_0(S)$.
3. Construct the $*$-order by checking if $I \leq_* J$ or $J \leq_* I$ for every pair (I, J).

4. For all downward closed subsets Λ of $\mathcal{G}_0(S)$:

   (a) consider $\Delta := \Lambda \cup \mathcal{F}_0^v(S)$;
   (b) check if $I \cap J \in \Delta$ for all $I, J \in \Lambda$;
   (c) if this condition holds, $\Delta = \mathcal{F}_0^*(S)$ for some star operation $*$.

This algorithm has been implemented in GAP, using the functions of the package `numericalsgps` [3, 29].

To calculate explicitly $\Xi(n)$ (for some $n \geq 2$), we can use Theorem 4.1 and Proposition 5.7 to limit the calculation to a finite number of semigroups, and the estimates in Sects. 5–7 to greatly shrink the number of semigroups.

1. Find the maximal m such that $\omega\left(\left\lceil\frac{m-1}{2}\right\rceil\right) \leq n$ (call it M);
2. For $m = 3$, calculate how many binomial coefficients $\binom{a}{b}$ satisfy $a+b \equiv 1 \bmod 3$ and $\binom{a}{b} \leq n$.
3. For $4 \leq m \leq M$, find all numerical semigroups S of multiplicity m with $g(S) \leq n-1$.
4. For every such semigroup S:

   (a) for every $a \in \mathbb{N} \setminus S$, bound $\omega_i(Q_a)$ by using Proposition 5.7, Theorem 7.1 or an explicit calculation;
   (b) if their sum is strictly larger than n, by Corollary 5.5 we have $|\mathrm{Star}(S)| > n$;
   (c) if the sum is at most n, calculate explicitly $|\mathrm{Star}(S)|$.

Remark 9.1

(a) The number of numerical semigroups of multiplicity m and genus up to $n-1$ grows polynomially, and M grows very slowly with n (as a double logarithm of n, by [27, Proposition 8.2]/Theorem 5.8 – for example, if $n = 7000$ we have only $M = 7$).
(b) Those semigroups can be found efficiently by solving linear inequalities, using the so-called *Kunz polytope* of S (see [10, 20]).
(c) Step 4 of the algorithm is very flexible, because it allows to use any kind of estimate on $|\mathrm{Star}(S)|$ before calculating it explicitly. For example, it is possible to use first Proposition 5.7 to obtain a quick estimate, and then, for those semigroups whose estimate is below n, calculate explicitly all of the sets Q_a (which is slower, but gives a better bound). It can also be used with other estimates, not necessarily depending on Q_a.

Using this algorithm, I calculated $\Xi(n)$ and $\Xi_m(n)$ for all $n \leq 150$, and $\Xi_m(n)$ for $m \in \{3, 5, 7\}$ and for all $n \leq 2000$ (for $m = 4$ and $m = 6$, the fact that m is not prime introduces linear families, which slow down considerably the calculation). Tables 2 and 3 show these values, and Table 4 lists those semigroups for $m(S) > 3$.

Table 2 $\Xi(n)$ for $n \leq 150$

n	$\Xi(n)$	$\Xi_3(n)$	$\Xi_4(n)$	$\Xi_5(n)$	$\Xi_6(n)$	$\Xi_7(n)$
10	8	7	1	0	0	0
20	18	14	4	0	0	0
30	27	22	4	1	0	0
40	40	31	6	3	0	0
50	46	37	6	3	0	0
60	57	46	8	3	0	0
70	69	54	9	6	0	0
80	76	60	10	6	0	0
90	83	67	10	6	0	0
100	93	75	11	7	0	0
110	101	82	12	7	0	0
120	111	90	13	8	0	0
130	122	98	15	9	0	0
140	131	105	17	9	0	0
150	141	112	17	12	0	0

Table 3 $\Xi_m(n)$ for $n \leq 2000$ and $m \in \{3, 5, 7\}$

n	$\Xi_3(n)$	$\Xi_5(n)$	$\Xi_7(n)$
100	75	7	0
200	148	13	0
300	220	16	0
400	290	21	0
500	361	21	0
600	431	22	0
700	500	22	0
800	570	22	0
900	639	24	0
1000	709	24	0
1100	776	25	0
1200	845	25	1
1300	914	25	1
1400	982	28	1
1500	1050	28	1
1600	1120	28	1
1700	1186	29	1
1800	1257	30	1
1900	1326	30	1
2000	1393	30	1

Table 4 Numerical semigroups with few star operations (with $|\mathrm{Star}(S)|$ in parentheses)

$m(S)=4$, $\lvert\mathrm{Star}(S)\rvert \leq 150$		
• $\langle 4,5,7\rangle$ (7) • $\langle 4,5,6,7\rangle$ (14) • $\langle 4,5,11\rangle$ (14) • $\langle 4,7,9\rangle$ (15) • $\langle 4,9,11\rangle$ (31) • $\langle 4,6,7,9\rangle$ (32)	• $\langle 4,6,9,11\rangle$ (51) • $\langle 4,7,17\rangle$ (57) • $\langle 4,11,13\rangle$ (63) • $\langle 4,6,11,13\rangle$ (71) • $\langle 4,6,13,15\rangle$ (91) • $\langle 4,7,10,13\rangle$ (105)	• $\langle 4,6,15,17\rangle$ (111) • $\langle 4,13,15\rangle$ (127) • $\langle 4,7,13\rangle$ (129) • $\langle 4,6,17,19\rangle$ (131) • $\langle 4,7,9,10\rangle$ (131)
$m(S)=5$, $\lvert\mathrm{Star}(S)\rvert \leq 2000$		
• $\langle 5,6,7,9\rangle$ (21) • $\langle 5,6,13\rangle$ (31) • $\langle 5,6,7\rangle$ (32) • $\langle 5,7,16\rangle$ (63) • $\langle 5,7,13\rangle$ (65) • $\langle 5,6,8\rangle$ (68) • $\langle 5,8,9,11\rangle$ (96) • $\langle 5,7,8\rangle$ (117) • $\langle 5,8,19\rangle$ (127) • $\langle 5,8,11,12\rangle$ (141)	• $\langle 5,7,9\rangle$ (147) • $\langle 5,6,8,9\rangle$ (148) • $\langle 5,6,7,8,9\rangle$ (163) • $\langle 5,6,14\rangle$ (206) • $\langle 5,9,22\rangle$ (255) • $\langle 5,6,19\rangle$ (275) • $\langle 5,7,9,13\rangle$ (340) • $\langle 5,9,16\rangle$ (351) • $\langle 5,7,8,11\rangle$ (369) • $\langle 5,6,9,13\rangle$ (387)	• $\langle 5,9,12,13\rangle$ (400) • $\langle 5,7,11\rangle$ (539) • $\langle 5,7,8,9,11\rangle$ (824) • $\langle 5,8,11\rangle$ (867) • $\langle 5,11,28\rangle$ (1023) • $\langle 5,6,13,14\rangle$ (1331) • $\langle 5,8,9\rangle$ (1356) • $\langle 5,11,12,14\rangle$ (1363) • $\langle 5,7,23\rangle$ (1685) • $\langle 5,8,9,12\rangle$ (1726)
$m(S)=7$, $\lvert\mathrm{Star}(S)\rvert \leq 2000$		
• $\langle 7,8,9,19\rangle$ (1116)		

10 The Ring Version

Suppose D is an integral domain with quotient field K. A *star operation* on D is a map $*: \mathcal{F}(D) \longrightarrow \mathcal{F}(D)$ that is extensive, order-preserving, idempotent, satisfies $D = D^*$ and such that $x \cdot I^* = (xI)^*$ for all $x \in K$ and all $I \in \mathcal{F}(D)$ (where $\mathcal{F}(D)$ is the set of fractional ideals of D, i.e., of the D-submodules I of the quotient field K of D such that $xI \subseteq D$ for some $x \neq 0$).

The concepts of principal star operations and of the $*$-order can be introduced also for rings; however, in general, there is no set corresponding to $\mathcal{F}_0(S)$ (and so to $\mathcal{G}_0(S)$). Furthermore, we may have $*_I = *_J$ even if I, J are nondivisorial and $I \neq xJ$ for all x.

In this section, we want to study star operations on a class of domains which is close to numerical semigroups. In particular, we shall study domains R satisfying the following conditions:

- R is Noetherian, one-dimensional and local;
- its integral closure V is a discrete valuation ring (DVR);

- the conductor ideal $(R : V)$ is nonzero;
- the extension of residue fields $R/\mathfrak{m}_R \subseteq V/\mathfrak{m}_V$ induced by the extension $R \subseteq V$ is an isomorphism.

Note that, in the previous conditions, we could have dropped "one-dimensional" and "local", since they follow from the fact that the integral closure is a DVR. An equivalent characterization is that the domains we study are the one-dimensional local Noetherian domains that are analytically irreducible and residually rational.

From now on, fix a discrete valuation ring V, and denote by $\mathscr{R}(V)$ the domains of this form whose integral closure is V; R will be a domain in $\mathscr{R}(V)$ and $\mathfrak{m}$ its maximal ideal. We shall use $\mathbf{v}$ to denote the normalized valuation relative to V: then, the set $\mathbf{v}(R) := \{\mathbf{v}(r) \mid r \in R\}$ is a numerical semigroup.

The questions we want to answer in this case are the same as in the numerical semigroup case: is the number of rings in $\mathscr{R}(V)$ with exactly n star operations finite? how many have less than n star operations? How to bound $|\mathrm{Star}(R)|$, for $R \in \mathscr{R}(V)$? For $n = 1$, the answer is well-known: $|\mathrm{Star}(R)| = 1$ if and only if R is Gorenstein, which happens if and only if $\mathbf{v}(R)$ is symmetric, i.e., if and only if $|\mathrm{Star}(\mathbf{v}(R))| = 1$ [2, 14].

Define $\mathcal{F}_0(R) := \{I \in \mathcal{F}(R) \mid R \subseteq I \subseteq V\}$: then, every fractional ideal I is isomorphic to an element of $\mathcal{F}_0(R)$ (just take $x^{-1}I$, where $x \in I$ satisfies $\mathbf{v}(x) = \min \mathbf{v}(I)$). However, unlike the semigroup case, this ideal is not unique: that is, if $y \in I$ is another element of minimal valuation, it may be that $x^{-1}I \neq y^{-1}I$. In particular, we can have $*_{x^{-1}I} = *_{y^{-1}I}$ even if $x^{-1}I \neq y^{-1}I$. However, if I and J are in $\mathcal{F}_0(S)$ and not divisorial, then $*_I = *_J$ implies that $\mathbf{v}(I) = \mathbf{v}(J)$ [21, Proposition 6.4]. We can thus prove an analogue to Theorem 4.1.

If S is a numerical semigroup, a *canonical ideal* of S is a fractional ideal Ω such that $(\Omega - (\Omega - I)) = I$ for every fractional ideal I of S, or equivalently such that $*_{\Omega(S)}$ is the identity. Every canonical ideal is in the form $a + K(S)$, where $K(S) := \{t \in \mathbb{N} \mid F(S) - t \in S\}$ is sometimes called the *standard canonical ideal* of S [9, Section 5]. Likewise, if D is an integral domain, a canonical ideal of D is a fractional ideal Ω such that $(\Omega : (\Omega : I)) = I$ for every fractional ideal I. If $R \in \mathscr{R}(V)$, then R admits canonical ideals [15, Theorem 15.7], and if Ω is one of them then $\mathbf{v}(\Omega)$ is a canonical ideal of $\mathbf{v}(R)$ [9, Satz 5].

Proposition 10.1 *Let $R \in \mathscr{R}(V)$, and suppose that R is not Gorenstein. Then,* $|\mathrm{Star}(R)| \geq g(\mathbf{v}(R)) + 1$.

Proof Let $S := \mathbf{v}(R)$. Since R is not Gorenstein, S is not symmetric, and thus there is a $\tau \in T(S) \setminus \{F(S)\}$; let $\lambda := \min\{\tau, F(S) - \tau\}$. For any positive $a \in \mathbb{N}$, let $T_a := R \cup \{\phi \in V \mid \mathbf{v}(\phi) > a\}$; then, T_a is a ring in $\mathscr{R}(V)$ and $\mathbf{v}(T_a) = \mathbf{v}(R) \cup \{x \in \mathbb{N} \mid x > a\}$, so that $F(\mathbf{v}(T_a)) = a$. For every a, let Ω_a be a canonical ideal of T_a such that $\mathbf{v}(\Omega_a) = \{t \in \mathbb{N} \mid a - t \in \mathbf{v}(T_a)\}$ is the standard canonical ideal of $\mathbf{v}(T_a)$.

Let $x \in \mathbb{N} \setminus S$. We distinguish three cases.

If $x < \lambda$ and $\lambda - x \notin S$, let $I_x := R + \{\phi \in \Omega_\lambda \mid \mathbf{v}(\phi) > x\}$. Then, I_x is an R-module, and $\mathbf{v}(I_x) = \mathbf{v}(R) \cup \{t \in \mathbf{v}(\Omega_\lambda) \mid t > x\}$; in particular, $\lambda \notin \mathbf{v}(I_x)$, and thus $\mathbf{v}(I_x)$ is not divisorial over S, which implies that I_x is not divisorial over R [1, Lemma II.1.22].

If $x < \lambda$ and $\lambda - x \in S$, let $y := g(S) - \lambda + x = g(S) - (\lambda - x)$, and define $I_x := R \cup \{\phi \in V \mid \mathbf{v}(\phi) > y\}$. Then, $\mathbf{v}(I_x)$ is not divisorial since it contains $g(S)$ but not $g(S) - \lambda$, and so I_x is not divisorial.

If $x \geq \lambda$ and $x \neq g(S)$, let $I_x := \Omega_x$: then, I_x is not divisorial since otherwise $T_x = (\Omega_x : \Omega_x)$ would be divisorial, against the fact that $\mathbf{v}(T_x)$ contains $g(S)$ but not λ (if $x = g$, then Ω_x is not divisorial since otherwise S would be symmetric).

It is straightforward to see that $\mathbf{v}(I_x) \neq \mathbf{v}(I_y)$ for $x \neq y$; hence, each one generates a different star operation, and $|\mathrm{Star}(R)| \geq g(\mathbf{v}(R)) + 1$. □

We also note that Proposition 5.7 carries over to the domain case, and in particular $|\mathrm{Star}(R)| \geq \omega(\nu(\mathbf{v}(R)))$. We now prove an analogue of Theorem 4.2, but we have to add an important additional hypothesis.

Theorem 10.2 *Let V be a DVR with finite residue field.*

(a) Every $R \in \mathscr{R}(V)$ has only finitely many star operations.
(b) For every $n > 1$, the set $\{R \in \mathscr{R}(V) \mid 2 \leq |\mathrm{Star}(R)| \leq n\}$ is finite.

Proof The first claim is a special case of [8, Theorem 2.5]. (It follows, for example, from the fact that $\mathcal{F}_0(R)$ is finite.)

For the second claim, we see that if $2 \leq |\mathrm{Star}(R)| \leq n$, then $\mathbf{v}(R)$ is not symmetric and $g(\mathbf{v}(R)) \leq n - 1$; hence, there are only finitely many possible $\mathbf{v}(R)$. Furthermore, since the residue field of V is finite, for any S there are only finitely many R such that $\mathbf{v}(R) = S$ [21, Lemma 5.13(a)]; hence, there are only finitely many $R \in \mathscr{R}(V)$ with $|\mathrm{Star}(R)| \leq n$. The claim is proved. □

In the previous theorem, the restriction to a finite residue field is not really restricting, since otherwise $\mathrm{Star}(R)$ is very often infinite.

Proposition 10.3 *Let $R \in \mathscr{R}(V)$, and suppose that the residue field F of R is infinite; suppose also that R is not Gorenstein. If $m(\mathbf{v}(R)) > 3$, then* $\mathrm{Star}(R)$ *is infinite.*

Proof Let $A := (\mathfrak{m} : \mathfrak{m})$; then, A is a ring, and it is local since its integral closure is V. Since R is not Gorenstein, $\dim_F(A/\mathfrak{m}) > 2$ [2, Theorem 6.3]. If $\dim_F(A/\mathfrak{m}) \geq 4$, then $|\mathrm{Star}(R)| = \infty$ by [8, Corollary 2.8]. If $\dim_F(A/\mathfrak{m}) = 3$, then following [6] let N be the maximal ideal of A and let $B := (N : N)$; by [6, Theorem 2.15], if $\mathrm{Star}(R)$ is finite then $B = V$ and $\dim_F(B/\mathfrak{m}B) = 3$. By [16],

$$\dim_F(B/\mathfrak{m}B) = |\mathbf{v}(B) \setminus \mathbf{v}(\mathfrak{m}B)| = m(\mathbf{v}(R))$$

since $\mathfrak{m}B$ contains all elements of valuation $m(\mathbf{v}(R))$ or more. Hence, if $m(\mathbf{v}(R)) > 3$ then $\mathrm{Star}(R)$ is infinite, as claimed. □

We can also obtain an explicit version of Theorem 10.2.

Lemma 10.4 *Let F be a finite field of cardinality q, and let W be a vector space over F of dimension n. Then, W has at most $2^n q^{n(n-1)/2}$ vector subspaces.*

Proof The number of vector subspaces of W of dimension k is the *q-binomial coefficient* (or *Gaussian binomial coefficient*)

$$\binom{n}{k}_q := \frac{(q^n-1)(q^{n-1}-1)\cdots(q^{n-t+1}-1)}{(q^t-1)(q^{t-1}-1)\cdots(q-1)}$$

(see e.g. [28, Proposition 1.3.18] or [5, Chapter 13, Proposition 2.1]). Using the q-binomial theorem [28, Chapter 3, Exercise 45] with $y = z = 1$ we have

$$\sum_{k=0}^{n}\binom{n}{k}_q \leq \sum_{k=0}^{n} q^{k(k-1)/2}\binom{n}{k}_q = \prod_{k=0}^{n-1}(1+q^k) \leq 2^n q^{n(n-1)/2},$$

as claimed. □

Theorem 10.5 *There is a constant C such that, for all discrete valuation rings V with residue field F of finite cardinality q and for all n,*

$$\Xi_V(n) := |\{R \in \mathscr{R}(V) \mid 2 \leq |\mathrm{Star}(R)| \leq n\}| \leq C(4\varphi)^n q^{n(2n-1)}$$

where $\varphi := \frac{1+\sqrt{5}}{2}$ is the golden ratio.

Proof If $|\mathrm{Star}(R)| \leq n$, then by Theorem 10.2 we have $g(\mathbf{v}(R)) \leq n-1$, and by [32] there are at most $C'\varphi^{n-1}$ semigroups with this property, for some constant C'. If S is a numerical semigroup, then as in the proof of [21, Lemma 5.13(a)] the $R \in \mathscr{R}(V)$ such that $\mathbf{v}(R) = S$ correspond to certain F-vector subspaces of $V/\mathfrak{m}_V^{F(S)+1}$; since $F(S) \leq 2g(S)$, using Lemma 10.4 we see that each S gives at most $2^{2n} q^{n(2n-1)}$ rings. Hence,

$$\Xi_V(n) \leq C'\varphi^{n-1} \cdot 2^{2n} q^{n(2n-1)} = C(4\varphi)^n q^{n(2n-1)}$$

with $C := C'/\varphi$. □

In this bound, the term φ^n can be substituted by a better bound, using (the analogue of) Proposition 5.7; however, the main term is $q^{n(2n-1)}$, whose lowering hinges on a more precise grasp of how many rings correspond to a given semigroup.

In general, the cardinality of $\mathrm{Star}(R)$ does not depend only on $S = \mathbf{v}(R)$ and on the residue field of V, but also on the precise nature of R itself; as a consequence, while it is possible to calculate explicitly $|\mathrm{Star}(R)|$ for a fixed R, in general there will not be a general formula (valid for each R). Sometimes, however, knowing S and the residue field is everything we need.

Proposition 10.6 *Let V be a DVR with residue field F, and let $q := |F|$. Let $R \in \mathscr{R}(V)$. Then:*

(a) [8, Theorem 3.8] if $\mathbf{v}(R) = \langle 3, 4, 5\rangle$, then $|\mathrm{Star}(R)| = 3$;
(b) [8, Example 3.10] if $\mathbf{v}(R) = \langle 3, 5, 7\rangle$, then $|\mathrm{Star}(R)| = 4$;
(c) [25, Proposition 3.4] if $\mathbf{v}(R) = \langle 4, 5, 7\rangle$, then $|\mathrm{Star}(R)| = 2^{2q+3}$;
(d) [30, Corollary 4.1.2] if $\mathbf{v}(R) = \langle 4, 5, 6, 7\rangle$, then $|\mathrm{Star}(R)| = 2^{2q+1} + 2^{q+1} + 2$.

Remark 10.7

(a) If $q = \infty$, then the last two cases should be interpreted as saying that $\mathrm{Star}(R)$ is infinite.
(b) The proofs given in [8, Example 3.10] and [30, Corollary 4.1.2] for $\mathbf{v}(R) = \langle 3, 5, 7\rangle$ and $\mathbf{v}(R) = \langle 4, 5, 6, 7\rangle$ (respectively) were given only in the case $R = K[[S]]$. However, their proofs can be applied also to the general case.

References

1. Barucci, V., Dobbs, D.E., Fontana, M.: Maximality properties in numerical semigroups and applications to one-dimensional analytically irreducible local domains. Mem. Am. Math. Soc. **125**(598), 78 (1997)
2. Bass, H.: On the ubiquity of Gorenstein rings. Math. Z. **82**, 8–28 (1963)
3. Delgado, M., Garcia-Sanchez, P.A., Morais, J.: NumericalSgps, a package for numerical semigroups, version 1.0.1 (2015). http://www.fc.up.pt/cmup/mdelgado/numericalsgps/. Refereed GAP package
4. Gilmer, R.: Multiplicative Ideal Theory, Pure and Applied Mathematics, vol. 12. Marcel Dekker, New York (1972)
5. Graham, R.L., Grötschel, M., Lovász, L. (eds.): Handbook of Combinatorics, vols. 1 and 2. Elsevier, Amsterdam/MIT Press, Cambridge (1995)
6. Houston, E.G., Park, M.H.: A characterization of local Noetherian domains which admit only finitely many star operations: the infinite residue field case. J. Algebra **407**, 105–134 (2014)
7. Houston, E.G., Mimouni, A., Park, M.H.: Integral domains which admit at most two star operations. Commun. Algebra **39**(5), 1907–1921 (2011)
8. Houston, E.G., Mimouni, A., Park, M.H.: Noetherian domains which admit only finitely many star operations. J. Algebra **366**, 78–93 (2012)
9. Jäger, J.: Längenberechnung und kanonische Ideale in eindimensionalen Ringen. Arch. Math. (Basel) **29**(5), 504–512 (1977)
10. Kaplan, N.: Counting numerical semigroups by genus and some cases of a question of Wilf. J. Pure Appl. Algebra **216**(5), 1016–1032 (2012)
11. Kim, M.O., Kwak, D.J., Park, Y.S.: Star-operations on semigroups. Semigroup Forum **63**(2), 202–222 (2001)
12. Kleitman, D., Markowsky, G.: On Dedekind's problem: the number of isotone Boolean functions. II. Trans. Am. Math. Soc. **213**, 373–390 (1975)
13. Krull, W.: Idealtheorie. Springer, Berlin (1935)
14. Kunz, E.: The value-semigroup of a one-dimensional Gorenstein ring. Proc. Am. Math. Soc. **25**, 748–751 (1970)
15. Matlis, E.: 1-Dimensional Cohen-Macaulay rings. Lecture Notes in Mathematics, vol. 327. Springer, Berlin (1973)
16. Matsuoka, T.: On the degree of singularity of one-dimensional analytically irreducible Noetherian local rings. J. Math. Kyoto Univ. **11**, 485–494 (1971)

17. Okabe, A., Matsuda, R.: Semistar-operations on integral domains. Math. J. Toyama Univ. **17**, 1–21 (1994)
18. Park, M.H.: On the cardinality of star operations on a pseudo-valuation domain. Rocky Mountain J. Math. **42**(6), 1939–1951 (2012)
19. Rosales, J.C., García-Sánchez, P.A.: Numerical Semigroups. Developments in Mathematics, vol. 20. Springer, New York (2009)
20. Rosales, J.C., García-Sánchez, P.A., García-García, J.I., Branco, M.B.: Systems of inequalities and numerical semigroups. J. Lond. Math. Soc. (2) **65**(3), 611–623 (2002)
21. Spirito, D.: Star Operations on Numerical Semigroups. Commun. Algebra **43**(7), 2943–2963 (2015)
22. Spirito, D.: Star operations on numerical semigroups: the multiplicity 3 case. Semigroup Forum **91**(2), 476–494 (2015)
23. Spirito, D.: The sets of star and semistar operations on semilocal Prüfer domains. J. Commut. Algebra (2017). arXiv:1707.07507
24. Spirito, D.: Embedding the set of nondivisorial ideals of a numerical semigroup into $\mathbb{N}^n$. J. Algebra Appl. **17**, 1850205 (2018)
25. Spirito, D.: Star operations on Kunz domains. Int. Electron. J. Algebra **25**, 171–185 (2019)
26. Spirito, D.: Vector subspaces of finite fields and star operations on pseudo-valuation domains. Finite Fields Appl. **56**, 17–30 (2019)
27. Spirito, D.: Star operations on numerical semigroups: antichains and explicit results. J. Commut. Algebra **11**(3), 401–431 (2019)
28. Stanley, R.P.: Enumerative combinatorics. Volume 1. Cambridge Studies in Advanced Mathematics, vol. 49, 2nd edn. Cambridge University Press, Cambridge (2012)
29. The GAP Group: GAP—Groups, Algorithms, and Programming, Version 4.8.7 (2017)
30. White, B.: Star operations and numerical semigroup rings. Ph.D. Thesis, The University of New Mexico, 2014
31. Wiedemann, D.: A computation of the eighth Dedekind number. Order **8**(1), 5–6 (1991)
32. Zhai, A.: Fibonacci-like growth of numerical semigroups of a given genus. Semigroup Forum **86**(3), 634–662 (2013)

Torsion in Tensor Products over One-Dimensional Domains

Neil Steinburg and Roger Wiegand

Abstract Over a one-dimensional Gorenstein local domain R, let E be the endomorphism ring of the maximal of R, viewed as a subring of the integral closure $\overline{R}$. If there exist finitely generated R-modules M and N, neither of them free, whose tensor product is torsion-free, we show that E must be local with the same residue field as R.

1 Introduction

Finding interesting examples of non-zero, finitely generated modules M, N over a commutative Noetherian ring R, with $M \otimes_R N$ torsion-free (meaning that no non-zero element of $M \otimes_R N$ is killed by a regular element of R) is a non-trivial task. Of course there are boring examples: take one of the modules to be torsion-free and the other to be projective. Or, if R is not local, take $M = R/\mathfrak{m}$ and $N = R/\mathfrak{n}$, where $\mathfrak{m}$ and $\mathfrak{n}$ are distinct maximal ideals. A slightly less boring example is obtained by taking $R = \mathbb{Q}[[x, y]]/(xy)$ and $M = N = R/(x)$.

> Let R be a local domain, and let M and N be finitely generated modules, neither one of them free. Must $M \otimes_R N$ always have non-zero torsion?

N. Steinburg
Drake University, Des Moines, IA, USA
e-mail: neil.steinburg@drake.edu

R. Wiegand (✉)
University of Nebraska-Lincoln, Lincoln, NE, USA
e-mail: rwiegand@unl.edu

V. Barucci et al. (eds.), *Numerical Semigroups*, Springer INdAM Series 40,
https://doi.org/10.1007/978-3-030-40822-0_18

Again, the answer is "no", and here is the connection with numerical semigroups:

Let $R = k[[t^4, t^5, t^6]]$, $M = (t^4, t^5)$, and $N = (t^4, t^6)$. Then $M \otimes_R N$ is torsion-free [5, 4.3].

In fact, the only known examples where the question above has a negative answer are numerical semigroup rings. This leads to a (somewhat halfhearted, since it is probably false) conjecture:

Conjecture 1 Suppose R is a one-dimensional local domain whose integral closure $\overline{R}$ is finitely generated as an R-module. If there exist finitely generated modules M and N, neither of them free, with $M \otimes_R N$ torsion-free, then $\overline{R}$ is local, and the inclusion $R \subseteq \overline{R}$ induces an isomorphism on residue fields.

2 Some Evidence

In this section we will prove the result stated in the abstract, which gives some support (admittedly rather sketchy) for Conjecture 1.

Throughout, $(R, \mathfrak{m}, k)$ is a one-dimensional Gorenstein local domain, with maximal ideal $\mathfrak{m}$ and residue field $k = R/\mathfrak{m}$. We let K denote the quotient field of R. If I and J are non-zero R-submodules of K, we identify $\operatorname{Hom}_R(I, J)$ with the set $\{\alpha \in K \mid \alpha I \subseteq J\}$, via the isomorphism $\varphi \mapsto \frac{1}{a}\varphi(a)$, where a is a fixed but arbitrary nonzero element of I. In particular, we identify $\operatorname{End}_R \mathfrak{m}$ with the ring $E = \{\alpha \in K \mid \alpha\mathfrak{m} \subseteq \mathfrak{m}\}$. Then $R \subseteq E \subseteq \overline{R}$, where $\overline{R}$ is the integral closure of R in K. The next lemma is due to Bass [2].

Lemma 1 *Assume $\mathfrak{m}$ is not a principal ideal. Then E/R is a simple R-module, and E is minimally generated, as an R-module, by $\{1, y\}$, where y is an arbitrary element of $E \setminus R$.*

Proof Since $\mathfrak{m}$ is indecomposable, there is no surjection $\mathfrak{m} \twoheadrightarrow R$. (Such a surjection would split, giving a decomposition $\mathfrak{m} \cong R \oplus H$, with $H \neq 0$, as $\mathfrak{m}$ is not principal; but clearly $\mathfrak{m}$ is indecomposable, since R is a domain.) This gives the second equality in the display

$$\mathfrak{m}^* = \operatorname{Hom}_R(\mathfrak{m}, R) = \operatorname{Hom}_R(\mathfrak{m}, \mathfrak{m}) = E\,. \tag{1}$$

Dualizing the short exact sequence

$$0 \to \mathfrak{m} \to R \to k \to 0,$$

and using the fact that $k^* = 0$, we get an exact sequence

$$0 \to R^* \to \mathfrak{m}^* \to \operatorname{Ext}^1_R(k, R) \to 0\,.$$

But $\operatorname{Ext}^1_R(k, R) \cong k$, as R is one-dimensional and Gorenstein. The identification of $\operatorname{End}_R(\mathfrak{m})$ with E is compatible with the identification of R^* with R (via multiplications), and thus the last short exact sequence shows that $E/R \cong k$. The next assertion is clear from simplicity of E/R and the fact that 1 is part of a minimal generating set for E, as $1 \notin \mathfrak{m} = \mathfrak{m}E$. □

Lemma 2 *Let S be a subring of K containing R and finitely generated as an R-module. Let M and N be finitely generated S-modules such that $M \otimes_R N$ is torsion-free over R. Then the natural surjection $M \otimes_R N \twoheadrightarrow M \otimes_S N$ is an isomorphism.*

Proof We consult the following commutative diagram:

$$\begin{array}{ccccc}
M \otimes_R N & \xrightarrow{\delta} & K \otimes_R (M \otimes_R N) & \xrightarrow{\cong} & (K \otimes_R M) \otimes_K (K \otimes_R N) \\
\downarrow{\scriptstyle\alpha} & & \downarrow{\scriptstyle\beta} & & \downarrow{\scriptstyle\gamma} \\
M \otimes_S N & \xrightarrow{\varepsilon} & K \otimes_S (M \otimes_S N) & \xrightarrow{\cong} & (K \otimes_S M) \otimes_K (K \otimes_S N)
\end{array} \tag{2}$$

The map δ is injective because $M \otimes_R N$ is torsion-free. One checks (by clearing denominators) that a subset of an S-module is linearly independent over S if and only if it is linearly independent over R, and so its rank as an S-module equals its rank as an R-module. Thus $r := \dim_K(K \otimes_R M) = \dim_K(K \otimes_S M)$ and $s := \dim_K(K \otimes_R N) = \dim_K(K \otimes_S N)$. The surjective map γ is therefore an isomorphism, since its domain and target both have the same K-dimension, namely rs. From the diagram, we see that β must be an isomorphism too, and hence α is injective. □

Theorem 1 *Let $(R, \mathfrak{m}, k)$ be a Gorenstein local domain of dimension one, and let $E = \operatorname{End}_R(\mathfrak{m})$, viewed as a ring between R and its integral closure $\overline{R}$. Assume that there exist finitely generated modules M and N, neither of them free, such that $M \otimes_R N$ is torsion-free. Then E is local, and the inclusion $R \to E$ induces a bijection on residue fields.*

Proof If $\mathfrak{m}$ is a principal ideal, then R is a discrete valuation ring, and $R = E = \overline{R}$. Therefore we assume from now on that $\mathfrak{m}$ is not principal.

We begin with some reductions. We first get rid of free summands, by writing $M = M' \oplus R^m$ and $N = N' \oplus R^n$, where both M' and N' are non-zero, and neither has a non-zero free direct summand. Notice that $M' \otimes_R N'$, being a direct summand of $M \otimes_R N$, is torsion-free. Replacing M by M' and N by N', we may assume that neither M nor N has a non-zero free direct summand.

Next, we have a reduction that goes back to Auslander's 1961 paper [1]. Let $\top X$ denote the torsion submodule of a module X, and put $\bot X = X/(\top X)$. By [3, Lemma 2.2], $(\bot M) \otimes_R (\bot M)$ is torsion-free. Moreover, both $\bot M$ and $\bot N$ are nonzero, since otherwise $M \otimes_R N$ would be a non-zero torsion module. We claim that $\perp M$ has no non-zero free summand. For, suppose there is a surjection $\perp M \twoheadrightarrow R$. Composing this with the natural surjection $M \twoheadrightarrow \perp M$, we get a surjection $M \twoheadrightarrow R$, and hence $M \cong R \oplus L$, a contradiction. Similarly, $\perp N$ has no non-zero free summand. Replacing M and N by their reductions modulo torsion, we may assume that both M and N are non-zero torsion-free R-modules, and that neither M nor N has a non-zero free direct summand.

As in [2], we note that every homomorphism $M \to R$ has its image in $\mathfrak{m}$, and so $M^* = \mathrm{Hom}_R(M, \mathfrak{m})$, which has a natural E-module structure extending the R-module structure. Therefore M^{**} is also an E-module. Since R is Gorenstein and M is torsion-free (= maximal Cohen-Macaulay), the natural map $M \to M^{**}$ is an isomorphism, and hence M itself has an E-module structure compatible with the original R-module structure. By symmetry, N too has a compatible E-module structure. Lemma 2 shows that the natural surjection $M \otimes_R N \twoheadrightarrow M \otimes_E N$ is an isomorphism and, in particular, $M \otimes_E N$ is torsion-free.

Suppose, by way of contradiction, that E is not local, and put $A = E/\mathfrak{m}E$. This is a two-dimensional k-algebra, and it is not local and hence must be isomorphic to $k \times k$. Let e be the idempotent of A supported on first coordinate. Then neither e nor $1-e$ is a unit of A. Let $\overline{M} = M/\mathfrak{m}M$ and $\overline{N} = N/\mathfrak{m}N$. We claim that $e\overline{M} \neq 0$. For suppose $e\overline{M} = 0$. Lift e to an element $\tilde{e} \in E$. Then $\tilde{e}M \subseteq \mathfrak{m}M$. Moreover, $\tilde{e}M + (1-\tilde{e})M + \mathfrak{m}M = M$, and hence $(1-\tilde{e})M = M$ by Nakayama's Lemma. The Determinant Trick yields an element $a \in (1-\tilde{e})E$ such that $(1+a)M = 0$. But M is faithful as an R-module and hence as an E-module (clear denominators). Therefore $1+a = 0$, and hence $-1 \in (1-\tilde{e})E$. But then $-1 \in (1-e)A$, contradicting the fact that $1-e$ is not a unit. This proves the claim and shows that $e\overline{M} \neq 0$. By symmetry, $(1-e)\overline{N} \neq 0$, and hence $e\overline{M} \otimes_k (1-e)\overline{N} \neq 0$. However, the isomorphism $\alpha : M \otimes_R N \xrightarrow{\cong} M \otimes_E N$ induces an isomorphism $\overline{M} \otimes_k \overline{N} \xrightarrow{\cong} \overline{M} \otimes_A \overline{N}$, carrying the non-zero module $e\overline{M} \otimes_k (1-e)\overline{N}$ onto $e\overline{M} \otimes_A (1-e)\overline{N} = 0$, a contradiction. This completes the proof that E is local.

Let $\mathfrak{n}$ be the maximal ideal of E, and put $\ell = E/\mathfrak{n}$. Suppose $\dim_k \ell > 1$. The inclusion $\mathfrak{m}E \hookrightarrow \mathfrak{n}$ induces a surjection $E/\mathfrak{m}E \twoheadrightarrow E/\mathfrak{n} = \ell$. Since, by Lemma 1, $\dim_k(E/\mathfrak{m}E) = 2$, this surjection must be an isomorphism, and hence $\mathfrak{n} = \mathfrak{m}E = \mathfrak{m}$. Observe that the isomorphism $\alpha : M \otimes_R N \to M \otimes_E N$ induces an isomorphism

$$\overline{M} \otimes_k \overline{N} \xrightarrow{\cong} \overline{M} \otimes_\ell \overline{N}. \tag{3}$$

Put $u = \dim_\ell \overline{M}$ and $v = \dim_\ell \overline{N}$. Then $\dim_\ell(\overline{M} \otimes_\ell \overline{N}) = uv$, and hence $\dim_k(\overline{M} \otimes_\ell \overline{N}) = 2uv$. On the other hand, $\dim_k(\overline{M} \otimes_k \overline{N}) = (\dim_k \overline{M})(\dim_k \overline{N}) = (2u)(2v) = 4uv$. The isomorphism in (3) forces $4uv = 2uv$, and hence either $u = 0$ or $v = 0$, contradicting Nakayama's Lemma. This shows that $\dim_k \ell = 1$, and the proof is complete. □

One might hope, at least for a Gorenstein ring $(R, \mathfrak{m}, k)$ with finite integral closure $\overline{R}$, that E being local with residue field k would force $\overline{R}$ to be local with residue field k. Of course, Theorem 1 would then answer Conjecture 1 affirmatively. The next example dashes this hope.

Let k be a field and $D = k[X]_{(X)\cup(X-1)}$. Then D is a principal ideal domain with 2 maximal ideals. Let $A = k[T]/(T^2)$, $B = k[X]/(X^2)\times k[X]/(X-1)^2$, and define $i : A \hookrightarrow B$ by $i(a+bt) = (a+bx, a+b(x-1))$ where $a, b \in k$, and decapitalization of the indeterminates indicates passage to cosets. Let $\pi : D \twoheadrightarrow B$ be the composition of the natural projection $D \twoheadrightarrow D/(X^2(X-1)^2)$ and the isomorphism $D/(X^2(X-1)^2) \xrightarrow{\cong} B$ provided by the Chinese Remainder Theorem. Define R to be the pullback of i and π:

$$\begin{array}{ccc} R & \rightarrowtail & D \\ \downarrow & & \downarrow \pi \\ A & \overset{i}{\rightarrowtail} & B \end{array} \tag{4}$$

By [6, Proposition 3.1], $(R, \mathfrak{m}, k)$ is a local one-dimensional domain, $\overline{R} = D$, and $\overline{R}$ is finitely generated as an R-module. Furthermore, letting $\mathfrak{f}$ be the conductor, we have $A \cong R/\mathfrak{f}$ and $B \cong D/\mathfrak{f}$. Since the length of $\overline{R}/\mathfrak{f}$, namely 4, is twice the length of $R/\mathfrak{f}$, [2, Corollary 6.5] guarantees that R is Gorenstein. One checks that $E := \operatorname{End}_R(\mathfrak{m})$ is local, with residue field k, but $\overline{R}$ is not local.

This example cannot be promoted to a counterexample to Conjecture 1. To see this, first observe that B is generated by two elements as an A-module. It follows that $\overline{R} = D$ is two-generated as an R-module. Therefore R has multiplicity two [4, Theorem 2.1], and hence every ideal of the completion $\widehat{R}$ is two-generated. It follows that $\widehat{R}$ is a hypersurface and therefore, by the main theorem of [5], the tensor product of any two non-free finitely generated R-modules has non-zero torsion.

Acknowledgements The second-named author is grateful to the Simons Foundation for support for this research through Simons Collaboration Grant 426885.

Some of the material in this paper is taken from the first-named author's 2018 Ph.D. dissertation at the University of Nebraska.

The authors thank the anonymous referee for several helpful suggestions, which have significantly improved the exposition.

References

1. Auslander, M.: Modules over unramified regular local rings. Ill. J. Math. **5**, 631–647 (1961)
2. Bass, H.: On the ubiquity of Gorenstein rings. Math. Z. **82**, 8–28 (1963)

3. Celikbas, O., Wiegand, R.: Vanishing of Tor, and why we care about it. J. Pure Appl. Algebra **219**, 429–448 (2015)
4. Greither, C.: On the two generator problem for the ideals of a one-dimensional ring. J. Pure Appl. Algebra **24**, 265–276 (1982)
5. Huneke, C., Wiegand, R.: Tensor products of modules and the rigidity of Tor. Math. Ann. **299**, 449–476 (1994); Correction: Math. Ann. **338**, 291–293 (2007)
6. Wiegand, R., Wiegand, S.: Stable isomorphism of modules over one-dimensional rings. J. Algebra **107**, 425–435 (1987)

Almost Symmetric Numerical Semigroups with Odd Generators

Francesco Strazzanti and Kei-ichi Watanabe

Abstract We study almost symmetric semigroups generated by odd integers. If the embedding dimension is four, we characterize when a symmetric semigroup that is not complete intersection or a pseudo-symmetric semigroup is generated by odd integers. Moreover, we give a way to construct all the almost symmetric semigroups with embedding dimension four and type three generated by odd elements. In this case we also prove that all the pseudo-Frobenius numbers are multiple of one of them and this gives many consequences on the semigroup and its defining ideal.

Keywords Symmetric numerical semigroups · Pseudo-symmetric numerical semigroups · Almost symmetric numerical semigroups · Pseudo-Frobenius numbers · RF-matrices

1 Introduction

Numerical semigroups have been extensively studied in the last decades for several reasons, since they appears in many areas of mathematics like commutative algebra, algebraic geometry, number theory, factorization theory, combinatorics or coding theory. For instance, the connection with commutative algebra has greatly influenced the theory of numerical semigroups and it is not a coincidence that many invariants of numerical semigroups have the same name of well-known invariants in commutative algebra. One of the main results that constructed a bridge between these two areas is the celebrated theorem proved by Kunz [16] that establishes the equivalence between Gorenstein rings and symmetric numerical semigroups.

F. Strazzanti (✉)
Dipartimento di Matematica e Informatica, Università degli Studi di Catania, Catania, Italy

K.-i. Watanabe
Department of Mathematics, College of Humanities and Sciences, Nihon University, Setagaya-ku, Tokyo, Japan
e-mail: watanabe@math.chs.nihon-u.ac.jp

V. Barucci et al. (eds.), *Numerical Semigroups*, Springer INdAM Series 40,
https://doi.org/10.1007/978-3-030-40822-0_19

More precisely, if R is a one-dimensional analytically irreducible and residually rational noetherian local ring, then it is Gorenstein if and only if the associated value-semigroup (that is a numerical semigroup) is symmetric.

An important notion related to the symmetry of a numerical semigroup is given by the pseudo-symmetric property. The rings that correspond to the pseudo-symmetric semigroups are called Kunz rings by many authors and there is an extensive literature about them. See for instance the monograph [2] that also provides a dictionary between commutative algebra and numerical semigroup theory.

In 1997 Barucci and Fröberg [1] introduced the notion of almost symmetric numerical semigroup that generalizes both symmetric and pseudo-symmetric ones. Similarly, in the same paper they introduced almost Gorenstein ring as the correspondent notion in commutative algebra; of course, it generalizes Gorenstein and Kunz rings. The last definition is given in the one-dimensional analytically unramified local case, but recently it was extended in the one-dimensional and higher dimensional local case as well as in the graded context, see [9, 10].

On the other hand almost symmetric semigroups have been studied by many authors from several points of view. They are also one of the main tools used in [20] to construct one-dimensional Gorenstein local rings with decreasing Hilbert functions in some level, giving an answer to a commutative algebra problem known as Rossi Problem. There are also many generalizations of the almost symmetric semigroups in literature, see [5, 6, 14].

The purpose of this paper is to study the almost symmetric semigroups generated by odd integers, in particular when the embedding dimension is four. In this case, independently of the parity of the generators, Moscariello [17] proved that the type of the semigroup is at most three confirming a conjecture of T. Numata. This means that we can divide the almost symmetric semigroups with embedding dimension four in three classes: symmetric, pseudo-symmetric and having type three.

If $S = \langle n_1, \ldots, n_e \rangle$ is a numerical semigroup, we say that $k[S] := k[t^s \mid s \in S]$ is the numerical semigroup ring associated to S, where k is a field and t is an indeterminate. It is possible to present this ring as a quotient of a polynomial ring $k[S] \cong k[x_1, \ldots, x_e]/I_S$ and I_S is called the defining ideal of S. We set $\deg(x_i) = n_i$ for every $i = 1, \ldots, e$, thus I_S is homogeneous.

Assume now that S has embedding dimension four. In the case of symmetric and pseudo-symmetric numerical semigroups the defining ideal is known by Bresinsky [4] and Komeda [15]. The type three case has been recently studied in [8, 13], where the defining ideal is found using the notion of RF-matrix, introduced in [17].

We focus on the case where all the generators of S are odd. In particular, if S is symmetric but not complete intersection we characterize when this happens in terms of some numbers related to the defining ideal of S. Moreover, in the pseudo-symmetric case we connect this property to the rows of a suitable RF-matrix associated to S. If S is almost symmetric with type three, we prove that the set of the pseudo-Frobenius numbers of S is $\mathrm{PF}(S) = \{f, 2f, 3f\}$ for some integer f. This lead to the description of the generators of both S and I_S as well as the minimal free resolution of $k[x_1, \ldots, x_4]/I_S$ in terms of the numbers

$\alpha_i = \min\{\alpha \mid \alpha n_i \in \langle n_1, \ldots, \widehat{n_i}, \ldots, n_4\rangle\}$ for $i = 1, \ldots, 4$, where n_1, n_2, n_3 and n_4 are the minimal generators of S. This allows us to construct all such semigroups and gives examples of numerical semigroups in which PF(S) has this particular shape, which have been studied in [11].

The structure of the paper is the following. In Sect. 2 we fix the notation and recall some useful definitions and results. In Sect. 3 we characterize when the generators of a symmetric numerical semigroup with embedding dimension four are all odd. In Sect. 4 we do the same in the pseudo-symmetric case. In the last section we consider the case of almost symmetric semigroups with embedding dimension four and type three. Here we prove Theorem 3 which gives the pseudo-Frobenius numbers and that allows to get Corollary 1, where the generators of S and I_S as well as the minimal free resolution of $k[x_1, \ldots, x_4]/I_S$ are described. Moreover, in Theorem 4 we give a way to construct all the almost symmetric semigroups with embedding dimension four and type three.

Several computations of the paper are performed by using the GAP system [21] and, in particular, the NumericalSgps package [7].

2 Basic Concepts

We denote by $\mathbb{N}$ the set of the natural numbers including 0. A numerical semigroup S is an additive submonoid of $\mathbb{N}$ such that $\mathbb{N}\setminus S$ is finite. Every numerical semigroup has a finite system of generators, i.e. there exist some positive integers $n_1, n_2, \ldots, n_s$ such that $S = \langle n_1, n_2, \ldots, n_s\rangle := \{\sum_{i=1}^{s} a_i n_i \mid a_i \in \mathbb{N} \text{ for } i = 1, \ldots, s\}$. Moreover, there exists a unique minimal system of generators $n_1, \ldots, n_e$ of S and the number e is called embedding dimension of S. The finiteness of $\mathbb{N}\setminus S$ is equivalent to $\gcd(n_1, \ldots, n_e) = 1$. If $S = \langle n_1, \ldots, n_e\rangle$, we denote by α_i the minimum integer such that $\alpha_i n_i = \sum_{j \neq i} a_j n_j$ for some non-negative integers $a_1, \ldots, a_e$.

The maximum of $\mathbb{Z}\setminus S$ is known as the Frobenius number of S and we denote it by F(S). We say that an integer $f \in \mathbb{Z}\setminus S$ is a pseudo-Frobenius number of S if $f + s \in S$ for every $s \in S\setminus\{0\}$. We denote the set of the pseudo-Frobenius numbers by PF(S) and we refer to its cardinality $t(S)$ as the type of S. Clearly F(S) is always a pseudo-Frobenius number, thus $t(S) \geq 1$.

Consider the injective map $\varphi : S \to \mathbb{Z}\setminus S$ defined by $\varphi(s) = \mathrm{F}(S) - s$. If φ is a bijection we say that S is symmetric, whereas if the image of φ is equal to $\mathbb{Z}\setminus S$ except for $\mathrm{F}(S)/2$ we say that S is pseudo-symmetric. It is not difficult to see that S is symmetric if and only if it has type 1 and it is pseudo-symmetric if and only if $\mathrm{PF}(S) = \{\mathrm{F}(S)/2, \mathrm{F}(S)\}$. Moreover, setting $g(S) = |\mathbb{N} \setminus S|$, S is symmetric (resp. pseudo-symmetric) if and only if $2g(S) = \mathrm{F}(S) + 1$ (resp. $2g(S) = \mathrm{F}(S) + 2$). We say that S is almost symmetric if and only if $2g(S) = \mathrm{F}(S) + t(S)$. There exists a useful characterization of the almost symmetric property due to H. Nari [18, Theorem 2.4]: if $\mathrm{PF}(S) = \{f_1 < f_2 < \cdots < f_t = \mathrm{F}(S)\}$, a numerical semigroup is almost symmetric if and only if $f_i + f_{t-i} = \mathrm{F}(S)$ for every $i = 1, \ldots, t-1$.

If $f \in \mathrm{PF}(S)$, then $f + n_i \in S$ for every i and, thus, there exist $\lambda_{i1}, \ldots, \lambda_{ie} \in \mathbb{N}$ such that $f + n_i = \sum_{j=1}^{e} \lambda_{ij} n_j$. Since $f \notin S$, λ_{ii} has to be equal to zero. For every $i, j = 1, \ldots, e$, set $a_{ii} = -1$ and $a_{ij} = \lambda_{ij}$ if $i \neq j$. Following [17] we say that the matrix $\mathrm{RF}(f) = (a_{ij})$ is a row-factorization matrix of f, briefly RF-matrix. Note that there could be several RF-matrices of f and that $f = \sum_{j=1}^{e} a_{ij} n_j$ for every i. For instance, consider the numerical semigroup $S = \langle 8, 10, 11, 13 \rangle$ that has embedding dimension four and is symmetric, because $\mathrm{PF}(S) = \{25\}$. The following are both RF-matrices of $\mathrm{F}(S) = 25$:

$$\begin{pmatrix} -1 & 0 & 3 & 0 \\ 3 & -1 & 1 & 0 \\ 2 & 2 & -1 & 0 \\ 1 & 3 & 0 & -1 \end{pmatrix}, \quad \begin{pmatrix} -1 & 2 & 0 & 1 \\ 0 & -1 & 2 & 1 \\ 0 & 1 & -1 & 2 \\ 2 & 0 & 2 & -1 \end{pmatrix}.$$

3 Symmetric Semigroups

We start by studying the symmetric numerical semigroups with embedding dimension four. If the semigroup is not complete intersection, there is a theorem proved by Bresinsky [4] that gives much information on the semigroup and its defining ideal. We state it following [3, Theorem 3]. By convention, if j is an integer not included between 1 and 4, we set $a_j = a_i$ and $b_j = b_i$ with $j \equiv i \pmod 4$ and $1 \leq i \leq 4$.

Theorem 1 *Let S be a numerical semigroup with 4 minimal generators. Then, S is symmetric and not complete intersection if and only if there are integers a_i and b_i with $i \in \{1, \ldots, 4\}$, such that $0 < a_i < \alpha_{i+1}$ and $0 < b_i < \alpha_{i+2}$ for all i,*

$$\alpha_1 = a_1 + b_1, \; \alpha_2 = a_2 + b_2, \; \alpha_3 = a_3 + b_3, \; \alpha_4 = a_4 + b_4 \tag{1}$$

and

$$\begin{aligned} n_1 &= \alpha_2\alpha_3 a_4 + a_2 b_3 b_4, \; n_2 = \alpha_3\alpha_4 a_1 + a_3 b_4 b_1, \\ n_3 &= \alpha_1\alpha_4 a_2 + a_4 b_1 b_2, \; n_4 = \alpha_1\alpha_2 a_3 + a_1 b_2 b_3. \end{aligned} \tag{2}$$

In this case $I_S = (f_1, f_2, f_3, f_4, f_5)$, where

$$f_1 = x_1^{\alpha_1} - x_3^{b_3} x_4^{a_4}, \quad f_2 = x_2^{\alpha_2} - x_1^{a_1} x_4^{b_4}, \quad f_3 = x_3^{\alpha_3} - x_1^{b_1} x_2^{a_2},$$
$$f_4 = x_4^{\alpha_4} - x_2^{b_2} x_3^{a_3}, \quad f_5 = x_1^{a_1} x_3^{a_3} - x_2^{a_2} x_4^{a_4}.$$

In this section we denote by a_i and b_i the integers that appear in the previous theorem.

Theorem 2 *Let S be a symmetric numerical semigroup minimally generated by $n_1, \dots, n_4$ and assume that S is not complete intersection. The following conditions are equivalent:*

1. *Every n_i is odd.*
2. *One of the following holds:*
   a. *All the α_i's and the a_i's are odd;*
   b. *There is exactly one index i_0 for which α_{i_0} is even. Moreover, a_{i_0} and a_{i_0-1} are odd, while the other a_i's are even;*
   c. *All the α_i's are even and all the a_i's are odd.*

Proof Using the equalities (1) and (2) it is easy to see that the conditions a, b and c imply that all the generators are odd.

Conversely, assume first that all the α_i's are odd and suppose by contradiction that a_1 is even. Since n_2 is odd, a_3 and b_4 are odd by (2). Therefore, $a_4 = \alpha_4 - b_4$ and $b_3 = \alpha_3 - a_3$ are even. Then n_1 should be even by (2). A contradiction!

Assume now that there is at least one α_i even. Without loss of generality, we can assume that α_1 is even. Since n_3 and n_4 are odd, the equalities in (2) imply that a_4, b_1, b_2, a_1 and b_3 are odd.

Assume first that a_2 is even. Then, the first equality in (2) implies that α_2 and α_3 are odd, so $a_3 = \alpha_3 - b_3$ is even. Moreover, since n_2 is odd, α_4 is odd by (2). Hence, we are in the case b.

Assume now that a_2 is odd. Then, $\alpha_2 = a_2 + b_2$ is even and it follows from the first equality in (2) that a_2 and b_4 are odd. In particular, $\alpha_4 = a_4 + b_4$ is even and, again by (2), a_3 is odd. Finally, we get that $\alpha_3 = a_3 + b_3$ is even and, then, we are in the case c. □

Example 1 We note that all the cases of the previous theorem can occur. All the following semigroups are symmetric but not complete intersections.

(a) Let $S = \langle 13, 17, 23, 19 \rangle$. In this case

$$f_1 = x_1^5 - x_3^2 x_4, \qquad f_2 = x_2^3 - x_1 x_4^2, \qquad f_3 = x_3^3 - x_1^4 x_2,$$

$$f_4 = x_4^3 - x_2^2 x_3, \qquad f_5 = x_1 x_3 - x_2 x_4,$$

in particular $\alpha_1 = 5, \alpha_2 = \alpha_3 = \alpha_4 = 3$ and $a_1 = a_2 = a_3 = a_4 = 1$.

(b) Let $S = \langle 13, 17, 33, 25 \rangle$. We have

$$f_1 = x_1^7 - x_3^2 x_4, \qquad f_2 = x_2^3 - x_1^2 x_4, \qquad f_3 = x_3^3 - x_1^5 x_2^2,$$

$$f_4 = x_4^2 - x_2 x_3, \qquad f_5 = x_1^2 x_3 - x_2^2 x_4,$$

therefore $\alpha_1 = 7$ and $\alpha_2 = \alpha_3 = 3$ and $\alpha_4 = 2$. Moreover, $a_1 = a_2 = 2$ and $a_3 = a_4 = 1$.

(c) Let $S = \langle 5, 7, 11, 9\rangle$. Then

$$f_1 = x_1^4 - x_3x_4, \quad f_2 = x_2^2 - x_1x_3, \quad f_3 = x_3^2 - x_1^3x_2,$$
$$f_4 = x_4^2 - x_2x_4, \quad f_5 = x_1x_3 - x_2x_4$$

and, thus, $\alpha_1 = 4, \alpha_2 = \alpha_3 = \alpha_4 = 2$ and $a_1 = a_2 = a_3 = a_4 = 1$.

Example 2 Unfortunately, in Theorem 2 it is not possible to characterize the parity of the generators by the parity of the α_i's, in fact we cannot eliminate the conditions on the a_i's in a, b and c, as the following examples show. All the listed semigroups are symmetric, but not complete intersections.

(a) Consider the semigroup $S = \langle 90, 91, 97, 93\rangle$. Then

$$f_1 = x_1^{13} - x_3^{12}x_4^2, \quad f_2 = x_2^3 - x_1^2x_4, \quad f_3 = x_3^{13} - x_1^{13}x_2,$$
$$f_4 = x_4^3 - x_2^2x_3, \quad f_5 = x_1^2x_3 - x_2x_4^2,$$

and all the α_i are odd, but there is an even generator. In fact, a_1 and a_4 are even.

(b) Let $S = \langle 22, 23, 29, 57\rangle$. We have

$$f_1 = x_1^5 - x_3^3x_4, \quad f_2 = x_2^2 - x_1x_4^4, \quad f_3 = x_3^5 - x_1^4x_2,$$
$$f_4 = x_4^5 - x_2x_3^2, \quad f_5 = x_1x_3^2 - x_2x_4.$$

In this case $\alpha_1 = \alpha_3 = \alpha_4 = 5$ and α_2 is even. However a generator is even, since a_4 is odd.

(c) Let $S = \langle 5, 14, 22, 18\rangle$. We have

$$f_1 = x_1^8 - x_3x_4, \quad f_2 = x_2^2 - x_1^2x_4, \quad f_3 = x_3^2 - x_1^6x_2,$$
$$f_4 = x_4^2 - x_2x_3, \quad f_5 = x_1^2x_3 - x_2x_4,$$

in particular all the α_i's are even, but three generators of S are even. Note that a_1 is even.

4 Pseudo-Symmetric Semigroups

Let $S = \langle n_1, n_2, n_3\rangle$ be a non-symmetric numerical semigroup. In [12] it is proved that the defining ideal of S is generated by the maximal minors of the matrix

$$\begin{pmatrix} x_1^{\alpha} & x_2^{\beta} & x_3^{\gamma} \\ x_1^{\alpha'} & x_2^{\beta'} & x_3^{\gamma'} \end{pmatrix} \tag{3}$$

for some positive integers $\alpha, \beta, \gamma, \alpha', \beta', \gamma'$. Moreover, by Nari et al. [19, Corollary 3.3], S is pseudo-symmetric if and only if $\alpha = \beta = \gamma = 1$ or $\alpha' = \beta' = \gamma' = 1$. Without loss of generality we assume that $\alpha' = \beta' = \gamma' = 1$. In [19, (2.1.1) pag. 69] it is proved that $n_1 = (\beta+1)\gamma+1$, $n_2 = (\gamma+1)\alpha+1$ and $n_3 = (\alpha+1)\beta+1$. Hence, it follows easily that n_1, n_2 and n_3 are odd if and only if either α, β, γ are odd or α, β, γ are even.

Now let $S = \langle n_1, n_2, n_3, n_4 \rangle$ be a pseudo-symmetric 4-generated numerical semigroup. By Herzog and Watanabe [13, Theorem 4.3] $\mathrm{F}(S)/2$ has a unique RF-matrix and, for a suitable relabeling of the generators of S, we have

$$\mathrm{RF}(\mathrm{F}(S)/2) = \begin{pmatrix} -1 & \alpha_2 - 1 & 0 & 0 \\ 0 & -1 & \alpha_3 - 1 & 0 \\ \alpha_1 - 1 & 0 & -1 & \alpha_4 - 1 \\ \alpha_1 - 1 & a & 0 & -1 \end{pmatrix} \tag{4}$$

for some non-negative integer a.

Given $f \in \mathrm{PF}(S)$ and $\mathrm{RF}(f) = (a_{ij})$, we say that the i-th row is even (resp. odd) if $\sum_{j=1}^{4} a_{ij}$ is even (resp. odd).

Proposition 1 *Assume that $S = \langle n_1, \ldots, n_4 \rangle$ is pseudo-symmetric and has embedding dimension 4. Then, every n_i is odd if and only if one of the following conditions hold:*

1. $\mathrm{F}(S)/2$ *is odd and every row of* $\mathrm{RF}(\mathrm{F}(S)/2)$ *is odd;*
2. $\mathrm{F}(S)/2$ *is even and every row of* $\mathrm{RF}(\mathrm{F}(S)/2)$ *is even.*

Proof We can assume that the matrix (4) is the RF-matrix of $f := \mathrm{F}(S)/2$.

Suppose first that every n_i is odd and f is odd. By the first row of (4), $f = -n_1 + (\alpha_2 - 1)n_2$ and $(\alpha_2 - 1)$ has to be even, i.e. the first row is odd. The same argument works for the second row. The third row (and similarly the last one) $f = (\alpha_1 - 1)n_1 - n_3 + (\alpha_4 - 1)n_4$ yields immediately that $\alpha_1 - 1$ and $\alpha_4 - 1$ have the same parity and, thus, the row is odd. If f is even we can use the same argument.

Assume now that Condition *1* holds. By the first two rows it follows that $\alpha_2 - 1$, $\alpha_3 - 1$ are even and, then, n_1 and n_2 are odd. Using the last row we have $\alpha_1 - 1 + a$ even, thus $(\alpha_1 - 1)n_1 + an_2$ is even and n_4 has to be odd. In the same way the third row implies that also n_3 is odd.

Finally, assume that Condition *2* holds. By the first row we get that $\alpha_2 - 1$ is odd and then n_1 and n_2 have the same parity. By the second one follows that also n_3 has the same parity of n_1 and n_2. If they are even, the last row implies that $f = (\alpha_1 - 1)n_1 + an_2 - n_4$ and, since f is even, also n_4 is even. This is a contradiction because $\gcd(n_1, n_2, n_3, n_4) = 1$, therefore, n_1, n_2 and n_3 are odd. Moreover, in the last row we have $\alpha_1 - 1 + a$ odd and, then, n_4 is odd. □

Remark 1 Let $S = \langle n_1, \ldots, n_4 \rangle$ be pseudo-symmetric and assume that $\mathrm{F}(S)/2$ and every n_i are odd. The previous proposition implies that α_2 and α_3 are odd. Moreover, α_1 and α_4 have the same parity, but we cannot determine if they are even or odd. In fact, if $S = \langle 15, 17, 35, 43\rangle$ we have $\mathrm{PF}(S) = \{53, 106\}$ and

$$\mathrm{RF}(53) = \begin{pmatrix} -1 & 4 & 0 & 0 \\ 0 & -1 & 2 & 0 \\ 3 & 0 & -1 & 1 \\ 3 & 3 & 0 & -1 \end{pmatrix},$$

whereas if $T = \langle 57, 61, 123, 163\rangle$, then $\mathrm{PF}(T) = \{431, 862\}$ and

$$\mathrm{RF}(431) = \begin{pmatrix} -1 & 8 & 0 & 0 \\ 0 & -1 & 4 & 0 \\ 4 & 0 & -1 & 2 \\ 4 & 6 & 0 & -1 \end{pmatrix}.$$

5 Almost Symmetric Semigroups with Type Three

Moscariello [17] proved that an almost symmetric numerical semigroup with embedding dimension four has type at most three. Therefore, to complete the picture we need to study the almost symmetric semigroups with type three. We start with an easy lemma that is probably known, but we include it for the reader's convenience.

Lemma 1 *Let $S = \langle n_1, \ldots, n_r\rangle$ and assume that $\alpha_1 = n_2$. Then $S = \langle n_1, n_2\rangle$.*

Proof If $T = \langle n_1, n_2\rangle$, then T is symmetric and $\mathrm{F}(T) = n_1n_2 - n_1 - n_2$. Suppose by contradiction that $n_3 \notin T$. Since T is symmetric, $\mathrm{F}(T) - n_3 \in T$, i.e. $\mathrm{F}(T) - n_3 = an_1 + bn_2$ for some non-negative integers a and b. Therefore $(n_2 - a - 1)n_1 = (b + 1)n_2 + n_3$ and, then, $\alpha_1 \le n_2 - a - 1$ gives a contradiction. □

Theorem 3 *Let $S = \langle n_1, n_2, n_3, n_4\rangle$ be an almost symmetric numerical semigroup with type three and assume that all the generators are odd. Then, its pseudo-Frobenius numbers are $\mathrm{PF}(S) = \{f, 2f, 3f\}$ for some integer f and, by a suitable change of order of n_1, n_2, n_3, n_4, there exists an RF-matrix of f and $2f$ of the following type:*

$$\begin{pmatrix} -1 & \alpha_2 - 1 & 0 & 0 \\ 0 & -1 & \alpha_3 - 1 & 0 \\ 0 & 0 & -1 & \alpha_4 - 1 \\ \alpha_1 - 1 & 0 & 0 & -1 \end{pmatrix}, \begin{pmatrix} -1 & \alpha_2 - 2 & \alpha_3 - 1 & 0 \\ 0 & -1 & \alpha_3 - 2 & \alpha_4 - 1 \\ \alpha_1 - 1 & 0 & -1 & \alpha_4 - 2 \\ \alpha_1 - 2 & \alpha_2 - 1 & 0 & -1 \end{pmatrix}.$$

Proof According to [8, Theorems 3.6 and 4.8], we distinguish four cases that in [8] are called UF1, UF2, nUF1 and nUF2. We will prove that only the last one is possible under our hypothesis. Let f and f' be the two pseudo-Frobenius numbers of S different from its Frobenius number.

Case UF1 In this case, by a suitable change of order of n_1, n_2, n_3, n_4, there exist RF-matrices of f and f' of the following type

$$\begin{pmatrix} -1 & \alpha_2-1 & 0 & 0 \\ \alpha_1-1 & -1 & 0 & 0 \\ \alpha_1-2 & 0 & -1 & 1 \\ 0 & \alpha_2-2 & 1 & -1 \end{pmatrix}, \quad \begin{pmatrix} -1 & 0 & 0 & \alpha_4-1 \\ 0 & -1 & 1 & \alpha_4-2 \\ b_{41}-1 & b_{32} & -1 & 0 \\ b_{41} & b_{32}-1 & 0 & -1 \end{pmatrix}$$

respectively and either $\alpha_2 = 2$ or $\alpha_4 = 2$.

If $\alpha_2 = 2$, the first two lines of the first matrix give $n_2 = f + n_1$ and $f + n_2 = (\alpha_1 - 1)n_1$. Hence, $2f = (\alpha_1 - 2)n_1$ and, since n_1 is odd, α_1 has to be even; consequently $f = (\alpha_1/2 - 1)n_1 \in S$ gives a contradiction.

Assume now that $\alpha_4 = 2$. The second matrix implies that $n_3 = n_2 + f'$ and $f' + n_3 = (b_{41} - 1)n_1 + b_{32}n_2$. Then,

$$\begin{aligned} 2n_3 &= n_3 + (f' + n_3) - f' = (n_2 + f') + ((b_{41} - 1)n_1 + b_{32}n_2) - f' \\ &= (b_{41} - 1)n_1 + (b_{32} + 1)n_2. \end{aligned} \tag{5}$$

Moreover, subtracting the first and the second rows of RF(f), we get $\alpha_1 n_1 = \alpha_2 n_2$. The previous lemma implies that $\alpha_1 < n_2$ and, then, $\gcd(n_1, n_2) = d > 1$. Since d is odd, in light of the equality (5) also n_3 is a multiple of d. Furthermore, $\alpha_4 = 2$ means that $2n_4 = \sum_{j=1}^{3} \alpha_{4j} n_j$ and, thus, also n_4 is a multiple of d; a contradiction.

Case UF2 By a suitable change of order of n_1, n_2, n_3, n_4, there exist RF-matrices of f and f' of the following type:

$$\begin{pmatrix} -1 & \alpha_2-1 & 0 & 0 \\ a_{21} & -1 & \alpha_3-2 & 0 \\ a_{21}-1 & 0 & -1 & 1 \\ 0 & \alpha_2-2 & \alpha_3-1 & -1 \end{pmatrix}, \quad \begin{pmatrix} -1 & 0 & 0 & \alpha_4-1 \\ 0 & -1 & \alpha_3-1 & \alpha_4-2 \\ a_{21}+b_{41} & 0 & -1 & 0 \\ b_{41} & \alpha_2-1 & 0 & -1 \end{pmatrix}.$$

respectively and either $\alpha_2 = 2$ or $\alpha_4 = 2$.

Assume first that $\alpha_2 = 2$. By subtracting the first and the last row in the first matrix we get

$$n_2 + n_4 = n_1 + (\alpha_3 - 1)n_3.$$

This implies that α_3 is even. Therefore, by adding the first two rows of the first matrix we have

$$2f = (a_{21} - 1)n_1 + (\alpha_3 - 2)n_3$$

and $a_{21} - 1$ has to be even. It follows that f is in the semigroup, that is a contradiction.

Assume now that $\alpha_4 = 2$. In this case $f' = n_4 - n_1$ is even, then $f = (\alpha_2 - 1)n_2 - n_1$ is odd and, thus, $\alpha_2 - 1$ is even. By adding the first and the last row of the second matrix we get

$$2f' = (b_{41} - 1)n_1 + (\alpha_2 - 1)n_2.$$

Again $b_{41} - 1$ has to be even and, then, f' is in the semigroup.

Case nUF1 By a suitable change of order of n_1, n_2, n_3, n_4, there exist RF-matrices of f and f' of the following type:

$$\begin{pmatrix} -1 & 0 & 0 & \alpha_4 - 1 \\ 0 & -1 & 1 & \alpha_4 - 2 \\ 0 & \alpha_2 - 1 & -1 & 0 \\ 1 & \alpha_2 - 2 & 0 & -1 \end{pmatrix}, \quad \begin{pmatrix} -1 & 1 & \alpha_3 - 2 & 0 \\ \alpha_1 - 1 & -1 & 0 & 0 \\ \alpha_1 - 2 & 0 & -1 & 1 \\ 0 & 0 & \alpha_3 - 1 & -1 \end{pmatrix}$$

respectively. Subtracting the first two rows of the first matrix we get $n_1 + n_3 = n_2 + n_4$, whereas subtracting the second and the third row we get $\alpha_2 n_2 = 2n_3 + (\alpha_4 - 2)n_4$ and, then, α_2 and α_4 have the same parity. In the same way, by subtracting the first two rows of the second matrix, we get that α_1 and α_3 have the same parity. By adding the first and the third row of the first matrix and using $n_1 + n_3 = n_2 + n_4$ we have

$$2f = -n_1 + (\alpha_2 - 1)n_2 - n_3 + (\alpha_4 - 1)n_4 = (\alpha_2 - 2)n_2 + (\alpha_4 - 2)n_4.$$

Since f is not in the semigroup, this implies that α_2 and α_4 are odd and, thus, f is odd by the first row.

If we do the same in the second matrix (with the second and the last row) we conclude that also f' is odd, that is a contradiction because $f + f'$ equals the Frobenius number that is odd.

Case nUF2 By a suitable change of order of n_1, n_2, n_3, n_4, there exist RF-matrices of f and f' of the following type:

$$\begin{pmatrix} -1 & \alpha_2 - 1 & 0 & 0 \\ 0 & -1 & \alpha_3 - 1 & 0 \\ 0 & 0 & -1 & \alpha_4 - 1 \\ \alpha_1 - 1 & 0 & 0 & -1 \end{pmatrix}, \quad \begin{pmatrix} -1 & \alpha_2 - 2 & \alpha_3 - 1 & 0 \\ 0 & -1 & \alpha_3 - 2 & \alpha_4 - 1 \\ \alpha_1 - 1 & 0 & -1 & \alpha_4 - 2 \\ \alpha_1 - 2 & \alpha_2 - 1 & 0 & -1 \end{pmatrix}$$

respectively. Since the sum of the first two rows of the first matrix is equal to the first row of the second matrix, it follows that $f' = 2f$. Hence, it is enough to recall that $\mathrm{F}(S) = f + f' = 3f$ by Nari's Theorem [18, Theorem 2.4]. □

Remark 2 Let $S = \langle n_1, n_2, n_3, n_4 \rangle$ be almost symmetric with type three and assume that all the generators are odd. By Theorem 3 the Frobenius number is equal to $3f$ and it is odd, so f is odd. Moreover, by a suitable change of order of n_1, n_2, n_3, n_4, we have $f = (\alpha_2-1)n_2-n_1 = (\alpha_3-1)n_3-n_2 = (\alpha_4-1)n_4-n_3 = (\alpha_1-1)n_1-n_4$. Therefore, α_1, α_2, α_3 and α_4 are odd.

Example 3 There are almost symmetric 4-generated semigroups with type three whose pseudo-Frobenius numbers have the structure of Theorem 3, even though some generators are even. For instance, if $S = \langle 4, 7, 10, 13 \rangle$, then $\mathrm{PF}(S) = \{3, 6, 9\}$. Moreover, also this semigroup is in the case nUF2, since

$$\mathrm{RF}(3) = \begin{pmatrix} -1 & 1 & 0 & 0 \\ 0 & -1 & 1 & 0 \\ 0 & 0 & -1 & 1 \\ 4 & 0 & 0 & -1 \end{pmatrix} \quad \text{and} \quad \mathrm{RF}(6) = \begin{pmatrix} -1 & 0 & 1 & 0 \\ 0 & -1 & 0 & 1 \\ 4 & 0 & -1 & 0 \\ 3 & 1 & 0 & -1 \end{pmatrix}$$

Note that $\alpha_2 = \alpha_3 = \alpha_4 = 2$ is even in this example.

By Theorem 3 in every row of $\mathrm{RF}(f)$ there is exactly one positive entry. Therefore, we immediately get the following corollary by Eto [8, Section 5.5] or [13, Lemma 5.4].

Corollary 1 *Let $S = \langle n_1, \ldots, n_4 \rangle$ be almost symmetric with type three and assume that n_i is odd for every $i = 1, \ldots, 4$. Then*

$$n_1 = (\alpha_2 - 1)(\alpha_3 - 1)\alpha_4 + \alpha_2, \qquad n_2 = (\alpha_3 - 1)(\alpha_4 - 1)\alpha_1 + \alpha_3,$$
$$n_3 = (\alpha_4 - 1)(\alpha_1 - 1)\alpha_2 + \alpha_4, \qquad n_4 = (\alpha_1 - 1)(\alpha_2 - 1)\alpha_3 + \alpha_1,$$

where $\alpha_1, \ldots, \alpha_4$ are odd and the defining ideal of S is $I_S = (x_1^{\alpha_1} - x_2^{\alpha_2-1}x_4, x_2^{\alpha_2} - x_3^{\alpha_3-1}x_1, x_3^{\alpha_3} - x_4^{\alpha_4-1}x_2, x_4^{\alpha_4} - x_1^{\alpha_1-1}x_3, x_1^{\alpha_1-1}x_2 - x_3^{\alpha_3-1}x_4, x_1x_4^{\alpha_4-1} - x_2^{\alpha_2-1}x_3)$. Moreover, setting $A = k[x_1, x_2, x_3, x_4]$, the minimal free resolution of A/I_S is

$$0 \longrightarrow A^3 \xrightarrow{\varphi_3} A^8 \xrightarrow{\varphi_2} A^6 \xrightarrow{\varphi_1} A \longrightarrow 0$$

where φ_1 is the obvious one and

$$\varphi_2 = \begin{pmatrix} x_3^{\alpha_3-1} & x_2 & 0 & 0 & 0 & 0 & x_4^{\alpha_4-1} & x_3 \\ x_1^{\alpha_1-1} & x_4 & 0 & 0 & x_4^{\alpha_4-1} & x_3 & 0 & 0 \\ 0 & 0 & x_1^{\alpha_1-1} & x_4 & x_2^{\alpha_2-1} & x_1 & 0 & 0 \\ 0 & 0 & x_3^{\alpha_3-1} & x_2 & 0 & 0 & x_2^{\alpha_2-1} & x_1 \\ -x_2^{\alpha_2-1} & -x_1 & x_4^{\alpha_4-1} & x_3 & 0 & 0 & 0 & 0 \\ 0 & 0 & 0 & 0 & -x_3^{\alpha_3-1} & -x_2 & x_1^{\alpha_1-1} & x_4 \end{pmatrix},$$

$$
{}^t\varphi_3 = \begin{pmatrix} 0 & x_3 & 0 & x_1 & 0 & -x_4 & 0 & -x_2 \\ x_4^{\alpha_4-1} & 0 & x_2^{\alpha_2-1} & 0 & -x_1^{\alpha_1-1} & 0 & -x_3^{\alpha_3-1} & 0 \\ -x_3 & -x_4^{\alpha_4-1} & -x_1 & -x_2^{\alpha_2-1} & x_4 & x_1^{\alpha_1-1} & x_2 & x_3^{\alpha_3-1} \end{pmatrix}.
$$

Example 4 Putting $(\alpha_1, \alpha_2, \alpha_3, \alpha_4) = (5, 3, 3, 3)$ in Corollary 1, we get the semigroup $S = \langle 15, 23, 27, 29 \rangle$. The set of its pseudo-Frobenius numbers is $\mathrm{PF}(S) = \{31, 62, 93\}$ and, then, S is almost symmetric with type three. According to Theorem 3 we have

$$
\mathrm{RF}(31) = \begin{pmatrix} -1 & 2 & 0 & 0 \\ 0 & -1 & 2 & 0 \\ 0 & 0 & -1 & 2 \\ 4 & 0 & 0 & -1 \end{pmatrix} \quad \text{and} \quad \mathrm{RF}(62) = \begin{pmatrix} -1 & 1 & 2 & 0 \\ 0 & -1 & 1 & 2 \\ 4 & 0 & -1 & 1 \\ 3 & 2 & 0 & -1 \end{pmatrix}.
$$

Obviously, this is the example with "smallest" generators.

Theorem 4 *Assume that* $\alpha_1, \alpha_2, \alpha_3, \alpha_4$ *are odd integers greater than* 1 *and let* n_1, n_2, n_3, n_4 *be as in Corollary 1. If* $\gcd(n_1, n_2, n_3, n_4) = 1$*, then* $S = \langle n_1, n_2, n_3, n_4 \rangle$ *is an almost symmetric semigroup generated by odd integers and it has type three. Moreover, all the* 4*-generated almost symmetric semigroups with type* 3 *and odd generators arise in this way.*

Proof Bearing in mind Corollary 1, it is easy to see that the ideal I_S contains

$$
J = (x_1^{\alpha_1} - x_2^{\alpha_2-1}x_4, x_2^{\alpha_2} - x_3^{\alpha_3-1}x_1, x_3^{\alpha_3} - x_4^{\alpha_4-1}x_2, x_4^{\alpha_4} - x_1^{\alpha_1-1}x_3,
$$
$$
x_1^{\alpha_1-1}x_2 - x_3^{\alpha_3-1}x_4, x_1x_4^{\alpha_4-1} - x_2^{\alpha_2-1}x_3).
$$

Let $A = k[x_1, x_2, x_3, x_4]$. Since $\gcd(n_1, n_2, n_3, n_4) = 1$, the k-vector space $A/(I_S + (x_1))$ has dimension $\dim_k A/(I_S + (x_1)) = \dim_k k[S]/(t^{n_1}) = n_1$. Moreover,

$$
J + (x_1) = (x_2^{\alpha_2-1}x_4, x_2^{\alpha_2}, x_3^{\alpha_3} - x_4^{\alpha_4-1}x_2, x_4^{\alpha_4}, x_3^{\alpha_3-1}x_4, x_2^{\alpha_2-1}x_3) \tag{6}
$$

and it is not difficult to see that $\dim_k A/(J + (x_1)) = n_1$. It follows that $A/(I_S + (x_1)) = A/(J + (x_1))$ and this implies $I_S = J$, see the last part of the proof of [13, Theorem 4.4].

We note that the socle of $A/(I_S + (x_1))$, defined as

$$
\mathrm{Soc}(A/(I_S + (x_1))) = \{y \in A/(I_S + (x_1)) \mid yx_2 = yx_3 = yx_4 = 0\},
$$

is generated by $y_1 = x_2^{\alpha_2-1}$, $y_2 = x_2^{\alpha_2-2}x_3^{\alpha_3-1}$, $y_3 = x_2^{\alpha_2-2}x_3^{\alpha_3-2}x_4^{\alpha_4-1} = x_2^{\alpha_2-3}x_3^{2\alpha_3-2}$. Therefore, the type of the ring $A/(I_S + (x_1))$ and, then, of S is three.

Moreover, the pseudo-Frobenius numbers of S are $f_i = \deg y_i - n_1$ for $i = 1, 2, 3$ and

$$\mathrm{F}(S) = f_3 = (\alpha_2 - 3)n_2 + (2\alpha_3 - 2)n_3 - n_1 =$$
$$= (\alpha_2 - 1)n_2 - n_1 + (\alpha_2 - 2)n_2 + (\alpha_3 - 1)n_3 - n_1 = f_1 + f_2,$$

since $x_2^{\alpha_2} - x_3^{\alpha_3 - 1}x_1 \in I_S$. This implies that S is almost symmetric with type three and, of course, it has embedding dimension four. The last statement of the theorem follows from Corollary 1. □

Example 5 Let n be a positive integer and set $\alpha_2 = \alpha_3 = \alpha_4 = 3, \alpha_1 = 3 + 2^n$. By Theorem 4 the semigroup

$$S_n = \langle 15,\ 15 + 2^{n+2},\ 15 + 2^{n+2} + 2^{n+1},\ 15 + 2^{n+2} + 2^{n+1} + 2^n \rangle.$$

is an almost symmetric semigroup with type three generated by four odd minimal generators. Moreover, using Theorem 3 it is easy to see that $\mathrm{PF}(S_n) = \{15 + 2^{n+3}, 2(15 + 2^{n+3}), 3(15 + 2^{n+3})\}$.

Remark 3 Table 1 shows the number of almost symmetric semigroups that are minimally generated by 3 or 4 odd generators less than 100, 150 and 200 respectively. These numbers are obtained using the GAP system [21] and, in particular, the NumericalSgps package [7]. In the table e denotes the embedding dimension of S, t denotes its type and c.i. stands for complete intersection.

It is not known if there is a bound for the type of an almost symmetric numerical semigroup with more than 4 generators in terms of the number of its generators. However, some computations suggest that in the case of 5 generators the type is at most 5. In Table 2 we show the number of almost symmetric semigroups generated by five odd integers.

Example 6 The type of an almost symmetric semigroup may be greater than its embedding dimension, even though all the generators are odd. For instance, the 7-generated semigroup $S = \langle 29, 33, 61, 65, 73, 81, 85\rangle$ is almost symmetric and its type is 12, in fact $\mathrm{PF}(S) = \{69, 77, 89, 93, 97, 101, 105, 109, 113, 125, 133, 202\}$.

Table 1 Number of almost symmetric semigroups with odd generators

Embedding dimension and type	Gen. ≤ 100	Gen. ≤ 150	Gen. ≤ 200
$e = 3$ and $t = 1$	2302	7978	18,751
$e = 3$ and $t = 2$	139	290	503
$e = 4$ and c.i.	596	4583	16,895
$e = 4$, $t = 1$ not c.i.	1927	7129	17,524
$e = 4$ and $t = 2$	595	1647	3481
$e = 4$ and $t = 3$	9	24	45

Table 2 Number of almost symmetric semigroups with odd generators

Embedding dimension and type	Gen. ≤ 100	Gen. ≤ 150	Gen. ≤ 200
$e = 5$ and c.i.	0	135	1199
$e = 5,\ t = 1$ not c.i.	3451	19,060	60,711
$e = 5$ and $t = 2$	1254	4592	11,489
$e = 5$ and $t = 3$	988	3582	8306
$e = 5$ and $t = 4$	359	970	1881
$e = 5$ and $t = 5$	2	4	6

Acknowledgements This work began when the second author was visiting the University of Catania and he would like to express his hearty thanks for the hospitality of Marco D'Anna. The first author was supported by INdAM, more precisely he was "titolare di una borsa per l'estero dell'Istituto Nazionale di Alta Matematica".

References

1. Barucci, V., Fröberg, R.: One-dimensional almost Gorenstein rings. J. Algebra **188**, 418–442 (1997)
2. Barucci, V., Dobbs, D.E., Fontana, M.: Maximality properties in numerical semigroups and applications to one-dimensional analytically irreducible local domain. Mem. Am. Math. Soc. **125**(599), 598 (1997)
3. Barucci, V., Fröberg, R., Şahin, M.: On free resolutions of some semigroup rings. J. Pure Appl. Algebra **218**(6), 1107–1116 (2014)
4. Bresinsky, H.: Symmetric semigroups of integers generated by 4 elements. Manuscripta Math. **17**(3), 205–219 (1975)
5. Chau, T.D.M., Goto, S., Kumashiro, S., Matsuoka, N.: Sally modules of canonical ideals in dimension one and 2-AGL rings. J. Algebra **521**, 299–330 (2019)
6. D'Anna, M., Strazzanti, F.: Almost canonical ideals and GAS numerical semigroups (submitted)
7. Delgado, M., García-Sánchez, P.A., Morais, J.: "NumericalSgps"—a GAP Package, Version 1.0.1. http://www.gap-system.org/Packages/numericalsgps.html
8. Eto, K.: Almost Gorenstein monomial curves in affine four space. J. Algebra **488**, 362–387 (2017)
9. Goto S., Matsuoka N., Phuong T.T.: Almost Gorenstein rings. J. Algebra **379**, 355–381 (2013)
10. Goto, S., Takahashi R., Taniguchi N.: Almost Gorenstein rings—towards a theory of higher dimension. J. Pure Appl. Algebra **219**, 2666–2712 (2015)
11. Goto, S., Kien, D.V., Matsuoka, N., Truong, H.L.: Pseudo-Frobenius numbers versus defining ideals in numerical semigroup rings. J. Algebra **508**, 1–15 (2018)
12. Herzog, J.: Generators and relations of abelian semigroups and semigroup rings. Manuscripta Math. **3**, 175–193 (1970)
13. Herzog, J., Watanabe, K.-i.: Almost symmetric numerical semigroups. Semigroup Forum **98**, 589–630 (2019)
14. Herzog J., Hibi T., Stamate D.I.: The trace of the canonical module. arXiv: 1612.02723
15. Komeda, J.: On the existence of Weierstrass points with a certain semigroup generated by 4 elements. Tsukuba J. Math. **6**(2), 237–270 (1982)
16. Kunz, E.: The value-semigroup of a one-dimensional Gorenstein ring. Proc. Am. Math. Soc. **25**, 748–751 (1970)

17. Moscariello, A.: On the type of an almost Gorenstein monomial curve. J. Algebra **456**, 266–277 (2016)
18. Nari, H.: Symmetries on almost symmetric numerical semigroups. Semigroup Forum **86**, 140–154 (2013)
19. Nari, H., Numata, T., Watanabe, K.-i.: Genus of numerical semigroups generated by three elements. J. Algebra **358**, 67–73 (2012)
20. Oneto, A., Strazzanti, F., Tamone G.: One-dimensional Gorenstein local rings with decreasing Hilbert function. J. Algebra **489**, 91–114 (2017)
21. The GAP Group: GAP—Groups, Algorithms, and Programming, Version 4.8.4 (2016). http://www.gap-system.org

Poincaré Series on Good Semigroup Ideals

Laura Tozzo

Abstract The Poincaré series of a ring associated to a plane curve was defined by Campillo, Delgado, and Gusein-Zade. This series, defined through the value semigroup of the curve, encodes the topological information of the curve. In this paper we extend the definition of Poincaré series to the class of good semigroup ideals, to which value semigroups of curves belong. Using this definition we generalize a result of Pol: under suitable assumptions, given good semigroup ideals E and K, with K canonical, the Poincaré series of $K - E$ is symmetric to the Poincaré series of E.

Keywords Poincaré series · Good semigroups · Symmetry

1 Introduction

Plane algebroid curves are determined by their value semigroups up to equivalence in the sense of Zariski, as shown by Waldi [14, 15]. Value semigroups are important invariants of curves also with regard to duality properties. Kunz [9] was the first to show that the Gorensteinness of an analytically irreducible and residually rational local ring corresponds to a symmetry of its numerical value semigroup. Waldi [14] gave a definition of symmetry for more branches, and showed that plane (hence Gorenstein) curves with two branches have symmetric value semigroups. Later Delgado [5] proved the analogue of Kunz' result for general algebroid curves: they are Gorenstein if and only if their value semigroup is symmetric. Campillo, Delgado and Kiyek [2] extended Delgado's result to analytically reduced and residually rational local rings with infinite residue field. D'Anna [4] then used the definition of symmetry given by Delgado to define a canonical semigroup ideal K_0, and showed that a fractional ideal $\mathcal{K}$ of R such that $R \subseteq \mathcal{K} \subseteq \overline{R}$ is canonical if and only if its

L. Tozzo (✉)
Department of Mathematics, Technische Universität Kaiserslautern, Kaiserslautern, Germany
e-mail: tozzo@mathematik.uni-kl.de

V. Barucci et al. (eds.), *Numerical Semigroups*, Springer INdAM Series 40,
https://doi.org/10.1007/978-3-030-40822-0_20

value semigroup coincides with K_0. Recently Pol [12] studied the value semigroup ideal of the dual of a fractional ideal over Gorenstein algebroid curves. In [8] the author together with Korell and Schulze gave a new definition of a canonical semigroup ideal K (see Definition 6) and extended D'Anna's and Pol's results to the larger class of admissible rings (see Definition 10). Moreover, one of the main results of [8] shows that Cohen–Macaulay duality (where the dual of a fractional ideal is obtained by applying the functor Hom$(-, \mathcal{K})$, with $\mathcal{K}$ a canonical ideal) and semigroup duality are compatible under taking values, if the ring is admissible. An admissible ring is in particular semilocal, and its value semigroup, as first observed by Barucci, D'Anna and Fröberg [1], satisfies particular axioms which define the class of good semigroups.

In this paper we analyze further the duality properties of good semigroups by showing symmetry properties of their Poincaré series. As a consequence of the more general result given by Stanley in [13], Gorenstein semigroup rings have symmetric Hilbert series. This is also equivalent to the value semigroup associated to the semigroup ring being symmetric. Adapting the concept of Hilbert series to value semigroups leads to the concept of Poincaré series. A definition of Poincaré series for a plane curve singularity was given by Campillo, Delgado and Gusein-Zade in [3], where they showed that it coincides with the Alexander polynomial, a complete topological invariant of the singularity. Moyano-Fernandez in [10], using a definition inspired by the above, analyzed the connection between univariate and multivariate Poincaré series of curve singularities and later on, together with Tenorio and Torres [11], they showed that the Poincaré series associated to generalized Weierstrass semigroups can be used to retrieve entirely the semigroup, hence highlighting the potential of Poincaré series. Later Pol [12] considered a symmetry problem on Gorenstein reduced curves. She proved that the Poincaré series of the Cohen–Macaulay dual of a fractional ideal $\mathcal{E}$ is symmetric to the Poincaré series of $\mathcal{E}$, therefore extending the symmetry known for Gorenstein curves to their fractional ideals. Pol's result strongly uses the fact that it is always possible to define a filtration on value semigroups (see Definition 3), as done first in [2]. To deal with this filtration an important tool is the distance $\mathrm{d}(E \backslash F)$ between two good semigroup ideals $E \subseteq F$ (see Definition 4). Using the notion of distance and the duality on good semigroups given in [8], we are able to generalize Pol's result to good semigroup ideals. We prove that, given good semigroup ideals E and K, with K canonical, the Poincaré series of $K - E$ is symmetric to the Poincaré series of E under suitable assumptions. In particular, the symmetry is true (without additional assumptions) whenever E is the value semigroup of a fractional ideal $\mathcal{E}$ of an admissible ring R.

2 Preliminaries

In this section we recall definitions and known results that will be needed in the rest of the paper.

Let $S \subseteq \overline{S}$ be a partially ordered cancellative commutative monoid, where $\overline{S}$ is a partially ordered monoid, isomorphic to $\mathbb{N}^s$ with its natural partial order. Then the group of differences D_S of $\overline{S}$ is isomorphic to $\mathbb{Z}^s$. In the following we always fix an isomorphism $D_S \cong \mathbb{Z}^s$, in order to talk about indexes $i \in \{1, \ldots, s\}$.

2.1 Good Semigroups and Their Ideals

The following where first defined in [6] and [4].

Definition 1 Let $E \subseteq \mathbb{Z}^s$. We define properties:

(E0) There exists an $\alpha \in \mathbb{Z}^s$ such that $\alpha + \mathbb{N}^s \subseteq E$.
(E1) If $\alpha, \beta \in E$, then $\min\{\alpha, \beta\} := (\min\{\alpha_i, \beta_i\})_{i \in I} \in E$.
(E2) For any $\alpha, \beta \in E$ and $j \in I$ with $\alpha_j = \beta_j$ there exists an $\epsilon \in E$ such that $\epsilon_j > \alpha_j = \beta_j$ and $\epsilon_i \geq \min\{\alpha_i, \beta_i\}$ for all $i \in I \setminus \{j\}$ with equality if $\alpha_i \neq \beta_i$.

Definition 2 We call S a *good semigroup* if properties (E0), (E1) and (E2) hold for $E = S$.

A *semigroup ideal* of a good semigroup S is a subset $\emptyset \neq E \subseteq D_S$ such that $E + S \subseteq E$ and $\alpha + E \subseteq S$ for some $\alpha \in S$.

If E satisfies (E1), we denote by $\mu^E := \min E$ its *minimum*.

If E satisfies (E1) and (E2), then we call E a *good semigroup ideal* of S. Note that any semigroup ideal of a good semigroup S automatically satisfies (E0).

If E and F are semigroup ideals of a good semigroup S, we define

$$E - F := \{\alpha \in D_S \mid \alpha + F \subseteq E\},$$

and we call

$$C_E := E - \overline{S} = \{\alpha \in D_S \mid \alpha + \overline{S} \subseteq E\}$$

the *conductor ideal* of E. If E is a semigroup ideal of S satisfying (E1), then we call $\gamma^E := \mu^{C_E}$ the *conductor* of E. We abbreviate $\gamma := \gamma^S$ and $\tau := \gamma - \mathbf{1}$, where $\mathbf{1} = (1, \ldots, 1) \in \mathbb{N}^s$.

Let S be a good semigroup. The set of good semigroup ideals of S is denoted by $\mathfrak{G}_S$.

Remark 1 Let S be a good semigroup. For any $E, F \in \mathfrak{G}_S$ and $\alpha \in D_S$ the following hold:

(a) $\alpha + E \in \mathfrak{G}_S$.
(b) $(\alpha + E) - F = \alpha + (E - F)$ and $E - (\alpha + F) = -\alpha + (E - F)$.
(c) $E - S = E$.

Definition 3 Let S be a good semigroup. For any $E \in \mathfrak{G}_S$, we define a decreasing filtration $E^\bullet$ on E by semigroup ideals

$$E^\alpha := \{\beta \in E \mid \beta \geq \alpha\}$$

for any $\alpha \in D_S$.

Remark 2 Let S be a good semigroup. For a semigroup ideal $E \in \mathfrak{G}_S$ we have $E = E^{\mu^E}$ and, by definition of conductor, $C_E = \gamma^E + \overline{S} = E^{\gamma^E}$.

2.2 Distance of Semigroup Ideals

Definition 4 Let $E \subseteq D_S$. Elements $\alpha, \beta \in E$ with $\alpha < \beta$ are called *consecutive* in E if $\alpha < \delta < \beta$ implies $\delta \notin E$ for any $\delta \in D_S$. For $\alpha, \beta \in E$, a chain

$$\alpha = \alpha^{(0)} < \cdots < \alpha^{(n)} = \beta \tag{1}$$

of points $\alpha^{(i)} \in E$ is said to be *saturated of length n* if $\alpha^{(i)}$ and $\alpha^{(i+1)}$ are consecutive in E for all $i \in \{0, \dots, n-1\}$. If E satisfies

(E4) For fixed $\alpha, \beta \in E$, any two saturated chains (1) in E have the same length n.

then we call $\mathrm{d}_E(\alpha, \beta) := n$ the *distance* of α and β in E.

Due to [4, Proposition 2.3], any $E \in \mathfrak{G}_S$ satisfies property (E4).

Definition 5 Let S be a good semigroup, and let $E \subseteq F$ be two semigroup ideals of S satisfying property (E4). Then we call

$$\mathrm{d}(F \backslash E) := d_F(\mu^F, \gamma^E) - \mathrm{d}_E(\mu^E, \gamma^E)$$

the *distance* between E and F.

The following was proved in [4, Proposition 2.7]:

Lemma 1 *If $E \subseteq F \subseteq G$ are semigroup ideals of a good semigroup S satisfying property* (E4), *then*

$$\mathrm{d}(G \backslash E) = \mathrm{d}(G \backslash F) + \mathrm{d}(F \backslash E).$$

Moreover, as proved by the author in [8, Proposition 4.2.6], distance can be used to check equality:

Proposition 1 *Let S be a good semigroup, and let $E, F \in \mathfrak{G}_S$ with $E \subseteq F$. Then $E = F$ if and only if* $\mathrm{d}(F \backslash E) = 0$.

2.3 Canonical Semigroup Ideals

The following definition is [8, Definition 5.11 in published version]:

Definition 6 Let S be a good semigroup. A *canonical ideal* K is a good semigroup ideal of S such that $K \subseteq E$ implies $K = E$ for any E with $\gamma^k = \gamma^E$.

Let $\alpha \in D_S$, $E \subseteq D_S$.

- $\Delta_i^E(\alpha) = \{\beta \in E \mid \beta_i = \alpha_i \text{ and } \beta_j > \alpha_j \text{ for all } j \neq i\}$;
- $\overline{\Delta}_i^E(\alpha) = \{\beta \in E \mid \beta_i = \alpha_i \text{ and } \beta_j \geq \alpha_j \text{ for all } j \neq i\}$;
- $\Delta^E(\alpha) = \cup_{i \in \{1,\ldots,s\}} \Delta_i^E(\alpha)$;
- $\overline{\Delta}^E(\alpha) = \cup_{i \in \{1,\ldots,s\}} \overline{\Delta}_i^E(\alpha)$.

We denote by $\mathbf{e}_i$ the i-th vector of the canonical basis of D_S. Then $\overline{\Delta}_i^E(\alpha) = \Delta_i^E(\alpha + \mathbf{e}_i - \mathbf{1})$.

Using [8, Proposition 5.18 in published version] and [4, Proposition 3.2] yields:

Proposition 2 *Let S be a good semigroup. Then K is a canonical ideal if and only if $K = \alpha + K_0$ for some $\alpha \in D_S$, where*

$$K_0 = \{\alpha \in D_S \mid \Delta^S(\tau - \alpha) = \emptyset\}$$

is a good semigroup ideal of S called normalized canonical ideal *of S. In particular, K_0 is the only canonical semigroup ideal with $\gamma^{K_0} = \gamma$.*

Lemma 2 *Let S be a good semigroup. If $E \in \mathfrak{G}_S$, then*

(a) $K_0 - E = \{\alpha \in D_S \mid \Delta^E(\tau - \alpha) = \emptyset\} \in \mathfrak{G}_S$;
(b) $\gamma^{K_0 - E} = \gamma - \mu^E$;
(c) $\mu^{K_0 - E} = \gamma - \gamma^E$.

Proof For part (a) see [4, Computation 3.3] and [8, Lemma 5.16 in published version]. Part (b) is proven in [8, Lemma 4.11 published version]. Part (c) follows by Korell et al. [8, Theorem 5.14 in published version]. In fact, $\mu^{K_0 - E} = \gamma - \gamma^{K_0 - (K_0 - E)} = \gamma - \gamma^E$. □

In the following, when we talk about *the* canonical semigroup ideal, we refer to K_0. To make notation easier, we will write K instead of K_0. Notice that by Remark 1 and Proposition 2 all the results hold as well for any K canonical, up to translation by a suitable α.

Definition 7 Let S be a good semigroup. Let $E \in \mathfrak{G}_S$ and K the canonical semigroup ideal. Then the *dual* of E is $K - E$.

Remark 3 Let S be a good semigroup, and $E \in \mathfrak{G}_S$. For all $\alpha \in D_S$ we have $E - D_S^\alpha = D_S^{\gamma - \alpha}$. In fact, Remark 1 implies:

$$E - D_S^\alpha = E - (\alpha + D_S) = -\alpha + (E - D_S) = -\alpha + \gamma + D_S = D_S^{\gamma - \alpha}.$$

This is in particular true for $E = K$.

The following theorem (see [8, Theorem 5.14 in published version]) shows that every good semigroup ideal coincides with its double dual (see Definition 7):

Theorem 1 *Let S be a good semigroup, $E \in \mathfrak{G}_S$, and let K be the canonical semigroup ideal. Then $K - (K - E) = E$.*

2.4 Value Semigroups

We now give a few definitions regarding rings, in order to make clear the connection between their value semigroups and good semigroups.

In the following, R is a commutative ring with 1, and Q_R its total ring of fractions. We always assume fractional ideals of R to be regular, i.e. to contain at least a regular element.

Definition 8 A *valuation ring* of Q_R is a subring $V \subsetneq Q_R$ such that the set $Q_R \setminus V$ is multiplicatively closed.

If $R \subseteq V$, we call V a *valuation ring over* R. We denote by $\mathfrak{V}_R$ the set of all valuation rings of Q_R over R.

A valuation ring V of Q with unique regular maximal ideal $\mathfrak{m}_V$ is called a *discrete valuation ring* if $\mathfrak{m}_V$ is the only regular prime ideal of V (see [7, Ch. I, (2.16) Def.]).

A *discrete valuation* of Q_R is a map $\nu \colon Q_R \twoheadrightarrow \mathbb{Z} \cup \{\infty\}$ satisfying

$$\nu(xy) = \nu(x) + \nu(y), \quad \nu(x + y) \geq \min\{\nu(x), \nu(y)\}$$

for any $x, y \in Q_R$. We refer to $\nu(x) \in \mathbb{Z} \cup \{\infty\}$ as the *value* of $x \in Q_R$ with respect to ν.

The following theorem is [7, Ch. II, (2.11) Thm.], and characterizes valuation rings over one-dimensional semilocal Cohen–Macaulay rings.

Theorem 2 *Let R be a one-dimensional semilocal Cohen–Macaulay ring. The set $\mathfrak{V}_R$ is finite and non-empty, and it contains discrete valuation rings only.*

Thanks to this theorem, we can give the following definition:

Definition 9 Let R be a one-dimensional semilocal Cohen–Macaulay ring, and let $\mathfrak{V}_R$ be the set of (discrete) valuation rings of Q_R over R with valuations

$$\nu = \nu_R := (\nu_V)_{V \in \mathfrak{V}_R} \colon Q_R \to (\mathbb{Z} \cup \{\infty\})^{\mathfrak{V}_R}.$$

To each fractional ideal $\mathcal{E}$ of R we associate its *value semigroup ideal*

$$\Gamma_{\mathcal{E}} := \nu(\{x \in \mathcal{E} \mid x \text{ is regular}\}) \subseteq \mathbb{Z}^{\mathfrak{V}_R}.$$

If $\mathcal{E} = R$, then the monoid Γ_R is called the *value semigroup* of R.

The following additional definitions are needed to make the value semigroup of a ring into a good semigroup.

Definition 10 Let R be a one-dimensional semilocal Cohen–Macaulay ring. Let $\widehat{R}$ denote its completion at the Jacobson radical and $\overline{R}$ its integral closure in its total ring of fractions Q_R.

(a) R is *analytically reduced* if $\widehat{R}$ is reduced or, equivalently, $\widehat{R_\mathfrak{m}}$ is reduced for all maximal ideals $\mathfrak{m}$ of R.
(b) R is *residually rational* if $\overline{R}/\mathfrak{n} = R/\mathfrak{n} \cap R$ for all maximal ideals $\mathfrak{n}$ of $\overline{R}$.
(c) R has *large residue fields* if $|R/\mathfrak{m}| \geq |\mathfrak{V}_{R_\mathfrak{m}}|$ for all maximal ideals $\mathfrak{m}$ of R.
(d) R is *admissible* if it is analytically reduced and residually rational with large residue fields.

The following was proven in [8, Corollary 3.14 in published version].

Proposition 3 *If R is admissible, then its value semigroup Γ_R is a good semigroup, and $\Gamma_\mathcal{E}$ is a good semigroup ideal for any fractional ideal $\mathcal{E}$ of R.*

Let R be an admissible ring, and let $\mathcal{E}$ be a fractional ideal of R. For any $\alpha \in D_S$ denote

$$\mathcal{E}^\alpha := \{x \in \mathcal{E} \mid \nu(x) \geq \alpha\}.$$

There is a clear link between filtrations of fractional ideals and filtrations of good semigroup ideals (see [8, Lemma 3.8 in published version]):

Lemma 3 *Let R be an admissible ring, and let $\mathcal{E}$ be a fractional ideal of R. Then $\mathcal{E}^\alpha$ is a (regular) fractional ideal of R and $(\Gamma_\mathcal{E})^\alpha = \Gamma_{\mathcal{E}^\alpha}$ for all $\alpha \in D_S$.*

The following was proven first by D'Anna [4, Proposition 2.2 in published version] and then extended in [8, Proposition 4.18 in published version].

Proposition 4 *Let R be an admissible ring, and let $\mathcal{E}, \mathcal{F}$ be two fractional ideals of R with $\mathcal{E} \subseteq \mathcal{F}$. Then*

$$\ell_R(\mathcal{F}/\mathcal{E}) = \mathrm{d}(\Gamma_\mathcal{F} \backslash \Gamma_\mathcal{E}),$$

where $\ell_R(\mathcal{F}/\mathcal{E})$ denotes the length of the quotient $\mathcal{F}/\mathcal{E}$ as R-module.

Finally, consider the Cohen–Macaulay dual of a fractional ideal $\mathcal{E}$, i. e. $\mathrm{Hom}(\mathcal{E}, \mathcal{K}) \cong \mathcal{K} : \mathcal{E}$ with $\mathcal{K}$ a canonical ideal. The following theorem (see [8, Theorem 5.27 in published version]) shows that the value semigroup ideal of the Cohen–Macaulay dual is the same as the dual of the value semigroup ideal (see Definition 7):

Theorem 3 *Let R be an admissible ring with canonical ideal $\mathcal{K}$. Then*

(a) $\Gamma_{\mathcal{K}:\mathcal{F}} = \Gamma_\mathcal{K} - \Gamma_\mathcal{F}$ for any fractional ideal $\mathcal{F}$ and
(b) $\mathrm{d}(\Gamma_\mathcal{K} - \Gamma_\mathcal{E} \backslash \Gamma_\mathcal{K} - \Gamma_\mathcal{F}) = \mathrm{d}(\Gamma_\mathcal{F} \backslash \Gamma_\mathcal{E})$ for any fractional ideals $\mathcal{E}, \mathcal{F}$ with $\mathcal{E} \subseteq \mathcal{F}$.

3 Distance and Duality

We now prove some technical results used in the coming section.

Lemma 4 *Let S be a good semigroup, $E \in \mathfrak{G}_S$, and $\alpha \in D_S$. Then* $\mathrm{d}(E^\alpha \backslash E^{\alpha+\mathbf{e}_i}) \leq 1$.

Proof We have the following:

$$\begin{aligned} \mathrm{d}(E^\alpha \backslash E^{\alpha+\mathbf{e}_i}) &= \mathrm{d}_{E^\alpha}(\mu^{E^\alpha}, \gamma^{E^{\alpha+\mathbf{e}_i}}) - \mathrm{d}_{E^{\alpha+\mathbf{e}_i}}(\mu^{E^{\alpha+\mathbf{e}_i}}, \gamma^{E^{\alpha+\mathbf{e}_i}}) \\ &= \mathrm{d}_{E^\alpha}(\mu^{E^\alpha}, \gamma^{E^{\alpha+\mathbf{e}_i}}) - \mathrm{d}_{E^\alpha}(\mu^{E^{\alpha+\mathbf{e}_i}}, \gamma^{E^{\alpha+\mathbf{e}_i}}) \end{aligned} \tag{2}$$

where the first equality is the definition of distance, and the second equality holds because a saturated chain between $\mu^{E^{\alpha+\mathbf{e}_i}}$ and $\gamma^{E^{\alpha+\mathbf{e}_i}}$ in $E^{\alpha+\mathbf{e}_i}$ is also saturated in E^α. Now observe that μ^{E^α} and $\mu^{E^{\alpha+\mathbf{e}_i}}$ are always comparable. In fact, by minimality of μ^{E^α} it has to be $\mu^{E^\alpha} = \min\{\mu^{E^\alpha}, \mu^{E^{\alpha+\mathbf{e}_i}}\} \leq \mu^{E^{\alpha+\mathbf{e}_i}}$. So (2) becomes

$$\mathrm{d}(E^\alpha \backslash E^{\alpha+\mathbf{e}_i}) = \mathrm{d}_{E^\alpha}(\mu^{E^\alpha}, \mu^{E^{\alpha+\mathbf{e}_i}}).$$

Now let $\mu^{E^\alpha} = \alpha^{(0)} < \cdots < \alpha^{(m)} = \mu^{E^{\alpha+\mathbf{e}_i}}$ be a saturated chain in E. Suppose $m \geq 2$. By minimality of $\mu^{E^{\alpha+\mathbf{e}_i}}$, we have that $\alpha^{(k)} \in \overline{\Delta}_i^E(\alpha)$ for all $k < m$. Consider $\alpha^{(0)}, \alpha^{(1)} \in E$. They have $\alpha_i^{(0)} = \alpha_i^{(1)} = \alpha_i$ and there exists a $j \neq i$ such that $\alpha_j^{(0)} < \alpha_j^{(1)} \leq \alpha_j^{(m)} = \mu_j^{E^{\alpha+\mathbf{e}_i}}$. We can apply property (E2) to $\alpha^{(0)}, \alpha^{(1)} \in E$ and obtain a $\beta \in E$ with $\beta_i > \alpha_i$ and $\beta_j = \min\{\alpha_j^{(0)}, \alpha_j^{(1)}\} = \alpha_j^{(0)}$. In particular, $\beta \in E^{\alpha+\mathbf{e}_i}$. Thus, by minimality of $\mu^{E^{\alpha+\mathbf{e}_i}}$, it has to be $\min\{\beta, \mu^{E^{\alpha+\mathbf{e}_i}}\} = \mu^{E^{\alpha+\mathbf{e}_i}}$. Then $\mu_j^{E^{\alpha+\mathbf{e}_i}} = \min\{\beta_j, \mu_j^{E^{\alpha+\mathbf{e}_i}}\} = \min\{\alpha_j^{(0)}, \mu_j^{E^{\alpha+\mathbf{e}_i}}\} = \alpha_j^{(0)} < \mu_j^{E^{\alpha+\mathbf{e}_i}}$. This is a contradiction. Hence the claim. □

Lemma 5 *Let S be a good semigroup, and let $E \in \mathfrak{G}_S$. Then $\mathrm{d}(E^\alpha \backslash E^{\alpha+\mathbf{e}_i}) = 1$ if and only if $\overline{\Delta}_i^E(\alpha) \neq \emptyset$.*

Proof Observe that by definition $E^\alpha = E^{\alpha+\mathbf{e}_i} \cup \overline{\Delta}_i^E(\alpha)$ and $E^{\alpha+\mathbf{e}_i} \cap \overline{\Delta}_i^E(\alpha) = \emptyset$. By Proposition 1, $\mathrm{d}(E^\alpha \backslash E^{\alpha+\mathbf{e}_i}) = 0$ if and only if $E^\alpha = E^{\alpha+\mathbf{e}_i}$, i.e. if and only if $\overline{\Delta}_i^E(\alpha) = \emptyset$. So the claim follows by Lemma 4. □

The following proposition characterizes the distance in terms of $\overline{\Delta}$-sets.

Proposition 5 *Let S be a good semigroup, $E \in \mathfrak{G}_S$, and $\alpha, \beta \in D_S$ with $\alpha \leq \beta$. Then $E^\beta \subseteq E^\alpha$.*

Let $\alpha = \alpha^{(0)} < \alpha^{(1)} < \cdots < \alpha^{(n)} = \beta$ be a saturated chain in D_S, with $\alpha^{(j+1)} = \alpha^{(j)} + \mathbf{e}_{i(j)}$ for any $j \in \{0, \ldots, n-1\}$. We have:

$$\mathrm{d}(E^\alpha \backslash E^\beta) = |\{j \in \{0, \ldots, n-1\} \mid \overline{\Delta}_{i(j)}^E(\alpha^{(j)}) \neq \emptyset\}|,$$

where $|-|$ denotes the cardinality.

Proof Using the additivity of the distance (see Lemma 1), our assumptions and Lemma 5 we get the following equalities:

$$\begin{aligned} \mathrm{d}(E^{\alpha}\backslash E^{\beta}) = \sum_{j=0}^{n-1} \mathrm{d}(E^{\alpha^{(j)}}\backslash E^{\alpha^{(j+1)}}) = \sum_{j=0}^{n-1} \mathrm{d}(E^{\alpha^{(j)}}\backslash E^{\alpha^{(j)}+\mathbf{e}_{i(j)}}) \\ = |\{j \in \{0,\dots,n-1\} \mid \overline{\Delta}^{E}_{i(j)}(\alpha^{(j)}) \neq \emptyset\}|. \end{aligned}$$

As a corollary, we obtain a way to compute the distance between two semigroup ideals.

Corollary 1 *Let S be a good semigroup. Let $E \subseteq F \in \mathfrak{G}_S$, and let $\mu^F = \alpha^{(0)} < \alpha^{(1)} < \cdots < \alpha^{(m)} = \mu^E < \cdots < \alpha^{(n)} = \gamma^E$ be a saturated chain in D_S. In particular, $\alpha^{(j+1)} = \alpha^{(j)} + \mathbf{e}_{i(j)}$ for any $j \in \{0,\dots,n-1\}$. Then*

$$\begin{aligned} \mathrm{d}(F\backslash E) = &|\{j \in \{0,\dots,n-1\} \mid \overline{\Delta}^{F}_{i(j)}(\alpha^{(j)}) \neq \emptyset\}| \\ &- |\{j \in \{m,\dots,n-1\} \mid \overline{\Delta}^{E}_{i(j)}(\alpha^{(j)}) \neq \emptyset\}| \end{aligned}$$

Proof By additivity of the distance (see Lemma 1) we have:

$$\mathrm{d}(F\backslash E) = \mathrm{d}(F\backslash C_E) - \mathrm{d}(E\backslash C_E) = \mathrm{d}(F^{\mu^F}\backslash F^{\gamma^E}) - \mathrm{d}(E^{\mu^E}\backslash E^{\gamma^E}).$$

The claim follows by Proposition 5. □

The following two lemmas are necessary to prove Proposition 6.

Lemma 6 *Let S be a good semigroup, and let $E \in \mathfrak{G}_S$. Let K be the canonical ideal of S. If $\overline{\Delta}^{K-E}_i(\tau-\alpha) \neq \emptyset$ then $\Delta^E_i(\alpha) = \emptyset$.*

Proof Let $\tau - \beta \in \overline{\Delta}^{K-E}_i(\tau-\alpha)$. Then

$$\begin{aligned} \tau_i - \beta_i &= \tau_i - \alpha_i, \\ \tau_j - \beta_j &\geq \tau_j - \alpha_j \text{ for all } j \neq i, \end{aligned}$$

and $\Delta^E(\beta) = \emptyset$ by Lemma 2(a). As $\beta_i = \alpha_i$ and $\beta_j \leq \alpha_j$, it follows $\Delta^E_i(\alpha) \subseteq \Delta^E_i(\beta) = \emptyset$. □

Lemma 7 *Let S be a good semigroup, $E \in \mathfrak{G}_S$, and $\alpha, \beta \in D_S$ with $\alpha \leq \beta$. Let K be the canonical ideal of S. Then:*

$$\mathrm{d}(E^{\alpha}\backslash E^{\beta}) \leq \mathrm{d}(D^{\alpha}_S \setminus D^{\beta}_S) - \mathrm{d}((K-E)^{\gamma-\beta} \setminus (K-E)^{\gamma-\alpha}).$$

Proof Let

$$\alpha = \alpha^{(0)} < \alpha^{(1)} < \cdots < \alpha^{(n)} = \beta$$

be a saturated chain in D_S, with $\alpha^{(j+1)} = \alpha^{(j)} + \mathbf{e}_{i(j)}$ for any $j \in \{0, \ldots, n-1\}$. Let us denote $J = \{0, \ldots, n-1\}$.

Set $\beta^{(j)} = \gamma - \alpha^{(n-j)}$. Then

$$\gamma - \beta = \beta^{(0)} < \beta^{(1)} < \cdots < \beta^{(n)} = \gamma - \alpha$$

is a saturated chain in D_S, and

$$\beta^{(j+1)} = \gamma - \alpha^{(n-(j+1))} = \gamma - (\alpha^{(n-j)} - \mathbf{e}_{i(n-(j+1))}) = \beta^{(j)} + \mathbf{e}_{i(n-(j+1))}.$$

By Proposition 5 we have $\mathrm{d}(E^\alpha \backslash E^\beta) = |\{j \in J \mid \overline{\Delta}^E_{i(j)}(\alpha^{(j)}) \neq \emptyset\}|$. Recall that $E = K - (K - E)$ by Proposition 1. Therefore we can apply Lemma 6 and obtain

$$\begin{aligned}
\mathrm{d}(E^\alpha \backslash E^\beta) &= |\{j \in J \mid \overline{\Delta}^E_{i(j)}(\alpha^{(j)}) \neq \emptyset\}| \\
&\leq |\{j \in J \mid \Delta^{K-E}_{i(j)}(\tau - \alpha^{(j)}) = \emptyset\}| \\
&= |\{j \in J \mid \Delta^{K-E}_{i(j)}(\gamma - \alpha^{(j)} - \mathbf{1}) = \emptyset\}| \\
&= |\{j \in J \mid \Delta^{K-E}_{i(j)}(\beta^{(n-j)} - \mathbf{1}) = \emptyset\}| \\
&= |\{j \in J \mid \overline{\Delta}^{K-E}_{i(j)}(\beta^{(n-(j+1))}) = \emptyset\}| \qquad (3) \\
&= n - |\{j \in J \mid \overline{\Delta}^{K-E}_{i(j)}(\beta^{(n-(j+1))}) \neq \emptyset\}| \\
&= n - |\{j \in J \mid \overline{\Delta}^{K-E}_{i(n-(j+1))}(\beta^{(j)}) \neq \emptyset\}| \\
&= n - \mathrm{d}((K-E)^{\gamma-\beta} \setminus (K-E)^{\gamma-\alpha}) \\
&= \mathrm{d}(D^\alpha_S \setminus D^\beta_S) - \mathrm{d}((K-E)^{\gamma-\beta} \setminus (K-E)^{\gamma-\alpha}).
\end{aligned}$$

Proposition 6 *Let S be a good semigroup, $E \in \mathfrak{G}_S$, and $\alpha, \beta \in D_S$ with $\alpha \leq \beta$. Let K be the canonical ideal of S. Then the following are equivalent:*

(i) $\mathrm{d}(E^\alpha \backslash E^\beta) = \mathrm{d}(D^\alpha_S \setminus D^\beta_S) - \mathrm{d}((K-E)^{\gamma-\beta} \setminus (K-E)^{\gamma-\alpha})$.

(ii) For all $\delta \in D_S$ such that $\alpha \leq \delta \leq \beta$ and for every $i \in \{1, \ldots, s\}$ such that $\delta + \mathbf{e}_i \leq \beta$,

$$\overline{\Delta}^E_i(\delta) \neq \emptyset \iff \Delta^{K-E}_i(\tau - \delta) = \emptyset.$$

(iii) For all $\delta \in D_S$ such that $\alpha \leq \delta \leq \beta$ and for every $i \in \{1, \ldots, s\}$ such that $\delta - \mathbf{e}_i \geq \alpha$,

$$\overline{\Delta}_i^{K-E}(\tau - \delta) \neq \emptyset \iff \Delta_i^E(\delta) = \emptyset.$$

Proof Let

$$\alpha = \alpha^{(0)} < \alpha^{(1)} < \cdots < \alpha^{(n)} = \beta$$

and

$$\gamma - \beta = \beta^{(0)} < \beta^{(1)} < \cdots < \beta^{(n)} = \gamma - \alpha$$

be as in Lemma 7. Let us denote again $J = \{0, \ldots, n-1\}$. Then, from the proof of Lemma 7 (see (3)) we have

$$\mathrm{d}(E^\alpha \backslash E^\beta) = \mathrm{d}(D_S^\alpha \setminus D_S^\beta) - \mathrm{d}((K-E)^{\gamma-\beta} \setminus (K-E)^{\gamma-\alpha})$$

if and only if

$$|\{j \in J \mid \overline{\Delta}_{i(j)}^E(\alpha^{(j)}) \neq \emptyset\}| = |\{j \in J \mid \Delta_{i(j)}^{K-E}(\tau - \alpha^{(j)}) = \emptyset\}|.$$

Since the first set is contained in the second by Lemma 6, we obtain

$$\{j \in J \mid \overline{\Delta}_{i(j)}^E(\alpha^{(j)}) \neq \emptyset\} = \{j \in J \mid \Delta_{i(j)}^{K-E}(\tau - \alpha^{(j)}) = \emptyset\}$$

In particular

$$\overline{\Delta}_{i(j)}^E(\alpha^{(j)}) \neq \emptyset \iff \Delta_{i(j)}^{K-E}(\tau - \alpha^{(j)}) = \emptyset.$$

Now let $\delta \in D_S$ be such that $\alpha \leq \delta \leq \beta$ and for every $i \in \{1, \ldots, s\}$, $\delta + \mathbf{e}_i \leq \beta$. Then it is always possible to find a saturated chain in D_S between α and β such that $\delta = \alpha^{(j)}$ and $i = i(j)$. Thus

$$\overline{\Delta}_i^E(\delta) \neq \emptyset \iff \Delta_i^{K-E}(\tau - \delta) = \emptyset.$$

Finally, observing that $E = K - (K - E)$ by Proposition 1, this is also equivalent to

$$\overline{\Delta}_i^{K-E}(\tau - \delta) \neq \emptyset \iff \Delta_i^E(\delta) = \emptyset.$$

if $\delta - \mathbf{e}_i \geq \alpha$ (i.e. $(\tau - \delta) + \mathbf{e}_i \leq \tau - \alpha$). □

The next corollary gives the necessary equivalent conditions for the main Theorem 4.

Corollary 2 *Let S be a good semigroup, $E \in \mathfrak{G}_S$, and $\alpha \in D_S$ with $\mu^E \leq \alpha \leq \gamma^E$. Let K be the canonical ideal of S. Then the following are equivalent:*

(i) $\mathrm{d}(D_S^{\mu^E} \setminus E) = \mathrm{d}((K - E) \setminus D_S^{\gamma - \mu^E})$.

(ii) $\mathrm{d}(E \setminus E^{\gamma^E}) = \mathrm{d}(D_S^{\mu^E} \setminus D_S^{\gamma^E}) - \mathrm{d}((K - E) \setminus (K - E)^{\gamma - \mu^E})$.

(iii) For every $i \in \{1, \ldots, s\}$ such that $\alpha + \mathbf{e}_i \leq \gamma^E$,

$$\overline{\Delta}_i^E(\alpha) \neq \emptyset \iff \Delta_i^{K-E}(\tau - \alpha) = \emptyset.$$

(iv) For every $i \in \{1, \ldots, s\}$ such that $\alpha - \mathbf{e}_i \geq \mu^E$,

$$\overline{\Delta}_i^{K-E}(\tau - \alpha) \neq \emptyset \iff \Delta_i^E(\alpha) = \emptyset.$$

Proof First of all observe that by additivity (see Lemma 1)

$$\mathrm{d}(D_S^{\mu^E} \setminus E) = \mathrm{d}(D_S^{\mu^E} \setminus D_S^{\gamma^E}) - \mathrm{d}(E \setminus D_S^{\gamma^E}).$$

As $D_S^{\gamma - \mu^E} = (K - E)^{\gamma - \mu^E}$ and $E^{\gamma^E} = D_S^{\gamma^E}$, (i) is equivalent to (ii). Now observe that by Lemma 2(c) and Remark 3, (ii) is the same as

$$\mathrm{d}(E^{\mu^E} \setminus E^{\gamma^E}) = \mathrm{d}(D_S^{\mu^E} \setminus D_S^{\gamma^E}) - \mathrm{d}((K - E)^{\gamma - \gamma^E} \setminus (K - E)^{\gamma - \mu^E}).$$

The claim follows then trivially from Proposition 6 with $\alpha = \mu^E$ and $\beta = \gamma^E$. □

Remark 4 Let R be an admissible ring and $\mathcal{E}$ a fractional ideal of R. Set $S = \Gamma_R$ and $E = \Gamma_{\mathcal{E}}$. Then Remark 3 and Proposition 3 imply Corollary 2(i).

4 Symmetry of the Poincaré Series

We now come to the main results of this paper. Let us first define the main objects of study, i.e. the Poincaré series.

For every $J \subseteq \{1, \ldots, s\}$, we denote $\mathbf{e}_J = \sum_{j \in J} \mathbf{e}_j$.

The following definition was given in [12]:

Definition 11 Let R be an admissible ring, and let $\mathcal{E}$ be a fractional ideal of R. We define

$$\ell_{\mathcal{E}}(\alpha) := \ell(\mathcal{E}^\alpha / \mathcal{E}^{\alpha + \mathbf{1}}), \quad \mathrm{L}_{\mathcal{E}}(\mathbf{t}) := \sum_{\alpha \in D_S} \ell_{\mathcal{E}}(\alpha) \mathbf{t}^\alpha,$$

where $\mathbf{t} = (t_1, \ldots, t_s)$, and $\mathbf{t}^\alpha = t_1^{\alpha_1} \cdots t_s^{\alpha_s}$.

The *Poincaré series of* $\mathcal{E}$ is

$$\mathrm{P}_{\mathcal{E}}(\mathbf{t}) := \mathrm{L}_{\mathcal{E}}(\mathbf{t}) \prod_{i=1}^{s} (t_i - 1).$$

We give an analogous definition for good semigroup ideals:

Definition 12 Let S be a good semigroup, and let $E \in \mathfrak{G}_S$. We define

$$\mathrm{d}_E(\alpha) := \mathrm{d}(E^{\alpha} \backslash E^{\alpha+\mathbf{1}}), \quad \mathrm{L}_E(\mathbf{t}) := \sum_{\alpha \in D_S} \mathrm{d}_E(\alpha)\mathbf{t}^{\alpha}.$$

The *Poincaré series of* E is

$$\mathrm{P}_E(\mathbf{t}) := \mathrm{L}_E(\mathbf{t}) \prod_{i=1}^{s} (t_i - 1).$$

Remark 5 Let R be an admissible ring, and let $\mathcal{E}$ be a fractional ideal of R. Then Lemma 3 and Proposition 4 yield $\mathrm{L}_{\Gamma_{\mathcal{E}}}(\mathbf{t}) = \mathrm{L}_{\mathcal{E}}(\mathbf{t})$, and in particular $\mathrm{P}_{\Gamma_{\mathcal{E}}}(\mathbf{t}) = \mathrm{P}_{\mathcal{E}}(\mathbf{t})$.

The Poincaré series can be written in a more compact fashion.

Lemma 8 *Let S be a good semigroup, and let $E \in \mathfrak{G}_S$. We define*

$$c_E(\alpha) := \sum_{J \subseteq \{1,\ldots,s\}} (-1)^{|J^c|} \mathrm{d}_E(\alpha - \mathbf{e}_J)$$

where J^c denotes the complement of J in $\{1, \ldots, s\}$. Then the Poincaré series can be written as

$$\mathrm{P}_E(\mathbf{t}) = \sum_{\alpha \in D_S} c_E(\alpha)\mathbf{t}^{\alpha}.$$

Proof Observe that

$$\begin{aligned}\prod_{i=1}^{s}(t_i-1) &= t_1 \cdots t_s + (-1)^1 \sum_{i_1<\cdots<i_{s-1}} t_{i_1} \cdots t_{i_{s-1}} + \cdots + (-1)^{s-1} \sum_{i=1}^{s} t_i + (-1)^s \\ &= \sum_{J \subseteq \{1,\ldots,s\}} (-1)^{|J^c|} \mathbf{t}^{\mathbf{e}_J}.\end{aligned}$$

Hence

$$\begin{aligned}\mathrm{P}_E(\mathbf{t}) &= \sum_{\alpha\in D_S} \mathrm{d}_E(\alpha)\mathbf{t}^{\alpha}\prod_{i=1}^{s}(t_i-1) = \sum_{\alpha\in D_S}\mathrm{d}_E(\alpha)\mathbf{t}^{\alpha}\sum_{J\subseteq\{1,\dots,s\}}(-1)^{|J^c|}\mathbf{t}^{\mathbf{e}_J}\\ &= \sum_{\alpha\in D_S}\sum_{J\subseteq\{1,\dots,s\}}(-1)^{|J^c|}\,\mathrm{d}_E(\alpha)\mathbf{t}^{\alpha+\mathbf{e}_J} =\\ &= \sum_{\alpha\in D_S}\sum_{J\subseteq\{1,\dots,s\}}(-1)^{|J^c|}\,\mathrm{d}_E(\alpha-\mathbf{e}_J)\mathbf{t}^{\alpha} = \sum_{\alpha\in D_S} c_E(\alpha)\mathbf{t}^{\alpha}.\end{aligned}$$

The next lemma is necessary to prove Proposition 7.

Lemma 9 *Let S be a good semigroup, $E \in \mathfrak{G}_S$, and $\beta \in D_S$. If $\beta_i + 1 < \mu_i^E$ or $\beta_i > \gamma_i^E$, then $\mathrm{d}_E(\beta) = \mathrm{d}_E(\beta + \mathbf{e}_i)$.*

Proof Let $\beta = \beta^{(0)} < \beta^{(1)} = \beta + \mathbf{e}_i < \cdots < \beta^{(s)} = \beta + \mathbf{1} < \beta^{(s+1)} = \beta + \mathbf{e}_i + \mathbf{1}$ be a saturated chain in D_S, where $\beta^{(j+1)} = \beta^{(j)} + \mathbf{e}_j$ for all $j \in \{1, \dots, s\} \setminus \{i\}$. Then by definition of $\mathrm{d}_E(\beta)$ and by additivity of the distance (see Lemma 1) we have

$$\mathrm{d}_E(\beta) = \mathrm{d}_E(E^{\beta}\backslash E^{\beta+\mathbf{1}}) = \sum_{j=0}^{s-1}\mathrm{d}_E(E^{\beta^{(j)}}\backslash E^{\beta^{(j+1)}}).$$

On the other hand we have

$$\mathrm{d}_E(\beta+\mathbf{e}_i) = \mathrm{d}_E(E^{\beta+\mathbf{e}_i}\backslash E^{\beta+\mathbf{e}_i+\mathbf{1}}) = \sum_{j=1}^{s}\mathrm{d}_E(E^{\beta^{(j)}}\backslash E^{\beta^{(j+1)}}).$$

Therefore

$$\begin{aligned}\mathrm{d}_E(\beta+\mathbf{e}_i) - \mathrm{d}_E(\beta) &= \mathrm{d}_E(E^{\beta^{(s)}}\backslash E^{\beta^{(s+1)}}) - \mathrm{d}_E(E^{\beta^{(0)}}\backslash E^{\beta^{(1)}})\\ &= \mathrm{d}_E(E^{\beta+\mathbf{1}}\backslash E^{\beta+\mathbf{e}_i+\mathbf{1}}) - \mathrm{d}_E(E^{\beta}\backslash E^{\beta+\mathbf{e}_i}).\end{aligned}$$

By Lemma 5 we know that

$$\mathrm{d}_E(E^{\beta}\backslash E^{\beta+\mathbf{e}_i}) = 1 \iff \overline{\Delta}_i^E(\beta) \neq \emptyset.$$

and

$$\mathrm{d}_E(E^{\beta+\mathbf{1}}\backslash E^{\beta+\mathbf{e}_i+\mathbf{1}}) = 1 \iff \overline{\Delta}_i^E(\beta+\mathbf{1}) \neq \emptyset.$$

If $\beta_i + 1 < \mu_i^E$, then also $\beta_i < \mu_i^E$, and therefore $\overline{\Delta}_i^E(\beta) = \overline{\Delta}_i^E(\beta + \mathbf{1}) = \emptyset$. This yields $\mathrm{d}_E(\beta + \mathbf{e}_i) - \mathrm{d}_E(\beta) = 0$. On the other hand, when $\beta_i > \gamma_i^E$, then also $\beta_i + 1 > \gamma_i^E$ and $\overline{\Delta}_i^E(\beta) \neq \emptyset$, $\overline{\Delta}_i^E(\beta + \mathbf{1}) \neq \emptyset$. This implies $\mathrm{d}_E(E^{\beta} \backslash E^{\beta + \mathbf{e}_i}) = \mathrm{d}_E(E^{\beta+\mathbf{1}} \backslash E^{\beta + \mathbf{e}_i + \mathbf{1}}) = 1$, and thus once again $\mathrm{d}_E(\beta + \mathbf{e}_i) - \mathrm{d}_E(\beta) = 0$. □

We can now prove that the Poincaré series of a good semigroup ideal is in fact a polynomial.

Proposition 7 *Let S be a good semigroup, and let $E \in \mathfrak{G}_S$. Then $\mathrm{P}_E(\mathbf{t})$ is a polynomial.*

Proof The goal is to prove that $c_E(\alpha) \neq 0$ only if $\mu^E \leq \alpha \leq \gamma^E$. Suppose there exists an i such that $\alpha_i < \mu_i^E$. Consider $J \subseteq \{1, \ldots, s\}$. It is not restrictive to consider $i \notin J$ (otherwise we can consider $J \setminus \{i\}$). Notice that if $\alpha - \mathbf{e}_{J \cup \{i\}} = \beta$, then $\alpha - \mathbf{e}_J = \beta + \mathbf{e}_i$. Since $\alpha_i < \mu_i^E$, then $\mu_i^E > (\alpha - \mathbf{e}_J)_i = (\beta + \mathbf{e}_i)_i = \beta_i + 1$. So by Lemma 9, we have

$$\mathrm{d}_E(\alpha - \mathbf{e}_{J \cup \{i\}}) = \mathrm{d}_E(\alpha - \mathbf{e}_J).$$

The same is true similarly if i is such that $\alpha_i > \gamma_i^E$. Therefore when $\alpha \notin \{\beta \mid \mu^E \leq \beta \leq \gamma^E\}$, for each $J \subseteq \{1, \ldots, s\}$ there exists a $J' \subset \{1, \ldots, s\}$ (either $J \cup \{i\}$ or $J \setminus \{i\}$) such that

$$\mathrm{d}_E(\alpha - \mathbf{e}_{J'}) = \mathrm{d}_E(\alpha - \mathbf{e}_J)$$

and $|J| = |J'| \pm 1$. Hence these terms annihilate each other in the sum

$$\sum_{J \subseteq \{1, \ldots, s\}} (-1)^{|J^c|} \mathrm{d}_E(\alpha - \mathbf{e}_J),$$

so that $c_E(\alpha) = 0$ for all $\alpha \notin \{\zeta \mid \mu^E \leq \zeta \leq \gamma^E\}$.

Thus $\mathrm{P}_E(\mathbf{t})$ is a polynomial. □

Finally, we are ready to prove our main theorem.

Theorem 4 *Let $S \subseteq \mathbb{N}^s$ be a good semigroup, γ its conductor, and let $E \in \mathfrak{G}_S$. Let K be the canonical ideal of S. If one of the equivalent conditions of Corollary 2 holds, then the Poincaré polynomials of E and $K - E$ are symmetric:*

$$\mathrm{P}_{K-E}(\mathbf{t}) = (-1)^{s+1} \mathbf{t}^{\gamma}\, \mathrm{P}_E\left(\frac{1}{\mathbf{t}}\right).$$

Proof By Lemma 8, $\mathrm{P}_{K-E}(\mathbf{t}) = \sum_{\alpha\in D_S} c_{K-E}(\alpha)\mathbf{t}^\alpha$, while

$$\begin{aligned}(-1)^{s+1}\mathbf{t}^\gamma\,\mathrm{P}_E\left(\frac{1}{\mathbf{t}}\right) &=(-1)^{s+1}\mathbf{t}^\gamma\sum_{\beta\in D_S} c_E(\beta)\mathbf{t}^{-\beta}\\ &=\sum_{\beta\in D_S}(-1)^{s+1}c_E(\beta)\mathbf{t}^{\gamma-\beta}\\ &=\sum_{\alpha\in D_S}(-1)^{s+1}c_E(\gamma-\alpha)\mathbf{t}^\alpha.\end{aligned}$$

Therefore the claim is equivalent to

$$c_{K-E}(\alpha) = (-1)^{s+1}c_E(\gamma-\alpha).$$

If $\alpha \notin \{\zeta \mid \mu^E \le \gamma-\zeta \le \gamma^E\} = \{\zeta \mid \gamma-\gamma^E \le \zeta \le \gamma-\mu^E\}$ then $c_{K-E}(\alpha) = c_E(\gamma-\alpha) = 0$ by proof of Proposition 7. So we can assume $\gamma-\gamma^E \le \alpha \le \gamma-\mu^E$.

Now let $\alpha = \gamma-\beta$. Then $\mu^E \le \beta \le \gamma^E$. As the equivalent conditions of Corollary 2 are satisfied, for any δ such that $\mu^E \le \delta \le \gamma^E$ with $\delta+\mathbf{e}_i \le \gamma^E$, $\overline{\Delta}_i^E(\delta) \neq \emptyset$ if and only if $\Delta_i^{K-E}(\tau-\delta) = \emptyset$. In particular, for any δ with $\mu^E \le \beta-\mathbf{1} \le \delta \le \beta \le \gamma^E$, $\overline{\Delta}_i^E(\delta) \neq \emptyset$ if and only if $\Delta_i^{K-E}(\tau-\delta) = \emptyset$. Hence by Proposition 6, $\mathrm{d}(E^{\beta-\mathbf{1}}\backslash E^\beta) = \mathrm{d}(D_S^{\beta-\mathbf{1}}\backslash D_S^\beta) - \mathrm{d}((K-E)^{\gamma-\beta}\backslash (K-E)^{\gamma-\beta+\mathbf{1}})$. Now recalling that $\alpha=\gamma-\beta$ we have $\mathrm{d}(E^{\gamma-\alpha-\mathbf{1}}\backslash E^{\gamma-\alpha}) = \mathrm{d}(D_S^{\gamma-\alpha-\mathbf{1}}\backslash D_S^{\gamma-\alpha}) - \mathrm{d}((K-E)^{\alpha}\backslash (K-E)^{\alpha+\mathbf{1}})$. As $\mathrm{d}(D_S^{\gamma-\alpha-\mathbf{1}}\backslash D_S^{\gamma-\alpha}) = d_{D_S}(\gamma-\alpha-\mathbf{1},\gamma-\alpha) = s$, this translates to

$$\mathrm{d}_{K-E}(\alpha) = s - \mathrm{d}_E(\gamma-\alpha-\mathbf{1}),$$

for any $\gamma-\gamma^E \le \alpha \le \gamma-\mu^E$ with $\alpha+\mathbf{1} \le \gamma-\mu^E$. Then

$$\begin{aligned}c_{K-E}(\alpha) = &\sum_{J\subseteq\{1,\ldots,s\}}(-1)^{|J^c|}\,\mathrm{d}_{K-E}(\alpha-\mathbf{e}_J)\\ =&(-1)^s\sum_{J\subseteq\{1,\ldots,s\}}(-1)^{|J|}(s-\mathrm{d}_E(\gamma-\alpha-\mathbf{1}+\mathbf{e}_J))\\ =&(-1)^s s\sum_{J\subseteq\{1,\ldots,s\}}(-1)^{|J|}+(-1)^{s+1}\sum_{J\subseteq\{1,\ldots,s\}}(-1)^{|J|}\,\mathrm{d}_E(\gamma-\alpha-\mathbf{1}+\mathbf{e}_J)\\ =&(-1)^s s\sum_{i=0}^{s}(-1)^i\binom{s}{i}+(-1)^{s+1}\sum_{J\subseteq\{1,\ldots,s\}}(-1)^{s+|J^c|}\,\mathrm{d}_E(\gamma-\alpha-\mathbf{e}_{J^c})\\ =&(-1)^s(1-1)^s+(-1)^{s+1}c_E(\gamma-\alpha)\\ =&(-1)^{s+1}c_E(\gamma-\alpha).\end{aligned}$$

Hence the claim. □

As a corollary, we obtain a generalization of Pol's result [12, Proposition 2.25].

Corollary 3 *Let R be an admissible ring, $\mathcal{E}$ a fractional ideal of R and $\mathcal{K}$ a canonical ideal of R such that $R \subseteq \mathcal{K} \subseteq \overline{R}$. Set $E = \Gamma_{\mathcal{E}}$ and $K = \Gamma_{\mathcal{K}}$. Then:*

$$\mathrm{P}_{K-E}(\mathbf{t}) = (-1)^{s+1}\mathbf{t}^{\gamma}\,\mathrm{P}_E\left(\frac{1}{\mathbf{t}}\right).$$

Proof It follows immediately from Remarks 4 and 5, and Theorem 4. □

Remark 6 Remark 4 shows that the equivalent conditions of Corollary 2 are true in the value semigroup case. It remains the question whether they are always satisfied. If not, they could represent a step forward in characterizing the class of value semigroups inside the bigger class of good semigroups.

References

1. Barucci, V., D'Anna, M., Fröberg, R.: Analytically unramified one-dimensional semilocal rings and their value semigroups. J. Pure Appl. Algebra **147**(3), 215–254 (2000)
2. Campillo, A., Delgado, F., Kiyek, K.: Gorenstein property and symmetry for one-dimensional local Cohen-Macaulay rings. Manuscripta Math. **83**(3–4), 405–423 (1994)
3. Campillo, A., Delgado, F., Gusein-Zade, S.M.: The Alexander polynomial of a plane curve singularity via the ring of functions on it. Duke Math. J. **117**(1), 125–156 (2003)
4. D'Anna, M.: The canonical module of a one-dimensional reduced local ring. Commun. Algebra **25**(9), 2939–2965 (1997)
5. Delgado de la Mata, F.: The semigroup of values of a curve singularity with several branches. Manuscripta Math. **59**(3), 347–374 (1987)
6. Delgado de la Mata, F.: Gorenstein curves and symmetry of the semigroup of values. Manuscripta Math. **61**(3), 285–296 (1988)
7. Kiyek, K., Vicente, J.L.: Resolution of curve and surface singularities in characteristic zero. In: Algebras and Applications, vol. 4. Kluwer Academic, Dordrecht (2004)
8. Korell, P., Schulze, M., Tozzo, L.: Duality on value semigroups. J. Commut. Algebra **11**(1), 81–129 (2019)
9. Kunz, E.: The value-semigroup of a one-dimensional Gorenstein ring. Proc. Am. Math. Soc. **25**, 748–751 (1970)
10. Moyano-Fernández, J.J.: Poincaré series for curve singularities and its behaviour under projections. J. Pure Appl. Algebra **219**(6), 2449–2462 (2015)
11. Moyano-Fernández, J.J., Tenório, W., Torres, F.: Generalized Weierstrass Semigroups and Their Poincaré Series. https://arxiv.org/pdf/1706.03733.pdf (2017)
12. Pol, D.: On the Values of Logarithmic Residues Along Curves. https://hal.archives-ouvertes.fr/hal-01074409 (2014), to appear on Ann. I. Fourier
13. Stanley, R.P.: Cohen-Macaulay complexes. In: Higher Combinatorics. Proceeding of NATO Advanced Study Inst., Berlin, 1976, pp. 51–62. Reidel, Dordrecht (1977)
14. Waldi, R.: Wertehalbgruppe und Singularität einer Ebenen Algebroiden Kurve. Universität Regensburg, Regensburg (1972, Dissertation)
15. Waldi, R.: On the equivalence of plane curve singularities. Commun. Algebra **28**(9), 4389–4401 (2000)

A Short Proof of Bresinski's Theorem on Gorenstein Semigroup Rings Generated by Four Elements

Kei-Ichi Watanabe

Abstract Let $H = \langle n_1, \ldots, n_4 \rangle$ be a numerical semigroup generated by four elements, which is symmetric and let $k[H]$ be the semigroup ring of H over a field k. H. Bresinski proved in Bresinsky (Manuscripta Math 17:205–219, 1975) that the defining ideal of $k[H]$ is minimally generated by three or five elements. We give a new short proof of Bresinski's Theorem using the structure theorem of Buchsbaum and Eisenbud on the minimal free resolution of Gorenstein rings of embedding codimension 3.

Keywords Gorenstein rings · Symmetric semigroups · Minimal free resolutions

1 Basic Concepts

Let $H = \langle n_1, \ldots, n_4 \rangle$ be a numerical semigroup generated by four elements. We denote

$$\mathrm{F}(H) = \max\{n \in \mathbb{Z} \mid n \notin H\}$$

the Frobenius number of H and $N = \sum_{i=1}^{4} n_i$. We call H symmetric if for every $n \in \mathbb{Z}$, $n \in H$ if and only if $\mathrm{F}(H) - n \notin H$. Let $k[H]$ be the semigroup ring of H over a field k and $S = k[x_1, \ldots, x_4]$ be the polynomial ring over k in the indeterminates $x_1, \ldots, x_4$. It is known by Kunz [4] that H is symmetric if and only $k[H]$ is Gorenstein. Let $\pi : S \to k[H]$ be the surjective k-algebra homomorphism with $\pi(x_i) = t^{n_i}$ for $i = 1, \ldots, n$. We consider S as a graded ring putting $\deg(x_i) = n_i$ so that π preserves the degree. We denote by I_H the kernel of π. If we assign to each x_i the degree n_i, then with respect to this grading, I_H is a homogeneous ideal,

K.-I. Watanabe (✉)
Department of Mathematics, College of Humanities and Sciences, Nihon University, Setagaya-ku, Tokyo, Japan
e-mail: watanabe@math.chs.nihon-u.ac.jp

V. Barucci et al. (eds.), *Numerical Semigroups*, Springer INdAM Series 40,
https://doi.org/10.1007/978-3-030-40822-0_21

generated by binomials. We denote by $\mu(I_H)$ the minimal number of generators of I_H. A binomial $\phi = \prod_{i=1}^e x_i^{\alpha_i} - \prod_{i=1}^e x_i^{\beta_i}$ belongs to I_H if and only if $\sum_{i=1}^e \alpha_i n_i = \sum_{i=1}^e \beta_i n_i$. A semigroup H is called a complete intersection if I_H is generated by a regular sequence. This condition is equivalent to say that $\mu(I_H) = 3$.

We define α_i to be the minimal positive integer such that

$$\alpha_i n_i = \sum_{j=1, j\neq i}^{4} \alpha_{ij} n_j. \tag{1}$$

Thus $f_i = x_i^{\alpha_i} - \prod_{j=1, j\neq i}^4 x_j^{\alpha_{ij}}$ $(i = 1, 2, 3, 4)$ is a minimal generator of I_H.

The purpose of this note is to give a short proof of Bresinski's Theorem;

Theorem 1.1 ([2]) *Assume that H is symmetric generated by four elements. If $k[H]$ is not a complete intersection, then $\mu(I_H) = 5$.* □

For the proof we let

$$F_\bullet = [\, 0 \to F_3 \to F_2 \xrightarrow{d_2} F_1 \xrightarrow{d_1} F_0 = k[H] \to 0\,]$$

be the graded minimal free resolution of $k[H]$ over S. Note that "H is symmetric" is equivalent to say "$k[H]$ is a Gorenstein ring". We denote $r = \mu(I_H) = \operatorname{rank} F_1$, $\phi_1, \ldots \phi_r$ be free basis of F_1 and we put $f_i = d_1(\phi_i) \in I_H$ so that $f_1, \ldots, f_r$ are minimal generators of I_H. We always assume that each f_i is a binomial.

Let us summarize known results about $F_\bullet$.

Theorem 1.2 ([3, 5])

(1) Since $k[H]$ is Gorenstein with a-invariant $a(k[H]) = \mathrm{F}(H)$, $F_3 \cong S(-\mathrm{F}(H) - N)$ and $F_\bullet$ is self-dual in the sense that there is an isomorphism $\mathrm{Hom}_S(F_\bullet, F_3) \cong F_\bullet$.

(2) r is an odd number.

(3) Let $M = (m_{ij})$ be the r by r matrix corresponding $d_2 : F_2 \to F_1$. Then we can choose the bases of F_2 and F_1 so that M is a skew-symmetric matrix.

(4) Let $\{e_1, \ldots, e_r\}$ be the free basis of F_2 so that $d_2(e_i) = \sum_{j=1}^r m_{ij}\phi_j$. Then if $M(i)$ denotes the $(r-1) \times (r-1)$ matrix obtained by deleting i-th row and i-th column of M, then f_i is obtained as the Pfaffian of $M(i)$ and $\deg(e_i) = \mathrm{F}(H) + N - \deg(f_i)$. Namely, $\mathrm{Det}(M(i)) = f_i^2$.

Note that if the i-th row of M is $(m_{i1}, \ldots, m_{ir})$, then we have

$$(*) \quad \sum_{i=1}^r m_{ij} f_j = 0.$$

2 The Proof

Now we will give a proof of Theorem 1.1 using Theorem 1.2.

Renumbering the minimal generators $\{f_1, \ldots, f_r\}$ of I_H, we can assume $f_p = x_p^{\alpha_p} - q_p$, where q_p is a monomial of $\{x_1, \ldots, x_4\} \setminus \{x_p\}$ $(p = 1, \ldots, 4)$. (We will show in Remark 2.3 that if some minimal generator of I_H is of the form $x_p^{\alpha_p} - x_q^{\alpha_q}$ then H is a complete intersection. Thus we can assume that $f_i = -f_j$ $(1 \le j < j \le 4)$ does not occur.) Hence, for $p \ge 5$, f_p is of the form

$$(**) \quad f_p = x_i^a x_j^b - x_k^c x_l^d \quad (a, b, c, d > 0) \quad (p \ge 5)$$

for some permutation $\{x_i, x_j, x_k, x_l\}$ of $\{x_1, x_2, x_3, x_4\}$.

Now we will show that $r = 5$. So, we assume $r \ge 7$ and get a contradiction. The next lemma will be the key of our proof.

Lemma 2.1 *If $s, t \ge 5$ and $s \ne t$, then $m_{s,t} = 0$.* □

Proof Assume $m_{s,t} \ne 0$ with $\deg(m_{s,t}) = h \in H_+$. Then we will have

$$\deg e_s = \mathrm{F}(H) + N - \deg(f_s) = h + \deg(f_t)$$

or

$$\mathrm{F}(H) + N = h + \deg f_s + \deg f_t.$$

Since $s, t \ge 5$, f_s, f_t are of the form (**) and we can take the expression

$$h + \deg f_s = \sum_{i=1}^{4} a_i n_i$$

so that 3 a_i's among 4 are positive. If some $a_j = 0$, then we can choose expression of $\deg f_t$ so that the coefficient of n_j is positive.

That means $\mathrm{F}(H) + N = h + \deg f_s + \deg f_t \ge_H N$, where we denote $a \ge_H b$ if $a - b \in H$. Then we get $\mathrm{F}(H) \in H$, a contradiction ! □

Corollary 2.2 $r \le 7$. □

Proof Assume $r \ge 9$. We know that f_1 is the Pfaffian of the matrix $M(1)$. Then by Lemma 2.1, we can see $\mathrm{Det}(M(1)) = 0$ because $m_{s,t} = 0$ if $s, t \ge 5$, □

Remark 2.3 Assume I_H has a minimal generator of the form $x_p^{\alpha_p} - x_q^{\alpha_q}$. Then we can assume f_p for $p \ge 4$ is of the form (**). Then above argument shows that $m_{s,t} = 0$ for $s, t \ge 4$. Now, if $r = 7$, then $\mathrm{Det}(M(1)) = 0$ since it contains a 4×4 0 matrix in it. Thus to show $r \le 5$ we can assume there is no minimal generator of I_H of type $x_p^{\alpha_p} - x_q^{\alpha_q}$. □

Proof *(Proof of Theorem 1.1)* Let us continue the proof of the Theorem 1.1. We assume $r = 7$ and deduce a contradiction.

We must have $f_1 = x_1^{\alpha_1} - q_1$ as the Pfaffian of $M(1)$. Now we know by Lemma 2.1 that if $s, t \geq 5$, $m_{s,t} = 0$. Let $N(1)$ be 3×3 matrix which is 2–4 rows and 5–7 columns of M. Then we must have $\mathrm{Det}(N(1)) = \pm(x_1^{\alpha_1} - q_1)$. That means, for every s, $2 \leq s \leq 4$, there should exists t with $5 \leq t \leq 7$ such that $m_{s,t}$ is a power of x_1. Namely, there should be at least three components that are a power of x_1.

Since the same should be true for $x_2, \ldots, x_4$, there should be $3 \times 4 = 12$ components in 1–4 rows and 5–7 columns. Namely we get

Claim Every (s, t) component of M with $1 \leq s \leq 4$ and $5 \leq t \leq 7$ is a power of some x_i and consequently $\neq 0$. □

On the other hand, assume, say, $f_1 = x_1^{\alpha_1} - x_2^b x_3^c$ with $b, c > 0$ and also $f_t = x_1^a x_4^d - x_2^{b'} x_3^{c'}$ for some $a, b', c', d > 0$ and $5 \leq t \leq 7$. Then $m_{1,t}$ should be 0, since otherwise

$$\mathrm{F}(H) + N = h + \deg f_1 + \deg f_t = h + an_1 + b'n_2 + c'n_3 + dn_4 \geq_H N,$$

which will lead to $\mathrm{F}(H) \in H$. A contradiction! Hence Claim 2 will lead to a contradiction.

Hence we get a contradiction from $r = 7$ and hence $\mu(I_H) = 5$ if H is not a complete intersection. □

Remark 2.4 If $r = 5$, we can show that I_H has no minimal generator of the form $x_p^{\alpha_p} - x_q^{\alpha_q}$. Indeed, assume that $f_1 = x_1^a - x_2^b$, $f_2 = x_3^c - q_3$, $f_3 = x_4^d - q_4$ for some monomials q_3, q_4 and f_4, f_5 are of the form (**). The above argument shows $m_{4,5} = m_{5,4} = 0$ and $\mathrm{Det}(M(1)) = (x_1^a - x_2^b)^2$. That means, $m_{i,j}$ are some power of x_1 or x_2 for $(i, j) = (2, 4), (2, 5), (3, 4), (3, 5)$. Then it is easy to see it is impossible to get a power of x_3 in $\mathrm{Det}(M(2))$, which contradicts $\mathrm{Det}(M(2)) = (x_3^c - q_3)^2$. □

Corollary 2.5 *If I_H has an element of type $x_p^{\alpha_p} - x_q^{\alpha_q}$ as a minimal generator, then H is a complete intersection.* □

Remark 2.6 If $r = 5$, we can deduce the form of M by our argument. For a monomial m of $\{x_1, \ldots, x_4\}$, let

$$\mathrm{supp}(m) = \{x_i \mid x_i \text{ divides } m\}.$$

We show that, if we put $f_i = x_i^{\alpha_i} - q_i$ $(i = 1, \ldots, 4)$ and $f_5 = q_5 - q_6$, then we can show that $\mathrm{supp}(q_i)$ $(i = 1, \ldots, 6)$ are all different and $\mathrm{supp}(q_i)$ consists of 2 variables. Also, if $f_i = x_i^{\alpha_i} - q_i$, $f_j = x_j^{\alpha_j} - q_j$ with $\mathrm{supp}(q_i) \cup \mathrm{supp}(q_j) = \{x_1, \ldots, x_4\}$, we show that $m_{i,j} = 0 = m_{j,i}$ and these are the only 0 of M except diagonals. Once we have proved these facts, we have exactly $16 = 5^2 - 5 - 4$ non-0 entries of M and they are powers of some x_i, hence we can deduce the matrix

$$\begin{pmatrix} 0 & -x_3^{\alpha_{43}} & 0 & -x_2^{\alpha_{32}} & -x_4^{\alpha_{24}} \\ x_3^{\alpha_{43}} & 0 & x_4^{\alpha_{14}} & 0 & -x_1^{\alpha_{31}} \\ 0 & -x_4^{\alpha_{14}} & 0 & -x_1^{\alpha_{21}} & -x_2^{\alpha_{42}} \\ x_2^{\alpha_{32}} & 0 & x_1^{\alpha_{21}} & 0 & -x_3^{\alpha_{13}} \\ x_4^{\alpha_{24}} & x_1^{\alpha_{31}} & x_2^{\alpha_{42}} & x_3^{\alpha_{13}} & 0 \end{pmatrix}$$

in Theorem 4 of [1].

Since $\mathrm{Det}(M(1)) = (x_1^{\alpha_1} - q_1)^2$, there should be at least a power of x_1 in rows 2, 3, 4, 5. Since $\mathrm{Det}(M(i)) = (x_i^{\alpha_i} - q_i)^2$ for $i = 2, 3, 4$, m_{i5} is a power of some x_k, $k \neq i$ and a power of every x_k $(1 \leq k \leq 4)$ should appear as some m_{5i}. Also we have

$$(***) \qquad \sum_{1=1}^{4} m_{i5} f_i = 0.$$

If, say, $\mathrm{supp}(q_1)$ has 3 variables, then we will have $m_{15} = 0$, since then we will have $\deg f_1 + \deg f_5 \geq_H N$. Then we must have $m_{51} = 0$, contradicting observation above. Thus we know that every q_i contains exactly 2 variables.

From (***) we know that if $m_{15} = x_j^p$, then q_1 is of the form $q_1 = x_j^s q_1'$, where we must have $\alpha_j = p + s$. Thus changing the order of variables, if necessary, we may assume $f_1 = x_1^{\alpha_1} - x_3^{\alpha_{13}} x_4^{\alpha_{14}}$ and $m_{51} = x_4^p$. Then we have $p + \alpha_{14} = \alpha_4$ since $m_{54} x_4^{\alpha_4}$ must cancel with $x_4^p x_3^{\alpha_{13}} x_4^{\alpha_{14}}$ and we must have $m_{54} = x_3^{\alpha_{13}}$. Then $x_3^{\alpha_{13}} q_4$ must cancel with $m_{53} x_3^{\alpha_3}$. Thus we have

$$q_1 = x_3^{\alpha_{13}} x_4^{\alpha_{14}}, q_2 = x_a^{\alpha_{21}} x_4^{\alpha_{24}}, q_3 = x_1^{\alpha_{31}} x_2^{\alpha_{32}}, q_4 = x_2^{\alpha_{42}} x_3^{\alpha_{43}}$$

and $m_{51} = x_4^{\alpha_{24}}, m_{52} = x_1^{\alpha_{31}}, m_{53} = x_2^{\alpha_{42}}, m_{54} = x_3^{\alpha_{13}}$ with $\alpha_1 = \alpha_{24} + \alpha_{21}, \alpha_2 = \alpha_{32} + \alpha_{42}, \alpha_3 = \alpha_{13} + \alpha_{43}, \alpha_4 = \alpha_{24} + \alpha_{14}$. We notice that $m_{13} = m_{24} = 0$ since $\deg f_1 + \deg f_3, \deg f_2 + \deg f_4 \geq_H N$ and we can fill in the other parts of M by $m_{i5} = m_{5i}$ and $\sum m_{ij} f_j = 0$. □

Acknowledgements The author thanks Kazufumi Eto for bringing him to this subject by his inspiring talk at "Singularity Seminar" at Nihon University. The main technique of this paper came out in the collaboration with Jürgen Herzog and the author is grateful to him for the collaboration. Also the author thanks the referee for valuable comments.

References

1. Barucci, V., Fröberg, R., Şahin, M.: On free resolutions of some semigroup rings. J. Pure Appl. Algorithm **218**, 1107–1116 (2014)
2. Bresinsky, H.: Symmetric semigroups of integers generated by 4 elements. Manuscripta Math. **17**, 205–219 (1975)

3. Buchsbaum, D.A., Eisenbud, D.: Algebra structures for finite free resolutions, and some structure theorems for ideals of codimension 3. Am. J. Math. **99**, 447–485 (1977)
4. Kunz, E.: The value-semigroup of a one-dimensional Gorenstein rings. Proc. Am. Math. Soc. **25**, 748–751 (1970)
5. Watanabe, J.: A note on Gorenstein rings of embedding codimension three. Nagoya Math. J. **50**, 227–232 (1973)

Printed in the United States
by Baker & Taylor Publisher Services

Printed in the United States
by Baker & Taylor Publisher Services